Antarctic Seaweeds

Iván Gómez • Pirjo Huovinen
Editors

Antarctic Seaweeds

Diversity, Adaptation and Ecosystem Services

Springer

Editors
Iván Gómez
Instituto de Ciencias Marinas y
Limnológicas, Facultad de Ciencias
Universidad Austral de Chile
Valdivia, Chile

Research Center Dynamics of High Latitude
Marine Ecosystems, (IDEAL)
Valdivia, Chile

Pirjo Huovinen
Instituto de Ciencias Marinas y
Limnológicas, Facultad de Ciencias
Universidad Austral de Chile
Valdivia, Chile

Research Center Dynamics of High Latitude
Marine Ecosystems, (IDEAL)
Valdivia, Chile

ISBN 978-3-030-39447-9 ISBN 978-3-030-39448-6 (eBook)
https://doi.org/10.1007/978-3-030-39448-6

This Springer imprint is published by the registered company Springer Nature Switzerland AG
The registered company address is: Gewerbestrasse 11, 6330 Cham, Switzerland

Preface

In 2002, Christian Wiencke and Margaret Clayton published the book *Antarctic Seaweeds* in the series Synopsis of the Antarctic Benthos edited by J.W. Wägele. To our knowledge, this volume represents the most recent and comprehensive guide dedicated exclusively to Antarctic macroalgae, which has made it an obligate reference for further studies. Nearly two decades later, we believe that it is timely and urgently needed to bring to light an update on the present state of the art of these important organisms in a rapidly changing world. Therefore, we invited various Antarctic researchers to contribute chapters covering recent advances in a variety of related topics. The list of contributors reflects well the currently active role of South American research groups in this field. Originally, the focus of the book was on the contributions from the South America; however, the invitation was expanded to some distinguished colleagues from other regions in order to improve the coverage of the thematics. It is clearly not possible to include all the topics, but we believe that a representative view of the current state of knowledge on the most relevant aspects is given, providing useful information for both undergraduate and postgraduate students as well as for scientific community. The book is organized in 5 parts with a total of 18 chapters. Part 1 gives a brief overview of the individual chapters and outlines the major gaps and challenges as well as the new directions in the study of Antarctic seaweeds. The following parts summarize the recent advances in diversity and biogeography (Part 2); physiology, productivity, and environmental responses (Part 3); biological interactions and ecosystem processes (Part 4); and chemical ecology of Antarctic seaweeds (Part 5). Many of the chapters discuss the topics in the context of environmental threats, especially climate change that is already affecting these ecosystems. Thus, unavoidably, there is some overlapping of these themes in different chapters, however, from distinct points of view and in other context.

We are grateful to all the colleagues who kindly accepted the invitation to contribute a chapter. We would also like to warmly thank our colleagues Chuck Amsler, Kai Bischof, Bernardo Broitman, Gabriela L. Campana, Marie-Laure Guillemin, Patrick Neal, Ellie Poulin, Martin Thiel, Nelson Valdivia, Christian Wiencke and Katharina Zacher for dedicating their time and expertise for peer review and improving the chapters of this book at their manuscript stages. We greatly acknowledge Prof. Christian Wiencke for contributing the foreword as one of the leading experts in polar seaweeds. Finally, we would like to thank the Universidad Austral de Chile, the Comisión Nacional de Investigación Científica y Tecnológica (CONICYT), and the Instituto Antártico Chileno (INACH) for permanent support of our research activities in the Antarctic and to the publisher for giving us the opportunity to make this volume. This publication is within the frame of the scientific program of the Research Center Dynamics of High Latitude Marine Ecosystems (IDEAL).

Valdivia, Chile

Iván Gómez
Pirjo Huovinen

Foreword

Seaweeds represent a group of photoautotrophic organisms of vital importance for the function of coastal ecosystems. They provide diverse habitats and breeding areas for uncountable numbers of organisms including crustaceans and fishes and represent an important food source not only for herbivores but also for detritivores feeding on degraded seaweed biomass. About 10% of the global oceanic production is based on seaweeds, a value similarly high as that of tropical rain forests. We knew for long that seaweed communities are well developed in the temperate regions of the world, but recent research shows that this is also true for the submarine seaweed forests of Antarctica: single species can attain biomasses of over 10 kg wet weight per square meter.

The exploration of Antarctic seaweeds began in 1817 with the expeditions of Gaudichaud, Bory, Montagne, Hooker, and Harvey. In a second phase of intensive research during the first two decades of the twentieth century, Gain, Skottsberg, and Kylin made important taxonomic studies on Antarctic seaweeds. In the 1960s, scuba diving investigations started and so, by the early 1980s, basic knowledge was available on taxonomy, geographic distribution, and depth zonation of Antarctic seaweeds. During the expeditions of Clayton and Wiencke in the 1980s and 1990s, numerous species of Antarctic seaweeds were isolated and taken in culture, allowing the description of algal life histories and the performance of physiological experiments in temperature-controlled rooms in the home laboratories. An important side product was a monograph on Antarctic seaweeds containing the first identification key.

These studies were a booster for scientists especially from South America, but also from Europe, Australia, and North America, to work further on the investigation of these interesting biotas growing in this remote, harsh, and unique environment. Purely descriptive surveys came to an end, and scientists applied the latest available methods to study distribution and biodiversity, metabolic adaptations to the extreme Antarctic environment characterized by low temperatures and long periods of darkness, the importance of the seaweeds as primary producers for Antarctic near-shore ecosystems, and the effects of global climate changes, in particular the increase of UV-B radiation due to stratospheric ozone depletion and the increase of the water temperatures in the Antarctic Peninsula region and their influence on zonation patterns and geographic distribution. Besides traditional studies, scientists used new approaches to study photosynthesis and carbon balance, genetic diversity, transcriptomic responses, and trophic interactions by experimental ecology and ecological network analysis, just to mention some.

To my knowledge, this is the first multi-authored book exclusively focused on Antarctic seaweeds and their role in coastal ecosystems in Antarctica with respect to their reaction to a changing environment from the metabolic, cellular, and organismic level to the level of communities. I am proud to say that I have supported many studies and interacted with almost all authors of the book. The book represents the present state of the art in this research area and as such will serve as an important baseline for future research.

Bremerhaven, Germany Christian Wiencke

Contents

Part I
Introduction

Chapter 1
Antarctic Seaweeds: Biogeography, Adaptation, and Ecosystem Services

Iván Gómez and Pirjo Huovinen

Abstract Seaweeds (macroalgae) represent the most striking benthic organisms in the Antarctic near-shore ecosystems. Their abundance, relevant roles as primary producers, and foundation organisms were recognized since the first Antarctic explorations. Furthermore, especially since the 1960s, improvements in the sub-aquatic survey techniques and laboratory facilities expanded considerably our knowledge on ecology, reproduction, and environmental adaptation of seaweeds whose biological processes determine much of the biogeochemical cycles in the Antarctic coastal systems. In recent years, the imminence of the climate change and the direct impact of human activities, which are affecting vast regions of the Antarctica, have highlighted the importance of seaweeds as central components shaping the structure, functions, and supporting services of benthic ecosystems under changing polar environment. The present book is aimed to put together the knowledge and experience gained in recent years by diverse research groups. Many of these research efforts have long tradition, while others have brought more recently important new approaches in the study of these organisms with benefits for the whole polar science. We believe that this initiative is timely and urgently needed in order to improve our scientific knowledge on these fascinating organisms. In this chapter, we describe the book's framework, summarizing the most important advances in areas related with diversity, biogeography, ecophysiology, biological interactions, and chemical ecology of Antarctic seaweeds. Finally, considerations regarding the major gaps and challenges as well as the new directions in the study of Antarctic seaweeds are outlined.

Keywords Antarctic · Climate change · Antarctic marine flora · Ecosystem functions

I. Gómez (✉) · P. Huovinen
Instituto de Ciencias Marinas y Limnológicas, Facultad de Ciencias, Universidad Austral de Chile, Valdivia, Chile

Research Center Dynamics of High Latitude Marine Ecosystems, (IDEAL), Valdivia, Chile
e-mail: igomezo@uach.cl; pirjo.huovinen@uach.cl

I. Gómez, P. Huovinen (eds.), *Antarctic Seaweeds*,
https://doi.org/10.1007/978-3-030-39448-6_1

1.1 Introduction: The Historical Context

In recent decades, important advances have been demonstrated in different areas of knowledge on Antarctic seaweeds, from organisms to ecosystems. However, in order to understand the roles and services of seaweed communities in Antarctica currently marked by climate change, it is essential to go back to the history of this endeavor. The first explorations of Antarctic seaweeds in the nineteenth century (Gaudichaud 1826; Hooker 1847) had already documented the exuberant presence of benthic seaweeds and recognized their importance for the coastal ecosystems in the Antarctic, especially around the Antarctic Peninsula. In his book *The Botany of the Antarctic Voyage of H.M. discovery ships "Erebus" and "Terror" in the years 1839–1843, under the command of Captain Sir James Clark Ross* (Fig. 1.1), one of most complete records of marine and terrestrial flora of the Southern Ocean, the British botanist Joseph D. Hooker disclosed much of the extraordinary conditions that characterize the habitat of many Antarctic seaweeds. Later, another important researcher, Karl Skottsberg, expanded this information from different Antarctic expeditions in the early twentieth century (e.g., Skottsberg 1907). During the 1960s and 1970s, descriptions based on scuba diving surveys carried out by Neushul (1965), Delépine et al. (1966), Zaneveld (1966), and Lamb and Zimmermann (1977), among others, confirmed this, highlighting the dominance of large endemic Desmarestiales at depths >10 m, where they occupy similar role as kelps as the dominant seaweed group in the Northern Hemisphere and the Arctic. The unique characteristics of the Antarctic marine flora reflect the complex biogeographic and evolutionary processes that followed the formation of the Antarctic Circumpolar Current (ACC) around 30–35 Ma and consequent full glaciation of the Antarctica (Clayton 1994). Diverse surveys across different sites in the Antarctic, including communities growing under ice shelves, expanded considerably our knowledge on vertical distribution, biomass, and diversity of seaweeds (Zielinski 1981, 1990; Amsler et al. 1995; Klöser et al. 1993, 1996; Brouwer et al. 1995).

Due to the harsh climatic conditions and logistic restrictions in Antarctica, advances in our knowledge on reproduction, phenology, and acclimation to the polar environment were only possible since the 1980s. Using cultured material, Moe and Henry (1982) described for the first time various aspects of the development of early phases of *Ascoseira mirabilis*. The first studies unraveling the seasonal development, life history, and physiological performance of Antarctic seaweeds were based on algae grown under cultivation conditions simulating the Antarctic light regime (Wiencke 1990). Based on these findings, two main growth strategies were defined: the season responders start growth and reproduction when environmental conditions are optimal in spring and summer, while the season anticipators develop during late winter and spring. Thereafter, the number of investigations focused on physiology of photosynthesis, growth, chemical ecology, etc., increased (revised in Wiencke 1996). A noticeable finding was that various endemic Antarctic brown algae, such as *Ascoseira mirabilis*, *Cystosphaera jacquinotii*, *Desmarestia anceps*, and *Himantothallus grandifolius*, exhibit thallus anatomical

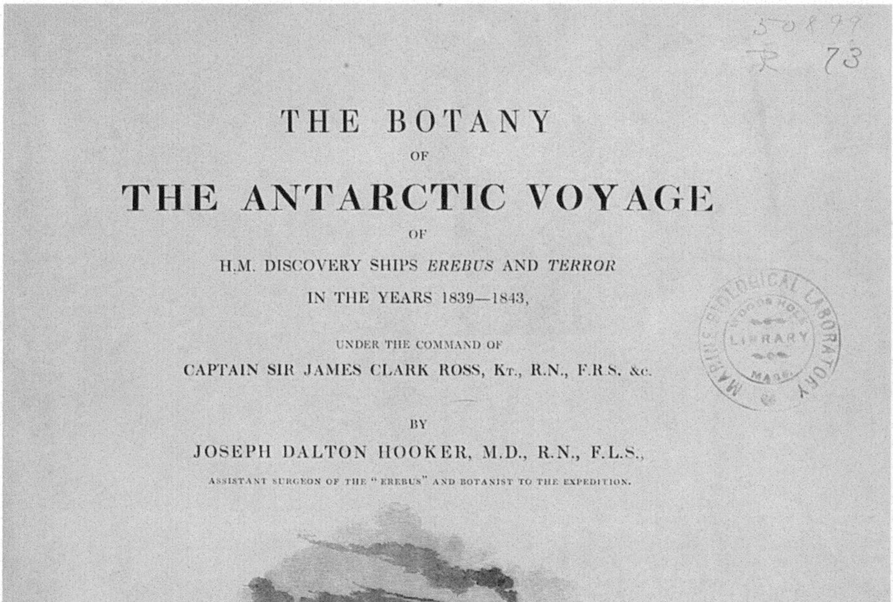

Fig. 1.1 Cover page of Hooker's publication describing the flora of the Southern Oceans

and functional characteristics resembling those of large kelps from the Northern Hemisphere (Drew and Hastings 1992; Gómez et al. 1995; Fig. 1.2). Here, the most remarkable morpho-functional adaptations of large Antarctic brown algae are their very low light demands for growth and photosynthesis and an efficient operation of light-independent carbon fixation (LICF) at the meristematic zones, which allow these organisms to display positive carbon balance at depth close to 30 m (Gómez et al. 1997). The knowledge on these structural and functional aspects of

Fig. 1.2 Large endemic brown algae are the most representative components of the Antarctic costal systems. (**a**) *Himantothallus grandifolius*, (**b**) *Desmarestia anceps*, (**c**) *Cystosphaera jacquinotii*. (Photos by Ignacio Garrido)

photosynthetic responses had important implications for understanding the biological interactions between seaweeds and their associated biota (Zacher et al. 2007; Iken et al. 2009; Amsler et al. 2011).

In the last decades, the Antarctic ozone depletion and associated increase in UV-B radiation, as well as the environmental shifts driven by climate change, oriented the research of Antarctic seaweeds. In this context, various studies have examined the effects of changing irradiance on different algal assemblages across Antarctica (Schwarz et al. 2003; Zacher et al. 2007; Huovinen and Gómez 2013; Clark et al. 2017; Deregibus et al. 2016). At an ecosystem level, seaweeds have been commonly recognized as important sentinels of climatic change in the Antarctic, highlighting the remarkable capacity of these organisms to adapt to new habitats (Quartino et al. 2013), and also providing some key ecological ecosystem properties that permit the maintenance of species richness and biomass (Valdivia et al. 2015). Through their ecosystem engineering functions, especially large endemic brown algae are able to minimize environmental variability enhancing the resilience of the whole system (Ortiz et al. 2017).

Despite these advances, much of the predictions related with adaptation and fate of Antarctic seaweeds are limited by scarce molecular evidence. From this perspective, the findings of an increasing number of cryptic species with Antarctic/sub-

Antarctic or even more vast joint distribution (van Oppen et al. 1993; Hommersand et al. 2009; Billard et al. 2015) challenge some traditional concepts related with the evolution and biogeographic patterns of the Antarctic marine flora (Crame 1992; Clayton 1994). According to current predictions, climatic anomalies, e.g., enhanced temperature, increased storms, and winds, will be able to break the ecological isolation of Antarctica and facilitate the arrival of temperate species (Fraser et al. 2018), with impacts on diversity and genetic configuration of local communities yet not well understood.

1.2 Antarctic Seaweeds in the Wake of Climate Change

The climate, oceanography, and related ecosystem processes in Antarctica and its surrounding oceanic system have been changing rapidly in the last decades (reviewed in Constable et al. 2014). Accelerated regional warming was reported especially in the WAP region almost 20 years ago (Vaughan et al. 2003). According to the IPCC scenarios, the mean annual air temperature in this region was predicted to increase by 1.4–5.8°C until 2100 (Clarke et al. 2007), although strong natural variability seems characteristic in this region (Turner et al. 2016). The surface waters of the Bellingshausen Sea have warmed by 1°C in summer since the 1950s (Meredith and King 2005), while Schloss et al. (2012) reported an increase of more than 2°C in winter sea surface temperature between 1991 and 2006 in Potter Cove (King George Island). This tendency and the possible effects on the polar system were recently highlighted in the last IPCC report (IPCC 2019). As a synthesis the report indicates that the Southern Ocean (area corresponding to 25% of world's oceans) has been warming at alarming rates, being responsible for 45–62% of the global ocean warming during the period 2005–2017. Although no clear overall trends in Antarctic sea ice cover were evident for the period 1979–2018, a strong decline has been observed recently (2016–2018), which can pose threats to the photosynthetic organisms due to unpredictable changes in the light regime (see Chap. 7 by Huovinen and Gómez). In the Arctic, massive ice-sheet losses, exceeding the rates of modeled estimations, have been observed (Bronselaer et al. 2018). Here, the role of albedo-reducing light-absorbing impurities in ice and snow fields exacerbating ice loss has been emphasized (Benning et al. 2014; Tedesco et al. 2016; Tedstone et al. 2017). Dark snow phenomenon has recently also been associated with decreased albedo in Maritime Antarctic (Huovinen et al. 2018). Recently, the active role of ice sheets and icebergs in the global carbon cycle has been recognized (reviewed by Barnes et al. 2018; Wadham et al. 2019) and can have important consequences for the adjacent marine realm in areas like Maritime Antarctic (Hood et al. 2015). Although various impacts of these changes are broadcasted for pelagic realms, their implications for the processes occurring in the Antarctic shallow benthos are much less known (Barnes and Conlan 2012; Constable et al. 2014).

The increasing number of volumes devoted to the present and projected impacts of global climate changes on the Southern Ocean and their different ecosystems (e.g., Bargagli 2005; Bergstrom et al. 2006; Rogers et al. 2012; Tin et al. 2014; Kanao et al. 2018) is a clear evidence of the importance of understanding their global consequences. Antarctica can be regarded as a natural laboratory where its physical environment brings the adaptation capacities of organisms to an extreme limit. In this context, seaweeds, as fundamental components of the Antarctic coastal systems, can give important insights into the structure and functioning of the biota in the new scenarios driven by climate change.

1.3 The Book

Based on recent quantitative, observational, and experimental evidences, this book updates the state of art about the diversity and geographic distribution of seaweeds as well as their biological interactions and responses to the environment, which is fundamental for understanding the coastal processes in a changing Antarctica. The main themes and the overall scientific framework discussed in the book can be summarized in Fig. 1.3.

1.3.1 Diversity and Biogeography

Compared to other biogeographical regions in the Southern Hemisphere, e.g., southern Australia, New Zealand, and the southern Chilean coast, the diversity of seaweeds in the Antarctic has been traditionally considered low. Based on Wiencke and Amsler (2014), the number of species is 124, showing high endemism (35%). In their chapter (Chap. 2), Oliveira et al. indicate that the richness of Antarctic seaweeds has been underestimated. Based on previous information and recent molecular surveys, the authors report a diversity of 151 species of which 85 are Rhodophyta, 32 Chlorophyta, and 34 Ochrophyta (most of them brown algae). Likewise, this update decreased the percentage of endemism to 24%. Overall, the increase in the number of catalogued species can be explained by improvements in the identification tools, e.g., the use of DNA barcoding, more complete gene databases, and more efficient approaches to detect, e.g., cryptic species. However, a conclusive outcome of this diversity is far from definitive: a lack of baseline datasets in order to accurately detect local loss of native species, or their replacement by alien assemblages, still persists. Thus, extending the geographical range and number of surveys, adjusting better the inventories of phylogenetic markers, and deepening the examination of less conspicuous algal groups, such as crustose and endophyte species, a hidden diversity normally overlooked, are suggested.

The Antarctic Circumpolar Current has defined the structure, diversity, and functioning of the biomes of the Southern Ocean. Fraser et al. (Chap. 3) make a comprehensive analysis of environmental and oceanographic conditions that characterize Antarctic from the sub-Antarctic regions, the dual role of ACC acting as an efficient

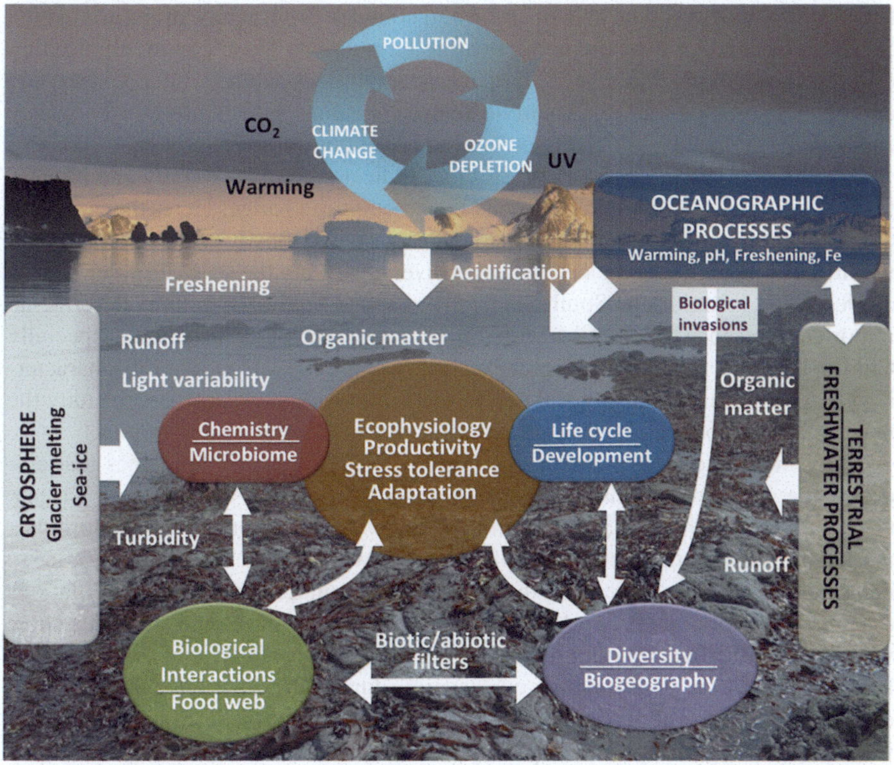

Fig. 1.3 Schematic presentation of the major drivers, organismal processes, and biological interactions of Antarctic seaweeds. The framework is based on the conclusions of the different chapters in this book

barrier and also as a bridge connecting marine assemblages and the requisites of organisms permitting their dispersal across these environmental gradients. Rafting of floating seaweeds driven by prevailing winds across the different fronts in the Southern Ocean appears as a central mechanism promoting transoceanic connections, not only of seaweeds but also invertebrates. The definitive establishment and persistence of new taxa in these zones will depend on different environmental filters, e.g., physical and biological constraints, and also on various organismal features related with reproductive viability, physiological capacities, etc.

Probably the extent of exchange of species and hence genetic fluxes between sub-Antarctic and Antarctic regions lie in the diversity of taxa that can be transported across long distances and their ability to remain alive during their journey. Macaya et al. (Chap. 4) indicate that a total of 39 species (3 Chlorophyta, 14 Ochrophyta, and 22 Rhodophyta) have been reported drifting, stranded or floating in Antarctica or crossing the Antarctic Polar Front (APF). Considering that many cold and cold-temperate species at both sides of the ACC show remarkable physiological adaptions to biotic and abiotic factors, e.g., grazing, UV radiation, and temperature, they could be able to arrive and colonize different locations around the Southern Ocean. An example is the floating large brown algae commonly used by different hitchhiking biota (e.g., barnacles, amphipods, algae). Interestingly, the authors suggest that various Antarctic seaweeds, some with floating or buoyancy capacity, have the physiological potential to travel out of the Antarctic.

In their chapter (Chap. 5), Pellizari et al. indicate that the diversity and biogeographic patterns of Antarctic seaweeds have begun to change. Here, the changing environmental scenarios in the Southern Ocean, related mostly with circulation and warming, will determine the new seaweed diversity. Using the seaweed assemblages of Deception Island in the South Shetlands as a case study, the authors describe an important presence of species with broad geographical distribution, especially Chlorophyceans, indicative of recent arrival. Apparently, areas like this characterized by peculiar physicochemical conditions could become key places to study the new Antarctic biodiversity, its biogeographic divergences and connections.

The Antarctic continental margins or peri-Antarctic islands are zones that evidence the long evolutionary history of seaweeds within the Southern Ocean. Guillemin et al. (Chap. 6) analyzed the sequences of mitochondrial and chloroplast markers in eight Antarctic species of green, brown, and red seaweeds in order to determine the genetic patterns in the context of the quaternary climatic oscillations (QCO). The haplotype network revealed that the studied Antarctic seaweeds show very low genetic diversity, and significant signatures are indicative of a recent population expansion after a massive constriction during the Last Glacial Maximum (20 Ka). Thus, the authors agree with a theory that this marine flora survived *in situ* in a unique refugium and subsequently recolonized the multiple postglacial open areas using the ACC as a predominant driving force.

In all, Antarctica is not a physically isolated continent, and in a scenario of increasing warming, the influx of marine organisms arriving, e.g., via rafting to its coasts, can find new opportunities for colonization, which finally will modify the local diversity (Fig. 1.3). Here, the examination of large-scale patterns of seaweeds may provide clues to evaluate aspects of endemism, biological corridors, and expansion of geographical distribution of various algal species. In this context, an account of the genetic footprints of past diversity can help to understand not only the large-scale processes that occurred along the evolution of the Antarctic flora, but also its future genetic structure.

1.3.2 Environment and Ecophysiology

Due to the harsh environmental conditions, the Antarctic has commonly been regarded as an inhospitable place for living organisms. Antarctic biota has adapted to these conditions and thrives in different types of habitats, some marked by extreme physical variability. However, the new environmental features as a consequence of regional warming and related phenomena occurring in the cryospheric realm, as well as direct anthropogenic pressures, are challenging the adaptive strategies of seaweeds in manners still not well understood.

Light is probably the most important environmental factor determining the phenology, spatial distribution, and productivity of Antarctic seaweeds. In Chap. 7, Huovinen and Gómez describe the underwater optics in the context of present and future variability and its importance for seaweed photobiology. The optical properties of the coastal waters, including their light absorbing and scattering components, define the underwater light environment at ecologically relevant depths (down to 40 m). Despite Antarctic seaweeds being regarded as shade-adapted organisms, they also show a striking capacity to acclimate to sudden increases in solar radiation. However, the natural variability in light regimes is being altered due to earlier sea ice breakup, enhanced runoff from the terrestrial and glacial melting, enhanced UV-B levels as a result of ozone depletion, etc. These new scenarios are accompanied by emergent stressors (e.g., local freshening, acidification, increasing contaminant load) whose influence on the underwater light climate in the Antarctic up to now is not well understood.

Probably one of the most striking signals of warming in the Antarctic is the retreat of glaciers, which is creating new ice-free habitats for benthic organisms. The question of how the future coastal scenarios driven by climate change will affect the colonization and fate of seaweeds was addressed by Quartino et al. (Chap. 8). In fact, the increased seaweed biomass will enhance the carbon flux and hence the organic matter towards the higher trophic levels. Due to some species attaining biomass values close to 10 kg m^{-2} wet weight, a strong impact on the coastal productivity can be expected. However, in these highly dynamic new habitats, reflected in the model system of Potter Cove in King George Island, seaweed colonization follows the sharp gradients set by the light penetration, which are strongly modified by enhanced sedimentation. Considering their great abundance and functional role as ecosystem engineers, benthic seaweeds can become important carbon sink in these systems. For instance, it has been estimated that seaweeds can account for a global net primary production of ca. 1.5 Tg C yr^{-1} (Krause-Jensen and Duarte 2016), thus forming part of the "blue carbon" components.

Low water transparency in the new ice-free areas affects the physiological performance of seaweeds in different ways. Deregibus et al. (Chap. 9), based on long-term records in areas nearby a retreating glacier at Potter Cove, describe the photosynthetic carbon balance of seaweeds (the gain of C in photosynthesis versus that lost in respiration) and its changes in relation with the light climate. Considering light requirements and photosynthetic efficiency estimated from P-E curves, the authors indicate that vertical distribution limits of some seaweed species changed as a result of enhanced turbidity. Accuracy of the carbon balance estimations requires a robust temporal set of solar irradiance data; thus, the importance of permanent *in situ* monitoring accounting for variations at short (hours, days) and long (monthly, inter-annual) timescale was highlighted.

The performance of seaweed populations under changing environmental regimes depends on the survivorship of their early reproductive stages. However, life cycle stages (e.g., spores, microscopic gametophytes, embryonic sporophytes, etc.) can be highly sensitive to environmental stressors. In Chap. 10, Navarro et al. make a thorough review of the aspects of the physiology of propagules of Antarctic seaweeds and how they respond to major physical factors, e.g., solar radiation, and temperature, considering present and future settings. UV effects are in many cases modified by temperature, showing interactions of factors. The response mechanisms and degree of tolerance of early developmental stages mirror those observed in the parental individuals. For example, differential responses to UV radiation determined in adult populations of congeneric and conspecific species from distinct depth zones (e.g., subtidal versus intertidal) or geographical origin (e.g., Antarctic versus sub-Antarctic) have also been observed in their propagules.

Antarctic seaweeds can be very abundant in terms of biomass and account by more than 50% of the coastal primary productivity, especially around the Antarctic Peninsula. Much of this ubiquity is strongly linked with efficient morpho-functional adaptations that have permitted these organisms to occupy niches characterized by sharp physical gradients. In Chap. 11, Gómez and Huovinen discuss the importance of the form and function of seaweeds. In general, the functional forms are well distributed along the major groups of Antarctic seaweeds: coarsely branched and leathery species, which can be regarded as the most robust and large-sized forms, represent 49% of the total number of species. In this group, endemic brown and red algae dominate, mainly growing at the subtidal zone. Filamentous, finely branched and foliose species (41%) belong mostly to green and red algae, common at shallow and intertidal sites, and are geographically widely distributed. Each of these morphs are integrated in different life strategies and hence distinct ecosystem functions. For example, perennial canopy-forming species show competitive abilities for light and substrate, but in general prevail less in sites subject to strong physical perturbation. Here, small colonizers and opportunistic species dominate in virtue of rapid metabolic adjustments and turnover rates.

The different chapters reveal that the abiotic environment of Antarctica is changing in extent that is already affecting several aspects of the physiology of marine biota in general and seaweeds in particular. The emergence of new habitats available as the glaciers retreat is modifying the composition, structure, and trophic relations of the benthic communities dominated by seaweeds. Apparently, a strategic factor underlying these responses is the ability of adult plants and their propagules to acclimate, via different functional traits, to the environmental shifts.

1.3.3 Ecological Functions

The ecological succession in Antarctic benthos determines the structure of the mature community and its biological network. Different types of positive and negative interactions between algal assemblages, invertebrates, fish, and microorganisms can be identified as the community develops. Based on *in situ* experiments, Campana et al. (Chap. 12) describe the successional stages and their biotic interrelations in a coastal site near Potter Cove. During the first three months, the incipient community is dominated by microorganisms and benthic diatoms, which apparently promote the development of small ephemeral filamentous green algae. The assemblage is successively enriched by the presence of various foliose and crustose forms of red algae and during late algal succession (after 4 years) by some perennial species of *Desmarestia*. The different components of the succession respond differently to environmental factors such as UV radiation, grazing, glacier retreat, and sea ice, and hence the structure and the biotic relations change dynamically within this early community.

Grazing is probably the most important biotic factor controlling the structure and composition of seaweed-dominated communities. In Antarctic benthic systems, the early successional stages dominated by small-sized seaweeds and periphyton represent excellent models to study how grazing modifies different ecological properties not only of native assemblages but also of alien species, whose arrival and establishment will be stimulated by climate warming. In Chap. 13, Valdivia determined by means of mathematical simulations the impact of mesograzers in sub-Antarctic and Antarctic sites connected by dispersal. *Ulva* sp. was regarded an alien species, being highly competitive in the Antarctic but not in the sub-Antarctic littoral. The results indicated that Antarctic mesograzers have a deterministic and marked effect on the biomass of the alien seaweeds; however, projected climate-change-driven shifts in temperature or pH could decrease the potential of, for example, amphipod grazers to control the development of invaders.

Antarctic seaweeds harbor complex and intricate microbiomes, which exert important influence on different molecular and biochemical processes of the algal host. Hitherto much of the coevolutionary processes of this association have been little studied. However, it is reasonable to argue that microbiota plays important functional roles in the ecology of Antarctic seaweeds. Gaitan-Spitia and Schmid (Chap. 14) review various aspects of structure, diversity, and functioning of Antarctic microbiomes and their implications for seaweeds. Members of phylum Actinobacteria show high diversity and persistence among different seaweed species, while Firmicutes are less represented. In general, the microbiomes associated to seaweeds are different from those found in the surrounding environment, which suggest that the bacterial composition is regulated by the seaweed host. Apparently, this feature reflects adaptive strategies to respond to multiple environmental conditions, e.g., antioxidation, antimicrobial activity, photoprotection, etc.

Seaweeds and microphytobenthos represent the basis of the Antarctic coastal food web. Because coastal areas can become highly perturbed, the dynamics and stability of the interspecific interrelations have fundamental influence on the whole benthic ecosystem at different spatial and temporal scales. Momo et al. (Chap. 15) determined that the food web at Potter Cove is based on 24 seaweed species and diverse other photosynthetic organisms, such as epiphytic and benthic diatoms and phytoplankton as well as their detritus. The system is also hyperconnected indicating multiple energy pathways. Considering extinction thresholds, this network can be regarded as relatively resilient to local losses of seaweed species. Similarly, using as a model Fildes Bay, a coastal system geographically close to Potter Cove, Ortiz et al. (Chap. 16) analyzed different keystone species complexes, which contribute importantly to the emergent network properties, such as growth, organization, development, maturity, and health of the ecosystem. The theoretical framework (based on network analysis, *ascendency*, and *loop analysis*) identified detritus, the phyto-zooplankton complex, sea stars, sea urchins, and seaweeds as the major components determining the overall structure and function of this system. Similar to Potter Cove, Fildes Bay appears to be a less developed system compared to other cold-temperate system, but being highly resilient to physical perturbations.

The described examples on the ecology of the coastal system in the Maritime Antarctic reveal complex biological interactions between, e.g., algae, microbiota, invertebrates, and fish, which are strongly regulated by the physical environment. In this scenario, seaweeds are identified as key components from the early stages of succession to the consolidate communities. Due to the strong influence of physical factors from terrestrial, freshwater, cryospheric and atmospheric processes, the structure, function, and trophic interrelations in these communities are in general very resistant to disturbances (Fig. 1.3). Thus, it seems that there are internal mechanisms operating at individual (e.g., efficient growth strategies, multiple anti-stress mechanisms) as well as at population and community (e.g., filters controlling native and alien species, high biological complexity based on species and biomass richness) levels, providing the system with a high resilience.

1.3.4 Chemical Ecology

The trophic relations in the Antarctic benthic system show a balance between consumption by herbivores and their deterrence. Amsler et al. (Chap. 17) review the recent advances in relationship between seaweeds and, e.g., amphipods, gastropods, and fish. Diverse halogenated monoterpenes and phlorotannins (phenolic compounds found in brown algae), and probably various other compounds, confer many species of Antarctic seaweeds unpalatability to different kinds of herbivores. Interestingly, the relationship between some seaweeds and various species of amphipods includes mutualism, in which chemically defended algae offer protection from, e.g., omnivorous fish, while amphipods reduce the biofouling and epiphytic load of the thalli.

Chemical defenses based on phlorotannins operate not only against grazing, but also form part of a wide suite of constitutive anti-stress mechanisms. In Chap. 18, Gómez and Huovinen summarize the different aspects that determine the synthesis and accumulation of these substances, which in some Antarctic brown algae can represent up to 12% of the dry weight. These compounds have different functions as grazing deterrents, reactive oxygen species (ROS) scavenging agents, and metal chelators and can be allocated in different thallus parts to optimize defense. Although phlorotannins are regarded as UV screening substances, no evidence on UV induction in Antarctic seaweeds has been reported. However, the antioxidant capacity increases substantially along with increasing phlorotannin concentrations in algal extracts, even in algae not naturally exposed to UV radiation.

The high prevalence of chemical defenses observed in Antarctic seaweeds is remarkable and suggests their central role in defining their dominance and biological interactions in the Antarctic coastal ecosystems (Fig. 1.3). Particularly, the constitutively high levels of phlorotannins measured in various dominant brown algae (e.g., *Desmarestia anceps*, *Himantothallus grandifolius*) open interesting questions about the activation of anti-stress mechanisms based on chemical substances with multiple primary and secondary functions. For seaweed assemblages subjected to climate-change-driven environmental shifts, such defenses could confer ecological advantages.

1.4 Gaps, Emerging Challenges, and Future Directions

The different chapters throughout this book update the current knowledge and provide novel insight into various aspects on diversity, ecophysiology, and ecology of Antarctic seaweeds, with particular emphasis on their responses to the changing polar environment. However, several gaps still persist and new questions require attention in the near future.

- *Long-term assessment:* Due to logistical constraints, research in Antarctica is normally restricted to the spring-summer season. This time frame clearly does not permit covering the entire environmental variability to which Antarctic organisms, especially annual and perennial species, are exposed. For example, many gaps exist on the metabolic performance of seaweeds (e.g., carbon and nutrient metabolism, use and remobilization of photoassimilates, etc.) during the long Antarctic winter. In fact, the few studies addressing photosynthesis in winter or under ice cover suggest that seaweeds are at their physiological limit during this period (Gutkowski and Maleszewski 1989; Drew and Hastings 1992; Schwarz et al. 2003). These studies should be complemented with long-term monitoring of annual and inter-annual physical fluctuations in order to delimit the ranges of acclimation and adaptation of organisms. Because most of the monitoring platforms deployed around the Antarctica are designed to record changes in the open ocean, long-term or real-time baseline information of near coastal processes is still very limited. In this context, the long-term observations focused on the impact of the retreating Fourcade Glacier in Potter Cove (King George Island) represent an important effort in gaining insights into the responses of benthos at ecological scales (Meredith et al. 2018; see Chap. 8 by Quartino et al. and Chap. 9 by Deregibus et al. and references therein).
- *Molecular ecology*: Although remarkable improvements in biomolecular tools have considerably expanded our capacities to record and elucidate the taxonomical status of Antarctic species (Held 2014), many seaweeds are still not well classified, are cryptic or due to their life form (e.g., epiphytes, endophytes or prostrates) remain undiscovered. Another important limitation challenging the efforts to expand not only the genetic inventories, but also the general knowledge

on Antarctic organisms, is that the surveyed areas are strongly biased towards some regions, especially around the Antarctic Peninsula and in sites in direct proximity to research stations, while other coasts, e.g., from the East Antarctic, have been scarcely visited (Mormède et al. 2014). Thus, it is assumed that in the near future, along with the advances in phylogeography and population genetics as well as in geographic coverage, the number of Antarctic seaweed species, both native and recently arrived, will increase (see Chap. 2 by Oliveira et al.).

There a considerable lags in our understanding of gene expression and regulation. This is probably one of the weakest areas in the study of seaweeds in general and Antarctic species in particular. Thus, use of molecular tools such as transcriptomic analysis will help identify the metabolic pathways and adaptive strategies that Antarctic seaweeds exhibit beyond their tolerance threshold. For example, recently high and constitutive gene expression of various physiological reactions, including photochemical and inorganic carbon utilization components, from RNA-Seq analysis was reported for the first time for an Antarctic endemic species (the brown alga *Desmarestia anceps*; Iñiguez et al. 2017). Clearly this type of techniques open new avenues for the identification of transcripts that are differentially expressed under different stress conditions. On the other hand, the new molecular tools together with improved physiological methodologies are fundamental to predict whether key Antarctic seaweeds exhibit the molecular machinery to respond to ongoing and near-future impacts of climate change.

- *Ontogenetic development and life cycle responses*: Developmental phases (e.g., spores, gametes, and embryonic sporophytes) are highly sensitive to environmental changes (reviewed in Chap. 10 by Navarro et al.). However, they are often overlooked due to their small size or because the logistical constraints associated with their isolation, culture, and experimentation in Antarctica (Wiencke 1988). Considering that the fate of these cells determine the structure and dynamics of further life phases, it is urgent to conduct research focused on the acquisition of stress tolerance capacity at different developmental stages and how this resilience is "transferred" over generations. Following important developments in the identification and visualization techniques in microalgae, e.g., fluorescence cell-based sensing and "*omics*" approaches (metabolomics, proteomics, genomics), it is now possible to quantify in real time the effects of different stressors on cellular structures of early stages of seaweeds. Thus, it will be possible to track the progressive expression of anti-stress mechanisms along the ontogeny or life cycle phases, an essential approach to understand the adjustments in response to environmental changes at an organismal level.
- *Direct anthropogenic impacts and interaction of multiple stressors*: Warming and ozone depletion are not the only threats to Antarctic biota. Among other concerns are ocean acidification and local decreases in salinity (freshening) due to enhanced melting of glaciers. Furthermore, increase of pollution in the Antarctic environment is generating new and not well-understood threats to these ecosystems. As the identification of sources, concentrations, and persistence of inorganic and organic pollutants poses considerable challenges (reviewed

in Caroli et al. 2001; Bargagli 2005), their effects on seaweeds and their communities are hitherto widely unknown. Moreover, many contaminants are reactive to other environmental factors (e.g., UV radiation), which may enhance their detrimental impact on biota. Because all these different variables are changing simultaneously, the research on the impact of their interactive effects (synergistic, antagonistic, additive, etc.) is challenging (see Chap. 7 by Huovinen and Gómez).

Finally, the contents of this book are in agreement with the increasing awareness of the importance of Antarctic and its biota in global processes and the urgency to improve our understanding on the role and sentinel responses of seaweeds to global climate change. We believe that a comprehensive account of the progress made in the last decades is timely and urgent in order to put into perspective how diversity, ecophysiological adaptations, and ecosystem relations of seaweeds will be molded in the future Antarctica.

References

Amsler CD, Rowley RJ, Laur DR, Quetin LB, Ross RM (1995) Vertical distribution of Antarctic Peninsular macroalgae: cover, biomass, and species composition. Phycologia 34:424–430

Amsler CD, Iken K, McClintock JB, Baker BJ (2011) Defenses of polar microalgae against herbivores and biofoulers. In: Wiencke C (ed) Biology of polar benthic algae. Walter de Gruyter GmbH & Co, KG, Berlin, New York, pp 101–120

Bargagli R (ed) (2005) Antarctic ecosystems: environmental contamination, climate change, and human impact. Springer, Berlin

Barnes DKA, Conlan KE (2012) The dynamic mosaic–disturbance and development of Antarctic benthic communities. In: Rogers AD, Johnston NM, Murphy EJ, Clarke A (eds) Antarctic ecosystems: an extreme environment in a changing world. Wiley, Chichester, UK

Barnes DKA, Fleming A, Sands CJ, Quartino ML, Deregibus D (2018) Icebergs, sea ice, blue carbon and Antarctic climate feedbacks. Philos Trans R Soc A 376 (2122): 20170176

Benning LG, Anesio AM, Lutz S, Tranter M (2014) Biological impact on Greenland's albedo. Nat Geosci 7:691

Bergstrom DM, Convey P, Huiskes AHL (eds) (2006) Trends in Antarctic terrestrial and limnetic ecosystems. Springer, Dordrecht, The Netherlands

Billard E, Reyes J, Mansilla A, Faugeron S, Guillemin M-L (2015) Deep genetic divergence between austral populations of the red alga *Gigartina skottsbergii* reveals a cryptic species endemic to the Antarctic continent. Polar Biol 38:2021–2034

Bronselaer B, Winton M, Griffies SM, Hurlin WJ, Rodgers KB, Sergienko OV, Stouffer RJ, Russell JL (2018) Change in future climate due to Antarctic meltwater. Nature 564:53–58

Brouwer PEM, Geilen EFM, Gremmen NJM, van Lent F (1995) Biomass, cover and zonation patterns of sublittoral macroalgae at Signy Island, South Orkney Islands, Antarctica. Bot Mar 38:259–270

Caroli S, Cescon P, Walton BT (eds) (2001) Environmental contamination in Antarctica, a challenge to analytical chemistry. Elsevier, Amsterdam, London

Clark GF, Stark JS, Palmer AS, Riddle MJ, Johnston EL (2017) The roles of sea-ice, light and sedimentation in structuring shallow Antarctic benthic communities. PLoS One 12(1):e0168391. https://doi.org/10.1371/journal.pone.0168391

Clarke A, Murphy EJ, Meredith MP, King JC, Peck LS, Barnes DKA, Smith RC (2007) Climate change and the marine ecosystem of the western Antarctic Peninsula. Philos Trans R Soc B 362:149–141

Clayton MN (1994) Evolution of the Antarctic benthic algal flora. J Phycol 30:897–904

Constable AJ, Melbourne -Thomas J, Corney SP, Arrigo KR, Barbraud C, Barnes DKA et al (2014) Climate change and Southern Ocean ecosystems. I: how changes in physical habitats directly affect marine biota. Glob Chang Biol 20:3004–3025. https://doi.org/10.1111/gcb.12623

Crame JA (1992) Evolutionary history of the polar regions. Hist Biol 6:37–60

Delépine R, Lamb JM, Zimmermann MH (1966) Preliminary report on the vegetation of the Antarctic Peninsula. Proc Int Seaweed Symp 5:107–116

Deregibus D, Quartino ML, Campana GL, Momo FR, Wiencke C, Zacher K (2016) Photosynthetic light requirements and vertical distribution of macroalgae in newly ice-free areas in Potter Cove, South Shetland Islands, Antarctica. Polar Biol 39:153–166

Drew EA, Hastings RM (1992) A year-round ecophysiological study of *Himantothallus grandifolius* (Desmarestiales, Phaeophyta) at Signy Island, Antarctica. Phycologia 31:262–277

Fraser CI, Morrison AK, Hogg AM, Macaya EC, van Sebille E, Ryan PG, Padovan A, Jack C, Valdivia N, Waters JM (2018) Antarctica's ecological isolation will be broken by storm-driven dispersal and warming. Nat Clim Change 8:704–708. https://doi.org/10.1038/s41558-018-0209-7

Gaudichaud C (1826) Botanique. In: Freycinet L (ed) Voyage autour du monde, entrepons par ordre du Roi, execute sur le corvettes de S.M. l 'Uranie' et la 'Physicienne', pendant les années 1817, 1818, 1819 et 1820. Pillet Aîné, Paris

Gómez I, Thomas DN, Wiencke C (1995) Longitudinal profiles of growth, photosynthesis and light independent carbon fixation in the Antarctic brown alga *Ascoseira mirabilis*. Bot Mar 38:157–164

Gómez I, Weykam G, Klöser H, Wiencke C (1997) Photosynthetic light requirements, daily carbon balance and zonation of sublittoral macroalgae from King George Island (Antarctica). Mar Ecol Prog Ser 148:281–293

Gutkowski R, Maleszewski S (1989) Seasonal changes of the photosynthetic capacity of the Antarctic macroalga *Adenocystis utricularis* (Bory) Skottsberg. Pol Biol 10:145–148

Held C (2014) Chapter 10.5. Phylogeography and population genetics. In: De Broyer C, Koubbi P, Griffiths HJ, Raymond B, d' U d'A C, Van de Putte AP, Danis B, David B, Grant S, Gutt J et al (eds) Biogeographic Atlas of the Southern Ocean. Scientific Committee on Antarctic Research, Cambridge, pp 437–440

Hommersand MH, Moe RL, Amsler CD, Fredericq S (2009) Notes on the systematics and biogeographical relationships of Antarctic and subAntarctic Rhodophyta with descriptions of four new genera and five new species. Bot Mar 52:509–534

Hood E, Battin TJ, Fellman J, O'Neel S, Spencer RGM (2015) Storage and release of organic carbon from glaciers and ice sheets. Nat Geosci 8:91–96

Hooker JD (1847) The botany of the Antarctic voyage of H. M. discovery ships *Erebus* and *Terror* in the years 1839–1843, vol. 1, Flora Antarctica, Part 1 Botany of Lord Auklands Group and Campbell's Island, Part 2 Botany of Fuegia, The Falklands, Kerguelens Land. Reeve Brothers, London

Huovinen P, Gómez I (2013) Photosynthetic characteristics and UV stress tolerance of Antarctic seaweeds along the depth gradient. Polar Biol 36:1319–1332

Huovinen P, Ramírez J, Gómez I (2018) Remote sensing of albedo-reducing snow algae and impurities in the Maritime Antarctica. ISPRS J Photogram Rem Sens 146:507–517

Iken K, Amsler CD, Amsler MO, McClintock JB, Baker BJ (2009) Field studies on deterrent roles of phlorotannins in Antarctic brown algae. Bot Mar 52:547–557

Iñiguez C, Heinrich S, Harms L, Gordillo FJL (2017) Increased temperature and CO_2 alleviate photoinhibition in *Desmarestia anceps*: from transcriptomics to carbon utilization. J Exp Bot 68(14):3971–3984

IPCC (2019) IPCC special report on the ocean and cryosphere in a changing climate. Pörtner H-O, Roberts DC, Masson-Delmotte V, Zhai P, Tignor M, Poloczanska E, Mintenbeck K, Nicolai M, Okem A, Petzold J, Rama B, Weyer N (eds)

Kanao M, Genti Toyokuni G, Yamamoto M-Y (eds) (2018) Antarctica–a key to global change. IntechOpen. https://doi.org/10.5772/intechopen.82197

Klöser H, Ferreyra G, Schloss I, Mercuri G, Laturnus F, Curtosi A (1993) Seasonal variation of algal growth conditions in sheltered Antarctic bays: the example of Potter Cove (King George Island, South Shetlands). J Mar Syst 4:289–301

Klöser H, Quartino ML, Wiencke C (1996) Distribution of macroalgae and macroalgal communities in gradients of physical conditions in Potter Cove, King George Island, Antarctica. Hydrobiologia 333:1–17

Krause-Jensen D, Duarte CM (2016) Substantial role of macroalgae in marine carbon sequestration. Nat Geosci 9(10):737

Lamb IM, Zimmermann MH (1977) Benthic marine algae of the Antarctic Peninsula. Ant Res Ser 23:130–229

Meredith MP, King JC (2005) Rapid climate change in the ocean west of the Antarctic Peninsula during the second half of the 20th century. Geophys Res Lett 32:L19604. https://doi.org/10.1029/2005GL024042

Meredith MP, Falk U, Bers AV, Mackensen A, Schloss IR, Ruiz Barlett E, Jerosch K, Silva Busso A, Abele D (2018) Anatomy of a glacial meltwater discharge event in an Antarctic cove. Phil Trans R Soc A 376:20170163. https://doi.org/10.1098/rsta.2017.0163

Moe R, Henry EC (1982) Reproduction and early development of *Ascoseira mirabilis* Skottsberg (Phaeophyta), with notes on Ascoseirales Petrov. Phycologia 21:55–56

Mormède S, Irisson JO, Raymond B (2014) Chapter 2.3. Distribution modelling. In: De Broyer C, Koubbi P, Griffiths HJ, Raymond B, Udekem d'Acoz C d', Van de Putte AP, Danis B, David B, Grant S, Gutt J et al (eds) Biogeographic Atlas of the Southern Ocean. Scientific Committee on Antarctic Research, Cambridge, pp 27–29

Neushul M (1965) Diving observations of subtidal Antarctic marine vegetation. Bot Mar 8:234–243

Ortiz M, Hermosillo-Nuñez B, González J, Rodríguez-Zaragoza F, Gómez I, Jordán F (2017) Quantifying keystone species complexes: Ecosystem-based conservation management in the King George Island (Antarctic Peninsula). Ecol Ind 81: 453-460

Quartino ML, Deregibus D, Campana GL, Edgar G, Latorre J, Momo FR et al (2013) Evidence of macroalgal colonization on newly ice-free areas following glacial retreat in Potter Cove (South Shetland Islands), Antarctica. PLoS One 8:e58223

Rogers AD, Johnston NM, Murphy EJ, Clarke A (eds) (2012) Antarctic ecosystems: an extreme environment in a changing world. Wiley, Chichester, UK

Schloss IR, Abele D, Moreau S, Demers S, Bers AV, Gonzalez O, Ferreyra GA (2012) Response of phytoplankton dynamics to 19-year (1991–2009) climate trends in Carlini Station (Antarctica). J Mar Syst 92:53–66. https://doi.org/10.1016/j.jmarsys.2011.10.006

Schwarz A, Hawes I, Andrew N, Norkko A, Cummings V, Thrush S (2003) Macroalgal photosynthesis near the southern global limit for growth, Cape Evans, Ross Sea, Antarctica. Polar Biol 26:789–799

Skottsberg K (1907) Zur Kenntnis der subantarktischen und antarktischen Meeresalgen. I. Phaeophyceen. In: Nordenskjöld O (ed) Wissenschaftliche Ergebnisse der Schwedischen Südpolar-Expedition 1901–1903, vol 4: 1. Lithographisches Institut des Generalstabs, Stockholm, pp 1–172

Tedesco M, Doherty S, Fettweis X, Alexander P, Jeyaratnam J, Stroeve J (2016) The darkening of the Greenland ice sheets: trends, drivers, and projections (1981–2100). Cryosphere 10:477–496. https://doi.org/10.5194/tc-10-477-2016

Tedstone AJ, Bamber JL, Cook JM, Williamson CJ, Fettweis X, Hodson AJ, Tranter M (2017) Dark ice dynamics of the south-west Greenland ice sheet. Cryosphere 11:2491–2506. https://doi.org/10.5194/tc-11-2491-2017

Tin T, Liggett D, Maher PT, Lamers M (2014) Antarctic futures: human engagement with the Antarctic environment. Springer, The Netherlands

Turner J, Lu H, White I, King JC, Phillips T, Hosking JS, Bracegirdle TJ, Marshall GJ, Mulvaney R, Deb P (2016) Absence of 21st century warming on Antarctic Peninsula consistent with natural variability. Nature 535:411–415

Valdivia N, Díaz MJ, Garrido I, Gómez I (2015) Consistent richness-biomass relationship across environmental stress gradients in a marine macroalgal-dominated subtidal community on the Western Antarctic Peninsula. Plos One 10(9):e0138582

Van Oppen MJH, Olsen JL, Stam WT, Van der Hoek C, Wiencke C (1993) Arctic–Antarctic disjunction in the benthic seaweeds *Acrosiphonia arcta* (Chlorophyta) and *Desmarestia viridis* (Phaeophyta) are of recent origin. Mar Biol 115:381–386

Vaughan DG, Marshall GJ, Connolley WM, Parkinson CL, Mulvaney R, Hodgson DA, King JC, Pudsey CJ, Turner J (2003) Recent rapid regional climate warming on the Antarctic Peninsula. Clim Change 60:243–274. https://doi.org/10.1023/A:1026021217991

Wadham JL, Hawkings JR, Tarasov L, Gregoire LJ, Spencer RGM, Gutjahr M, Ridgwell A, Kohfeld KE (2019) Ice sheets matter for the global carbon cycle. Nat Comm 10 (1)

Wiencke C (1988) Notes on the development of some benthic marine macroalgae of King George Island, Antarctica. Ser Cient INACh 37:23–47

Wiencke C (1990) Seasonality of brown macroalgae from Antarctica- a long-term culture study under fluctuating Antarctic daylengths. Polar Biol 10:589–600

Wiencke C (1996) Recent advances in the investigation of Antarctic macroalgae. Polar Biol 16:231–240

Wiencke C, Amsler CD, Clayton MN (2014) Chapter 5.1 Macroalgae. In: de Broyer C, Koubbi P, Griffiths HJ, Raymond B, Udekem d'Acoz Cd', Van de Putte PAP, Danis B, David B, Grant S, Gutt J et al (eds) Biogeographic atlas of the Southern Ocean. Scientific Committee on Antarctic Research, Cambridge, pp 66–73

Zacher K, Wulff A, Molis M, Hanelt D, Wiencke C (2007) UV radiation and consumer effects on a field-grown intertidal macroalgal assemblages in Antarctica. Glob Chang Biol 13:1201–1215

Zaneveld JS (1966) Vertical zonation of Antarctic and subAntarctic benthic marine algae. Antarctic J US 1(5):211–213

Zielinski K (1981) Benthic macroalgae of Admiralty Bay (King George Island, South Shetland Islands) and circulation of algal matter between the water and the shore. Pol Polar Res 2:71–94

Zielinski K (1990) Bottom macroalgae of the Admiralty Bay (King George Island, South Shetlands, Antarctica). Pol Polar Res 11:95–131

Part II
Diversity and Biogeography

Chapter 2
Diversity of Antarctic Seaweeds

Mariana C. Oliveira, Franciane Pellizzari, Amanda S. Medeiros, and Nair S. Yokoya

Abstract Antarctica is characterized by extremes of climate and biogeographic isolation from other continents by distance, high depths, and the Antarctic Circumpolar Current. Even under these harsh conditions, macroalgae thrive in different coastal ecosystems contributing to primary production and serving as habitat and food for a variety of species of marine fauna. However, it is known that the Antarctic marine flora presents low species richness compared to other biogeographical regions: until the past decade a number of 120 Antarctic seaweeds had been reported. On the other hand, long geographical isolation and extreme climatic and oceanographic conditions justify their high degree of endemism (ca. of 33–40%). A new compilation of the Antarctic seaweed diversity is presented in this chapter, reporting a list of 151 species cited to the entire Antarctica, comprising 85 Rhodophyta, 34 Ochrophyta (Phaeophyceae and Chrysophyceae), and 32 Chlorophyta with an endemism degree of 27%, lower than in previous reports. Molecular approaches based on different markers (ITS, UPA, COI-5P) are being used to assist species identification. The collection of marine specimens in Antarctica is expensive and still very difficult, and therefore, the occurrence for many species can become inaccurate. The difficult access to samples is another limitation, which could explain that most of the best known species are concentrated around scientific stations. Consequently, the macroalgal diversity in Antarctica and its distribution is probably underestimated. A better knowledge on this diversity and its distribution is urgent, as the region is facing significant climate changes that may drive shifts on the assemblages of macroalgae.

M. C. Oliveira (✉) · A. S. Medeiros
Department of Botany, Institute of Biosciences, University of São Paulo, São Paulo, SP, Brazil
e-mail: mcdolive@ib.usp.br

F. Pellizzari
Laboratório de Ficologia e Qualidade de Água Marinha, Universidade Estadual do Paraná, Campus Paranaguá, Paranaguá, PR, Brazil
e-mail: franciane.pellizzari@unespar.edu.br

N. S. Yokoya
Institute of Botany, São Paulo, SP, Brazil
e-mail: nyokoya@pq.cnpq.br

I. Gómez, P. Huovinen (eds.), *Antarctic Seaweeds*,
https://doi.org/10.1007/978-3-030-39448-6_2

Keywords Benthic algae · Diversity · Endemic species · Maritime Antarctica · Molecular taxonomy

2.1 The Antarctic Environment

Antarctica is characterized by extremes of climate, which makes its habitats and biogeographic context quite unique. It is isolated from other continents by distance, high depths, and oceanographic currents, mainly by the Antarctic Circumpolar Current (Barker and Thomas 2003) (Fig. 2.1). Broad variations in temperature and salinity can occur in the Antarctic habitat, with high incidence of alternating ultraviolet radiation with long periods of absence of light, freezing, and thawing cycles that determine highly variable physical mosaics (Wynn-Williams 1996; Vincent 2000; Clarke et al. 2005). However, few attempts have been made to provide sub-

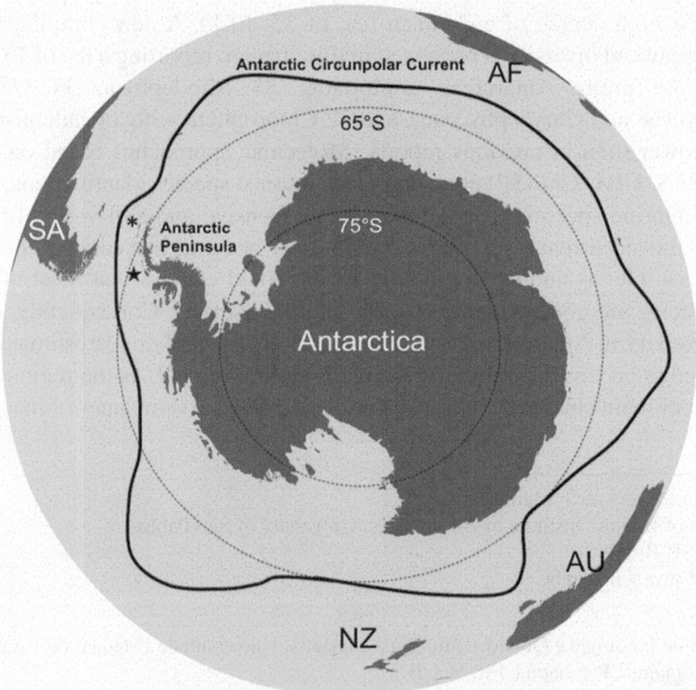

Fig. 2.1 Schematic map showing Antarctica and surroundings, Africa (AF), Australia (AU), New Zeeland (NZ), and South America (SA). King George Island (South Shetland Islands) is marked by ∗ and Adelaide Island is marked by ★. The Antarctic Circumpolar Current is represented by a black line and arrowheads. (https://freevectormaps.com/globes/antarctica/GLB-AN-01-0001)

stantial insight into the implications of these gradients of abiotic factors for the spatial distribution of biodiversity (Convey et al. 2014).

Low temperatures, high salinities, and the occurrence of long periods of ice cover in coastal regions can characterize the Southern Ocean. The seasonality of irradiance levels and photoperiod, as well as the ice cover, exert a strong effect on supralittoral, eulittoral, and sublittoral communities (Hempel 1987; Drew and Hastings 1992). The alternation of annual ice cycles impacts the physical and chemical environment in many ways and, consequently, the local biota as a whole. The effects of ice cover are most obvious in the intertidal zone and in the upper sublittoral due to the physical action of the displacement of ice blocks that remove sessile organisms, including macroalgae. Ice formation, as well as summer melting, significantly alters salinity and light penetration, exposing benthic organisms to extreme values of these parameters.

Even so, the Southern Ocean is known to afford living space for high abundance of benthic organisms compared to other regions of the world. Since the first studies carried out in the region, the high densities and relatively high diversity of benthic communities have been demonstrated (Clarke 1990).

2.2 Seaweeds in Antarctica: Definition and Importance

Similar to the terrestrial plants, seaweeds – or marine macroalgae – are photosynthetic organisms that form underwater forests on consolidated or hard substrate. In virtue of their diversity and biomass, seaweeds play a key role in sustaining the primary production in coastal ecosystems and are important food sources, especially to the rockfish, *Notothenia coriiceps*; echinoderms, e.g., *Odontaster validus* and *Sterechinus neumayeri*; and amphipods, e.g., *Gondogeneia antarctica*. Moreover, macroalgae serve as shelter and substrate for the growth and reproduction of several marine organisms (Nedzarek and Rakusa-Suszcewski 2004, see also Chap. 15 by Momo et al. and Chap. 16 by Ortiz et al.).

Antarctic seaweeds belong to the major divisions of algae: golden-brown and brown (Ochrophyta: Chrysophyceae and Phaeophyceae, respectively), green (Chlorophyta), and red algae (Rhodophyta). The morphology of Antarctic macroalgae varies from crustose forms or delicate filamentous forms of few centimeters to large foliaceous, terete, and leathery forms that may reach several meters long (Fig. 2.2). Some species deposit calcium carbonate in their cell walls, presenting a rigid thallus. These calcareous species provide substrate for other marine species (see also Chap. 11 by Gómez and Huovinen). Specific groups of algae are protected against predation by synthesizing chemical compounds that make them unpalatable (see Chap. 17 by Amsler et al.). The synthesis and accumulation of these bioactive compounds is the result of a long and complex evolutionary and ecological process. Some of these substances, especially for the species adapted to extreme polar ecosystems, may have biotechnological and pharmaceutical applications. The reproductive sexual and/or asexual cells of Antarctic seaweeds are well adapted to the

Fig. 2.2 Macromorphology of Antarctic seaweeds. (**a**) Crustose calcareous red alga *Clathromorphum* sp. (Rhodophyta). (Photo by F. Pellizari); (**b**) monostromatic saccate green alga *Monostroma hariotii* (Chlorophyta); (**c**) terete red alga *Plocamium* sp. (Rhodophyta). (Photos by A.S. Medeiros); (**d**) leathery brown alga *Himantothallus grandifolius* (Phaeophyceae), with an approximate length of 3.7 m. (Photo by E.C. Oliveira)

environmental conditions marked by low temperature and limited solar radiation. These cells, either an asexual stage, a gamete, or a zygote (after gametes fusion), will float within the marine currents and settle and develop on a suitable substrate, in a dynamic tuned with the seasonal Antarctic variability (see also Chap. 10 by Navarro et al.).

In the Southern Ocean, macroalgae together with the phytoplankton are responsible for 30–50% of the dissolved O_2 released in the marine habitats surrounded by the Southern Ocean (data retrieved from visibleearth.nasa.gov). In particular, the robust and perennial endemic brown algae (Phaeophyceae), such as *Himantothallus grandifolius* and *Desmarestia* spp., are well represented in terms of biomass and are the main primary producers in these areas (Amsler et al. 1995; Wiencke et al. 2014).

2.3 Seaweed Taxonomic Studies in Antarctica: Toward a New Species Compilation

Antarctic seaweed communities have been characterized by low diversity and high levels of endemism (Lamb and Zimmermann 1977; Wiencke et al. 2014). The first reports date back to the nineteenth century when seaweeds were collected during cruises of the corvettes Uranie and Physicienne (Gaudichaud 1826). However, the first compilations on diversity of Antarctic seaweeds along the Antarctic Peninsula were published much later by Skottsberg (1906, 1941, 1953, 1964), who is considered the pioneer of the Antarctic phycology. The first scuba diving surveys were conducted by Neushul (1959, 1961, 1963, 1965, 1968) and Skottsberg and Neushul (1960). Skottsberg (1964) estimated the occurrence of 96 species of Antarctic seaweeds (16 Chlorophyta, 19 Phaeophyceae, and 61 Rhodophyta). In later surveys, higher species richness was documented: 100 species (Papenfuss 1964; Moe 1985), 120 species (Clayton 1994), and 117–123 species (Wiencke and Clayton 2002). The most recent review on seaweed diversity for the entire Antarctica reported a total of 124 taxa, comprising 80 species of Rhodophyta, 27 species of Phaeophyceae, and 17 Chlorophyta (Wiencke et al. 2014).

Different researchers have broadly studied seaweed communities from the South Shetland Islands, mainly from the King George Island, including Zielinski (1990), Quartino et al. (2005), Oliveira et al. (2009), Valdivia et al. (2014), and Gómez et al. (2019). The macroalgal diversity from Deception Island and Livingston Island was studied by Ramírez (1982), Clayton et al. (1997), and Gallardo et al. (1999). Medeiros (2013) generated a set of sequences of macroalgal DNA barcodes and phylogenetic markers from Admiralty Bay (King George Island), contributing to a molecular database useful for future investigations on the diversity of Antarctic seaweeds. Molecular data were obtained for 8 species of Chlorophyta, 9 species of Phaeophyceae, and 14 species of Rhodophyta. *Prasiola* sp., *Protomonostroma rosulatum* (Chlorophyta), *Chordaria linearis* (Phaeophyceae), *Acanthococcus antarcticus*, and *Plumariopsis peninsularis* (Rhodophyta) are new records for Admiralty Bay, and *Callophyllis* sp. is possibly a new species for science.

Mystikou et al. (2014) analyzed the seaweed diversity in the Southwestern Antarctic Peninsula (Adelaide Island) over records of 35 years, reporting 41 species. This part of the Antarctic Peninsula is a key region affected by contemporary climate change, but has been rarely studied. Sanches et al. (2016) performed multivariate analyses of Antarctic and sub-Antarctic seaweed distribution patterns focusing on a new evaluation of the role of the Antarctic Circumpolar Current. Regarding species diversity, in this study the authors identified 129 and 145 macroalgal species, respectively, for the Southern Antarctic Circumpolar Front (SACF) and for the Polar Front (PF) and considering genera, 95 and 101, for the SACF and PF, respectively.

Pellizzari et al. (2017) recorded a total of 104 species of benthic marine algae along the South Shetland Islands (28 Phaeophyceae, 24 Chlorophyta, and 52 Rhodophyta), representing 82% of all seaweed taxa described in Antarctica. The authors also reported nine new records, mainly previously recorded at other latitudes (*Protomonostroma rosulatum, Monostroma grevillei, Cladophora coelothrix, Chaetomorpha irregularis, Dictyota decumbens, Asteronema ferruginea, Microzonia velutina, Cladodonta lyalli, Rhodophyllis centrocarpa*) and two putative new species of *Prasiola* sp. and *Callophyllis* sp. (see also Medeiros 2013).

Spatial variation in the species diversity was observed among the collecting sites: Livingston Island and King George Island showed the highest diversity. Deception Island, an area with geothermal activity and intense tourism, was dominated by opportunistic and broadly distributed filamentous green algae (see also Chap. 5 by Pellizzari et al.).

A new compilation of the Antarctic seaweed diversity is presented in this chapter, based primarily on data reported by Wiencke and Clayton (2002), Oliveira et al. (2009), Wiencke et al. (2014), Mystikou et al. (2014), and Pellizzari et al. (2017). This survey reports a species richness of 151 species, comprising 85 Rhodophyta (plus one uncertain species), 34 Phaeophyceae, and 32 Chlorophyta to the entire Antarctic region (Tables 2.1, 2.2, and 2.3, based on Guiry and Guiry (2019) for taxonomical nomenclature).

Antarctic marine flora can be characterized by a high number of endemic species, approximately 33% (Wiencke and Clayton 2002), 35% (Neushul 1968; Wulff et al. 2009; Wiencke and Amsler 2012), or 39.6% (Skottsberg 1964). However, this percentage decreases to 27.1% of species restricted to areas within the Antarctic Circumpolar Current, i.e., Antarctica and sub-Antarctic islands (with latitude higher than 55°S) in the compilation of the present chapter (Tables 2.1, 2.2, and 2.3). Considering the degree of endemism close to 27.1% calculated from a total of 41 endemic species, Rhodophyta represents 61% of endemic taxa, Phaeophyceae 29.3%, and Chlorophyta 9.7%. However, if we consider separately and proportionally (total group richness/group endemic richness), the algal group with the highest endemism degree is Phaeophyceae (35.3%), followed by Rhodophyta (29.4%), and Chlorophyta (12.5%).

Large brown seaweeds of the order Laminariales (kelps) are common structuring species in cold-temperate coasts in the southern and northern hemisphere and also in the Arctic, while in Antarctica this order is substituted by the Desmarestiales,

Table 2.1 Compiled taxa list of Chlorophyta

Species	Antarctica[a]	South Shetland Islands[b]	King George Island[c]	Adelaide Island[d]
Acrosiphonia arcta (Dillwyn) Gain 1912	X	X	X	
Blidingia minima (Nägeli ex Kützing) Kylin 1947	X	X		
Chaetomorpha irregularis (Zaneveld) Cormaci, Furnari & Alongi 2014 (as *Lola irregularis*) **(E)**	X	X		
Chaetomorpha mawsonii Lucas 1919	X	X		
Cladophora coelothrix Kützing 1843		X		
Cladophora repens Harvey 1849	X			
Endophyton atroviride O'Kelly in Ricker 1987	X	X		
Entocladia maculans (AD Cotton) Papenfuss 1964	X			
Lambia antarctica (Skottsberg) Delépine 1967 **(E)**	X	X		
Monostroma grevillei (Thuret) Wittrock 1866		X		
Monostroma hariotii Gain 1911	X	X	X	X
Prasiola sp. (SP428305)		X		
Prasiola crispa (Lightfoot) Kützing 1843	X	X	X	
Protomonostroma rosulatum Vinogradova **(E)**		X		
Protomonostroma undulatum (Wittrock) Vinogradova 1969			X	
Pseudothrix groenlandica (Agardh) Hanic & SC Lindstrom 2008 (as *Capsosiphon groenlandicus*)				X
Rhizoclonium ambiguum (Hooker & Harvey) Kützing 1849		X		
Rhizoclonium riparium (Roth) Harvey 1849		X		
Spongomorpha pacifica (Montagne) Kützing 1854 (as *Acrosiphonia pacifica*)	X	X		
Ulothrix australis Gain 1911 **(E)**	X	X	X	
Ulothrix flacca (Dillwyn) Thuret in Le Jolis 1863	X	X		
Ulothrix subflaccida Wille 1901	X			
Ulothrix zonata (Weber & Mohr) Kützing 1833		X		
Ulva sp. (foliose)				X
Ulva compressa Linnaeus 1753	X	X		
Ulva hookeriana (Kützing) HS Hayden, Blomster, Maggs, Silva, Stanhope & Waaland 2003 (as *Enteromorpha bulbosa*)	X	X	X	

(continued)

Table 2.1 (continued)

Species	Antarctica[a]	South Shetland Islands[b]	King George Island[c]	Adelaide Island[d]
Ulva intestinalis Linnaeus 1753	X	X		
Ulva lactuca Linnaeus 1753	X			
Ulva rigida C.Agardh 1823	X			
Ulvella leptochaete (Huber) Nielsen, O'Kelly & Wysor in Nielsen et al. 2013				X
Ulvella viridis (Reinke) Nielsen, O'Kelly & Wysor in Nielsen et al. 2013 (as *Entocladia viridis*)	X	X		
Urospora penicilliformis (Roth) Areschoug 1866	X	X	X	X

Based on [a]Wiencke and Clayton (2002) and Wiencke et al. (2014), [b]Pellizzari et al. (2017), [c]Oliveira et al. (2009), and [d]Mystikou et al. (2014)
E = endemic species (restricted to Antarctic and sub-Antarctica)

Table 2.2 Compiled species list of Ochrophyta (Phaeophyceae and Chrysophyceae)

Species	Antarctica[a]	South Shetland Islands[b]	King George Island[c]	Adelaide Island[d]
Adenocystis utricularis (Bory) Skottsberg 1907	X	X	X	X
Antarctosaccion applanatum (Gain) Delépine **(E)**	X	X		X
Ascoseira mirabilis Skottsberg 1907 **(E)**	X	X	X	
Asteronema ferruginea (Harvey) Delépine & Asensi 1975		X		
Australofilum incommodum (Skottsberg) AF Peters 2003		X		
Chordaria linearis (Hooker & Harvey) Cotton 1915	X	X		
Cystosphaera jacquinotii (Montagne) Skottsberg 1907 **(E)**	X	X	X	
Desmarestia anceps Montagne 1842	X	X	X	
Desmarestia antarctica Moe & Silva 1989 **(E)**	X	X	X	
Desmarestia chordalis Hooker & Harvey 1845			X	
Desmarestia confervoides (Bory) Ramírez & Peters 1993	X			
Desmarestia menziesii Agardh 1848 **(E)**	X	X	X	X
Dictyota decumbens (Ricker) Hörnig, Schnetter & Prud'homme van Reine 1992		X		

(continued)

Table 2.2 (continued)

Species	Antarctica[a]	South Shetland Islands[b]	King George Island[c]	Adelaide Island[d]
Ectocarpus constanciae Hariot 1887		X		
Ectocarpus siliculosus (Dillwyn) Lyngbye 1819		X		
Elachista antarctica Skottsberg 1953 **(E)**	X	X	X	X
Geminocarpus austrogeorgiae Skottsberg 1907 **(E)**	X	X	X	X
Geminocarpus geminatus (Hooker & Harvey) Skottsberg 1907	X	X	X	X
Halopteris corymbosa (Dickie) Draisma, Prud'homme & Kawai 2010	X			
Halopteris obovata (Hooker & Harvey) Sauvageau 1904	X	X		
Haplogloia moniliformis RW Ricker 1987		X	X	
Himantothallus grandifolius (A Gepp & E Gepp) Zinova 1959 **(E)**	X	X	X	X
Leptonematella falklandica (Skottsberg) MJ Wynne 1969	X			
Lithoderma antarcticum Skottsberg 1953 **(E)**	X			X
Microzonia australe (Levring) Camacho & Fredericq 2018 (as *Syringoderma australe*) **(E)**		X		
Microzonia velutina (Harvey) Agardh 1894		X		
Petalonia fascia (Müller) Kuntze 1898	X	X		
Petroderma maculiforme (Wollny) Kuckuck 1897	X	X		
Phaeurus antarcticus Skottsberg 1907 **(E)**	X	X	X	
Pylaiella littoralis (Linnaeus) Kjellman 1872	X	X	X	X
Ralfsia australis Skottsberg 1921	X	X		
Scytosiphon lomentaria (Lyngbye) Link 1833	X	X		
Scytothamnus fasciculatus (Hooker & Harvey) Cotton 1915	X	X		
Utriculidium durvillei Skottsberg 1907	X		X	

[a]Wiencke and Clayton (2002) and Wiencke et al. (2014), [b]Pellizzari et al. (2017), [c]Oliveira et al. (2009), and [d]Mystikou et al. (2014)
E = endemic species (restricted to Antarctic and sub-Antarctica)

Table 2.3 Compiled taxa list of Rhodophyta

Species	Antarctica[a]	South Shetland Islands[b]	King George Island[c]	Adelaide Island[d]
Acanthococcus antarcticus Hooker & Harvey 1845 **(E)**	X	X		
Ahnfeltia plicata (Hudson) Fries 1836	X			
Antarcticothamnion polysporum Moe & Silva 1979 **(E)**	X			
Antarctocolax lambii Skottsberg 1953 **(E)**	X			
Ballia callitricha (C Agardh) Kützing 1843	X	X	X	X
*Ballia sertularioides (*Suhr) Papenfuss 1940	X			
Bangia fuscopurpurea (Dillwyn) Lyngbye 1819	X	X		
Callophyllis sp. 1		X		
Callophyllis sp. 2				X
Callophyllis atrosanguinea (JD Hooker & Harvey) Hariot 1887	X	X	X	
Callophyllis pinnata Setchell & Swezy 1923			X[e]	
Callophyllis tenera Agardh 1849	X			
Callophyllis variegata (Bory) Kützing 1843	X	X		
Carlskottsbergia antarctica (Hooker & Harvey) Athanasiadis 2018 (as *Lithophyllum antarcticum*)				X
Ceramium involutum Kützing 1849	X			
Cladodonta lyallii (Hooker & Harvey) Skottsberg 1923		X		
Clathromorphum sp.				X
Clathromorphum obtectulum (Foslie) Adey 1970	X	X		
Curdiea racovitzae Hariot in De Wildemann 1900 **(E)**	X	X	X	X
Delisea pulchra (Greville) Montagne 1844	X	X	X	
Erythrotrichia carnea (Dillwyn) Agardh 1883	X			
Falklandiella harveyi (Hooker) Kylin 1956 (as *Dasyptilon harveyi*)	X			
Gainia mollis Moe 1985 **(E)**	X	X		
Georgiella confluens (Reinsch) Kylin 1956 **(E)**	X	X	X	
Gigartina skottsbergii Setchell & Gardner 1936	X	X	X	
Gracilariopsis longissima (Gmelin) Steentoft, Irvine & Farnham 1995 (as *Gracilaria verrucosa*)	X			

(continued)

Table 2.3 (continued)

Species	Antarctica[a]	South Shetland Islands[b]	King George Island[c]	Adelaide Island[d]
Griffithsia antarctica Hooker & Harvey in Hooker 1847	X			
Gymnogongrus antarcticus Skottsberg 1953 **(E)**	X	X	X	
Gymnogongrus turquetii Hariot 1907 **(E)**	X	X	X	
Hildenbrandia lecannellieri Hariot 1887	X	X	X	
Hydrolithon sp.	X			
Hymenocladia sp.				X
Hymenocladiopsis prolifera (Reinsch) Wynne 2004 (as *H. crustigena*) **(E)**	X	X		X
Iridaea cordata (Turner) Bory de Saint-Vincent 1826	X	X	X	X
Iridaea mawsonii Lucas 1919*	X	X		
Leptophytum coulmanicum (Foslie) Adey 1970 **(E)**	X			
Lithothamnion granuliferum Foslie 1905	X	X		
Meiodiscus concrescens (KM Drew) Gabrielson in Gabrielson et al. 2000 (as *Audouinella concrescens*)	X			
Mesophyllum sp.				X
Microrhinus carnosus (Reinsch) Skottsberg 1923	X	X		
Myriogramme livida (Hooker & Harvey) Kylin 1924	X			
Myriogramme manginii (Gain) Skottsberg 1953 **(E)**	X	X		X
Myriogramme smithii (Hooker & Harvey) Kylin 1924	X	X		X
Nereoginkgo adiantifolia Kylin in Kylin & Skottsberg 1919 **(E)**	X	X		
Neuroglossum delesseriae (Reinsch) Wynne 1997 (as *N. ligulatum*) **(E)**	X	X	X	
Notophycus fimbriatus Moe 1986 **(E)**	X	X	X	X
Pachymenia orbicularis (Zanardini) Setchell & Gardner 1934	X	X	X	
Palmaria decipiens (Reinsch) Ricker 1987	X	X	X	X
Palmaria georgica (Reinsch) Ricker 1987	X	X		
Pantoneura plocamioides Kylin 1919 **(E)**	X	X		X
Paraglossum lancifolium (Agardh) Agardh (as *Delesseria lancifolia*)	X	X	X	
Paraglossum salicifolium (Reinsch) Showe in Fredericq & Hommersand 2012 (as *Delesseria salicifolia*)	X	X		X

(continued)

Table 2.3 (continued)

Species	Antarctica[a]	South Shetland Islands[b]	King George Island[c]	Adelaide Island[d]
Peyssonnelia harveyana P Crouan & H Crouan ex Agardh 1851	X			
Phycodrys antarctica (Skottsberg) Skottsberg 1923 **(E)**	X	X		X
Phycodrys austrogeorgica Skottsberg 1923 **(E)**	X	X		X
*Phycodrys quercifolia (*Bory) Skottsberg 1922	X	X		
Phyllophora abyssalis Skottsberg in Kylin & Skottsberg 1919 **(E)**				X
Phyllophora ahnfeltioides Skottsberg in Kylin & Skottsberg 1919 **(E)**	X	X		
Phyllophora antarctica A Gepp & ES Gepp 1905 **(E)**	X			X
Leptophytum foecundum (Kjellmann) Adey	X			
Phymatolithon lenormandii (Areschoug) Adey 1966	X			
Picconiella plumosa (Kylin) De Toni 1936	X	X	X	
Plocamium cartilagineum (Linnaeus) Dixon 1967	X	X	X	X
Plocamium secundatum (Kützing) Kützing 1866	X	X	X	X
Plocamium hookeri Harvey in Hooker & Harvey 1845 **(E)**	X	X		X
*Plumariopsis eatonii (*Dickie) De Toni 1903	X			
Plumariopsis peninsularis Moe & Silva 1983 **(E)**	X	X		
Polysiphonia abscissa Hooker & Harvey 1845	X			
Pseudolithophyllum sp.		X		
Pterothamnion antarcticum (Kylin) Moe & Silva 1980	X			
Pterothamnion simile (Hooker & Harvey) Nägeli 1862	X			
Ptilonia magellanica (Montagne) Agardh 1852	X			
Porphyra plocamiestris Ricker 1987 **(E)**		X	X	X
Porphyra umbilicalis Kützing 1843	X			
Porphyra woolhouseae Harvey 1863			X	
Pyropia endiviifolia (A Gepp & E Gepp) Choi & Hwang in Sutherland et al. 2011 (as *Porphyra endiviifolia*) **(E)**	X	X	X	

(continued)

.

Table 2.3 (continued)

Species	Antarctica[a]	South Shetland Islands[b]	King George Island[c]	Adelaide Island[d]
Rhodochorton purpureum (Lightfoot) Rosenvinge 1900 (as *Audouinella purpurea*)	X			
Rhodophyllis centrocarpa (Montagne) Wynne		X		
Rhodymenia coccocarpa (Montagne) Wynne 2007 (as *Rhodymenia subantarctica*)	X	X		X
Rubrointrusa membranacea (Magnus) Clayden & Saunders 2010 (as *Audouinella membranacea*)	X			
Sarcodia sp.				X
Sarcodia montagneana (Hooker & Harvey) Agardh 1852	X	X		
Sarcothalia circumcincta (Agardh) Hommersand in Hommersand et al. 1993	X			
Sarcothalia papillosa (Bory) Leister in Hommersand, Guiry, Fredericq & Leister 1993	X	X		

Adapted from [a]Wiencke and Clayton (2002) and Wiencke et al. (2014), [b]Pellizzari et al. (2017), [c]Oliveira et al. (2009), [c]Mystikou et al. (2014), [d]Mystikoy et al. (2014) and [e]Yoneshigue-Valentin et al. (2013)

E = endemic species (restricted to Antarctic and sub-Antarctica)

*This species is considered uncertain in Algaebase (https://www.algaebase.org/GuiryandGuiry 2019), and therefore was not considered in the calculations for total and endemic species.

e.g., *Himantothallus grandifolius* and various *Desmarestia* spp. (Moe and Silva 1977; Clayton 1994). The fact that Laminariales do not reproduce at temperatures below 0°C and/or the competition with Desmarestiales could explain the absence of kelps in Antarctica (Peters and Breeman 1993).

2.4 Molecular Taxonomy for the Study of Antarctic Seaweed Diversity

Some Antarctic macroalgae exhibit phenotypic plasticity, which leads to identification problems; therefore, it is important to use molecular markers (e.g., DNA barcoding) as an additional tool to gain accuracy in the classification (Medeiros 2013). However, there are some difficulties in the applicability of this technique, due to the occurrence of variability in the region of the primers, especially for *cox*1 DNA barcode, and the lack of effectiveness of this marker in the case of Chlorophyta. The proposal of a quick, easy, and low-cost method for inventory of seaweed biodiversity shows advantages over other techniques, since standardization of a DNA barcode in the identification and report of new records or new species facilitates the exchange of information among laboratories around the world (Le Gall and Saunders 2010).

Medeiros (2013) has successfully used three DNA barcodes for the identification of the macroalgal assemblages of Admiralty Bay, despite some limitations. The three used DNA barcodes (UPA, *cox*1, and *tuf*A), although presented different rates of divergence, were consistent with other markers used in the analyses. The main advantage found in the amplification of the UPA marker (Sherwood et al. 2010) was the universality of the primers, since a single pair of primers was used for the three seaweed groups, whereas different combinations of primers were required to amplify the *cox*1. However, the low variability in UPA sequences may underestimate the species diversity, while *cox*1 presents a higher level of divergence, being more suitable as a specific marker. The absence of universal primers can limit the amplification in some groups of algae. In other instance, *tuf*A showed to be very promising for green algae, since it was possible to amplify this marker with a single set of primers and the levels of divergence found were relatively higher than for UPA and comparable to those found for *cox*1 in red and brown algae. Finally, the use of the UPA can be a fast and efficient tool for biodiversity monitoring, mainly of cryptogenic and cryptic seaweed species from Antarctica. Medeiros (2013) conclude that due to the analysis of only a few specimens per species in Admiralty Bay, it was not possible to establish a reliable limit between the values of intraspecific and interspecific divergences (barcode gap) for the distinct genera, which according to Meier et al. (2008) is necessary for successful species identification.

Thus, further studies are needed to establish these limits in order to use the DNA barcoding technique extensively in the identification and surveys of macroalgae diversity from Antarctica. Although larger molecular markers such as *rbc*L, ITS and SSUrDNA, which have more sequences available for comparison in global database, are relevant to obtain. In addition, the use of GenBank data for molecular species identification purposes is not totally reliable, since accurate species identification is not always guaranteed and identification based on morphological characters is still indispensable.

Seaweed species from Admiralty Bay, King George Island, exhibit affinities with seaweeds from South America and New Zealand, as well as with cold and polar regions from the northern hemisphere. In Hommersand et al. (2009), *rbc*L data showed that the affinity between the Antarctic and South American Rhodophyta species is complex, since the species groups are phylogenetically distant. Still, according to Hommersand et al. (2009), there is a possibility that particularities in the Antarctic climate caused a faster evolution of the *rbc*L gene, compared to temperate waters from South America. This fact may justify the high divergence found by Medeiros (2013) in sequences of *rbc*L for the species *Iridaea cordata* and *Plocamium* aff. *cartilagineum* obtained in Antarctica, when compared to the same species sampled in Chile. Finally, the author demonstrated that the DNA barcoding tool, together with other markers, proved to be a very suitable approach for large-scale application in biodiversity and conservation studies, providing information for global database that would combine molecular, morphological, and distributional data (see also Chap. 5 by Pellizzari et al.).

2.5 Seaweed Distribution in Antarctica

Seaweeds are conspicuous components of subtidal benthic communities; however, their abundance and diversity vary strongly depending on the habitat characteristics (Klöser et al. 1996). In many cases the hard-bottom substrata are often covered by crustose red algae (e.g., *Hildenbrandia lecannellieri* and species of the order Corallinales). Pebbles and boulders of granite, as well as volcanic outcrops, often form the hard substrate. These formations are common on King George Island, Penguin Island, Nelson Island, and Robert Island. Elephant Island, Livingston Island, and Half Moon Island are formed of volcanic bedrocks and of pebbles in the intertidal and subtidal zones. The slope and exposure to waves, ice melting, and abrasion vary among Antarctic islands and the continental coast influencing directly the recruitment of seaweeds (Pellizzari et al. 2017).

Large brown algae of the order Desmarestiales (e.g., *Desmarestia anceps*, *D. menziesii*, and *Himantothallus grandifolius*) are dominant in the benthic communities of Western Antarctica, growing in substrata below the zone affected by ice scouring down to ≥30-m water depth (Wiencke and Clayton 2002). On the other hand, red algae are dominant in the benthic communities at higher latitudes, as in the Ross Sea. Many macroalgae may survive, after detached from the original substrate, during long periods as free-floating thalli. Recent studies (synopsis in Fraser et al. 2018) report that these robust free-floating specimens may also transport hitchhiker species by rafting, including smaller epiphyte seaweed species and several marine invertebrates. These floating seaweed masses anchor other species, being recognized as vectors of introduction of alien/cryptogenic taxa around Antarctica (see also Chap. 3 by Fraser et al. and Chap. 4 by Macaya et al.).

Intertidal seaweeds are filamentous (e.g., the green algae *Urospora penicilliformis* and *Ulothrix* spp.), foliose (e.g., *Monostroma hariotii, Pyropia endiviifolia*), or saccate (e.g., *Adenocystis utricularis*), or possess perennial basal attachments (e.g., *Palmaria decipiens*). Intertidal species are well developed in the milder climates of the Antarctic maritime islands, especially in the South Shetland and the South Orkney archipelagos (Wiencke and Clayton 2002; Oliveira et al. 2009). These organisms display a suite of morpho-functional adaptations to cope with a high variability of physical factors, such as temperature, ice scour, light, salinity, etc. (Gómez et al. 2019, see also Chap. 11 by Gómez and Huovinen).

In the Maritime Antarctica, which compress the Antarctic Peninsula and nearby islands, the substrate availability, milder climate, and less marked seasonality permit the colonization and persistence of a higher diversity of marine organisms (Amsler et al. 1995). Here, more than 90% of all Antarctic species have been reported (Clayton 1994). Species richness decreases in the East Antarctic Peninsula (EAP), and only 7 and 17 species occurred, respectively, on the coasts of the Ross

Sea (Zaneveld 1966) and Terra Nova Bay (Cormaci et al. 1992). Generally, an inverse relationship between species diversity and latitude is observed in Antarctic seaweeds (Wiencke and Clayton 2002). A total of 104 taxa was identified in South Shetland Islands (ca. 60°S) by Pellizzari et al. 2017, a species number higher than in Adelaide Island (67°S) with 41 taxa (Mystikou et al. 2014) and Terra Nova Bay (Ross Sea, above latitude 70°S) with 17 taxa (Cormaci et al. 1992). Only few species grow in latitudes above 76°S, such as the red algae *Iridaea cordata*, *Phyllophora antarctica*, *Phycodrys antarctica*, and *Hildenbrandia lecannellieri*, the green alga *Monostroma hariotii*, and the brown alga *Desmarestia menziesii* (Wiencke and Clayton 2002).

2.6 Concluding Remarks: Gaps and Prospects for the Future

The knowledge on the seaweed diversity throughout Antarctica is essential, as they are fundamental as primary producers and their composition affects the structure of the ecosystems. Species of seaweeds are bioindicators of environmental changes, including climate change, invasive species, and seawater pollution, among others. Therefore, knowledge on seaweed diversity throughout Antarctica is urgently needed.

There are some clear gaps in the knowledge on seaweed diversity from the Antarctic region, especially in deeper water. As described above, species richness tends to increase with the number of collections in a wide range of locations and in different environments, but collection in Antarctica is expensive and difficult. Furthermore, as mentioned above, little is known on the diversity fluctuations of the marine flora and how those are influenced by environmental factors or anthropogenic activities. Regular monitoring of sites should be done, especially of those that are subjected to human activities such as those from scientific bases or touristic visitation (Hughes and Ashton 2017).

In Antarctica, some of the strongest environmental gradients on the planet can be found, therefore providing an ideal study ground to test hypotheses related with environmental variability and its impact on biodiversity. The most important physical driver of Antarctic marine communities is the oceanographic boundary of the Polar Front. At smaller spatial scales, ice cover, ice scour, and salinity gradients are clearly important determinants of diversity at habitat and community level. However, stochastic and extreme events remain an important driving force, particularly in the context of local extinction, colonization, or recolonization of Antarctic biota (Convey et al. 2014).

Biogeographic barriers are known to be fundamental in macroecological and evolutionary processes. Ocean circulation, considering present and past patterns of continental drift, can isolate or connect many groups of marine organisms, including seaweeds (see also Chap. 3 by Fraser et al. and Chap. 5 by Pellizzari et al.). These benthic organisms release spores and propagules as planktonic stages that drift with currents and/or tides and have been a sensible indicator of changes in biogeographic distribution patterns (Sanches et al. 2016, see also Chap. 10 by

Navarro et al.). According to Pellizzari et al. (2017) and Fraser et al. (2018), the Antarctic Circumpolar Current, previously considered a biogeographic barrier, may become a new pathway for biota interconnectivity and deserves further investigation. Mystikou et al. (2014) discuss that Antarctic seaweeds display plasticity and adaptability in response to extreme environmental conditions such as low temperatures and limited light availability (Wiencke and Amsler 2012). Thus, it is relevant to examine how environmental alterations, such as those caused by climate change, will be affecting algal seasonality, richness, depth zonation, and latitudinal distribution.

The higher seaweed diversity observed in the South Shetland Islands (Pellizzari et al. 2017), a transitional area, bring the imminent need of long-term biological and abiotic monitoring in order to establish conservation guidelines across the Antarctic and sub-Antarctic zones, especially upon increasing tourism, global climate, and oceanographic changes.

Moreover, laboratory culture techniques associated with morphological and molecular analyses could reveal the hidden diversity of Antarctic marine flora, especially with respect to small species (e.g., epiphytes and endophytes), including potential alien/cryptogenic species. Integrative taxonomical studies are needed to less known groups, e.g., crustose algae, and molecular studies are essential to clarify the phylogenetic and biogeographic relationships of Antarctic seaweeds.

Finally, a reference baseline database of seaweed diversity is urgently necessary. This should integrate different inventories based on morphology and life cycles with molecular data from fast, high-throughput, and low-cost methods, such as DNA barcodes and/or metagenomics (Oliveira et al. 2018). Hence, researchers will be enabled to effectively monitor environmental changes and help in the conservation of this unique environment.

Acknowledgments We thank PROANTAR (Brazilian Antarctic Program 557030/2009-9, 407588/2013-2 and 442258/2018-2), INCT Criosfera 2, Brazilian Navy (Polar Ship Almirante Maximiano–H41), Brazilian Air Force, MMA (Ministry of Environment), MCTIC (Ministry of Science, Technology and Innovation), and CNPq (National Council of Research and Development; for grant 310672/2016-3 to NSY and 301491/2013-5 to MCO). This study was financed in part by the Coordenação de Aperfeiçoamento de Pessoal de Nível Superior – Brasil (CAPES) – Finance Code 001.

References

Amsler CD, Rowley RJ, Laur DR, Quetin LB, Ross RM (1995) Vertical distribution of Antarctic peninsular macroalgae: cover, biomass and species composition. Phycologia 34:424–430

Barker PF, Thomas E (2003) Origin, signature and palaeoclimatic influence of the Antarctic Circumpolar Current. Earth Sci Rev 66:143–162

Clarke DL (1990) Arctic Ocean ice-cover; geologic history and climatic significance. In: Grantz A, Johnson L, Sweeney JL (eds) The Arctic Ocean region. The Geological Society of America, Boulder, CO, pp 53–62

Clarke A, Barnes KA, Hodgson DA (2005) How isolated is Antarctica. Trends Ecol Evol 20:1–3

Clayton MN (1994) Evolution of Antarctic benthic algal flora. J Phycol 30:897–904

Clayton MN, Wiencke C, Klöser H (1997) New records of temperate and sub-Antarctic marine benthic macroalgae from Antarctica. Polar Biol 17:141–149. https://doi.org/10.1007/s003000050116

Convey P, Chown SL, Clarke A, Barnes DKA, Schiaparelli S, Wall DH (2014) The spatial structure of Antarctic biodiversity. Ecol Monogr 84:203–244

Cormaci M, Furnari G, Scammacca B (1992) The benthic algal flora of Terra Nova Bay (Ross Sea, Antarctica). Bot Mar 35:541–552

Drew EA, Hastings RM (1992) A year-round ecophysiological study of *Himantothallus grandifolius* (Desmarestiales, Phaeophyta) at Signy Island, Antarctica. Phycologia 31:262–277

Fraser CI, Morrison AK, McC Hogg A, Macaya EC, Van Sebille E, Ryan PG, Padovan A, Jack C, Valdivia N, Waters JM (2018) Antarctica's ecological isolation will be broken by storm-driven dispersal and warming. Nat Clim Change Lett 8:704–708. https://doi.org/10.1038/s41558-018-0209-7

Gallardo T, Pérez-Ruzafa IM, Flores-Moya A, Conde F (1999) New collections of benthic marine algae from Livingston and Deception Islands (South Shetland Islands) and Trinity Island (Bransfield Strait), Antarctica. Bot Mar 42:61–69. https://doi.org/10.1515/bot.1999.009

Gaudichaud C (1826) Botanique. In: Freycinet I (ed) Voyage au tur de monde…executé sur les corvettes de SM l'Uranie et la Physicienne pendant les années 1817–1820. Pillet Ainé, Paris

Gómez I, Navarro NP, Huovinen P (2019) Bio-optical and physiological patterns in Antarctic seaweeds: a functional trait based approach to characterize vertical zonation. Prog Oceanogr 174:17–27

Guiry MD, Guiry GM (2019) AlgaeBase. World-wide electronic publication. National University of Ireland, Galway. http://www.algaebase.org, searched on 24 August 2019

Hempel GV (1987) Die Polarmeere – ein biologischer Vergleich. Polarforschung 57:173–189

Hommersand MH, Moe RL, Amsler CD, Fredericq S (2009) Notes on the systematics and biogeographical relationships of Antarctic and sub-Antarctic Rhodophyta with descriptions of four new genera and five new species. Bot Mar 52:509–534. https://doi.org/10.1515/bot.2009.081

Hughes KA, Ashton GV (2017) Breaking the ice: the introduction of biofouling organisms to Antarctica on vessel hulls. Aquat Conserv Mar Freshw Ecosyst 27:158–164

Klöser H, Quartino ML, Wiencke C (1996) Distribution of macroalgae and macroalgal communities in gradients of physical conditions in Potter Cove, King George Island, Antarctica. Hydrobiologia 333:1–17

Lamb I, Zimmermann MH (1977) Benthic marine algae of the Antarctic Peninsula. In: Pawson DL (ed) Biology of the Antarctic Seas V. American Geophysical Union, Antarctic Research Series. 23 (4), United States. 229 p. ISSN: 0066-4634

Le Gall L, Saunders GW (2010) DNA barcoding is a powerful tool to uncover algal diversity: a case study of the Phyllophoraceae (Gigartinales, Rhodophyta) in the Canadian flora. J Phycol 46:374–389

Medeiros AS (2013) Macroalgae diversity of Admiralty Bay, King George Island, Antarctic Peninsula based on DNA barcoding and other molecular markers. MSci Thesis, São Paulo University, p. 164. http://www.teses.usp.br/teses/disponiveis/41/41132/tde-24032014-090801/pt-br.php

Meier R, Zhang G, Ali F (2008) The use of mean instead of smallest interspecific distances exaggerates the size of the "Barcoding Gap" and leads to misidentification. Syst Biol 57:809–822

Moe RL (1985) *Gainia* and Gainiaceae: a new genus and family of crustose marine Rhodophyceae from Antarctica. Phycologia 24:419–428

Moe RL, Silva PC (1977) Antarctic marine flora: uniquely devoid of kelps. Science 196:1206–1208

Mystikou A, Peters AF, Asensi AO, Fletcher KI, Brickle P, Van West P et al (2014) Seaweed biodiversity in the south-western Antarctic Peninsula: surveying macroalgal community composition in the Adelaide Island/Marguerite Bay region over a 35-year time span. Polar Biol 37:1607–1619. https://doi.org/10.1007/s00300-014-1547-1

Nedzarek A, Rakusa-Suszcewski S (2004) Decomposition of macroalgae and the release of nutrients in Admiralty Bay, King George Island, Antarctica. Polar Biosci 17:16–35
Neushul M (1959) Biological collecting in Antarctic waters. Veliger 2:15–17
Neushul M (1961) Diving in Antarctic water. Polar Rec 10:83–88
Neushul M (1963) Reproductive morphology of Antarctic kelps. Bot Mar 5:19–24
Neushul M (1965) Diving observation of sub-tidal Antarctic marine vegetation. Bot Mar 8:234–243
Neushul M (1968) Benthic marine algae. Antarctic Map Folio Ser 10:9–10
Oliveira EC, Absher TM, Pellizzari FM, Oliveira MC (2009) The seaweed flora of Admiralty Bay, King George Island, Antarctic. Polar Biol 32:1639–1647. https://doi.org/10.1007/s00300-009-0663-9
Oliveira MC, Repetti SI, Iha C, Jackson CJ, Díaz-Tapia P, Lubiana KMF, Cassano V, Costa JF, Cremen MCM, Marcelino VR, Verbruggen H (2018) High-throughput sequencing for algal systematics. Eur J Phycol 53:256–272. https://doi.org/10.1080/09670262.2018.1441446
Papenfuss GF (1964) Catalogue and bibliography of Antarctic and Subantarctic benthic marine algae. Antarct Res Ser 1:1–76
Pellizzari F, Silva MC, Silva EM, Medeiros A, Oliveira MC, Yokoya NS et al (2017) Diversity and spatial distribution of seaweeds in the South Shetland Islands, Antarctica: an updated database for environmental monitoring under climate change scenarios. Polar Biol 40:1671–1685. https://doi.org/10.1007/s00300-017-2092-5
Peters AF, Breeman AM (1993) Temperature tolerance and latitudinal range of brown algae from temperate Pacific South America. Mar Biol 115:143–150
Quartino ML, Zaixso HE, Boraso de Zaixso AL (2005) Biological and environmental characterization of marine macroalgal assemblages in Potter Cove, South Shetland Islands, Antarctica. Bot Mar 48:187–197. https://doi.org/10.1515/BOT.2005.029
Ramírez ME (1982) Catálogo de las algas marinas del Territorio Chileno Antártico. Ser Cient INACh 29:39–67
Sanches PF, Pellizzari FM, Horta PH (2016) Multivariate analyses of Antarctic and sub-Antarctic seaweed distribution patterns: an evaluation of the role of the Antarctic Circumpolar Current. J Sea Res 110:29–38. https://doi.org/10.1016/j.seares.2016.02.002
Sherwood AR, Kurihara A, Conklin KY, Sauvage T, Presting GG (2010) The Hawaiian Rhodophyta biodiversity survey (2006–2010): a summary of principal findings. BMC Plant Biol 10(258):1–29
Skottsberg C (1906) Observations on the vegetations of the Antarctic Sea. Botaniska Studier tillagnade F.R. Kjelman, Uppsala, pp 245–264
Skottsberg C (1941) Communities of marine algae in subantarctic and Antarctic waters. K Sven Velenslapsakad Handl Tredje Serien 19:1–92
Skottsberg C (1953) On two collections of Antarctic marine algae. Arkiv Botanik 2:531–566
Skottsberg C (1964) Antarctic phycology. In: Carrick R, Holgate M, Prevost J (eds) Biologie Antarctique. 1st SCAR Symposium Hermann, Paris, pp 147–154
Skottsberg C, Neushul M (1960) *Phyllogigas* and *Himantothallus*. Antarctic Phaeophyceae Bot Mar 2:164–173
Valdivia N, Diaz MJ, Holtheuer J, Garrido I, Huovinen P, Gómez I (2014) Up, down, and all around: scale-dependent spatial variation in rocky-shore communities of Fildes Peninsula, King George Island, Antarctica. PLoS One 9(6):e100714. https://doi.org/10.1371/journal.pone.0100714
Vincent WF (2000) Evolutionary origins of Antarctic microbiota: invasion, selection and endemism. Antarct Sci 12:374–385
Wiencke C, Amsler CD (2012) Seaweeds and their communities in polar regions. In: Wiencke C, Bischof K (eds) Seaweed biology: novel insights into ecophysiology, ecology and utilization. Springer, Germany, pp 265–291
Wiencke C, Clayton MN (2002) Antarctic seaweeds. In: Wägele JW (ed) Synopses of the Antarctic benthos, vol 9. A.R.G. Gantner Verlag, Rugell, Lichtenstein, p 239

Wiencke C, Amsler CD, Clayton MN (2014) Chapter 5.1. Macroalgae. In: DeBroyer C, Koubbi P, Griffiths HJ, Raymond B, Udekem d'Acoz CD (eds) Biogeographic atlas of the Southern Ocean. Scientific Committee on Antarctic Research, Cambridge, pp 66–73

Wulff A, Iken K, Quartino ML, Al-Handal A, Wiencke C, Clayton MN (2009) Biodiversity, biogeography and zonation of marine benthic micro–and macroalgae in the Arctic and Antarctic. Bot Mar 52:491–507. https://doi.org/10.1515/bot.2009.072

Wynn-Williams DD (1996) Response of pioneer soil microalgal colonists to environmental change in Antarctica. Microb Ecol 31:177–188

Yoneshigue-Valentin Y, Silva IB, Fujii MT, Yokoya NS, Pupo D, Guimarães SMPB, Martins AP, Sanches PF, Pereira DC, Dalto AG, Souza JMC, Pereira CMP, Pellizzari F, Colepicolo P (2013) Marine macroalgal diversity in Admiralty Bay, King George Island, South Shetlands Islands, Antarctica. Ann Act Rep:140–148

Zaneveld JS (1966) Vertical zonation of Antarctic and Subantarctic benthic marine algae. Antarct J US 1:211–213

Zielinski K (1990) Bottom macroalgae of Admiralty Bay (King George Island, South Shetland Island, Antarctic). Polar Res 11:95–131

Chapter 3
Biogeographic Processes Influencing Antarctic and sub-Antarctic Seaweeds

Ceridwen I. Fraser, Adele Morrison, and Pamela Olmedo Rojas

Abstract Antarctica has long been seen as biologically isolated, surrounded by the vast Southern Ocean and its circumpolar oceanographic currents and fronts and home to many endemic species. New evidence demonstrates, however, that buoyant seaweeds can cross perceived oceanographic barriers in the Southern Ocean to reach Antarctic coasts. These macroalgal rafts can carry diverse passengers, including marine invertebrates and other, non-buoyant seaweeds. The stark differences between Antarctic and sub-Antarctic near-coastal ecosystems are therefore more probably the result of environmental differences than physical isolation. Modelling indicates that algal rafts from the sub-Antarctic could reach Antarctic coasts every month, providing an ongoing influx of marine propagules that are poised to colonise as the climate warms. In this chapter, we review the following: (i) the evidence for the isolation of Antarctica, (ii) the oceanographic processes that can hinder or promote passive dispersal into Antarctic waters and (iii) the characteristics of organisms that could be rafting to Antarctic coasts with buoyant macroalgae.

Keywords Antarctic circumpolar current · Connectivity · Dispersal · Polar front · Stokes drift

C. I. Fraser (✉) · P. Olmedo Rojas
Department of Marine Science, University of Otago, Dunedin, New Zealand

A. Morrison
Research School of Earth Sciences, and ARC Centre of Excellence for Climate Extremes, Australian National University, Acton, ACT, Australia
e-mail: adele.morrison@anu.edu.au

I. Gómez, P. Huovinen (eds.), *Antarctic Seaweeds*,
https://doi.org/10.1007/978-3-030-39448-6_3

3.1 Antarctica's Place in the World: An Isolated Continent?

Antarctica has long been seen as biologically isolated from the rest of the world (Clarke et al. 2005; Fraser et al. 2012; Convey et al. 2014; Chown et al. 2015). The opening of the Drake Passage between South America and Antarctica, around 41 million years ago (Scher and Martin 2006), and the subsequent onset of the Antarctic Circumpolar Current (AAC), around 30 Ma (Scher et al. 2015), geographically isolated Antarctica. Along with declines in atmospheric CO_2 in the Cenozoic, the onset of the ACC enhanced cooling and glaciation of the Antarctic (DeConto and Pollard 2003), resulting in marked reductions in diversity in some marine groups (Clarke 1990). The ACC is the world's largest ocean current, about 23,000 km long and up to 2000 km wide in some areas (Constantin and Johnson 2016); connects the Indian, Pacific and Atlantic Oceans; and extends up to 4–5-km depth (Marynets 2019). There are several well-recognised circumpolar thermal fronts in the Southern Ocean, including the Subantarctic Front and the Antarctic Polar Front (see next section). These oceanic features, encircling Antarctica, enhance the stark environmental differences between Antarctic and more northern (sub-Antarctic or temperate) ecosystems. Biologically, the Southern Ocean with its strong, eastward-flowing ACC and circumpolar fronts represents a major hurdle for some organisms to cross (Patarnello et al. 1996; Fraser et al. 2012; Gonzalez-Wevar et al. 2012), particularly for passive dispersers. The observed high-level endemism in the Southern Ocean marine biota (Barnes et al. 2006) and the relatively low diversity and high endemism of terrestrial organisms in Antarctica have largely been considered to result from long periods of evolution in isolation (Fraser et al. 2012). However, we are increasingly recognising that the Southern Ocean is not an impermeable biological barrier; dispersal events into the Antarctic do occur (Clarke et al. 2005; Barnes et al. 2006) and at higher frequencies than previously thought (Fraser et al. 2017; Fraser et al. 2018b). The distinctive Antarctic biota is thus probably a result of adaptation to extreme environmental conditions, rather than an inability of passively dispersing organisms to reach the region.

3.1.1 Adaptations of Terrestrial Organisms to Antarctic Conditions

Only around 0.2–0.3% of Antarctic land is currently ice-free (Convey and Stevens 2007; Burton-Johnson et al. 2016). Antarctic terrestrial organisms, restricted to this ice-free land, have apparently persisted in Antarctica for millions of years and have adapted to the extreme conditions including low temperatures, low winter photoperiod and scarce food supply (Convey and Stevens 2007). Antarctic terrestrial biodiversity is restricted mainly to microinvertebrates, bryophytes and lichens; only two vascular plants occur, both restricted to the maritime Antarctica

(Ochyra et al. 2008). Lichens and mosses are the dominant flora of Antarctica – there are ca 386 and 111 species of lichens and mosses identified, respectively (Øvstedal and Lewis Smith 2001; Ochyra et al. 2008). Lichens have been described as the organisms best adapted to Antarctic conditions (Ochyra et al. 2008) and have been reported to be photosynthetically active under suboptimal temperatures (Kappen 2000). Antarctic mosses have also shown remarkable physiological adaptations to both water availability and solar radiation. For example, mosses can survive under long periods of both desiccation and submergence (Wasley et al. 2006), including being frozen by glaciers for hundreds (Cannone et al. 2017) and even thousands (Roads et al. 2014) of years, despite water availability having been shown to be a key factor underpinning the abundance and distribution of Antarctic terrestrial organisms (Kennedy 1993; Convey and Stevens 2007; Robinson et al. 2018). The Antarctic terrestrial fauna is dominated by microarthropods, such as springtails and mites (Convey and Stevens 2007). Glycerol in cell membranes might help arthropods inhabiting polar conditions to survive freezing temperatures (Teets and Denlinger 2014). Additionally, the small size of terrestrial organisms could assist with minimising moisture loss in windy environments and maximising the use of limited nutrients (Kappen et al. 1995).

3.1.2 Adaptations of Marine Organisms to Antarctic Conditions

The contemporary marine biota of Antarctica has been strongly shaped by glacial cycles (Allcock and Strugnell 2012; Fraser et al. 2012) and mainly comprises fish, macroalgae and filter feeders such as sponges, small crustaceans, molluscs and anemones (Clayton 1994; Griffiths 2010). Some taxa such as decapods, sharks and skates are poorly represented in Antarctica (Aronson and Blake 2015). Up to 90% of the Antarctic fishes and marine invertebrates (Barnes et al. 2006) and 35% of macroalgae (Clayton 1994; Gómez 2015) are endemic to the region. Whereas some Antarctic algal species are also present in nearby northern areas (such as *Adenocystis utricularis, Gigartina skottsbergii, Monostroma hariotii* or *Iridaea cordata*, which are found in both Antarctica and South America: Gómez 2015), others are unique to the region but are closely related to taxa elsewhere. Sunlight is highly limited in the Antarctic marine environment, but Antarctic macroalgae can survive at low temperatures, photosynthesise under very low light conditions and can store organic compounds to use in dark periods (Wiencke et al. 2007). Other marine species also show adaptations to their extreme environment. Antarctic notothenioid fish have developed antifreeze glycoproteins, which lower the internal freezing point of most of their fluids, preventing freezing in sub-zero Antarctic waters (DeVries 1988).

3.1.3 Evidence for Dispersal of Organisms into the Antarctic

Transoceanic dispersal mainly occurs either by active dispersal (e.g. swimming and flying) or through passive dispersal such as through transport with ocean or wind currents, or 'hitch-hiking' with larger animals, driftwood or seaweed (Muñoz et al. 2004; Gillespie et al. 2012; Moon et al. 2017). Some microbes appear to have reached Antarctica via aerial dispersal with wind (Vincent 2000), and small arthropods also appear capable of dispersing with wind (Hawes et al. 2007), although long-distance dispersal of arthropods across the Southern Ocean via this mechanism seems unlikely (Pugh 2003). Recently, Fraser et al. (2018b) reported that southern bull kelp, *Durvillaea antarctica*, which grows in the sub-Antarctic but not in the Antarctic, had travelled south across the Southern Ocean. These were the longest biological rafting events ever recorded, >20,000 km, and were apparently driven by strong winds and storms that pushed the kelp across Southern Ocean fronts (see below). Similarly, in the last couple of decades, king crabs (*Neolithodes yaldwyni*) have been found on the Antarctic continental shelf (Smith et al. 2012) and might represent invasions from deeper water, although there has been some debate as to whether the crabs are new immigrants (Thatje et al. 2005) or have been long-term residents of the Antarctic and have simply gone undetected (Griffiths et al. 2013).

With global warming, many organisms are migrating poleward (Hickling et al. 2006; Chen et al. 2011; Fraser et al. 2012). Invasive (anthropogenically transported) and naturally dispersing non-native species represent a major threat to Antarctic biota (Frenot et al. 2005; Chown et al. 2012; Chown et al. 2015; Duffy et al. 2017). Understanding how permeable the Southern Ocean 'barrier' is to dispersal of different sorts of organisms will help us to understand how Antarctic biodiversity might be affected by future colonisations (Fraser et al. 2017; Wauchope et al. 2019).

3.2 Physical Oceanographic Processes Influencing Movement of Seaweeds into or out of the Antarctic

There has been a long-held perception that the fronts of the ACC act as oceanographic barriers preventing movement of passively dispersing marine biota southward into Antarctic waters (Hunt et al. 2016). Fronts are sharp transitions in temperature, salinity and biogeochemical properties and align with the narrow, eastward currents of the ACC. Traditionally, three primary fronts have been identified in the Southern Ocean – the Subantarctic Front, the Polar Front and the Southern Antarctic Circumpolar Current Front (Orsi et al. 1995). Traversing southward towards Antarctica, a front appears as a near step change from warmer, saltier waters in the north to colder, fresher waters on the southern side. The separation of warm and cold waters across the ACC fronts has cultivated the conceptual picture that there is very little mixing or transport across the fronts. However, this is an overly simplistic view of circulation in the Southern Ocean. Recent analysis of higher-resolution observations shows that the ACC has a more complicated structure with numerous fronts that merge and split around Antarctica (Sokolov and Rintoul 2009). Mesoscale variability is also rich in the Southern Ocean, and eddies and jets that meander in time can move objects away from the direction of the main currents (Lehahn et al. 2011). An additional process that has recently been recognised as extremely important for dispersal of floating surface material is non-linear advection by surface waves, known as Stokes drift (Fraser et al. 2018b; Dobler et al. 2019). Below we expand upon the most important processes influencing movement of seaweeds in the Southern Ocean: the mean northward drift of Ekman transport, transient north and south movement by eddies and storm-driven Stokes drift, in addition to the predominantly zonal movement by the large-scale horizontal ocean currents.

3.2.1 *Ekman Transport*

The westerly winds over the Southern Ocean, also known as the 'Roaring Forties', are the strongest average winds on the planet, with annual average speeds up to 10 m s^{-1} (Lin et al. 2018). The winds blow from west to east over a wide latitude band covering approximately 30–65°S. Due to the Coriolis effect, which deflects motion to the left in the Southern Hemisphere, the eastward wind stress on the ocean drives a northward surface movement referred to as Ekman transport. Ekman transport in the Southern Ocean extends to a depth of around 100 m and is surface intensified. Observed northward speeds are up to ~2 cm s^{-1} at the surface (Lenn and Chereskin 2009), which in isolation would result in floating objects moving more than 600 km, or 6° of latitude northward in 1 year. Although the Southern Ocean fronts provide a visible separation between the sub-Antarctic islands and Antarctica, it is the constant northward drift of Ekman transport that represents the biggest obstacle to southward movement of floating seaweeds.

3.2.2 *Eddies*

The Southern Ocean fronts are dynamically unstable, forming ubiquitous mesoscale eddies and transient meanders in the ACC jets (Thompson Andrew 2008). Drifting objects can be trapped inside coherent ring-like eddies and transported over long distances away from the large-scale ocean currents (Lehahn et al. 2011). In the Southern Ocean, eddies are ~10–100 km across and commonly last for longer than 4 months, over which time they can travel north or south by 5° or more of latitude (Chelton et al. 2011). A drifting object would be unlikely to remain trapped for the entire lifetime of an eddy, as eddies 'leak' as they stretch and interact with other ocean circulation features (d'Ovidio et al. 2013). However, seaweeds may encounter a series of eddies over time, resulting in a net northward or southward transport. Fraser et al. (2018b) showed that the inclusion of mesoscale variability was essential for modelled virtual particles to drift sufficiently southward to reach Antarctica.

3.2.3 Wave-Driven Stokes Drift

High wind speeds associated with atmospheric storms over the Southern Ocean generate an intense wave climate and make it consistently the roughest ocean on earth (Young 1999). The non-linear nature of surface ocean waves results in a net advection of floating objects in the direction of the waves, known as Stokes drift. While the average direction of waves in the Southern Ocean is eastward, the same as the average wind direction, atmospheric storms result in large variability of the wind direction and associated wave direction. Seaweeds can be transported southwards across fronts by large, individual storm events, during which the Stokes drift is frequently as large as 0.3 m s^{-1} (Rascle and Ardhuin 2013). Fraser et al. (2018b) showed that the influence of wave-driven Stokes drift is necessary for floating objects to drift sufficiently southward to reach Antarctica (Fig. 3.1).

3.2.4 Surface Currents

The large-scale horizontal circulation in the open Southern Ocean is predominantly eastward and therefore cannot in isolation drive floating objects southward. Nonetheless, the ocean circulation is important to consider, as it largely determines the locations where seaweeds approach Antarctica. While eddy processes and

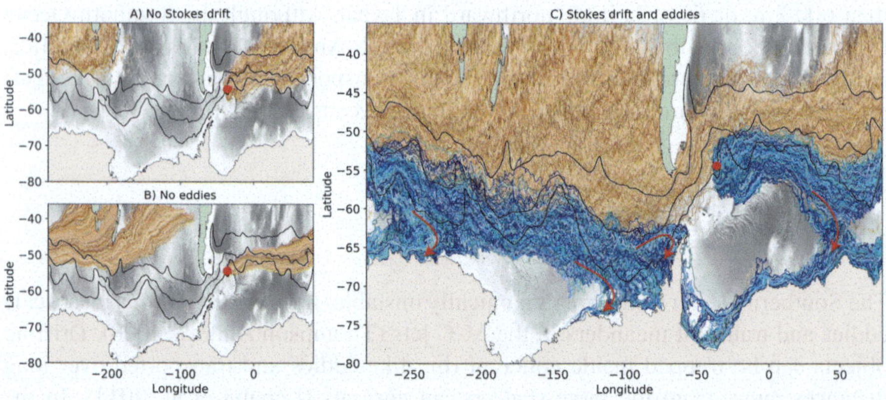

Fig. 3.1 Simulated drift particle trajectories from South Georgia, as described in Fraser et al. (2018a, b). More than 4.2 million virtual particles were released from South Georgia (red dot) and advected for 3 years with simulated surface ocean velocities where (**a**) eddies, but not Stokes drift, were included; (**b**) Stokes drift, but not eddies, was included; and (**c**) both Stokes drift and eddies were included. Blue lines show trajectories for all particles located on the Antarctic shelf after 3 years. Brown lines show trajectories for a random sample (0.1%) of particles not satisfying this condition. Black lines show positions of major fronts. Red arrows show the four major pathways of approach to Antarctica, two aligned with southward deviations of the ACC and two associated with the southward movement of the Weddell and Ross Gyres

Stokes drift are responsible for moving floating objects southward across the ACC fronts, the large-scale meanders of the ACC and the subpolar gyres set the most likely pathways of approach to Antarctica (Fraser et al. 2018b).

The eastward ACC dominates the ocean circulation in the Southern Ocean. The ACC plays an important role in dispersing species zonally around the Southern Ocean (Waters 2008) and in rapidly moving drifting objects to longitudes where shorter routes to Antarctica are possible. The circumpolar path of the ACC is steered by the underlying bathymetry and continental gateways. There are two close approaches of the ACC to the Antarctic continental slope – along the West Antarctic Peninsula and offshore from Wilkes Land in East Antarctica. In the recent Lagrangian study of Fraser et al. (2018b), these two locations were preferred routes for southward drifting particles to first make contact with the continent.

Two additional locations with high influx of floating objects to the Antarctic coast are associated with the subpolar gyres (Fraser et al. 2018b). The Weddell and Ross Gyres are clockwise circulation features between Antarctica and the ACC, driven by Ekman divergence at the intersection of the westerly winds to the north and easterly winds to the south. At the eastern edge of the Weddell and Ross Gyres (~50°E and ~120°W, respectively), the flow is southward towards Antarctica. If seaweeds are driven sufficiently southward across the ACC fronts through a combination of eddy and wave-driven processes (Fig. 3.2), the subpolar gyres can connect dispersal pathways from the southern edge of the ACC to the Antarctic coast. Once seaweeds reach the coast of Antarctica, circumpolar connectivity around the coastline is enabled by the westward Antarctic Coastal and Antarctic Slope Currents.

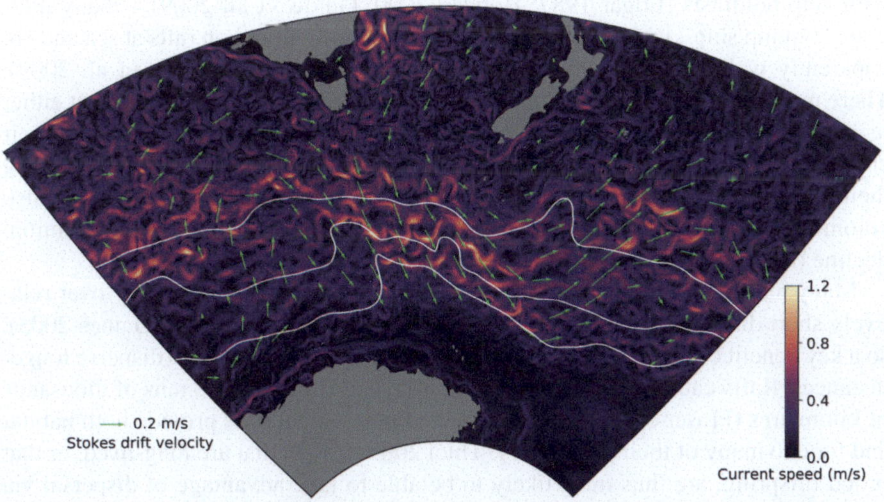

Fig. 3.2 Illustration of eddies and Stokes drift in the Southern Ocean. White lines show long-term average fronts in the Southern Ocean. Purple colours show a snapshot of surface ocean current speed, and green arrows show a snapshot of the additional surface velocity from wave-driven Stokes drift. (All data are from sources listed in Fraser et al. 2018b)

3.3 Hitch-Hiking to the Antarctic: Passengers on Seaweed Rafts

Buoyant macroalgae have long been recognised as potential vectors for the transport of diverse other taxa (Thiel and Gutow 2005), including invertebrates (e.g. Helmuth et al. 1994; Miranda and Thiel 2008; Nikula et al. 2010) and non-buoyant seaweeds (e.g. Edgar 1987; Fraser et al. 2013; López et al. 2018). Indeed, entire coastal communities can potentially travel long distances at sea via rafting with seaweeds (Fraser et al. 2011). With new evidence that buoyant macroalgae can traverse perceived oceanic 'barriers' and reach the Antarctic coast (Fraser et al. 2018b), rafting events have the potential to enable diverse non-native species to reach Antarctica. Currently, these taxa seem unable to establish, perhaps because the frequent ice scour of rocky shores in Antarctica precludes establishment of the large kelps these species require for habitat and food (Fraser et al. 2009; Fraser et al. 2018b). Antarctic coastal waters are warming, however, and some parts of the Antarctic Peninsula could be comparable to present-day sub-Antarctic environments by the end of the century (Griffiths et al. 2017), facilitating establishment of non-native species brought by kelp rafts.

3.3.1 Characteristics of Rafting Species

Shortly after detachment, there is a decline in the number of invertebrates associated with kelp holdfasts (Edgar 1987; Hobday 2000; Gutow et al. 2009) – many individuals 'jump ship', including those that are commonly found on rafts at sea and are apparently well suited to rafting (Miranda and Thiel 2008; Gutow et al. 2009). There may be bet-hedging benefits for those species that have individuals that either leave the detached kelp or remain, with departing individuals enhancing their chances of remaining in suitable kelp habitats, and remaining individuals enhancing their chances of colonising new territory (Gutow et al. 2009). Some species also colonise already detached, drift macroalgae, enhancing raft diversity after the initial decline (Edgar 1987).

Many marine invertebrates have pelagic larvae that can disperse only over relatively short distances (hundreds to thousands of metres: Kinlan and Gaines 2003), so a key benefit of dispersing with macroalgal rafts is the capacity to disperse longer distances. Rafts can endure for hundreds of days, travelling up to tens of thousands of kilometres (Fraser et al. 2011, 2018b), and macroalgal rafts provide both habitat and food to many of their passengers (Thiel 2003a). Taxa that are long-lived, or that brood offspring, are thus most likely to be able to take advantage of dispersal via long-distance rafting journeys – direct-developing organisms have been noted to make up an increasing proportion of the rafting community the longer a raft is at sea (Thiel 2003a). Examples of such taxa include peracarid crustaceans (amphipods, isopods and tanaids) and pulmonate gastropods (Thiel and Gutow 2005). In some

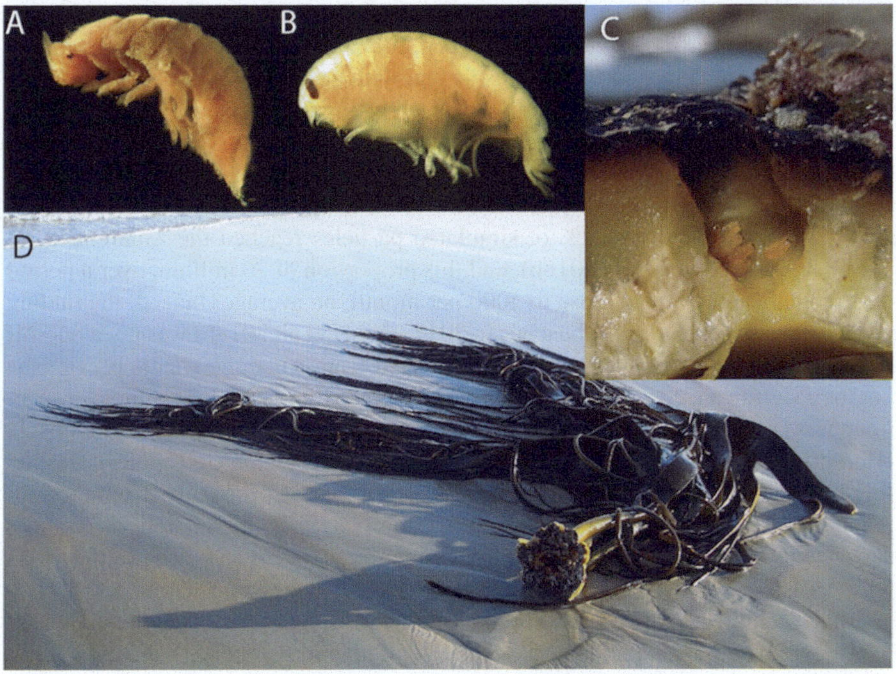

Fig. 3.3 Peracarid crustaceans including isopods and amphipods can brood their young within kelp holdfasts, and are well suited to survive long rafting journeys. (**a**) The boring isopod *Limnoria stephenseni*, (**b**) the amphipod *Parawaldeckia kidderi*, (**c**) a group of *Parawaldeckia* amphipods in a tunnel in a dissected holdfast of *Durvillaea antarctica* bull kelp, (**d**) beached *Durvillaea antarctica* bull kelp raft, with holdfast. (Photos by Ceridwen Fraser)

cases, these bore into kelp holdfasts, creating tunnels in which they are largely protected from exposure to harsh conditions in the open ocean and in which they can nurture subsequent generations (e.g. the isopod genus *Limnoria:* Thiel 2003b) (Fig. 3.3). Some other taxa that have been found associated with macroalgal rafts that have dispersed long distances in the Southern Hemisphere include chitons, limpets, bivalves, echinoderms and pycnogonids (Fraser et al. 2011), highlighting the potential for macroalgal rafts to carry diverse new taxa to the Antarctic.

3.3.2 *Processes Affecting Establishment of New Taxa in the Antarctic*

Simulated dispersal of surface particles (e.g. buoyant macroalgal rafts) from sub-Antarctic locations suggests that the vast majority are driven eastward and eventually northward, away from Antarctica. The proportions of particles able to reach the Antarctic coast in simulations were relatively small – from 0.0001% to 0.1915% (Fraser et al. 2018b). Estimates based on empirical surveys of kelp at sea suggest, however, that there are tens of millions of macroalgal rafts adrift in the Southern Ocean at any time; for *Durvillaea antarctica*, for example, there are estimated to be about 70 million detached individuals at sea, with around 20 million of those still bearing intact holdfasts (Smith 2002). If such estimates are correct and assuming most rafts complete their journeys within 3 years (Fraser et al. 2018b), there could be between a few and several thousand sub-Antarctic-origin *D. antarctica* rafts reaching the Antarctic coast each month. For example, for releases from South Georgia, approximately 0.19% of simulated particles reached the Antarctic coast within 3 years (Fraser et al. 2018b), and this proportion of 70 million, over a period of 36 months, amounts to close to 4000 per month, on average. Indeed, the finding of *D. antarctica* on Antarctic beaches in 2017 (Fraser et al. 2018b) was not a one-off event; *D. antarctica* samples had been collected from Antarctic waters previously, but their significance was not immediately recognised. In 1989, for example, a specimen was collected from Arthur Harbour at Anvers Island off the Antarctic Peninsula (Herbarium of the University of California Berkeley, accession UC157330). Despite this evidence for long-term, persistent transport of marine propagules from the sub-Antarctic to Antarctica, the large kelps that dominate the sub-Antarctic near-shore ecosystems (such as *D. antarctica* and *Macrocystis pyrifera*) and many of the invertebrates associated with them have not yet been established in Antarctica.

There could be a range of factors that influence the chance of establishment of these large macroalgae and their associated epifauna in Antarctica. The most obvious of these is ice scour – neither *D. antarctica* nor *M. pyrifera* currently occurs in areas heavily affected by iceberg scour or sea ice (Fraser et al. 2009; Macaya and Zuccarello 2010; Fraser 2012). Most Antarctic coasts are likely to be affected by ice scour long into the future, but warming near the Antarctic Peninsula over coming decades could make shallow marine environments in that region somewhat comparable to contemporary sub-Antarctic environments by the end of this century (Griffiths et al. 2017). If areas along the Peninsula, for example, bays and channels, are partly protected from ice scour in this warmer environment, some large sub-Antarctic macroalgae – and, subsequently, their associated epibionts – might have an opportunity to establish.

Another consideration is the viability of dispersing seaweeds and other taxa on reaching Antarctica. The viability of kelp rafts reaching Antarctica has yet to be tested, but the two *D. antarctica* specimens found at King George Island in 2017 (Fraser et al. 2018b) were both male and had mature antheridia that appeared likely to be viable. Furthermore, prior research indicates that the duration of survival and viability of rafting macroalgae is greater in cooler, high-latitude waters versus warmer, lower-latitude waters (Rothäusler et al. 2009). Establishment of dispersing macroalgae can also be influenced by the presence or absence of competing conspecific (Waters et al. 2013; Fraser et al. 2018a) or heterospecific (Valentine and Johnson 2004) seaweeds. For Antarctic intertidal ecosystems, however, such competition seems unlikely to play a major role in preventing the establishment of sub-Antarctic-origin macroalgae, as Antarctic rocky shores generally are not densely blanketed in algal cover (Fig. 3.4). Nonetheless, should ice scour become less frequent in some areas, competitive processes could influence which macroalgae can establish and survive on exposed rocky shores.

Fig. 3.4 Antarctic rocky shores are starkly different to sub-Antarctic shores. In the former, frequent ice scour removes all but the hardiest organisms, and seaweeds tend to be small and/or have biphasic life cycles. In the sub-Antarctic, large brown macroalgae create habitat for diverse other taxa including invertebrates and other seaweeds. (**a**, **b**) show sub-Antarctic shores dominated by *Durvillaea antarctica* bull kelp; (**c**, **d**) show Antarctic shores lacking large seaweeds; (**e**, **f**) show small, hardy seaweeds that can occur in both the sub-Antarctic and Antarctic: *Adenocystis utricularis* (**e**) and *Bostrychia intricata* (**f**). White scale bars represent approximately 20 cm. (Photos A, B, E and F by Ceridwen Fraser, sub-Antarctic Marion Island and New Zealand; C by Carlos Olavarria; D by Emma Newcombe, South Shetland Islands, Antarctica) (Figure modified from Fraser et al. 2012).

3.4 Concluding Remarks

Antarctica has long been considered biologically isolated, but we now recognise that the Southern Ocean and its fronts are not impermeable barriers to southward dispersal of drifting objects. While there are no direct large-scale ocean currents crossing the fronts, there are a number of both oceanic and directly wind-driven processes that can move floating objects such as buoyant seaweed rafts north or south across the fronts. Fraser et al. (2018b) showed that the combination of both transient mesoscale variability and wave-driven Stokes drift is necessary for such frontal traverses. Once floating objects have crossed to the southern edge of the ACC, the large-scale horizontal ocean currents determine the locations of final approach to Antarctica. Of the four primary regions where objects approach the Antarctic coast (Fig 3.1), the region considered most feasible for future establishment of sub-Antarctic macroalgae is the Antarctic Peninsula and nearby islands, where warming is forecast to be most rapid. Should such establishment occur, there could be drastic changes to Antarctic near-shore marine ecosystems, as drift macroalgae can transport diverse other taxa including invertebrates and other seaweeds.

Acknowledgements We thank Richard Moe for alerting us to the existence of the *D. antarctica* specimen from Antarctica in the UC Berkeley herbarium and Erasmo Macaya for discussions. CIF was funded by a Rutherford Discovery Fellowship from the Royal Society of New Zealand (RDF-UOO1803), AKM was supported by an Australian Research Council DECRA Fellowship (DE170100184), and POR was supported by a Strategic Doctoral Scholarship from the Division of Sciences at the University of Otago.

References

Allcock AL, Strugnell JM (2012) Southern Ocean diversity: new paradigms from molecular ecology. Trends Ecol Evol 27:520–528

Aronson RB, Blake DB (2015) Global climate change and the origin of modern benthic communities in Antarctica. Integr Comp Biol 41:27–39

Barnes DKA, Hodgson DA, Convey P, Allen CS, Clarke A (2006) Incursion and excursion of Antarctic biota: past, present and future. Glob Ecol Biogeogr 15:121–142

Burton-Johnson A, Black M, Fretwell P, Kaluza-Gilbert J (2016) An automated methodology for differentiating rock from snow, clouds and sea in Antarctica from Landsat 8 imagery: a new rock outcrop map and area estimation for the entire Antarctic continent. Cryosphere 10:1665–1677

Cannone N, Corinti T, Malfasi F, Gerola P, Vianelli A, Vanetti I et al (2017) Moss survival through in situ cryptobiosis after six centuries of glacier burial. Sci Rep 7:4438

Chelton DB, Schlax MG, Samelson RM (2011) Global observations of nonlinear mesoscale eddies. Prog Oceanogr 91:167–216

Chen I-C, Hill JK, Ohlemüller R, Roy DB, Thomas CD (2011) Rapid range shifts of species associated with high levels of climate warming. Science 333:1024–1026

Chown SL, Huiskes AHL, Gremmen NJM, Lee JE, Terauds A et al (2012) Continent-wide risk assessment for the establishment of nonindigenous species in Antarctica. Proc Natl Acad Sci USA 109:4938–4943

Chown SL, Clarke A, Fraser CI, Cary SC, Moon KL, McGeoch MA (2015) The changing form of Antarctic biodiversity. Nature 522:431–438

Clarke A (1990) Temperature and evolution: Southern Ocean cooling and the Antarctic marine fauna. In: Kerry KR, Hempel G (eds) Antarctic ecosystems. Springer, Berlin, Heidelberg, pp 9–22

Clarke A, Barnes DKA, Hodgson DA (2005) How isolated is Antarctica? Trends Ecol Evol 20:1–3

Clayton MN (1994) Evolution of the Antarctic marine benthic algal flora. J Phycol 30:897–904

Constantin A, Johnson RS (2016) An exact, steady, purely azimuthal flow as a model for the Antarctic Circumpolar Current. J Phys Oceanogr 46:3585–3594

Convey P, Stevens MI (2007) Antarctic biodiversity. Science 317:1877–1878

Convey P, Chown SL, Clarke A, Barnes DKA, Bokhorst S, Cummings V, Ducklow HW, Frati F et al (2014) The spatial structure of Antarctic biodiversity. Ecol Monogr 84:203–244

d'Ovidio F, De Monte S, Penna AD, Cotté C, Guinet C (2013) Ecological implications of eddy retention in the open ocean: a Lagrangian approach. J Phys A Math Theor 46:254023

DeConto RM, Pollard D (2003) Rapid Cenozoic glaciation of Antarctica induced by declining atmospheric CO_2. Nature 421:245–249

DeVries AL (1988) The role of antifreeze glycopeptides and peptides in the freezing avoidance of antarctic fishes. Comp Biochem Physiol B Comp Biochem 90:611–621

Dobler D, Huck T, Maes C, Grima N, Blanke B, Martinez E, Ardhuin F (2019) Large impact of Stokes drift on the fate of surface floating debris in the South Indian Basin. Mar Pollut Bull 148:202–209

Duffy GA, Coetzee BWT, Latombe G, Akerman AH, McGeoch MA, Chown SL (2017) Barriers to globally invasive species are weakening across the Antarctic. Divers Distrib 23:982–996

Edgar GJ (1987) Dispersal of faunal and floral propagules associated with drifting *Macrocystis pyrifera* plants. Mar Biol 95:599–610

Fraser CI (2012) The impacts of past climate change on sub-Antarctic nearshore ecosystems. Pap Proc R Soc Tasman 146:89–93

Fraser CI, Nikula R, Spencer HG, Waters JM (2009) Kelp genes reveal effects of subantarctic sea ice during the Last Glacial Maximum. Proc Natl Acad Sci U S A 106:3249–3253

Fraser CI, Nikula R, Waters JM (2011) Oceanic rafting by a coastal community. Proc R Soc B 278:649–655

Fraser CI, Nikula R, Ruzzante DE, Waters JM (2012) Poleward bound: biological impacts of Southern Hemisphere glaciation. Trends Ecol Evol 27:462–471

Fraser CI, Zuccarello GC, Spencer HG, Salvatore LC, Garcia GR, Waters JM (2013) Genetic affinities between trans-oceanic populations of non-buoyant macroalgae in the high latitudes of the Southern Hemisphere. PLoS One 8:e69138

Fraser CI, Kay GM, Plessis MD, Ryan PG (2017) Breaking down the barrier: dispersal across the Antarctic Polar Front. Ecography 40:235–237

Fraser CI, Davies ID, Bryant D, Waters JM (2018a) How disturbance and dispersal influence intraspecific structure. J Ecol 106:1298–1306

Fraser CI, Morrison AK, Hogg AM, Macaya EC, van Sebille E, Ryan PG, Padovan A, Jack C, Valdivia N, Waters JM (2018b) Antarctica's ecological isolation will be broken by storm-driven dispersal and warming. Nat Clim Change 8:704–708

Frenot Y, Chown SL, Whinam J, Selkirk PM, Convey P, Skotnicki M, Bergstrom DM (2005) Biological invasions in the Antarctic: extent, impacts and implications. Biol Rev 80:45–72

Gillespie RG, Baldwin BG, Waters JM, Fraser CI, Nikula R, Roderick GK (2012) Long-distance dispersal: a framework for hypothesis testing. Trends Ecol Evol 27:47–56

Gómez I (ed) (2015) Flora marina antártica: patrimonio de biodiversidad. Ediciones Kultrún, Valdivia, Chile, p 243

González-Wevar CA, Díaz A, Gerard K, Cañete JI, Poulin E (2012) Divergence time estimations and contrasting patterns of genetic diversity between Antarctic and southern South America benthic invertebrates. Rev Chil Hist Nat 85:445–456

Griffiths HJ (2010) Antarctic marine biodiversity–what do we know about the distribution of life in the Southern Ocean? PLoS One 5:e11683

Griffiths HJ, Whittle RJ, Roberts SJ, Belchier M, Linse K (2013) Antarctic crabs: invasion or endurance? PLoS One 8:e66981

Griffiths HJ, Meijers AJS, Bracegirdle TJ (2017) More losers than winners in a century of future Southern Ocean seafloor warming. Nat Clim Change 7:749–754
Gutow L, Giménez L, Boos K, Saborowski R (2009) Rapid changes in the epifaunal community after detachment of buoyant benthic macroalgae. J Mar Biol Assoc UK 89:323–328
Hawes TC, Worland MR, Convey P, Bale JS (2007) Aerial dispersal of springtails on the Antarctic Peninsula: implications for local distribution and demography. Antarct Sci 19:3–10
Helmuth B, Veit RR, Holberton R (1994) Long-distance dispersal of a subantarctic brooding bivalve (*Gaimardia trapesina*) by kelp rafting. Mar Biol 120:421–426
Hickling R, Roy DB, Hill JK, Fox R, Thomas CD (2006) The distributions of a wide range of taxonomic groups are expanding polewards. Glob Chang Biol 12:450–455
Hobday AJ (2000) Persistence and transport of fauna on drifting kelp (*Macrocystis pyrifera* (L.) C. Agardh) rafts in the Southern California Bight. J Exp Mar Biol Ecol 253:75–96
Hunt GL, Drinkwater KF, Arrigo K, Berge J, Daly KL, Danielson S, Daase M, Hop H, Isla E et al (2016) Advection in polar and sub-polar environments: impacts on high latitude marine ecosystems. Prog Oceanogr 149:40–81
Kappen L (2000) Some aspects of the great success of lichens in Antarctica. Antarct Sci 12:314–324
Kappen L, Sommerkorn M, Schroeter B (1995) Carbon acquisition and water relations of lichens in polar regions—potentials and limitations. The Lichenologist 27:531–545
Kennedy AD (1993) Water as a limiting factor in the Antarctic terrestrial environment: a biogeographical synthesis. Arct Alp Res 25:308–315
Kinlan BP, Gaines SD (2003) Propagule dispersal in marine and terrestrial environments: a community perspective. Ecology 84:2007–2020
Lehahn Y, d'Ovidio F, Lévy M, Amitai Y, Heifetz E (2011) Long range transport of a quasi isolated chlorophyll patch by an Agulhas ring. Geophys Res Lett 38.
Lenn Y-D, Chereskin TK (2009) Observations of Ekman currents in the Southern Ocean. J Phys Oceanogr 39:768–779
Lin X, Zhai X, Wang Z, Munday DR (2018) Mean, variability, and trend of Southern Ocean wind stress: role of wind fluctuations. J Climate 31:3557–3573
López BA, Macaya EC, Rivadeneira MM, Tala F, Tellier F, Thiel M (2018) Epibiont communities on stranded kelp rafts of *Durvillaea antarctica* (Fucales, Phaeophyceae)—Do positive interactions facilitate range extensions? J Biogeogr 45:1833–1845
Macaya EC, Zuccarello GC (2010) Genetic structure of the giant kelp *Macrocystis pyrifera* along the southeastern Pacific. Mar Ecol Prog Ser 420:103–112
Marynets K (2019) On the modeling of the flow of the Antarctic Circumpolar Current. Monatsh Math 188:561–565
Miranda L, Thiel M (2008) Active and passive migration in boring isopods *Limnoria* spp. (Crustacea, Peracarida) from kelp holdfasts. J Sea Res 60:176–183
Moon KL, Chown SL, Fraser CI (2017) Reconsidering connectivity in the sub-Antarctic. Biol Rev 92:2164–2181
Muñoz J, Felicísimo ÁM, Cabezas F, Burgaz AR, Martínez I (2004) Wind as a long-distance dispersal vehicle in the Southern Hemisphere. Science 304:1144–1147
Nikula R, Fraser CI, Spencer HG, Waters JM (2010) Circumpolar dispersal by rafting in two subantarctic kelp-dwelling crustaceans. Mar Ecol Prog Ser 405:221–230
Ochyra R, Lewis Smith RI, Bednarek-Ochyra H (2008) The illustrated moss flora of Antarctica. Cambridge University Press, Cambridge, p 685
Orsi AH, Whitworth T, Nowlin WD (1995) On the meridional extent and fronts of the Antarctic Circumpolar Current. Deep Sea Res I Oceanogr Res Papers 42:641–673
Øvstedal DO, Lewis Smith RI (2001) Lichens of Antarctica and South Georgia: a guide to their identification and ecology. Cambridge University Press, Cambridge, UK, p 413
Patarnello T, Bargelloni L, Varotto V, Battaglia B (1996) Krill evolution and the Antarctic ocean currents: evidence of vicariant speciation as inferred by molecular data. Mar Biol 126:603–608
Pugh PJA (2003) Have mites (*Acarina*: Arachnida) colonised Antarctica and the islands of the Southern Ocean via air currents? Polar Rec 39:239–244

Rascle N, Ardhuin F (2013) A global wave parameter database for geophysical applications. Part 2: model validation with improved source term parameterization. Ocean Model 70:174–188
Roads E, Longton RE, Convey P (2014) Millennial timescale regeneration in a moss from Antarctica. Curr Biol 24:R222–R223
Robinson SA, King DH, Bramley-Alves J, Waterman MJ, Ashcroft MB, Wasley J, Turnbull JD (2018) Rapid change in East Antarctic terrestrial vegetation in response to regional drying. Nat Clim Change 8:879–884
Rothäusler E, Gómez I, Hinojosa IA, Karsten U, Tala F, Thiel M (2009) Effect of temperature and grazing on growth and reproduction of floating Macrocystis spp. (Phaeophyceae) along a latitudinal gradient. J Phycol 45:547–559
Scher HD, Martin EE (2006) Timing and climatic consequences of the opening of Drake Passage. Science 312:428–430
Scher HD, Whittaker JM, Williams SE, Latimer JC, Kordesch WEC, Delaney ML (2015) Onset of Antarctic Circumpolar Current 30 million years ago as Tasmanian Gateway aligned with westerlies. Nature 523:580–583
Smith SDA (2002) Kelp rafts in the Southern Ocean. Glob Ecol Biogeogr 11:67–69
Smith CR, Grange LJ, Honig DL, Naudts L, Huber B, Guidi L, Domack E (2012) A large population of king crabs in Palmer Deep on the west Antarctic Peninsula shelf and potential invasive impacts. Proc R Soc Lond [Biol] 279:1017–1026
Sokolov S, Rintoul SR (2009) Circumpolar structure and distribution of the Antarctic Circumpolar Current fronts: 1. Mean circumpolar paths. J Geohys Res Oceans 114:C11018
Teets NM, Denlinger DL (2014) Surviving in a frozen desert: environmental stress physiology of terrestrial Antarctic arthropods. J Exp Biol 217:84–93
Thatje S, Anger K, Calcagno JA, Lovrich GA, Pörtner H-O, Arntz WE (2005) Challenging the cold: crabs reconquer the Antarctic. Ecology 86:619–625
Thiel M (2003a) Rafting of benthic macrofauna: important factors determining the temporal succession of the assemblage on detached macroalgae. Hydrobiologia 503:49–57
Thiel M (2003b) Reproductive biology of *Limnoria chilensis*: another boring peracarid species with extended parental care. J Nat Hist 37:1713–1726
Thiel M, Gutow L (2005) The ecology of rafting in the marine environment. II. The rafting organisms and community. Oceanogr Mar Biol 43:279–418
Thompson AF (2008) The atmospheric ocean: eddies and jets in the Antarctic Circumpolar Current. Philos Trans Royal Soc A 366:4529–4541
Valentine JP, Johnson CR (2004) Establishment of the introduced kelp *Undaria pinnatifida* following dieback of the native macroalga *Phyllospora comosa* in Tasmania, Australia. Mar Freshw Res 55:223–230
Vincent WF (2000) Evolutionary origins of Antarctic microbiota: invasion, selection and endemism. Antarct Sci 12:374–385
Wasley J, Robinson SA, Lovelock CE, Popp M (2006) Some like it wet biological characteristics underpinning tolerance of extreme water stress events in Antarctic bryophytes. Funct Plant Biol 33:443–455
Waters JM (2008) Driven by the West Wind Drift? a synthesis of southern temperate marine biogeography, with new directions for dispersalism. J Biogeogr 35:417–427
Waters JM, Fraser CI, Hewitt GM (2013) Founder takes all: density-dependent processes structure biodiversity. Trends Ecol Evol 28:78–85
Wauchope HS, Shaw JD, Terauds A (2019) A snapshot of biodiversity protection in Antarctica. Nat Commun 10:946
Wiencke C, Clayton MN, Gómez I, Iken K, Lüder UH, Amsler CD, Karsten U, Hanelt D, Bischof K, Dunton K (2007) Life strategy, ecophysiology and ecology of seaweeds in polar waters. Rev Environ Sci Bio/Tech 6:95–126
Young IR (1999) Seasonal variability of the global ocean wind and wave climate. Int J Climatol 19:931–950

Chapter 4
Detached Seaweeds as Important Dispersal Agents Across the Southern Ocean

Erasmo C. Macaya, Fadia Tala, Iván A. Hinojosa, and Eva Rothäusler

Abstract After detachment from their substratum, many seaweeds can float or drift at the mercy of currents and winds, thereby facilitating their dispersal and connectivity. In Antarctica only one species possess floating structures (gas-filled vesicles), the brown seaweed *Cystosphaera jacquinotti*. However, other species such as *Durvillaea antarctica* and *Macrocystis pyrifera* that form abundant forests around the sub-Antarctic islands can also remain at the sea surface once detached, providing a potential dispersal mechanism not only for the seaweeds but also for the associated biota. Additionally, recent reports indicate that floating *D. antarctica* can reach the Antarctic continent from sub-Antarctic regions. Herein, we collect

E. C. Macaya (✉)
Laboratorio de Estudios Algales (ALGALAB), Departamento de Oceanografía, Universidad de Concepción, Concepción, Chile

Millennium Nucleus Ecology and Sustainable Management of Oceanic Island (ESMOI), Coquimbo, Chile

Research Center Dynamics of High Latitude Marine Ecosystems (IDEAL), Valdivia, Chile
e-mail: emacaya@oceanografia.udec.cl

F. Tala
Departamento de Biología Marina, Facultad de Ciencias del Mar, Universidad Católica del Norte, Coquimbo, Chile

Centro de Investigación y Desarrollo Tecnológico en Algas, Facultad de Ciencias del Mar, Universidad Católica del Norte (CIDTA-UCN), Coquimbo, Chile
e-mail: ftala@ucn.cl

I. A. Hinojosa
Millennium Nucleus Ecology and Sustainable Management of Oceanic Island (ESMOI), Coquimbo, Chile

Departamento de Ecología, Facultad de Ciencias y Centro de Investigación en Biodiversidad y Ambientes sustentables (CIBAS), Universidad Católica de la Santísima Concepción, Concepción, Chile
e-mail: ihinojosa@ucsc.cl

E. Rothäusler
Centro de Investigaciones Costeras - Universidad de Atacama (CIC - UDA), Avenida Copayapu 485, Copiapó, Chile
e-mail: eva.rothausler@uda.cl

I. Gómez, P. Huovinen (eds.), *Antarctic Seaweeds*,
https://doi.org/10.1007/978-3-030-39448-6_4

information about floating and drifting seaweeds in Antarctica, but also their biology, physiology, and distribution within the sub-Antarctic sources. Up to now, only a few species have been recorded floating in Antarctica, and mainly during the first Antartic explorations. So far, most of the studies on detached seaweeds only highlight their importance, when already stranded and serving as carbon sources for benthic communities. However, some seaweed species are able to handle present sea surface conditions in Antarctica and thus in the future when higher temperatures, less ice and more available substrate are available, they might be able to frequently travel and colonize this region, thereby representing an important dispersal mode.

Keywords Floating seaweed · Connectivity · Drifting · Rafting · Stranding

The two most striking vegetable productions of this island are a noble seaweed, called Sargassum jacquinotii, and a Lichen. The first of these was not found attached, but floating in the ocean among the ice, by which it was sometimes much mutilated. Though belonging to a highly variable order, it is a perfectly distinct as well as conspicuous species. (James Clark Ross 1847)

4.1 Introduction

The description above by the British Royal Navy Officer Sir James Ross onboard of the *HMS Erebus* during the Antarctic exploration (January 1843) it is one of the first records of floating seaweeds in Antarctica. In his report, Ross also mention the surgeon of the HMS Chanticleer, William Webster, who was the first naturalist in collecting and describing such species at Deception Island: "*the most common seaweed was found floating. It was of a pale chocolate colour, stem and branches flat, quarter of an inch in breadth, leaves equitant, thin, delicate, four of five inches long, and at the base of each was a spherical air-cell the size of a grape. The mode of reproduction appeared to be from a cluster of buds appended to the terminal branches*" Webster (1834).

The seaweed observed by Webster and Ross is known today as *Cystosphaera jacquinotti* and is endemic to the Antarctic. This brown seaweed species possesses "pneumatocysts" (the size of a grape, according to Webster's description) which are floating structures and allow the seaweed to remain at the surface once being detached from their primary substratum (Fig. 4.1). Although there are abundant records of floating seaweeds around the world (e.g. Thiel and Gutow 2005a; Rothäusler et al. 2012; Macaya et al. 2016 and references therein), polar regions appear poorly represented. This may be due to the harsh environmental conditions onboard of a research vessel, the limited accessibility to many regions, and the isolation of Antarctica from the rest of the Southern Ocean. That is probably why most of the information comes from microalgal aggregates (Cefarelli et al. 2011; Assmy et al.

Fig. 4.1 The endemic *Cystosphaera jacquinotti,* black arrows indicate the gas filled structures "pneumatocysts". Right, a floating raft observed in Fildes Bay, King George Island. (Photos by Erasmo C. Macaya)

2013; Katlein et al. 2015; Belt et al. 2018). As a result, few records of floating seaweeds have been reported in Antarctica, indicating only their importance as drifting and stranding seaweeds, because they can serve as a carbon source for benthic communities (e.g. Fischer and Wiencke 1992; Norkko et al. 2004; Braeckman et al. 2019; see also Chap. 8 by Quartino et al.). However recent records of non-native positively buoyant kelps (*Durvillaea antarctica* and *Macrocystis pyrifera*) that stranded on South Shetland Islands opened the discussion regarding the isolation of Antarctica (Fraser et al. 2018; Avila et al. 2020). This is especially important in the context of future warming scenarios, where long-distance dispersal might facilitate connectivity with sub-Antarctic seaweed sources. In fact, long-distance dispersal and connectivity within these vast areas of open oceans have been already demonstrated for *D. antarctica* and *Macrocystis pyrifera*, with single haplotypes having a wide sub-Antarctic distribution (Fraser et al. 2009; Macaya and Zuccarello 2010). At the same time, these seaweeds serve as important dispersal vehicles for associated flora and fauna (Nikula et al. 2010; Wichmann et al. 2012; Nikula et al. 2013; Cumming et al. 2014; Macaya et al. 2016), considering the large number of floating *D. antarctica* rafts in the Southern Ocean (Smith 2002) (see also Chap. 3 by Fraser et al.). In this chapter we give information about floating, drifting and stranded seaweeds in Antarctica in relation to their abundance and distribution. We also address whether the extreme environmental conditions (irradiance and temperature) can affect seaweed physiology, growth and thus survival mainly in floating conditions. In addition, we give potential sources of seaweeds that may have the chance to travel into Antarctica.

4.2 Detached Seaweeds in Antarctica

A total of 39 seaweeds species, including 3 Chlorophyta, 14 Ochrophyta, and 22 Rhodophyta–have been recorded drifting, stranding or floating in Antarctica or crossing the Antarctic Polar Front APF (Table 4.1). From the 95 records included herein, 75% reported about observations or studies of drifting seaweeds, followed by a 17% of floating seaweeds, while only an 8% reported about stranded seaweeds. To our knowledge, so far only brown but no red or green seaweed species have been found floating in Antarctica. The brown seaweeds species include *Cystosphaera jacquinotti*, *Himantothallus grandifolius*, *Desmarestia anceps* and *Adenocystis utricularis*. Among them, only *C. jacquinotti* possesses floating structures and is also the most frequent species observed. For instance, abundant seaweed pieces were recorded floating during 1908–1909 in the Neumayer and Peltier Channels as well as in the Bransfield and Gerlache Straits (Gain 1912). Fertile floating individuals were observed in Harmony Cove (Neushul 1963) and pieces up to 90 cm in length were spotted among ice in Cockburn Island (Ross 1847). In contrast, *A. utricularis* owes buoyancy when mature thalli become filled with air and individuals have been found floating abundantly within the Fildes Bay (E. Macaya pers. obs). Floating thalli but also holdfasts of *H. grandifolius* were observed within Admiralty Bay (Zemko et al. 2015). Large numbers of floating kelps such as *Macrocystis pyrifera* and *Durvillaea antarctica* were found floating south of the APF (Fraser et al. 2017) and floating *D. antarctica* pieces were recorded in the interior pass of Deception Island (Gain 1912). Several records of floating seaweed were also carried out by James Cook during his voyage towards the South Pole during 1770's, he described floating seaweeds at latitude 70°S, longitude 108°W: "*We continued our course to the south, and passed a piece of weed covered with barnacles, which a brown albatross was picking off*". Those rafts might have been afloat for a while, because of the barnacle presence (Thiel and Gutow 2005a).

Among the drifting red seaweeds, most of the species belong to the order Ceramiales and Gigartinales, with seven and five species recorded, respectively. Attached to the spines of the sea urchins *Sterechnus neumayeri* (Amsler et al. 1999; Schwarz et al. 2003) or accumulated in hollows, a bundant biomass of drifting *Phyllophora antarctica* were observed at 15–30 m depth in Cape Evans, McMurdo Sound (Norkko et al. 2004). Interestingly, *S. neumayeri* uses drift pieces of *P. antarctica* and *Iridaea cordata* as defense against its main predator, the anemone *Isotealia antarctica*. At the same time, this retention of drifting seaweed pieces indirectly supports to maintain the seaweed populations, if pieces are fertile and still within the photic zone (Amsler et al. 1999). The most common drifting brown seaweeds belong to the order Desmarestiales. For instance individuals of *Desmarestia*

Table 4.1 Detached seaweeds reported drifting, floating or stranded in Antarctica. Floating kelps crossing the Antarctic Polar Front are also included

Phylum/order	Species	Algal-status	Locality	Reference
Chlorophyta				
Prasiolales	*Prasiola crispa*	Drift	Fildes Bay, King George Island	(Fischer and Wiencke 1992)
Ulothricales	*Monostroma hariotii*	Drift	Admiralty Bay, King George Island	(Oliveira et al. 2009)
			Fildes Bay, King George Island	(Fischer and Wiencke 1992)
		Stranded	Fildes Bay, King George Island	(Westermeier et al. 1992)
Ulvales	*Ulva intestinalis*	Drift	Whalers Bay, Deception Island	(Clayton et al. 1997)
Ochrophyta				
Ascoseirales	*Ascoseira mirabilis*	Drift	Fildes Bay, King George Island	(Müller et al. 1990; Sato et al. 1992; Tada et al. 1996)
		Drift	Admiralty Bay, King George Island	(Zielinski 1990)
Chrysomeridales	*Antarctosaccion applanatum*	Drift	South Bay, Livingston Island	(Gallardo et al. 1999)
Desmarestiales	*Desmarestia anceps*	Drift	Fildes Bay, King George Island	(Fischer and Wiencke 1992; Tada et al. 1996)
			DeLaca Island	(Amsler et al. 2012)
			Signy Island, South Orkney Islands	(Richardson 1979; Brouwer et al. 1995)
			Admiralty Bay, King George Island	(Zielinski 1990)
		Floating, stranded	Bransfield Strait; Deception Island; King George Island	(Gain 1912)
	Desmarestia antarctica	Drift	Deception Island	(Lastra et al. 2014)
			Fildes Bay, King George Island	(Fischer and Wiencke 1992)
	Desmarestia menziesii	Drift	Deception Island	(Lastra et al. 2014)
			Signy Island, South Orkney Islands	(Brouwer 1996)

(continued)

Table 4.1 (continued)

Phylum/order	Species	Algal-status	Locality	Reference
			Fildes Bay, King George Island	(Fischer and Wiencke 1992)
			Admiralty Bay, King George Island	(Zielinski 1990)
	Desmarestia sp.	Stranded	Deception Island	(Gain 1912)
	Himantothallus grandifolius	Floating	Admiralty Bay, King George Island	(Zemko et al. 2015)
		Drift	Admiralty Bay, King George Island	(Zielinski 1981, 1990)
			Signy Island, South Orkney Islands	(Brouwer 1996)
Ectocarpales	*Adenocystis utricularis*	Drift,	Fildes Bay, King George Island	(Westermeier et al. 1992; Tada et al. 1996)
		Floating	Fildes Bay, King George Island	(Erasmo Macaya pers. obs.)
	Chordaria linearis	Drift	Admiralty Bay, King George Island	(Oliveira et al. 2009)
	Petalonia fascia	Drift	Whalers Bay, Deception Island	(Clayton et al. 1997)
	Scytosiphon lomentaria	Drift	Whalers Bay, Deception Island	(Clayton et al. 1997)
	Utriculidium durvillei		Fildes Bay, King George Island	(Müller et al. 1992)
Fucales	*Cystosphaera jacquinotii*	Floating	Deception Island	(Hooker 1844)
			Cockburn Island	(Ross 1847)
			Antarctic Peninsula	(Montagne 1842; Skottsberg 1907)
			Half Moon Island	(Neushul 1963)
			Fildes Bay, King George Island	(Gain 1912; Macaya E. Personal Observation; Weykam et al. 1996)
			Gerlache Strait	(De Wildeman 1935)
			South Bay, Doumer Island	Macaya E. Personal Observation
		Drift	Admiralty Bay, King George Island	(Zielinski 1981, 1990; Oliveira et al. 2009)

(continued)

Table 4.1 (continued)

Phylum/order	Species	Algal-status	Locality	Reference
		Drift	Ardley Bay, King George Island	(Fischer and Wiencke 1992)
		Drift, stranded	Half Moon Island	(Neushul 1965)
	Durvillaea antarctica	Floating, stranded	Deception Island, King George Island	(Gain 1912)
		Stranded	Livingtone Island	(Avila et al. 2020)
		Floating	South of New Zealand 65°S	(Hooker 1844)
			South of the Antarctic Polar Front	(Fraser et al. 2017)
		Drift	Livingston and Elephant Islands	(Pellizzari et al. 2017)
		Stranded	Fildes Bay, King George Island	(Fraser et al. 2018)
Laminariales	*Macrocystis pyrifera*	Floating	South of the Antarctic Polar Front	(Fraser et al. 2017)
		Stranded	Deception Island	(Avila et al. 2020)
Rhodophyta				
Balliales	*Ballia callitricha*	Drift	South Bay, Livingston Island	(Gallardo et al. 1999)
Bangiales	*Porphyra plocamiestris*	Drift	Cuverville Island	(Clayton et al. 1997)
Bonnemaisoniales	*Delisea pulchra*	Drift	Admiralty Bay, King George Island	(Oliveira et al. 2009)
Ceramiales	*Georgiella confluens*	Drift	Fildes Bay, King George Island	(Fischer and Wiencke 1992)
	Myriogramme manginii	Drift	Livingston Island	(Gallardo et al. 1999)
			Deception Island	(Gain 1912)
			Signy Island, South Orkney Islands	(Brouwer 1996)
	Neuroglossum delesseriae	Drift	Fildes Bay, King George Island	(Fischer and Wiencke 1992)
	Pantoneura plocamioides	Drift	Signy Island, South Orkney Islands	(Brouwer 1996)
			South Bay, Livingston Island	(Gallardo et al. 1999)

(continued)

Table 4.1 (continued)

Phylum/order	Species	Algal-status	Locality	Reference
	Paraglossum lancifolium	Drift	Admiralty Bay, King George Island	(Oliveira et al. 2009)
	Phycodrys antarctica	Drift	Fildes Bay, King George Island	(Fischer and Wiencke 1992)
	Phycodrys quercifolia	Drift	Fildes Bay, King George Island	(Weykam et al. 1996)
Gigartinales	*Cystoclonium obtusangulum*	Drift	Byers Peninsula, Livingston Island	(Hommersand et al. 2009)
	Gigartina skottsbergii	Drift	Fildes Bay, King George Island	(Fischer and Wiencke 1992)
			Admiralty Bay, King George Island	(Oliveira et al. 2009)
	Iridaea cordata	Drift	Cape Evans, McMurdo Sound	(Schwarz et al. 2003)
			Fildes Bay, King George Island	(Fischer and Wiencke 1992)
			McMurdo Sound	(Amsler et al. 1999)
			Admiralty Bay, King George Island	(Oliveira et al. 2009)
		Stranded	Fildes Bay, King George Island	(Westermeier et al. 1992)
	Notophycus fimbriatus	Drift	Admiralty Bay, King George Island	(Oliveira et al. 2009)
	Phyllophora antarctica	Drift	Cape Evans, McMurdo Sound, Ross Island	(Miller and Pearse 1991; Schwarz et al. 2003; Norkko et al. 2004)
			McMurdo Sound	(Amsler et al. 1999)
Gracilariales	*Curdiea racovitzae*	Drift	Fildes Bay, King George Island	(Fischer and Wiencke 1992)
Halymeniales	*Pachymenia orbicularis*	Drift	Admiralty Bay, King George Island	(Oliveira et al. 2009)
Palmariales	*Palmaria decipiens*	Drift	Half Moon Island	(Neushul 1965)
			Foster Bay, Deception Island	(Lastra et al. 2014)
			Fildes Bay, King George Island	(Fischer and Wiencke 1992)

(continued)

Table 4.1 (continued)

Phylum/order	Species	Algal-status	Locality	Reference
		Drift, stranded	Admiralty Bay, King George Island	(Zielinski 1981; Westermeier et al. 1992; Oliveira et al. 2009)
Plocamiales	*Trematocarpus antarcticus*	Drift	Fildes Bay, King George Island	(Fischer and Wiencke 1992)
	Plocamium cartilagineum	Drift	Fildes Bay, King George Island	(Fischer and Wiencke 1992)
			Admiralty Bay, King George Island	(Oliveira et al. 2009)
			Signy Island, South Orkney Islands	(Brouwer 1996)
	Plocamium hookeri	Drift	South Bay, Livingston Island	(Gallardo et al. 1999)
Rhodymeniales	*Hymenocladiopsis prolifera*	Drift	South Bay, Livingston Island	(Gallardo et al. 1999)

anceps and *D. menziesii* were observed at 10 to 40 m depth in Admiralty Bay, where icebergs provoked their detachment (Zielinski 1990). Huge drifting patches of *D. anceps* with up to 60 m^3 were recorded between 5 and 20 m depth in Fildes Bay. These, detached individuals represent an important habitat for a variety of invertebrates such as gastropods (*Nacella concinna*), giant isopods (*Glyptonotus antarcticus*) and gammaridean amphipods (Tada et al. 1996) (Fig. 4.2).

Similarly, high quantities of drifting *D. anceps* detached by ice and wave action accumulated in hollows at Signy Island, South Orkney (Brouwer 1996). When detached this species continued to grow and remained healthy for over 44 weeks, without changing its palatability (Brouwer 1996; Amsler et al. 2012). To date, only 3 species of drifting green seaweeds have been reported in Antarctica, namely: *Monostroma hariotti*, *Prasiola crispa* and the non-native *Ulva intestinalis* (Fischer and Wiencke 1992; Clayton et al. 1997; Oliveira et al. 2009). Large inflated thalli of *U. intestinalis* were found drifting in Whalers Bay (Deception Island), which is probably a recent introduction, caused by shipping activities (Clayton et al. 1997), also this species has been reported as a common fouling organism in other latitudes (Blomster et al. 1998). However, introduction via floating dispersal cannot be ruled out because *U. intestinalis* can form massive floating mats, that are known to tolerate severe winter conditions (Bäck et al. 2000) together with a rapid acclimation to changes in salinity, nutrients and light (Cohen and Fong 2004). Similarly, free floating *Ulva linza* acclimated rapidly to changing light conditions by developing effective mechanisms to cope with excessive irradiation (Häder et al. 2001).

Detached seaweeds washed ashore (Fig. 4.3) have been studied at different sites with in Antarctica. Lastra et al. (2014) estimated 1545 tons of seaweeds that arrived

Fig. 4.2 Drifting seaweeds accumulated in hollows, Fildes Bay, King George Island. (Photo by Ignacio Garrido)

to the shore in Foster Bay (Deception Island) within the period from November to July, and only 3% of the detached seaweeds accumulated as wrack along the intertidal. In Admiralty Bay, 1643 tons of stranded seaweeds were estimated between February and November 1979, and especially after storms (Rakusa-Suszczewski 1980; Zielinski 1981). Stranded seaweeds remnants decompose with contact to air, and within 20 d, more than 50% were transformed to organic matter (Rakusa-Suszczewski 1980), a process that becomes accelerated by winds and waves (Zielinski 1981). The wrack deposition can be highly variable with more than 50% of replacement at each tidal cycle (Lastra et al. 2014), probably caused by the interaction between wind direction and coastal orientation (Rakusa-Suszczewski 1980; Zielinski 1981; Rakusa-Suszczewski 1995; Lastra et al. 2014). Slow decomposition rates due to the extreme environment, however facilitates the longevity of detached seaweeds and constitute an important carbon source for higher trophic levels via the detrital food web (Fischer and Wiencke 1992; Schwarz et al. 2003; see also Chap 8 by Quartino et al.).

Fig. 4.3 Stranded seaweeds at Fildes Bay, King George Island. Low left, *Desmarestia antarctica* as epiphyte of stranded *Curdiea racovitzae*. (Photos by Erasmo C. Macaya)

4.3 Abiotic Factors Influencing Floating Seaweeds

Solar radiation (280–700 nm) and sea surface temperature (SST) are the main abiotic factors that determine the persistence and thus dispersal ability of floating seaweeds (Hobday 2000; Rothäusler et al. 2009, 2012; Graiff et al. 2013; Tala et al. 2016, 2019). In Antarctica and sub-Antarctic islands, there is a strong seasonal light regime, especially at latitudes >70°S, where several months of darkness in winter and of complete daylight in summer with up to 1700 μmol m^{-2} $s^{-1}$of PAR, 44 W m^{-2} of UV-A, and 2.3 W m^{-2} of UV-B can prevail (Quartino et al. 2005; Zacher et al. 2007, 2009; see also Chap 7 by Huovinen and Gómez). Also SST varies from warmer sub-Antarctic waters (4 °C to 14 °C) to colder, icier Antarctic waters (−1.8 °C and 2.2 °C off the Antarctic Peninsula) (Drew and Hastings 1992; Klöser et al. 1993; Kang et al. 2002; Mélice et al. 2003). Thus, depending on the latitude or season when detached, and/or when occasionally become pushed through the APF, seaweeds are confronting extreme different conditions during rafting.

In situ experiments carried out with two tethered kelp species in Tierra del Fuego at 54°S, showed for the temperate *M. pyrifera* that its growth capacity is favored by high light levels (~1000 μmol m^{-2} s^{-1}) and moderate SSTs (~10 °C) in summer but the high growth rates cannot be maintained in low light (~700 μmol m^{-2} s^{-1}) combined with low SST (~6 °C) in winter (Tala et al. 2016). In contrast, the sub-Antarctic *D. antarctica* showed a reverse pattern with positive growth in winter despite the low light and temperature conditions. Apparently, the potential for rafting dispersal in *D. antarctica* appears to be less dependent on seasonal variations in environmental conditions (Tala et al. 2019). Clearly, sub-Antarctic floating seaweeds perform well in cooler waters and under low solar radiation than their temperate counterparts (Rothäusler et al. 2009; Graiff et al. 2013; Tala et al. 2016, 2019).

Seaweeds that are distributed in sub-Antarctic islands and Tierra del Fuego such as *D. antarctica* and *M. pyrifera* typically grow in environments with large fluctuations in seawater temperature and have thus a broad performance breadth (Eggert 2012). Consequently, if floating conspecifics have the chance to become pushed through the APF (Fraser et al. 2017; Fraser et al. 2018) they possibly can acclimate to cold-water temperatures more easily and thereby contributing to long distance dispersal (Fraser et al. 2018). However, *M. pyrifera* may fail to establish in the colder, icier Antarctic environment because it was shown that they lost biomass at 54°S in winter probably due to low light availability (Tala et al. 2016). Hence, abiotic conditions in polar waters >60 °S might temporarily, especially in winter, be unfavorable for some cold-temperate floating seaweeds. This might be the reason why *M. pyrifera* (> 1 m) has been reported in coexistence with the endemic Antarctic *H. grandifolius* under the extreme environmental conditions in South Georgia (Barnes et al. 2006). At the same latitude (54°S), but at Tierra del Fuego, floating *D. antarctica* did not show biomass losses in winter and steadily grew, which was not the case in summer (Tala et al. 2016). As a result its floating time exceeded 200 d in winter but in summer it dropped to 90 d (Tala et al. 2019). This suggests that *D. antarctica* may survive and be successful when entering into icier SSTs at the APF. Indeed, two mature thalli of *D. antarctica* were found stranded at the Antarctic Peninsula (62°S) in summer 2017 with goose barnacles attached (Fraser et al. 2018). Molecular studies showed that these seaweed pieces originated from sub-Antarctic Islands and therefore the dispersal distances represented the longest rafting events ever recorded, with trajectories between 20,000 and 25,000 km (Fraser et al. 2018). This suggests that these pieces started their journey as rafts consisting of several individuals because it was estimated that they travelled a maximum of 2 years and a minimum of 489 d in order to reach Antarctica from their origins (Fraser et al. 2018). During this very long pelagic journey, rafts experienced substancial fluctuations in the abiotic and biotic environment, thereby probably losing biomass.

It is known that nutrient abundance stimulates seaweed growth and thus it has been inferred to affect the persistence time of floating seaweeds (Rothäusler et al. 2012 and references therein). When seaweeds become occasionally pushed through the APF and float in the open waters of the Southern Ocean, which are defined as high nutrient environments, their growth and thus survival is unlikely to be restricted. The same is true when seaweeds arrive to the coastal waters of the Antarctic Peninsula, where high levels of nitrate and phosphate are present at the sea surface throughout the year (reviewed in Zacher et al. 2009). There is still, however, very little information available on the impact of nutrients on floating seaweeds.

4.4 Biotic Factors Affecting Floating Seaweeds

Large numbers of hitchhiking organisms have been reported from floating seaweeds (Khalaman and Berger 2006; Vandendriessche et al. 2007a; Wichmann et al. 2012; Abé et al. 2013; Macaya et al. 2016; López et al. 2018). Many of these organisms are grazers e.g. amphipods and isopods that mostly feed on vegetative parts from the rafts, thereby removing photosynthetic tissue, and thus contribute to their demise (Vandendriessche et al. 2007b; Rothäusler et al. 2009; Rothäusler et al. 2018). But also grazing by e.g. snails and sea urchins and fouling by bryozoans, goose barnacles, and epiphytic seaweeds can limit their persistence at the sea surface (Rothäusler et al. 2011b; Graiff et al. 2016; Rothäusler et al. 2018; see also Chap. 17 by Amsler et al. about algal-grazers interactions in Antarctic seaweeds). Especially, the overgrowth with sessile organisms can reduce buoyancy, because epibionts contribute to an increase in the specific gravity of the floating seaweed (Graiff et al. 2016). These biotic interactions are particularly important for seaweed rafts because they will directly affect the persistence (and dispersal potential) of rafts at the sea surface.

The destructive effect of herbivorous grazers on their rafts is dependent on water temperature (Vandendriessche et al. 2007a; Rothäusler et al. 2009, 2018). In mesocosms but also in field studies it was shown that low temperatures (5–15 °C) slowed amphipod and isopod consumption on floating *M. pyrifera*, *Fucus vesiculosus,* and *Ascophyllum nodosum,* and that these seaweeds can compensate for grazer induced tissue losses (Vandendriessche et al. 2007a; Rothäusler et al. 2009, 2018). For instance, at 5 °C, *F. vesiculosus* and *A. nodosum* kept in mesocosms gained 3 times their initial weight while afloat, but the weight gain was less in the presence of isopods (Vandendriessche et al. 2007b). This is probably because herbivore metabolism, which is strongly controlled by temperature (O'Connor 2009), slows down at low SSTs and seaweed growth is favored under benign environmental conditions. Similarly, in cold sub polar waters around Iceland (64°N), Ingólfsson (1998) reported that fronds of *A. nodosum*, which were accidentally left afloat for 43 d, did not show any signs of decay and were still in perfect condition. Consequently, floating seaweeds are expected to be less grazed under cold SSTs.

Rafts serve as substratum for many sessile epibionts, including e.g. encrusting bryozoans and goose barnacles (Hinojosa et al. 2006; Rothäusler et al. 2011a; Graiff et al. 2016). Bryozoan growth and thus additional weight generally increase with seaweed floating time, thereby negatively affecting seaweed persistence (Rothäusler et al. 2011b for *M. pyrifera* at 30°S), which can finally result in their decay and sinking (Graiff et al. 2016). However, recently, it was shown that maximum bryozoan cover on tethered *M. pyrifera* (30°S) was reached earlier in spring/summer than in autumn/winter (Graiff et al. 2016). In a study done with natural floating rafts of *M. pyrifera* (30°S), the colonization by bryozoans increased with decreasing latitudes, which coincides with warmer SSTs (Rothäusler et al. 2011a). This implies that bryozoan growth is slowed down at lower SSTs. Hence, seaweeds carrying epibionts and traveling in cold and icy sub-Antarctic or Antarctic waters may stay afloat for substantially longer periods, which is also underlined by the fact that *D. antarctica* thalli beached at Antarctica had goose barnacles attached (Fraser et al. 2018; E. Macaya Pers. Obs.). Preliminary observations on floating rafts of *C. jacquinotti* within the Antarctic Peninsula revealed very few epibionts (E. Macaya pers. obs.).

Moreover, there is evidence from molecular studies, showing that hitchhiking organisms, such as crustaceans, mollusks, seaweed parasites, and non-buoyant seaweeds, can travel vast distances (Nikula et al. 2010, 2013; Fraser and Waters 2013; Fraser et al. 2013; Boo et al. 2014; Guillemin et al. 2014; Macaya et al. 2016). Some of them even have a circum sub-Antarctic distribution, as has been reported for two holdfast dwelling peracarid crustaceans (Nikula et al. 2010). However, there are no studies from high latitudes determining whether these hitchhikers affect the floating persistence of their rafts. Probably, under favorable conditions such as prevailing in Tierra del Fuego, sub-Antarctic islands, and Antarctica, seaweeds can continuously grow, and thus provide a long lasting substratum and food source for their associated hitchhikers, which in turn may have a decreased metabolic rate. Particularly, high abundances of amphipods (up to 300,000 individuals m^2) have been recorded in benthic Antarctic seaweeds stands, such as *Desmarestia antarctica, D. menziesii* and *D. anceps* (Huang et al. 2007; see also Chap. 17 by Amsler et al.), some of these species have been found floating or drifting (Table 4.1) and might represent a dispersal vehicle for associated hitchhikers.

4.5 Physiology of Floating and Drifting Seaweeds: Traspassing Thermal Barriers

Our knowledge about the physiological, reproductive and growth performance of Antarctic seaweeds comes mainly from benthic populations and/or from seaweeds kept under experimental conditions. Similarly, the performance of floating seaweeds at high latitudes has been almost exclusively studied under mesocosm conditions. Generally, in Antarctic benthic seaweeds, photosynthetic and bio-optical properties as well as UV and/or temperature stress tolerances are linked to their vertical distribution and biogeographic affinity (Huovinen and Gómez 2013; Gómez et al. 2019). These seaweeds are defined as shade-adapted, and hence are photosynthetically and metabolically prepared to cope with extreme irradiance conditions but also metabolically with very low temperatures (Gómez et al. 2009; Huovinen and Gómez 2013; Rautenberger et al. 2015; Gómez et al. 2019). Hence, during spring/summer, when irradiances, day length, and light penetration are suddenly increasing, they can become stressful for many benthic seaweeds but also for their community associated (Zacher et al. 2009). This is especially the case for subtidal species, which are growing in a more stable environment compared to intertidal species.

Upon detachment, floating seaweeds suffer a sudden environmental change from the benthic (shaded) to the pelagic enviroment (sunny). Thus, depending on the season and latitude when detached, floating seaweeds but also their associated hitchhikers might respond differently (e.g. growth, physiological acclimation and reproduction) (Rothäusler et al. 2012; Macaya et al. 2016 and references therein). At high latitudes (54°S), floating rafts of *M. pyrifera* and *D. antarctica* displayed different short-term physiological adjustments to floating conditions depending on the season, which are apparently associated to their biogeographic affinity and growth strategies (Tala et al. 2016). During winter, *M. pyrifera* lost biomass although seaweeds maintained their maximum quantum yield (F_v/F_m) and increased their pigment levels, presumably because of light limitation and to support some biomass production. Opposite responses such as positive biomass changes, photoregulation, decreasing F_v/F_m and pigment levels were found during summer (Tala et al. 2016). *Durvillaea antarctica* might support better the winter conditions that prevail at high latitudes because positive biomass changes were found during the first three months afloat and only small variations in F_v/F_m, pigment, and phlorotannin levels. However, in summer, tissue deterioration and antioxidant activity increased, which were reflected in biomass loss (Tala et al. 2016, 2019). Therefore, at high latitudes, *D. antarctica* can stay longer afloat (>200 days) during winter than during summer (90 days).

4.5.1 Out of Antarctic: Is it Physiologically Feasible?

Even though there is no specific information about the physiological and reproductive responses of floating seaweed in and out Antarctica, important transoceanic routes have been identified through phylogeographic studies at high latitudes (Fraser et al. 2010; Coyer et al. 2011; Fraser et al. 2013; Boo et al. 2014; Moon et al. 2017), confirming the capacities of floating and non-buoyant seaweeds to acclimate during long journeys (see also Chap. 3 by Fraser et al.). The successful dispersal of floating seaweeds over short or long distances requires an efficient physiological acclimation (e.g. photosynthetic efficiency adjust, high antioxidant activity, production of defensive metabolites) not only to maintain growth and reproductive capacity, but also to deal with the herbivore pressure and epibiont load. When rafting is successful, seaweeds can extend their distributional ranges according to their physiological and/or reproductive tolerances and/or abiotic and biotic interactions (Thiel and Gutow 2005b; Nikula et al. 2010; Coyer et al. 2011; Waters et al. 2013; Batista et al. 2018; López et al. 2018).

Performance and tolerance of benthic seaweed populations can give us an indication of how they may respond under floating or drifting conditions. The optimum temperatures for photosynthesis in Antarctic seaweeds range between 5 and 15 °C (Wiencke et al. 1993; Eggert and Wiencke 2000), thus floating seaweeds entering Antarctic from Tierra del Fuego or sub-Antarctic sources might be able to survive. Temperature requirements for growth and survival of 15 Antarctic red seaweed species were studied by Bischoff-Bäsmann and Wiencke (1996). Consequently, if eurythermal species (with a broad performance breadth) become pushed through the APF they can acclimate more easily to more temperate temperatures more easily and can support large distance dispersal. Hence an "out of Antarctic" scenario, could be supported at least by the high environmental tolerance showed by some cold-adapted Antarctic seaweeds. In fact, various species display different mechanisms to support elevated temperatures and high UVR /PAR conditions (Huovinen and Gómez 2013; Rautenberger et al. 2015). In experimental conditions, UVR tolerance is improved in sensitive species modulated by temperature via an efficient damage repair of the photosynthetic apparatus instead of increasing photo-protective metabolites or antioxidant activity (Rautenberger et al. 2015). Therefore, chemically based tolerance mechanisms are not inducible as has been reported for some Desmarestiales (Flores-Molina et al. 2016). In fact, high constitutive concentrations of soluble and insoluble phlorotannins (a type of phenolics found mainly in brown seaweeds) have been determined irrespective of the UV levels at which these seaweeds are exposed in their habitat (Rautenberger et al. 2015). In brown seaweeds, phlorotannins have different biological functions such as UV shielding, ROS scavenging, herbivore deterrence, and cell wall formation. Therefore, species that maintain high phlorotannin concentrations are better prepared to face variable environmental conditions (see also Chap. 18 by Gómez and Huovinen). However, due to an extended period of low light during the polar winter, and turbid seawater during summer and autumn following sea-ice melting and phytoplankton blooms, Antarctic seaweeds are strongly shade adapted, which could impair their physiological performance during rafting and limiting the migration to lower latitudes.

The endemic floating species *Cystosphaera jacquinotii* is commonly found between 5 to 30 m depth, usually in more exposed areas and disappear south of 66°S (Neushul 1963; Wulff et al. 2009). *Adenocystis utricularis* with its vesicle saccate-like morphology are abundant in the intertidal (Valdivia et al. 2014) and have also been recorded floating (see above). Photosynthetic characteristics showed that *C. jacquinotti* has a low saturating irradiance (E_k = 37 μmol photons m^{-2} s^{-1}) and stress tolerance to UVR and enhanced temperature, while *A. utricularis* showed high saturating irradiance (E_k = 137 μmol photons m^{-2} s^{-1}) and stress tolerance (Gómez et al. 2019). Both saturating irradiance levels match with the subtidal and intertidal prevailing light conditions. Similar patterns were detected in seaweeds from King George Island, with saturation irradiances of 55 μmol m^{-2} s^{-1} in subtidal and 120 μmol m^{-2} s^{-1} in eulittoral species (Huovinen and Gómez 2013). For instance, photosynthesis of *Desmarestia antarctica*, *D. anceps* and *Gymnogongrus antarcticus* was sensitive when exposed to higher light conditions, especially to UVB (Bischof et al. 1998). In this context, seaweeds from subtidal environments that form drifting patches can deteriorate quickly, thus contributing to an important carbon source for higher trophic levels.

High concentrations of soluble phlorotannins are found in *C. jacquinotii* (>5% of DW in both vegetative and reproductive tissues), acting as UVR shield and herbivore deterrent (Iken et al. 2007; Huovinen and Gómez 2015; see also Chap. 17 by

Amsler et al. and Chap. 18 by Gómez and Huovinen), thereby contributing to a higher floating persistence.

Environmental factors that affect seaweed physiology and growth under benthic and floating conditions, can also modify the reproductive capacity. Considering the short-life and dispersal distances of seaweed propagules (Santelices 1990), the ability to remain reproductively competent during rafting and when arriving on a new site is crucial for a successful dispersal and colonization. Long floating times together with high temperature and radiation conditions decreas the availability of propagules and can increas the disintegration of reproductive tissues such as shown for *D. antarctica* (Tala et al. 2016, 2019), *M. pyrifera* (Macaya et al. 2005; Hernández-Carmona et al. 2006; Rothäusler et al. 2011b) and *Hormosira banksii* (McKenzie and Bellgrove 2008). Rafting and long distance dispersal might be facilitated in monoecious species or in dioecious species (e.g. *D. antarctica* and *C. jacquinotti*) when male and female individuals are traveling together and arriving to a new suitable habitat (Lizée-Prynne et al. 2016).

Presumably the establishment of non-native species within the Antarctic shores is prevented by the physiological and reproductive tolerance ability to extreme environmental conditions rather than transport. Low temperatures, tides and wave action, ice-scouring, salinity changes, sediment and detritus accumulation may limit the establishment of foreign taxa on hard substrates (Campana et al. 2009). Biotic interactions associated to local herbivores and competition can also determine the colonization success of the colonization. However, under a warming climate scenario, more sub-Antarctic and temperate species might be able to grow and survive at higher latitudes. However, if the substrate availability is low at the new site of arrival or the the habitat is highly diverse, only strong competitors may have the chance to establish (Stachowicz et al. 2002). The density-dependent blocking mechanism of conspecifics has been described to prevent the colonization of floating *D. antarctica,* due to low space for new recruits and the low contribution of new gametes in relation to those produced by local individuals (Waters et al. 2013). Future warming scenarios might facilitate colonization by floating seaweeds, reducing competition with local species and increasing substrate availability.

Acknowledgements We thank the editors for inviting us to contribute to this review. Financial support was received through Fondap-IDEAL grant 15150003 to ECM and FONDECYT grant 1131023 to FT and ECM, CONICYT PAI (Chile) 79160069 to FT.

References

Abé H, Komatsu T, Kokubu Y, Alabsi N, Rothäusler E, Shishido H, Yoshizawa S, Ajisaka T (2013) Invertebrate fauna associated with floating *Sargassum horneri* (Fucales: Sargassaceae) in the East China Sea. Spec Diver 18(1):75–85

Amsler CD, McClintock JB, Baker BJ (1999) An antarctic feeding triangle: defensive interactions between macroalgae, sea urchins, and sea anemomes. Mar Ecol Prog Ser 183:105–114

Amsler CD, Mcclintock JB, Baker BJ (2012) Palatability of living and dead detached Antarctic macroalgae to consumers. Antarct Sci 24(6):589–590

Assmy P, Ehn JK, Fernández-Méndez M, Hop H, Katlein C, Sundfjord A, Bluhm K, Daase M, Engel A, Fransson A, Granskog MA, Hudson SR, Kristiansen S, Nicolaus M, Peeken I, Renner AHH, Spreen G, Tatarek A, Wiktor J (2013) Floating ice-algal aggregates below melting Arctic Sea ice. PLoS One 8(10):e76599

Avila C, Angulo-Preckler C, Martín-Martín RP, Figuerola B, Griffiths HJ, Waller CL (2020) Invasive marine species discovered on non–native kelp rafts in the warmest Antarctic island. Sci Rep 10:1639

Bäck S, Lehvo A, Blomster J (2000) Mass occurrence of unattached *Enteromorpha intestinalis* on the Finnish Baltic Sea Coast. Ann Bot Fenn 37:155–161

Barnes DKA, Linse K, Waller C, Morely S, Enderlein P, Fraser KPP, Brown M (2006) Shallow benthic fauna communities of South Georgia Island. Polar Biol 29:223–228

Batista MB, Batista AA, Franzan SP, Simionatto PP, Lima ST, Velez-Rubio G, Scarabino F, Camacho O, Schmitz C, Martinez A (2018) Kelps' long-distance dispersal: role of ecological/oceanographic processes and implications to marine forest conservation. Diversity 10(11)

Belt ST, Brown TA, Smik L, Assmy P, Mundy CJ (2018) Sterol identification in floating Arctic sea ice algal aggregates and the Antarctic sea ice diatom *Berkeleya adeliensis*. Org Geochem 118:1–3

Bischof K, Hanelt D, Wiencke C (1998) UV-radiation can affect depth-zonation of Antarctic macroalgae. Mar Biol 131:597–605

Bischoff-Bäsmann B, Wiencke C (1996) Temperature requirements for growth and survival of Antarctic Rhodophyta. J Phycol 32(4):525–535

Blomster J, Maggs CA, Stanhope MJ (1998) Molecular and morphological analysis of *Enteromorpha intestinalis* and *E. compressa* (Chlorophyta) in the British isles. J Phycol 34(2):319–340

Boo GH, Mansilla A, Nelson W, Bellgrove A, Boo SM (2014) Genetic connectivity between trans-oceanic populations of *Capreolia implexa* (Gelidiales, Rhodophyta) in cool temperate waters of Australasia and Chile. Aquat Bot 119:73–79

Braeckman U, Pasotti F, Vázquez S, Zacher K, Hoffmann R, Elvert M, Marchant H, Buckner C, Quartino ML, Mác Cormack W, Soetaert K, Wenzhöfer F, Vanreusel A (2019) Degradation of macroalgal detritus in shallow coastal Antarctic sediments. Limnol Oceanogr 64(4):1423–1441

Brouwer PEM (1996) Decomposition in situ of the sublittoral Antarctic macroalga *Desmarestia anceps* Montagne. Polar Biol 16(2):129–137

Brouwer PEM, Geilen EFM, Gremmen NJM, van Lent F (1995) Biomass, cover and zonation patterns of sublittoral macroalgae at Signy Island, South Orkney Islands, Antarctica. Bot Mar 38:259–270

Campana GL, Zacher K, Fricke A, Molis M, Wulff A, Quartino ML, Wiencke C (2009) Drivers of colonization and succession in polar benthic macro- and microalgal communities. Bot Mar 52(6):655–667

Cefarelli AO, Vernet M, Ferrario ME (2011) Phytoplankton composition and abundance in relation to free-floating Antarctic icebergs. Deep-Sea Res II Top Stud Oceanogr 58:1436–1450

Clayton MN, Wiencke C, Klöser H (1997) New records of temperate and sub Antarctic marine benthic macroalgae from Antarctica. Polar Biol 17(2):141–149

Cohen RA, Fong P (2004) Physiological responses of a bloom-forming green macroalga to short-term change in salinity, nutrients, and light help explain its ecological success. Estuaries 27(2):209–216

Coyer JA, Hoarau G, Van Schaik J, Luijckx P, Olsen JL (2011) Trans-Pacific and trans-Arctic pathways of the intertidal macroalga *Fucus distichus* L. reveal multiple glacial refugia and colonizations from the North Pacific to the North Atlantic. J Biogeogr 38(4):756–771

Cumming RA, Nikula R, Spencer HG, Waters JM (2014) Transoceanic genetic similarities of kelp-associated sea slug populations: long-distance dispersal via rafting? J Biogeogr 41(12):2357–2370

De Wildeman É (1935) Observations sur la algues rapporlées par l'expedition animctiq'ue de la "Belgica". Résullanls du voyage de la "Belgica" en 897-1899. Ansvers, p. 47
Drew E, Hastings R (1992) A year-round ecophysiological study of *Himantothallus grandifolius* (Desmarestiales, Phaeophyta) at Signy Island, Antarctica. Phycologia 31(3–4):262–277
Eggert A (2012) Seaweed responses to temperature. In: Wiencke C, Bischof K (eds) Seaweed biology: novel insights into ecophysiology, ecology and utilization. Springer-Verlag, Berlin, pp 47–66
Eggert A, Wiencke C (2000) Adaptation and acclimation of growth and photosynthesis of five Antarctic red algae to low temperatures. Polar Biol 23(9):609–618
Fischer G, Wiencke C (1992) Stable carbon isotope composition, depth distribution and fate of macroalgae from the Antarctic Peninsula region. Polar Biol 12(3):341–348
Flores-Molina MR, Rautenberger R, Muñoz P, Huovinen P, Gómez I (2016) Stress tolerance of the endemic Antarctic brown alga Desmarestia anceps to UV radiation and temperature is mediated by high concentrations of phlorotannins. Photochem Photobiol 92(3):455–466
Fraser CI, Waters JM (2013) Algal parasite *Herpodiscus durvillaeae* (Phaeophyceae: Sphacelariales) inferred to have traversed the Pacific Ocean with its buoyant host. J Phycol 49(1):202–206
Fraser CI, Nikula R, Spencer HG, Waters JM (2009) Kelp genes reveal effects of subantarctic sea ice during the Last Glacial Maximum. P Natl Acad Sci USA 106(9):3249–3253
Fraser CI, Thiel M, Spencer HG, Waters JM (2010) Contemporary habitat discontinuity and historic glacial ice drive genetic divergence in Chilean kelp. BMC Evol Biol 10:203
Fraser CI, Zuccarello GC, Spencer HG, Salvatore LC, Garcia GR, Waters JM (2013) Genetic affinities between trans-oceanic populations of non-buoyant macroalgae in the high latitudes of the southern hemisphere. PLoS One 8(7):e69138
Fraser CI, Kay GM, Md P, Ryan PG (2017) Breaking down the barrier: dispersal across the Antarctic Polar Front. Ecography 40(1):235–237
Fraser CI, Morrison AK, Hogg AM, Macaya EC, van Sebille E, Ryan PG, Padovan A, Jack C, Valdivia N, Waters JM (2018) Antarctica's ecological isolation will be broken by storm-driven dispersal and warming. Nat Clim Chang 8(8):704
Gain L (1912) La flore algologique des régions antarctiques et subantarctiques.—Deuxième Expédition Antarctique Française (1908–1910) commandée par le Dr Jean Charcot: sciences naturelles. Sci Nat Doc Sci-Paris:156–218
Gallardo T, Pérez-Ruzafa IM, Flores-Moya A, Conde F (1999) New collections of Benthic marine algae from Livingston and Deception Islands (South Shetland Islands) and Trinity Island (Bransfield Strait). Bot Mar 42:61–69
Gómez I, Wulff A, Roleda MY, Huovinen P, Karsten U, Quartino ML, Dunton K, Wiencke C (2009) Light and temperature demands of marine benthic microalgae and seaweeds in polar regions. Bot Mar 52(6):593–608
Gómez I, Navarro NP, Huovinen P (2019) Bio-optical and physiological patterns in Antarctic seaweeds: a functional trait based approach to characterize vertical zonation. Prog Oceanogr 174:17–27
Graiff A, Karsten U, Meyer S, Pfender D, Tala F, Thiel M (2013) Seasonal variation in floating persistence of detached *Durvillaea antarctica* (Chamisso) Hariot thalli. Bot Mar 56(1):3–14
Graiff A, Pantoja JF, Tala F, Thiel M (2016) Epibiont load causes sinking of viable kelp rafts: seasonal variation in floating persistence of giant kelp *Macrocystis pyrifera*. Mar Biol 163(9):191
Guillemin M-L, Valero M, Faugeron S, Nelson W, Destombe C (2014) Tracing the trans-pacific evolutionary history of a domesticated seaweed (*Gracilaria chilensis*) with archaeological and genetic data. PLoS One 9(12):e114039
Häder D-P, Lebert M, Helbling EW (2001) Effects of solar radiation on the Patagonian macroalga *Enteromorpha linza* (L.) J. Agardh — Chlorophyceae. J Photochem Photobiol B Biol 62(1):43–54
Hernández-Carmona G, Hughes B, Graham MH (2006) Reproductive longevity of drifting kelp *Macrocystis pyrifera* (Phaeophyceae) in Monterey Bay, USA. J Phycol 42(6):1199–1207

Hinojosa IA, Boltaña S, Macaya E, Ugalde P, Valdivia N, Vásquez N, Newman W, Thiel M (2006) Geographic distribution and description of four pelagic barnacles along the south east Pacific coast of Chile – a zoogeographical approximation. Rev Chil Hist Nat. 78: 603–614.

Hobday A (2000) Age of drifting *Macrocystis pyrifera* (L.) C. Agardh rafts in the Southern California Bight. J Exp Mar Biol Ecol 253:97–114

Hommersand MH, Moe RL, Amsler CD, Fredericq S (2009) Notes on the systematics and biogeographical relationships of Antarctic and sub-Antarctic Rhodophyta with descriptions of four new genera and five new species. Bot Mar 52(6): 509–534

Hooker JD (1844) The botany of the Antarctic voyage of H. M. discovery ships Erebus and Terror in the years 18391843, vol. 1, Flora Antarctica, Part 1 Botany of Lord Auklands Group and Campbell's Island, Part 2 Botany of Fuegia, The Falklands, Kerguelens Land. Reeve Brothers, London

Huang Y, Amsler M, McClintock J, Amsler C, Baker B (2007) Patterns of gammarid amphipod abundance and species composition associated with dominant subtidal macroalgae along the western Antarctic Peninsula. Polar Biol 30:1417–1430

Huovinen P, Gómez I (2013) Photosynthetic characteristics and UV stress tolerance of Antarctic seaweeds along the depth gradient. Polar Biol 36(9):1319–1332

Huovinen P, Gómez I (2015) UV Sensitivity of vegetative and reproductive tissues of two Antarctic brown algae is related to differential allocation of phenolic substances. Photochem Photobiol 91(6):1382–1388

Iken K, Amsler CD, Hubbard JM, McClintock JB, Baker BJ (2007) Allocation patterns of phlorotannins in Antarctic brown algae. Phycologia 46(4):386–395

Ingólfsson A (1998) Dynamic of macrofaunal communities of floating seaweed clumps off western Iceland: a study of patches on the surface of the sea. J Exp Mar Biol Ecol 23:119–137

Kang J-S, Kang S-H, Lee JH, Lee S (2002) Seasonal variation of microalgal assemblages at a fixed station in King George Island, Antarctica, 1996. Mar Ecol Prog Ser 229:19–32

Katlein C, Fernández-Méndez M, Wenzhöfer F, Nicolaus M (2015) Distribution of algal aggregates under summer sea ice in the Central Arctic. Polar Biol 38:719–731

Khalaman V, Berger VY (2006) Floating seaweeds and associated fauna in the White Sea. Oceanology 46:827–833

Klöser H, Ferreyra G, Schloss I, Mercuri G, Laturnus F, Curtosi A (1993) Seasonal variation of algal growth conditions in sheltered Antarctic bays: the example of Potter Cove (King George Island, South Shetlands). J Mar Syst 4(4):289–301

Lastra M, Rodil IF, Sánchez-Mata A, García-Gallego M, Mora J (2014) Fate and processing of macroalgal wrack subsidies in beaches of Deception Island, Antarctic Peninsula. J Sea Res 88:1–10

Lizée-Prynne D, López B, Tala F, Thiel M (2016) No sex-related dispersal limitation in a dioecious, oceanic long-distance traveller: the bull kelp *Durvillaea antarctica*. Bot Mar 59(1):39–50

López BA, Macaya EC, Rivadeneira MM, Tala F, Tellier F, Thiel M (2018) Epibiont communities on stranded kelp rafts of *Durvillaea antarctica* (Fucales, Phaeophyceae)–do positive interactions facilitate range extensions? J Biogeogr 45(8):1833–1845

Macaya EC, Zuccarello GC (2010) DNA Barcoding and genetic divergence in the giant kelp *Macrocystis* (Laminariales). J Phycol 46(4):736–742

Macaya EC, Boltana S, Hinojosa IA, Macchiavello JE, Valdivia NA, Vásquez NR, Buschmann AH, Vásquez JA, Vega JMA, Thiel M (2005) Presence of sporophylls in floating kelp rafts of Macrocystis spp. (Phaeophyceae) along the Chilean Pacific coast. J Phycol 41(5):913–922

Macaya EC, López B, Tala F, Tellier F, Thiel M (2016) Float and raft: role of buoyant seaweeds in the phylogeography and genetic structure of non-buoyant associated flora. In: Hu ZM, Fraser C (eds) Seaweed phylogeography. Springer, pp 97–130

McKenzie PF, Bellgrove A (2008) Dispersal of *Hormosira banksii* (Phaeophyceae) via detached fragments: reproductive viability and longevity. J Phycol 44(5):1108–1115

Mélice J, Lutjeharms J, Rouault M, Ansorge I (2003) Sea-surface temperatures at the sub-Antarctic islands Marion and Gough during the past 50 years. S Afr J Sci 99(7–8):363–366

Miller KA, Pearse JS (1991) Ecological studies of seaweeds in McMurdo Sound, Antarctica. Am Zool 31:35–48

Montagne JFC (1842) Prodromus generum specierumque phycearum novarum, in itinere ad polum antarcticum...ab illustri Dumont d'Urville peracto collectarum, notis diagnosticis tantum huc evulgatarum, descriptionibus verò fusioribus nec non iconibus analyticis iam iamque illustrandarum. Paris apud Gide, editorem, Parisiis, pp. 1–16

Moon KL, Chown SL, Fraser CI (2017) Reconsidering connectivity in the sub-Antarctic. Biol Rev 92(4):2164–2181

Müller DG, Westermeier R, Peters A, Boland W (1990) Sexual reproduction of the Antarctic brown alga *Ascoseira mirabilis* (Ascoseirales, Phaeophyceae). Bot Mar 33:251–255

Müller DG, Ramírez ME, Westermeier R (1992) *Utriculidium durvillei* (Bory?) Skottsberg en isla Rey Jorge, Antártica. Ser Cien INACH 42:47–52

Neushul M (1963) Reproductive morphology of Antarctic kelps. Bot Mar 5:19–24

Neushul M (1965) Diving observations of subtidal Antarctic marine vegetation. Bol Mar 8(2–4):234–243

Nikula R, Fraser CI, Spencer HG, Waters JM (2010) Circumpolar dispersal by rafting in two subantarctic kelp-dwelling crustaceans. Mar Ecol Prog Ser 405:221–230

Nikula R, Spencer HG, Waters JM (2013) Passive rafting is a powerful driver of transoceanic gene flow. Biol Lett 9(1):20120821

Norkko A, Thrush SF, Cummings VJ, Funnell GA, Schwarz A-M, Andrew NL, Hawes I (2004) Ecological role of Phyllophora antarctica drift accumulations in coastal soft-sediment communities of McMurdo Sound, Antarctica. Polar Biol 27(8):482–494

O'Connor MI (2009) Warming strengthens an herbivore-plant interaction. Ecology 90:388–398. https://doi.org/10.1890/08-0034.1

Oliveira EC, Absher TM, Pellizzari FM, Oliveira MC (2009) The seaweed flora of Admiralty Bay, King George Island, Antarctic. Polar Biol 32(11):1639–1647

Pellizzari F, Silva MC, Silva EM, Medeiros A, Oliveira MC, Yokoya NS, Pupo D, Rosa LH, Colepicolo P (2017) Diversity and spatial distribution of seaweeds in the South Shetland Islands, Antarctica: an updated database for environmental monitoring under climate change scenarios. Polar Biol 40:1671–1685

Quartino ML, Zaixso HE, Boraso de Zaixso AL (2005) Biological and environmental characterization of marine macroalgal assemblages in Potter Cove, South Shetland Islands, Antarctica. Bot Mar 48(3):187–197

Rakusa-Suszczewski S (1980) Environmental conditions and the functioning of Admiralty Bay (South Shetland Islands) as part of the near shore Antarctic ecosystem. Pol Polar Res 1:11–27

Rakusa-Suszczewski S (1995) Flow of matter in the Admiralty Bay area, King George Island, Maritime Antarctic. Proc NIPR Symp Polar Biol 8:101–113

Rautenberger R, Huovinen P, Gómez I (2015) Effects of increased seawater temperature on UV tolerance of Antarctic marine macroalgae. Mar Biol 162(5):1087–1097

Richardson MG (1979) The distribution of Antarctic marine macro-algae related to depth and substrate. Br Antarct Surv Bull 49:1–13

Ross JC (1847) A voyage of discovery and research in the Southern and Antarctic Regions, during the Years 1839–43, vol 2. John Murray

Rothäusler E, Gómez I, Hinojosa IA, Karsten U, Tala F, Thiel M (2009) Effect of temperature and grazing on growth and reproduction of floating *Macrocystis* spp.(Phaeophyceae) along a latitudinal gradient. J Phycol 45(3):547–559

Rothäusler E, Gómez I, Hinojosa IA, Karsten U, Miranda L, Tala F, Thiel M (2011a) Kelp rafts in the Humboldt Current: interplay of abiotic and biotic factors limit their floating persistence and dispersal potential. Limnol Oceanogr 56(5):1751–1763

Rothäusler E, Gomez I, Hinojosa IA, Karsten U, Tala F, Thiel M (2011b) Physiological performance of floating giant kelp *Macrocystis pyrifera* (Phaeophyceae): latitudinal variability in the effects of temperature and grazing. J Phycol 47(2):269–281

Rothäusler E, Gomez I, Karsten U, Tala F, Thiel M (2011c) Physiological acclimation of floating *Macrocystis pyrifera* to temperature and irradiance ensures long-term persistence at the sea surface at mid-latitudes. J Exp Mar Biol Ecol 405(1–2):33–41

Rothäusler E, Gutow L, Thiel M (2012) Floating seaweeds and their communities. In: Wiencke C, Bischof K (eds) Seaweed biology: novel insights into ecophysiology, ecology and utilization. Springer-Verlag, Berlin, pp 359–380

Rothäusler E, Reinwald H, López BA, Tala F, Thiel M (2018) High acclimation potential in floating *Macrocystis pyrifera* to abiotic conditions even under grazing pressure–a field study. J Phycol 54(3):368–379

Santelices B (1990) Patterns of reproduction dispersal and recruitment in seaweeds. Oceanogr Mar Biol Ann Rev 28:177–276

Sato T, Sakurai H, Takasaki A, Watanabe K, Hirano Y (1992) Underwater observation of Antarctic fishes and invertebrates with a note on collecting and transportation techniques for research and exhibition in the aquarium. Ser Cient INACH 42:95–103

Schwarz A-M, Hawes I, Andrew N, Norkko A, Cummings V, Thrush S (2003) Macroalgal photosynthesis near the southern global limit for growth; Cape Evans, Ross Sea, Antarctica. Polar Biol 26(12):789–799

Skottsberg K (1907) Zur Kenntnis der subantarktischen und antarktischen Meeresalgen. I. Phaeophyceen. In: Nordenskjöld O (ed) Wissenschaftliche Ergebnisse der Schwedischen Südpolar-Expedition 1901–1903, vol 4: 1. Lithographisches Institut des Generalstabs, Stockholm, pp 1–172

Smith SDA (2002) Kelp rafts in the Southern Ocean. Glob Ecol Biogeogr 11(1):67–69

Stachowicz JJ, Fried H, Osman RW, Whitlatch RB (2002) Biodiversity, invasion resistance, and marine ecosystem function: reconciling pattern and process. Ecology 83(9):2575–2590

Tada S, Sato T, Sakurai H, Arai H, Kimpara I, Kodama M (1996) Benthos and fish community associated with clumps of submerged drifting algae in Fildes Bay, King George Island, Antarctica. Proc NIPR Symp Polar Biol 9:243–251

Tala F, Gómez I, Luna-Jorquera G, Thiel M (2013) Morphological, physiological and reproductive conditions of rafting bull kelp (*Durvillaea antarctica*) in northern-central Chile (30º S). Mar Biol 160(6):1339–1351

Tala F, Velásquez M, Mansilla A, Macaya EC, Thiel M (2016) Latitudinal and seasonal effects on short-term acclimation of floating kelp species from the South-East Pacific. J Exp Mar Biol Ecol 483:31–41

Tala F, López BA, Velásquez M, Jeldres R, Macaya EC, Mansilla A, Ojeda J, Thiel M (2019) Long-term persistence of the floating bull kelp *Durvillaea antarctica* from the South-East Pacific: potential contribution to local and transoceanic connectivity. Mar Environ Res 149:67–79

Thiel M, Gutow L (2005a) The ecology of rafting in the marine environment. I. The floating substrata. Oceanogr Mar Biol Ann Rev 42:181–263

Thiel M, Gutow L (2005b) The ecology of rafting in the marine environment. II. The rafting organisms and community. Oceanogr Mar Biol Ann Rev 43:279–418

Valdivia N, Díaz MJ, Holtheuer J, Garrido I, Huovinen P, Gómez I (2014) Up, down, and all around: scale-dependent spatial variation in rocky-shore communities of Fildes Peninsula, King George Island, Antarctica. PLoS One 9(6)

Vandendriessche S, Messiaen M, O'Flynn S, Vincx M, Degraer S (2007a) Hiding and feeding in floating seaweed: floating seaweed clumps as possible refuges or feeding grounds for fishes. Estuar Coast Shelf Sci 71(3–4):691–703

Vandendriessche S, Vincx M, Degraer S (2007b) Floating seaweed and the influences of temperature, grazing and clump size on raft longevity—a microcosm study. J Exp Mar Biol Ecol 343(1):64–73

Waters JM, Fraser CI, Hewitt GM (2013) Founder takes all: density-dependent processes structure biodversity. Trends Ecol Evol 28(2):78–85

Webster W (1834) Narrative of a voyage to the southern Atlantic Ocean in the years 1828, 29, 30, performed in HM sloop Chanticleer under the command of the late Captain Henry Foster, FRS & c. vol II. London, Richard Bentley, 1–398

Westermeier R, Gómez I, Rivera PJ, Müller DG (1992) Macroalgas marinas antárticas: distribución, abundancia y necromasa en isla Rey Jorge, Shetland del Sur, Antártica Ser. Cient INACH 42:21–34

Weykam G, Gómez I, Wiencke C, Iken K, Klöser H (1996) Photosynthetic characteristics and C:N ratios of macroalgae from King George Island (Antarctica). J Exp Mar Biol Ecol 204:1–22

Wichmann CS, Hinojosa IA, Thiel M (2012) Floating kelps in Patagonian Fjords: an important vehicle for rafting invertebrates and its relevance for biogeography. Mar Biol 159(9):2035–2049

Wiencke C, Rahmel J, Karsten U, Weykam G, Kirst GO (1993) Photosynthesis of marine macroalgae from Antarctica: light and temperature requirements. Bot Acta 106(1):78–87

Wulff A, Iken K, Quartino M, Al-Handal A, Wiencke C, Clayton M (2009) Biodiversity, biogeography and zonation of marine benthic micro-and macroalgae in the Arctic and Antarctic. Bot Mar 56:491–507

Zacher K, Roleda MY, Hanelt D, Wiencke C (2007) UV effects on photosynthesis and DNA in propagules of three Antarctic seaweeds (*Adenocystis utricularis*, *Monostroma hariotii* and *Porphyra endiviifolium*). Planta 225(6):1505–1516

Zacher K, Rautenberger R, Hanelt D, Wulff A, Wiencke C (2009) The abiotic environment of polar marine benthic algae. Bot Mar 52(6):483–490

Zemko K, Pabis K, Siciński J, Błażewicz-Paszkowycz M (2015) Diversity and abundance of isopod fauna associated with holdfasts of the brown alga *Himantothallus grandifolius* in Admiralty Bay, Antarctic. J Pol Polar Res 36(4):405–415

Zielinski K (1981) Benthic macroalgae of Admiralty Bay (King George Island, South Shetland Islands) and circulation of algal matter between the water and the shore. J Pol Polar Res 2(3–4):71–94

Zielinski K (1990) Bottom macroalgae of the Admiralty Bay (King George Island, South Shetlands, Antarctica). J Pol Polar Res 11:95–131

Chapter 5
Biogeography of Antarctic Seaweeds Facing Climate Changes

Franciane Pellizzari, Luiz Henrique Rosa, and Nair S. Yokoya

Abstract The seaweed biogeography and diversity in remote areas, such as Antarctica, should be reassessed considering the population shifts induced by global changes. This chapter addresses the hypothesis that ecological isolation can be disrupted and that biogeographical distribution of some species could be altered by thermohaline changes, which in turn would alter the dispersal patterns of macroalgae driven by severe meteorological and oceanographic events. Algal growth and distribution are limited by physical and biological processes, acting as sensitive bioindicators of changes or abrupt oscillations in the environmental regimes. In addition, Antarctica represents a natural laboratory highly susceptible to the climate changes, and the monitoring of their ecosystems may help to predict their potential effects beyond the Southern Ocean. Another fundamental issue is to understand the increase in species richness due to the cryptic and alien species, considering shifts in their biogeographic distribution. The large-scale patterns of some of these species reported for Antarctica may provide clues to reevaluate aspects of endemism, biological corridors, ecotone, and expansion of geographical distribution of algal assemblages facing climate changes, reinforcing the hypothesis that these isolated ecosystems will become gradually more connected.

Keywords Alien species · Biogeographic processes · Deception Island · Endemism · Seaweed diversity

---

F. Pellizzari (✉)
Universidade Estadual do Paraná, Campus Paranaguá, Laboratório de Ficologia e Qualidade de Água Marinha, Paranaguá, Brazil
e-mail: franciane.pellizzari@unespar.edu.br

L. H. Rosa
Universidade Federal de Minas Gerais, Departamento de Microbiologia, Belo Horizonte, Brazil
e-mail: ihrosa@icb.ufmg.br

N. S. Yokoya
Instituto de Botânica, Núcleo de Pesquisa em Ficologia, São Paulo, Brazil
e-mail: nyokoya@hotmail.com

I. Gómez, P. Huovinen (eds.), *Antarctic Seaweeds*,
https://doi.org/10.1007/978-3-030-39448-6_5

5.1 The Abiotic Setting of the Southern Ocean

The Southern Ocean (SO) biota distribution is the result of major geological, oceanographic, and climate changes during the last 50 Ma, and there is a paradox between the broad distributions of some species and their inherently poor dispersal capacity, although marine biogeographic studies supported by molecular studies are increasingly revealing examples of this paradox and indicating that long-distance dispersal of macroalgae is possible (Fraser et al. 2013).

The seaweed processes in the Antarctic and sub-Antarctic regions are strongly limited and correlated with environmental parameters (Wiencke and Amsler 2012). In contrast with the Arctic Ocean, the Southern Ocean has no land bridge to temperate regions since the late Mesozoic and has been further separated from the southern continents by the Antarctic Circumpolar Current (ACC) since ca. 26 Ma (Kirst and Wiencke 1995). The eastward movement of the ACC or the West Wind Drift (WWD) had strongly defined the oceanography in the Southern Hemisphere and consequently the diversity, biogeography, and ecology of seaweeds in all the adjacent coastal regions (Orsi et al. 1995; Huovinen and Gómez 2012). Thus, biogeographic barriers and ecological corridors are fundamental concepts to understand macroecological and evolutionary processes, since that ocean circulation, considering present and past patterns of continental drift, may isolate or connect groups of marine organisms, including seaweeds.

The Southern Ocean encircling Antarctica, whose natural boundary is the Antarctic Convergence, is known as the most extreme environment on the Earth. Considering ocean circulation in the SO, the ACC is permeated by the Southern Antarctic Circumpolar Front, located between the Polar Front (50°–60°S) and surroundings of the Antarctic coastline, representing the disjunction of Antarctic waters and resulting in distinct thermal stratifications for each zone (Orsi et al. 1995; Sanches et al. 2016). In the southern hemisphere, two circumpolar fronts define the boundaries of the cold-temperate region (Fig. 5.1a): (1) the Antarctic Polar Front (APF or Antarctic Convergence), the southern limit, characterized by cold surface waters (ranging 3–5°C), and (2) the Subtropical Front (STF or Subtropical Convergence), the northern limit that separates subtropical water in the north from sub-Antarctic water, showing ca. of the 10°C and 15°C surface temperature, during winter and summer, respectively (Huovinen and Gómez 2012). Thus the ACC, being the largest and deepest current in the world, was considered in several past studies as a physical barrier that could limit passive dispersal of new species to Antarctica (Wells et al. 2011; Pritchard et al. 2012; Sanches et al. 2016).

In a macroscale, Longhurst (2007) identified four "biogeochemical" provinces in the Southern Ocean considering physical and hydrological properties of these currents: South Subtropical Convergence, sub-Antarctic water ring, Antarctic, and Polar Southern provinces. On a regional scale, the dynamics of ice formation and/or its presence as drifting sheets or icebergs, substrate type, and current regime are the main factors that set biogeographical boundaries for biota.

Fig. 5.1 (**a**) Antarctica contextualized in the Southern Ocean, showing continental proximities and major oceanographic fronts: subtropical front (STF), Antarctic Polar Front (APF), and Antarctic Circumpolar Current (ACC). (**b**) Detailed map indicating the geographical position of the SSI South Shetland Islands, including Deception, WAP Western Antarctic Peninsula, and EAP Eastern Antarctic Peninsula. (Photo by Franciane Pellizzari)

5.2 Biogeographic Patterns

Due to the isolation, the SO presents higher endemism of seaweeds, although lower richness than some sub-Antarctic islands and continental areas from the Southern Atlantic and Pacific. The Eastern Antarctic Peninsula (EAP) and southern regions from the Western Antarctic Peninsula (WAP) present low species richness when compared to the South Shetland Islands (SSI – Fig. 5.1b) (Pellizzari et al. 2017) (see also Chap. 2 by Oliveira et al.). However, Wulff et al. (2009), comparing the biodiversity, biogeography, and zonation of seaweeds from the Arctic and Antarctic regions, reported that the diversity and number of endemic species are higher in Antarctica than in the Arctic. Wiencke and Clayton (2002) mentioned that about a third of the 120 species recorded to the Antarctic region are endemic (Phaeophyceae 44%, Rhodophyta 32%, and Chlorophyta 18%). These differences are explained by the biogeographical histories of both regions (Wiencke and Clayton 2002).

Temperature is the most important factor controlling biogeographic distribution of macroalgae, and then shifts in their distribution are inevitable during periods of global changes. However, other factors must be considered. For example, rafting of floating organisms and objects appears to have a critical role in facilitating long-distance dispersal and structuring intertidal ecosystems in Antarctica (Fraser et al. 2013). Many species that lack obvious transoceanic dispersal ability (e.g., brooding invertebrates, small non-buoyant macroalgae, and terrestrial vertebrates) are nonetheless evidently able to disperse long distances attached in floating wood or detached kelp (see also Chap. 3 by Fraser et al.). Also, aiming to avoid generalized biogeography of the SO benthos, Griffiths et al. (2009) studied distributional data of mollusks, cheilostome, cyclostome, and bryozoans (species richness, rates of endemism, patterns of radiation) in 29 provinces in the South American, South African, Tasmanian, New Zealand, sub-Antarctic, and Antarctic regions. The authors reported high species numbers in New Zealand, Tasmania, and South Africa and low species richness in South America. In contrast, no difference was found in richness between the east and west parts of the Southern Ocean (called "single functional unit"). Besides evidence of strong faunal links between the Antarctica and South America, biogeographical regions in the SO differ depending upon the class of organism, which is being considered. Linse et al. (2006) indicated that patterns of diversity and endemism were very different between the bivalves and gastropods. Moreover, Pierrat et al. (2013), in a comparative study among echinoids, bivalves, and gastropods from the Antarctic, sub-Antarctic, and cold-temperate areas, reported a significant advantage of the BSN (bootstrapped spanning network) procedure in the identification of faunal similarities among biogeographical regions and transitional areas, considering the following faunal provinces: (1) New Zealand, (2) southern South America, (3) east sub-Antarctic islands, (4) West Antarctica, and (5) East Antarctica. Strong faunal relationships perfectly match the flows of the ACC and Antarctic Coastal currents, suggesting strong connections and groupings between bioregions. However, Antarctic and sub-Antarctic regions are regarded as distinct biogeographical regions, with patterns driven by a small number of widely distributed species (Griffiths and Waller 2016). Gutt et al. (2016) studied macroepibenthic communities in Weddell Sea, Bransfield Strait, and Drake Passage and suggested that large-scale patterns in pelagic organisms segregate three eco-regions possibly correlated with ice presence and depth. Keith et al. (2011), using a geographically weighted regression (GWR) model, determined that some environmental conditions could predict genus richness in macroalgae and confirmed the hypothesis that environmental tolerances influence species distributions more strongly at higher latitudes, whereas biotic interactions play a more prominent role in the tropics. This chapter discusses in that extent shifts in seaweed diversity and distribution in the SO may be also associated to the rapid climatic changes, resulting in higher connections or disruption of these previously established provinces.

5.3 Seaweed Assemblages: Are Antarctic Seaweed Diversity and Richness Changing?

Shifts in seaweed distribution were important during periods of changing temperatures as demonstrated in the geological past during the last glaciation (Crame 1994). Similar changes may be expected during the current period of global warming with a significant impact on marine communities. In addition, Antarctic macroalgal communities will presumably expand north and also southward due to less strong ice abrasion and improved light conditions, and the new ice-free areas appearing due to glacier retreat will be probably colonized by seaweeds (Quartino et al. 2013; Wiencke et al. 2014 and studies summarized in Pellizzari et al. 2017) (see also Chap. 8 by Quartino et al.).

Climate changes in Antarctic Peninsula are resulting in shifts of marine populations (summarized in Ducklow et al. 2013). These authors suggested that macroalgal assemblages show lower biomass and diversity in the southern region of WAP when compared to the northern assemblages. Additionally, they mentioned that the assemblages of the northern WAP seem to be expanding to the south as the annual sea ice declines.

Physico-chemical changes, mainly due to increased surface seawater temperatures (SST) in the SO, result in a potential driver to shifts in seaweed assemblages (Müller et al. 2009; Sangil et al. 2012; Wernberg et al. 2016; Pellizzari et al. 2017). Based on a multivariate analysis of Antarctic and sub-Antarctic seaweed distribution focusing in an evaluation of the role of the ACC, Sanches et al. (2016) found differences in the distributional patterns of species and genera in relation to the zones and sections connected with South America (1: north), influenced by the Ross Sea Gyre (2: Western Antarctica Peninsula) and by the Weddell Sea Gyre (3: Eastern Antarctica Peninsula). The predicted latitudinal gradient of species richness was confirmed, and two main biogeographical clusters were defined: (1) subtropical and sub-Antarctica and (2) Antarctic Peninsula and surroundings of continental Antarctic. These differences in diversity patterns between both regions suggest the existence of a species distributional flux. This may result either from natural dispersion or due to global changes, suggesting that Antarctica may not be as isolated and as suggested in Chap 3 of this book and by Fraser et al. (2019).

Recently, high seaweed richness among the South Shetland Islands (SSI), highlighted by six new records for the area, four cryptogenic species and two putative new species, was reported (Pellizzari et al. 2017). It is noteworthy that the 104 species of benthic marine algae listed by these authors represent ~82% of all seaweeds described to Antarctica. The majority of these new records are cryptic species, represented by small specimens and inconspicuous forms, and their occurrence may easily pass unnoticed until a large-scale sampling and a seasonal program have been implemented (Fig. 5.2) (see also Chap. 2 by Oliveira et al.). Dubrasquet et al. (2018) studied the diversity of red algae and their distribution along the WAP and SSI through a molecular-assisted revision, identifying significant differences among assemblages of SSI and northern part of the WAP (at ≈63°S), central part of the

Fig. 5.2 Activities carried out within the frame of the Antarctic Program supported by the Brazilian Navy. (Photos by Franciane Pellizzari)

WAP (at ≈64°S), and central-southern part of the WAP (at ≈67°S). Apparently, these subregions do not correspond to the classical bioregions reported in previous studies, suggesting that the Bransfield Strait cannot be regarded as a strong barrier for red macroalgae.

The South Shetland Islands and interface islands between Antarctic and sub-Antarctic regions were previously reported to have transitional biota features, which are most likely related with the composition and distribution of benthic marine communities (Pellizzari et al. 2017). The seaweed richness found in the SSI (ca. 60°S) was higher than in Adelaide Island and Terra Nova Bay (Ross Sea, above latitude 70°S), according to Cormaci et al. (1992) and Mystikou et al. (2014). In addition, Pellizzari et al. (2017) considered Livingston and King George islands as hotspots among the SSI.

Recent changes in the composition and structure of algal assemblages from Antarctica are featured by new records, including green algae and cyanobacteria identified unprecedentedly for the area, followed by detection of cryptogenic species and reduction of endemism (Pellizzari et al. 2017; see also Chap. 2 by Oliveira et al.). Cyanobacteria are opportunistic organisms that can inhabit diversified and extreme environments, and they have been regarded as indicators of changes in seawater pH and temperature (Sangil et al. 2012). Antarctic cyanobacteria are conspicuous and well studied in lake assemblages (Taton et al. 2006). However, incipient information in marine ecosystems is available. Pellizzari et al. (2017) identified six species of filamentous Cyanobacteria as epiphytes on large seaweeds, occurring along the intertidal mats of Deception Island. They also reported new record of opportunistic Chlorophyta, including a cryptogenic species. It has been suggested that general diversity may be decreasing due to the effects of climate changes (summary in Pibernat et al. 2007; Wiencke and Amsler 2012). However, this pattern is not reported in some studies (synopsis in Pellizzari et al. 2017), considering that seaweed communities seem to be in "transit" for a higher connectivity and distribution expansion. Further taxonomical studies supported by molecular analyses and biogeographical models are needed to infer more consistent information regarding effects of global changes in seaweed diversity.

5.4 The Physiological Bases of Macroalgal Shifts

Antarctic seaweed species are adapted to extreme environmental conditions; however, they can be highly sensitive to abrupt changes in meteorological and oceanographical settings, suggesting that the distribution, diversity, and richness of these organisms can be used as suitable sentinels of the effects of the climate change.

In general, Antarctic macroalgae are very well adapted to low seawater temperatures. Kirst and Wiencke (1995) suggested that changes in lipid composition of membranes are related to genetic adaptation to extreme conditions. Moreover, endemic Antarctic species show high photosynthetic efficiency and high P:R ratios at low temperatures (0–5°C), which allow them to exhibit high growth rates and positive carbon balance at depths close to 30–40-m locations (Gómez et al. 1997; Wiencke et al. 2007) (see also Chap. 9 by Deregibus et al.). Additionally, endemic species exhibit upper survival temperatures (USTs) between 9°C and 13°C. The red alga *Georgiella confluens* exhibits even lower temperature demands, growing only at 0°C, but not at 5°C, and exhibits an UST of 11°C (Wiencke et al. 2014). In contrast, few endemic Antarctic species have higher temperature demands. For example, the brown alga *Ascoseira mirabilis* grows up to 10°C but exhibits a low UST similar to the above-mentioned species. The red algae *Gymnogongrus antarcticus* and *Phyllophora ahnfeltioides* grow up to temperatures of 10°C or even 15°C, but exhibit considerably higher USTs of 19°C or 22°C, respectively. So, the latter three species could theoretically occur even at temperate sites (Bischoff-Bäsmann and Wiencke 1996). The highest temperature demands of Antarctic species were demonstrated in Antarctic cold-temperate green algae. These species grow up to 10°C, 15°C, or 20°C and exhibit USTs between 19 and 26 (28)°C (Wiencke et al. 2014). These temperature requirements reflect the preferential occurrence of these species along supra- and eulittoral zones, where temperatures may vary considerably.

Responses to climate changes are particularly rapid and strong in marine habitats especially in high latitudes and in the intertidal and shallow subtidal zones, where species are often growing at their upper temperature tolerance limits (Hoegh-Guldberg and Bruno 2010; Sorte et al. 2010). In this context, temperature requirements for growth and survival of endemic Antarctic species and species distributed in the Antarctic cold-temperate region have been extensively tested in laboratory. However, results from laboratory experiments should be interpreted with caution since growth and survival of seaweeds in their natural habitat depend on a complex interaction of physical, chemical, and biological factors. In fact, warming at coastal Antarctic waters can alter light regime due to increased turbidity and shortening in the ice cover period, with consequences not well understood for algae adapted to short periods of favorable light conditions and extended periods of darkness (Wiencke et al. 2007) (see also Chap. 7 by Huovinen and Gómez). On the other hand, the inflow of melt water in summer, due to enhanced glacier melting or increased terrestrial runoff, has also considerable effects on the salinity and temperature regime in inshore waters down to 20-m depth (Klöser et al. 1993; Zacher et al. 2009). In other instance, inorganic nutrient concentrations may be high and in general not limiting for seaweeds at any time of the year in the Antarctica (Sanches et al. 2016).

It is well known that species that fails to acclimatize physiologically or evolve genetically to increasing temperature will probably move into another habitats or become extinct (Parmesan 2006; Thomas 2010; Jueterbock et al. 2013). According to Guo et al. (2005), the most effective and informative method of predicting declines or disappearances and/or expansion of nonnative species is by monitoring boundary conditions of very isolated populations, as is the case in Antarctic seaweed assemblages. In all, large-scale climate-driven changes have recently been observed in maritime Antarctic (Macaya et al. 2016; Pellizzari et al. 2017; López et al. 2018; Fraser et al. 2019; Diaz Tapia et al. 2018), which can favor the proliferation and extend the growing season of warm affinity species, inducing local dominance of introduced species and causing adverse effect on cold-adapted species (López et al. 2018).

5.5 Deception Island: A Case Study of Opportunistic, Alien, Cryptic and Cryptogenic Species

Deception Island (62°57'S; 60°38'W), South Shetland Islands (SSI – Fig. 5.1b), affords unique ecosystems essential to understanding the impacts of global warming due to conditions of a highly temperature-sensitive environment. This island is also an excellent model to work in a semi-enclosed environment that can be monitored with long-term instrumentation while being free from the blocking effects of the prevalence of ice and icebergs even during the winter (Smith et al. 2003). The ATCM XXXV (2012) defines Deception Island as an ecosystem of high relevance associated to a unique terrestrial flora, including at least 18 species that have not been recorded elsewhere in Antarctica (e.g., the communities associated with geothermal areas of the Antarctic flowering plant, pearlwort, *Colobanthus quitensis*). All these areas are under conservation strategies (classified in ASPAs and ASMAs). Due to its scientific importance, the area offers a rare opportunity to study the effects

Fig. 5.3 (**a**) Deception Island (SSI) aerial view. (Photo: DAE, Brazilian Navy); (**b**) hydrothermal vents; (**c**) volcanic fumaroles; (**d**) *Spongomorpha arcta* (Chlorophyta); (**e**) filamentous Cyanobacteria epiphytized by diatoms, opportunistic taxa dominating the benthic algae assemblages in the area; (**f**) lake along the island cone. (Photos by Franciane Pellizzari)

of environmental changes, in order to contrast them with the dynamics and recovery from natural disturbance. On the other side, the long history of human activity (since c.1820), including exploration sailing, whaling, war aviation, and scientific research (e.g., the Norwegian whaling station; the British secret base of the World War II, "Base B"), makes this island a site of high historical value. The island was previously a whaling station, and the presence of research stations (Argentina and Spain) is also since decades ago. About 11 localities were designated as Antarctic Specially Protected Area (ASPA).

Deception Island is a circular-shaped volcano, with a linear glaciated coastline, forming semi-enclosed environments showing intense geothermal activity. Black sand beaches and outcrops of volcanic rock form the hard substrate (Fig. 5.3a–c, f). Deception Island exhibits widely varying microclimates with a diameter of ca. 12 km; the center of the island was formed by a huge eruption that flooded the sea to form a large bay (Port Foster). The bay has a narrow entrance (Neptune's Bellows), and shortly after the pass, there is a cove (Whalers Bay).

The fumarolic emissions and thermal springs generally occur along a principal fracture encompassing Fumarole Bay, Telefon Bay, and Pendulum Cove. Temperatures of fumarolic discharges as high as 110°C have been recorded depending on tidal cycle. These fumarolic emissions and thermal springs along the main fracture result in temperature and pH anomalies (Fig. 5.3a–c), and the bottom water temperatures in shallow areas could rise to 8°C (Pellizzari et al. 2017).

Thus, mainly due to the unique physical and chemical features besides long human presence history, including current tourism, Deception Island (Antarctic Specially Managed Area No. 4) is exceptionally vulnerable to the impacts of introduced nonnative species. The human visitation, the mild climate (compared to other polar areas), protected and safest natural harbor, and the presence of geothermal heated sites may catalyze more introduced marine and terrestrial species than any other Antarctic locations.

The diversity of macroalgae from Deception was early described by Clayton et al. (1997), Gallardo et al. (1999), and Pellizzari et al. (2017). Two putative new species, *Prasiola* sp. and *Callophyllis* sp., occur in Deception Island and deserve elucidation due to their cryptic features (Pellizzari et al. 2017). Besides, the widely distributed species *Ulvella viridis*, *Spongomorpha arcta*, and *Rhizoclonium ambiguum* are also conspicuous in the island. Pellizzari et al. (2017) also reported the predominance of filamentous, opportunistic, and widely distributed or cosmopolitan cyanobacteria and green algae, high diatom epiphytism, and it was one of the locations where the bipolar and cryptogenic species were reported (Fig. 5.3d, e). *Monostroma grevillei*, whose introduction possibilities lie in the anthropic activity (whaling and/or tourism), is also a bipolar species, although with some degree of latitudinal disruption. Clayton (1994) reported the first occurrence of *Petalonia fascia* for Deception Island and suggested that ships could introduce the species, mentioning also that the intertidal and subtidal seaweeds from Port Foster belong probably to widely distributed cold-adapted species. At least 20 seaweed species in Antarctica are broadly distributed, e.g., the red alga *Plocamium cartilagineum*, the brown alga *Petalonia fascia*, and the green alga *Ulothrix flacca* (Wiencke and Clayton 2002). It is possible that such species may be recent invaders from temperate regions (Clayton 1994). Two decades later, Pellizzari et al. (2017) observed the intertidal diversity permeated by filamentous algae belonging mainly to Cyanobacteria and typical eurythermal Chlorophyta such as *Cladophora albida* and *Rhizoclonium riparium*. Following McCarthy et al. (2019), many Northern Hemisphere species have become invasive in southern temperate sites, such as Port Phillip Bay (Melbourne, Australia). Pathways to the Southern Hemisphere are clearly viable for some species of algae, polychaetes, and sponges that can inhabit polar waters. Besides, several bipolar species have been reported, including microorganisms, macrofauna, and macroalgae, for example, *M. grevillei* (Pellizzari et al. 2017). Other studies have demonstrated the presence of cryptic species in Antarctica (Billard et al. 2015 and synopsis in Dubrasquet et al. 2018), and Deception Island is one of the target places where cryptic and hidden species must be investigated.

Crustose calcareous algae belonging to Corallinaceae (P*seudolithophyllum* sp. and *Lithophyllum subantarcticum*) and Hapalidiaceae (*Clathromorphum obtectulum*, *Lithothamnion granuliferum*, *Synarthrophyton patena*, and two unidentified species of the *Phymatolithon*/*Mesophyllum* complex) have been reported along the Shetland Islands (listed in Pellizzari et al. 2017) (Fig. 5.4a, f). However, in Deception Island, these calcareous algae are absent, probably due to lower pH values, confirming the island as a natural laboratory to study responses in species facing extreme abiotic changes. Moreover, the macroalgae from Deception Island host a rich

Fig. 5.4 Some species sampled along the SSI. (**a**) *Clathromorphum* (calcareous Rhodophyta), (**b**) fertile *Iridaea cordata*, (**c**) transverse section of *I. cordata*, (**d**) Delesseriaceae, (**e**) cystocarp of *Myriogramme mangini*, (**f**) *Nacella concinna* (the Antarctic limpet) grazing over benthic algae. (Photos by Franciane Pellizzari)

associated fungal diversity composed of endemic, cold-adapted, or cosmopolitan taxa. Fungal species of the genera *Aspergillus*, *Cryptococcus*, *Debaryomyces*, *Meyerozyma*, *Penicillium*, and *Pseudogymnoascus* have been isolated from algae such as *Adenocystis utricularis*, *Monostroma hariotii*, and *Adenocystis* sp. and *Pyropia endiviifolia* (Godinho et al. 2013; Furbino et al. 2014). The studies also emphasize that the fate of endemic or cold-adapted macroalgal species under changing Antarctic environment will affect also their associated microbiota.

5.6 Reevaluating Eco-Regions, Isolation, and Endemism in the Southern Ocean

According to Wiencke and Tom Dieck (1990), many endemic Antarctic species show a lower range of thermal tolerance compared to Arctic species. Arctic species have been isolated for a much less time than Antarctic species. However, some of these species were also listed in Falkland Islands (Clayton 2003), South Georgia (Wells et al. 2011), and Macquarie Island (Ricker 1987), e.g., *Desmarestia menziesii*, *G. confluens*, and *Myriogramme manginii*. Ramírez and Santelices (1991) reported *D. anceps* for Chile. *Monostroma hariotii* Gain was registered in Argentinean Patagonia (Boraso de Zaixso 2003, 2013), South Georgia (Wells et al.

2011), and the Falkland Islands (Clayton 2003). *Iridaea cordata* (Fig. 5.4b, c) and *Geminocarpus geminatus* also occurs in the sub-Antarctic islands and Tierra del Fuego. *Ballia callitricha* and *Adenocystis utricularis* also occur in New Zealand and Australia, suggesting higher connectivity among the assemblages and as already discussed in this chapter.

Studies using molecular markers (Hommersand et al. 2009; Medeiros 2013; Billard et al. 2015; Ocaranza-Barrera et al. 2018; Dubrasquet et al. 2018) have revealed that cryptic species previously listed in both maritime Antarctica and South America, e.g., *Plocamium*, *Ulothrix*, *Gigartina*, and *Iridaea*, are in fact distinct taxonomic entities. John et al. (1994) reported 127 species of benthic algal flora from South Georgia (54°S). Comparing the results of this sub-Antarctic island with the list from the SSI (Pellizzari et al. 2017, Fig. 5.5a–d), 40% of the species co-occurred in both locations, suggesting high similarity between marine flora of distinct eco-regions or provinces (Sanches et al. 2016). Wells et al. (2011) described the intertidal and subtidal benthic seaweed diversity of South Georgia revealing that 63% of these species list co-occur in the SSI. *Palmaria decipiens* and *Iridaea cordata* (Fig. 5.4), conspicuous Antarctic red algae, occur from the Ross Sea to sub-Antarctic islands (Wiencke and Clayton 2002). However, *I. cordata* was also

Fig. 5.5 (**a**) King George Island (SSI) threatened by global changes. (**b**) Weddell seal in a drifted bed of conspicuous *Palmaria decipiens*, (**c**) *Iridaea cordata*, *Adenocystis utricularis*, *Ascoseira mirabilis*, (**d**) *Monostroma hariotii*. *(Photos by Franciane Pellizzari)*

reported in Chile and recently in Argentinean Patagonia (Ramírez and Santelices 1991; Boraso de Zaixso 2013), and *P. decipiens* was also recorded in New Zealand (Nelson 2012), Macquarie Island (Ricker 1987), and South Georgia (Wells et al. 2011).

Advances in molecular techniques have allowed estimating more precisely phylogenetic relationships, levels of differentiation, and divergence time between populations from these continents (Ocaranza-Barrera et al. 2018). According to Papenfuss (1964), the seaweed flora of the sub-Antarctic islands and the Antarctic region totaled 550 species. Approximately 75% of these species were distributed in the sub-Antarctic islands (Clayton 1994). Wulff et al. (2009) suggested that, after the endemics, the second largest seaweed group in Antarctic includes those occurring both in the Antarctic region and on sub-Antarctic islands and Tierra del Fuego (see also Chap. 2 by Oliveira et al.). Some of these species may be examples of distributional extension northward; besides few studies are supported by molecular data (Hommersand et al. 2009; Fraser et al. 2013; Medeiros 2013; Billard et al. 2015; Pellizzari et al. 2017; Ocaranza-Barrera et al. 2018). Fraser et al. (2013) studied the genetic affinities between transoceanic populations of *Adenocystis utricularis* and *Bostrychia intricata* in southern Chile, New Zealand, and several sub-Antarctic islands (disjoint regions connected oceanographically by the ACC). The authors observed divergent clades for both species, but close phylogenetic relationships – even shared haplotypes – among populations separated by large oceanic distances. Despite not being particularly buoyant, *A. utricularis* and *B. intricata* showed genetic signatures of recent dispersal across vast oceanic distances, presumably by attachment to floating substrata (e.g., wood, buoyant large macroalgae, or still, hull fouling, or ballast water), although the genetic data indicate the possibility of cryptic species within both taxa (see also Chap. 4 by Macaya et al.).

For algae present on both sides of the ACC, e.g., *Gigartina skottsbergii*, two hypotheses have been discussed: (1) they correspond to recent immigrants from adjacent continents, or (2) they have evolved in situ surviving the effects of the last glaciations. The haplotype networks of *G. skottsbergii* suggested that there is some evolutionary divergence between populations and a disjoint distribution, a pattern that exceeds its dispersal capacity (Billard et al. 2015; see also Chap 6 by Guillemin et al.). In accordance to Hommersand et al. (2009), the monophyletic clades of *G. skottsbergii* may correspond to two cryptic species: (1) *G. skottsbergii* distributed in the sub-Antarctic islands and in the southern coast of Chile and (2) a new species, still to be formally described and named, that occurs in the Antarctic Peninsula, South Shetland, and South Orkney Islands. The divergence time between these two cryptic species suggested that algae with limited dispersal capabilities were able to cross the Scotia Sea after separation of the continents, potentially via a stepping stone process through the volcanic arc of islands. This strongly supports the idea that the last Quaternary glaciations induced marked bottlenecks that were followed by rapid colonization events (Billard et al. 2015; see also Chap. 6 by Guillemin et al.).

Regarding other connections in the SO, the seaweed assemblage of Southern Australia (Victoria-Tasmania Region) has been featured by high species richness

(Kerswell 2006) – endemic (40–77%) and temperate (17–45%) species – whereas species with tropical (4.5– 9.7%) and polar (1.5–5.1%) affinities are much less represented. Wernberg et al. (2011a) used a seaweed database from herbarium records, sampled in Australia since the 1940s, and documented changes in communities' distribution in Indian and Pacific Oceans, related with rapid warming over the past five decades. For the coasts of New Zealand, regarded as areas of high seaweed diversity (Kerswell 2006), a recent list suggested that among the 770 currently known seaweed recorded, 265 are endemic (34%) and 22 alien (2.9%). Rhodophyta accounts around 60% of the total taxa. Overall, a reduction in endemism (see also Chap. 2 by Oliveira et al.) and higher species "transit" among assemblages from Antarctica, sub-Antarctic, and Southern America have been suggested, probably requiring increasing seawater temperatures over a longer period of time (Sjøtun et al. 2015).

Although temperature increases may not directly affect the latitudinal distribution of Antarctic key structural seaweeds in the short term (Müller et al. 2009), they may modify the presence of these species indirectly via changes in sea ice dynamics, which enhances the hard substrate availability, besides changes on thermohaline and circulation patterns. Stammerjohn et al. (2008) reported that the sea ice cover along the southern portion of the Western Antarctic Peninsula (WAP) advances nearly 2 months later in the winter and retreats approximately 1 month earlier in the spring. In this context, the duration of the sea ice season around the Antarctic Peninsula decreased by 4 days per year from 1979/1980 to 2011/2012 (Hughes and Ashton 2017). However, the sea ice surrounding Antarctica has increased in extent and concentration from the late 1970s, when satellite-based measurements began, until 2015. While this increased ice cover trend is modest, it is surprising given the general warming of the global climate (Fig. 5.6a, b) and despite some climate models – that incorporate a better understanding of the processes affecting the region – usually simulating a tendency of sea ice decrease in the mid and long term. In contrast, in the Artic, the sea ice cover has exhibited pronounced declines over the same period, consistent with global climate model simulations. However, and due to the oscillating patterns of the sea ice cover recently in the Antarctic shelves, this random behavior in distinct Antarctic areas has presented an enigma for global climate change science. In addition, significant differences between the Southern and Arctic Oceans resulted from SST opposite anomalies (Fig. 5.6c).

Recent data suggest that the SO seems to be barely sensitive to greenhouse gases (GHG), resulting in a current lower warming levels of SST than observed in other areas (NASEM 2017). However, changes in the deep ocean can have important effects due to significant upwelling (NASEM 2017). Other factors that may play a role in this conundrum include tropical Pacific and Atlantic teleconnections, variability in the wind, and ocean circulation that circumnavigate Antarctica. However, an understanding of the mechanisms and processes driving sea ice variability, as well SST trends, is limited by the lack of long and homogenous records.

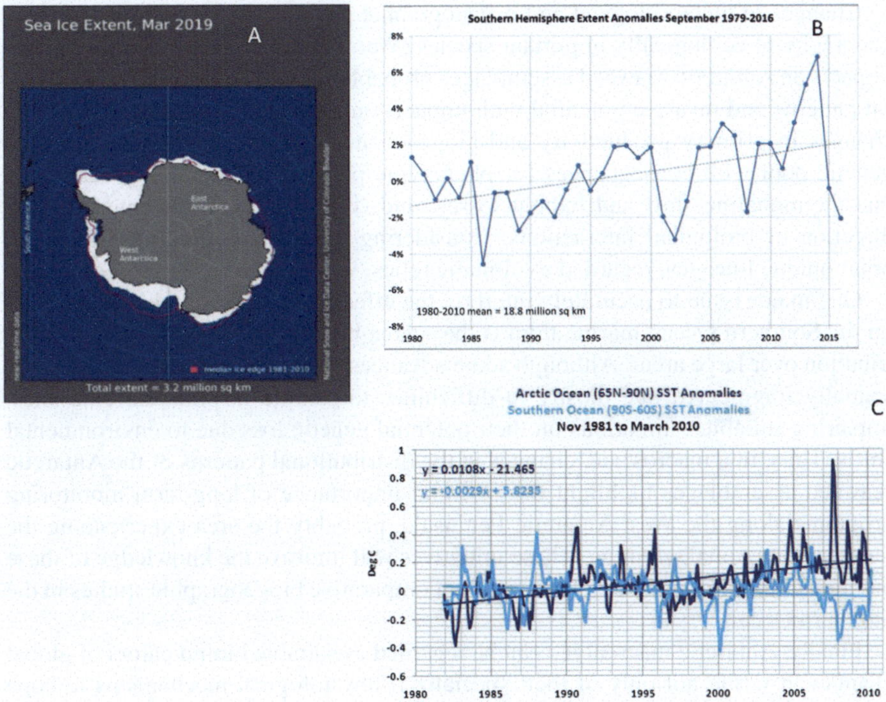

Fig. 5.6 (**a**) Antarctic sea ice extent: line is the median of ice edge between 1981 and 2010 and the white contour, in 2019. A decrease has been observed since January 2016. (Image Source: National Snow & Ice Data Center http://nsidc.org/data/seaice_index). (**b**) Southern Ocean sea ice extends anomalies. (Between 2010 and 2016 the trend line is slightly positive http://nsidc.org/data/seaice_index). (**c**) SST anomalies of the Arctic and Southern Oceans. (Source: NOAA NOMADS http://nomad1.ncep.noaa.gov/cgi-bin/pdisp_sst.sh)

5.7 Concluding Remarks: Prospects for the Future Marine Flora of the Southern Ocean

Shifts in species distribution are being reported worldwide as a result of changes in connectivity patterns, mediated by anthropic activity, climate changes, and increased species "transit" even towards remote areas. Dispersal refugia for Antarctic and sub-Antarctic seaweeds (see Billard et al. 2015) are also concepts that contribute to the better understanding of the dispersal mechanisms and recolonization processes during biogeographic expansion in the Southern Ocean and connections (see also Chap. 6 by Guillemin et al.). Thus, seaweed assemblages in Antarctica, and in the Southern Hemisphere, will be modified not only by large-scale geological and paleoclimatic processes, but also by long-distance dispersal events (see also Chap. 3 by Fraser et al.).

Changes in meteorological and oceanographic patterns are rearranging the biogeography of ecologically important species (Wernberg et al. 2010, 2011a, b). The impacts on Antarctic seaweed assemblages probably will include (i) shifts in dispersal patterns and invasive potential with impacts on local richness and diversity, (ii) changes in primary productivity and biogeochemical fluxes, (iii) changes in the genetic pool with consequences for phenotypic plasticity and adaptive responses due thermohaline shift and habitat losses, and (iv) changes in the intensity and direction of biological interactions. Considering these projections, the threat to polar unique lineages, regarded as climatic relics, will increase.

One major issue to accurately interpret the effects of abiotic and biotic changes on the Southern Ocean marine flora is the scarce basic knowledge of seaweed distribution over large areas. Although some advances have been made, the studies are spatially fragmented due mostly to difficulties to sample in polar/extreme sites, impairing suitable estimations on their potential genetic loss due to environmental shifts. Thus, in a macroscale, changes in the distributional patterns of the Antarctic seaweed assemblages highlight the essential importance of long-term monitoring programs along the West Antarctic Peninsula, probably the area experiencing the most rapid regional warming. These initiatives will improve the knowledge of these natural laboratories and models for further comparative biogeographic studies in the Southern Ocean.

Finally, Antarctic macroalgae can be regarded as suitable bioindicators of global changes in virtue not only of their specialized physiological mechanisms to cope with low temperatures and limited light conditions, but also by their high adaptability to extreme conditions in their habitat. These remote assemblages can provide essential clues to understand the resilience potential of organisms living in the edge of their adaptive windows and are fundamental to reinforce the need for maintaining global database aiming to integrate and normalize abiotic/biotic metadata (see also Chap. 2 by Oliveira et al.).

Acknowledgements The authors thank PROANTAR (Brazilian Antarctic Program 557030/2009-9, 407588/2013-2 and 442258/2018-2), INCT Criosfera 2, Brazilian Navy (Polar Ship Almirante Maximiano–H41), Brazilian Air Force, MMA (Ministry of Environment), MCTIC (Ministry of Science, Technology and Innovation), CNPq (National Council of Research and Development), Fundação Araucária–Government of Paraná, CAPES, FAPEMIG, and FNDC.

References

ATCM XXXV (2012) Final report. Deception Island management package: management plan for Antarctic specially managed area No 4 Deception Island, South Shetland Islands, Antarctica

Billard E, Reyes J, Mansilla A, Faugeron S, Guillemin M-L (2015) Deep genetic divergence between austral populations of the red alga *Gigartina skottsbergii* reveals a cryptic species endemic to the Antarctic continent. Polar Biol 38:2021. https://doi.org/10.1007/s00300-015-1762-4

Bischoff-Bäsmann B, Wiencke C (1996) Temperature requirements for growth and survival of Antarctic Rhodophyta. J Phycol 32:525–535

Boraso de Zaixso AL (2003) Algas marinas de la Patagonia: una guía ilustrada. Fundación de Historia Natural Félix de Azara, Argentina

Boraso de Zaixso AL (ed) (2013) Elementos para el estudio de las macroalgas de Argentina, Con colaboración de J.M. Zaixso. 1a ed. – Comodoro Rivadavia, Universitaria de la Patagonia. ISBN 978-987-1937-14-1

Clayton MN (1994) Evolution of the Antarctic marine benthic algal flora. Rev J Phycol 30:897–904

Clayton MN (2003) Falkland Islands seaweed survey, Shackleton Scholaship Fundation, Monash University, Victoria 3800, Australia

Clayton MN, Wiencke C, Klöser H (1997) New records of temperate and sub-Antarctic marine benthic macroalgae from Antarctica. Polar Biol 17:141–149. https://doi.org/10.1007/s003000050116

Cormaci M, Furnari G, Scammacca B (1992) The benthic algal flora of Terra Nova Bay (Ross Sea, Antarctica). Bot Mar 35:541–552. https://doi.org/10.1515/botm.1992.35.6.541

Crame JA (1994) Evolutionary history of Antarctica. In: Hempel G (ed) Antarctic science: global concerns. Springer, Berlin, pp 188–214

Díaz Tapia P, Maggs C, Macaya EC, Verbruggen H (2018) Widely distributed red algae often represent hidden introductions, complexes of cryptic species or species with strong phylogeographic structure. J Phycol 54:829–839. https://doi.org/10.1111/jpy.12778

Dubrasquet H, Reyes J, Sanchez RP, Valdivia N, Guillemin ML (2018) Molecular-assisted revision of red macroalgal diversity and distribution along the Western Antarctic Peninsula and South Shetland Islands. Cryptogamie Algol 39(4):409–429. https://doi.org/10.7872/crya/v39.iss4.2018.409

Ducklow HW, Fraser WR, Meredith MP, Stammerjohn SE, Doney SC, Martinson DG, Sailley SF, Schofield OM, Steinberg DK, Venables HJ, Amsler CD (2013) West Antarctic Peninsula: an ice-dependent coastal marine ecosystem in transition. Oceanography 26:190–203. https://doi.org/10.5670/oceanog.2013.62

Fraser CI, Zuccarello GC, Spencer HG, Salvatore LC, Garcia GR et al (2013) Genetic affinities between trans-oceanic populations of non-buoyant macroalgae in the high latitudes of the southern Hemisphere. PLoS One 8(7):e69138. https://doi.org/10.1371/journal.pone.0069138

Fraser CI, Morrison AK, McC Hogg A, Macaya EC, van Sebille E, Ryan PG, Padovan A, Jack C, Valdivia N, Waters JM (2019) Antarctica's ecological isolation will be broken by storm-driven dispersal and warming. Nat Clim Change Lett 42:475–483. https://doi.org/10.1038/s41558-018-0209-7

Furbino LE, Godinho VM, Santiago IF, Pellizzari FM, TMA A, Zani CL, PAS J, Romanha AJ, AGO C, LHVG G, Rosa CA, Minnis AM, Rosa LH (2014) Diversity patterns, ecology and biological activities of fungal communities associated with the endemic macroalgae across the Antarctic Peninsula. Microb Ecol 67:775–787

Gallardo T, Pérez-Ruzafa IM, Flores-Moya A, Conde F (1999) New collections of benthic marine algae from Livingston and Deception Islands (South Shetland Islands) and Trinity Island (Bransfield Strait), Antarctica. Bot Mar 42:61–69. https://doi.org/10.1515/bot.1999.009

Godinho VM, Furbino LE, Santiago IF, Pellizzari FM, Yokoya N, Rosa LH (2013) Diversity and bioprospecting of fungal communities associated with endemic and cold-adapted macroalgae in Antarctica. ISME 7:1434–1451

Gómez I, Weykam G, Klöser H, Wiencke C (1997) Photosynthetic light requirements, metabolic carbon balance and zonation of sublittoral macroalgae from King George Island (Antarctica). Mar Ecol Prog Ser 149:281–293

Griffiths HJ, Waller CL (2016) The first comprehensive description of the biodiversity and biogeography of Antarctic and Sub-Antarctic intertidal communities. J Biogeogr 43(6):1143–1155. https://doi.org/10.1111/jbi.12708

Griffiths HJ, Barnes DKA, Linse K (2009) Towards a generalized biogeography of the Southern Ocean benthos. J Biogeogr 36:162–177

Guo KM, Taper M, Schoenberger B, Brandle J (2005) Spatial-temporal population dynamics across species range: from centre to margin. Oikos 108:47–57

Gutt J, Alvaro MC, Barco A, Böhmer A, Bracher A, David B et al (2016) Macroepibenthic communities at the tip of the Antarctic Peninsula, an ecological survey at different spatial scales. Polar Biol 39:829–849. https://doi.org/10.1007/s00300-015-1797-6

Hoegh-Guldberg O, Bruno JF (2010) The impact of climate change on the world's marine ecosystems. Science 328:1523–1528

Hommersand MH, Moe RL, Amsler CD, Fredericq S (2009) Notes on the systematics and biogeographical relationships of Antarctic and sub-Antarctic Rhodophyta with descriptions of four new genera and five new species. Bot Mar 52:509–534. https://doi.org/10.1515/bot.2009.081

Hughes KA, Ashton GV (2017) Breaking the ice: the introduction of biofouling organisms to Antarctica on vessel hulls. Aquatic Conserv Mar Freshw Ecosyst 27:158–164. https://doi.org/10.1002/aqc.2625

Huovinen P, Gómez I (2012) Cold-temperate seaweed communities of the Southern Hemisphere. In: Wiencke C, Bischof K (eds) Seaweed biology, novel insights into ecophysiology, ecology and utilization. Ecological studies 219. Springer, Berlin Heidelberg, pp 293–314. https://doi.org/10.1007/978-3-642-28451-9_14

John DM, Pugh PJA, Tittley I (1994) Observations on the benthic marine algal flora of South Georgia: a floristic and ecological analysis. Bull Nat Hist Museum 24:101–114

Jueterbock A, Tyberghein L, Verbruggen H, Coyer JA, Olsen JL, Hoarau G (2013) Climate change impact on seaweed meadow distribution in the North Atlantic rocky intertidal. Ecol Evol 3(5):1356–1373. https://doi.org/10.1002/ece3.541

Keith SA, Kerswell AP, Connolly SR (2011) Global diversity of marine macroalgae: environmental conditions explain less variation in the tropics. Glob Ecol Biogeogr 23:517–529

Kerswell AP (2006) Global biodiversity patterns of benthic marine algae. Ecology 87:2479–2488

Kirst GO, Wiencke C (1995) Ecophysiology of polar algae. J Phycol 31:181–199

Klöser H, Ferreyra G, Schloss I, Mercuri G, Laturnus F, Curtosi A (1993) Seasonal variation of algal growth conditions in sheltered Antarctic bays: the example of Potter Cove (King George Island, South Shetlands). J Mar Syst 4:289–301

Linse K, Griffiths HJ, Barnes DKA, Clarke A (2006) Biodiversity and biogeography of Antarctic and sub-Antarctic mollusca. Deep-Sea Res II 53:985–1008

Longhurst A (ed) (2007) Ecological geography of the sea, 2nd edn. Academic Press, London, p 390

López BA, Macaya EC, Rivadeneira MM, Tala F, Tellier F, Thiel M (2018) Epibiont communities on stranded kelp rafts of *Durvillaea antarctica* (Fucales, Phaeophyceae). Do positive interactions facilitate range extensions? J Biogeogr 45:1833–1845. https://doi.org/10.1111/jbi.13375

Macaya EC, López B, Tala F, Tellier F, Thiel M (2016) Float and raft: role of buoyant seaweeds in the phylogeography and genetic structure of non-buoyant associated flora. In: Hu ZM, Fraser C (eds) Seaweed phylogeography. Springer, Dordrecht, pp 97–130. https://doi.org/10.1007/978-94-017-7534-2_4

McCarthy AH, Peck LS, Hughes KA, Aldridge DC (2019) Antarctica: the final frontier for marine biological invasions. Glob Chang Biol 25(7): 2221–2241. https://doi.org/10.1111/gcb.14600

Medeiros AS (2013) Macroalgae diversity of Admiralty Bay, King George Island, Antarctic Peninsula based on DNA barcoding and other molecular markers. Thesis. São Paulo University. http://www.teses.usp.br/teses/disponiveis/41/41132/tde-24032014-090801/

Müller R, Laepple T, Bartsch I, Wiencke C (2009) Impact of oceanic warming on the distribution of seaweeds in polar and cold-temperate waters. Bot Mar 52:617–638. https://doi.org/10.1515/bot.2009.080

Mystikou A, Peters AF, Asensi AO, Fletcher KI, Brickle P, van West P, Convey P, Küpper FC (2014) Seaweed biodiversity in the south-western Antarctic Peninsula: surveying macroalgal community com- position in the Adelaide Island/Marguerite Bay region over a 35-year time span. Polar Biol 37:1607–1619. https://doi.org/10.1007/s00300-014-1547-1

NASEM (2017) National academies of sciences, engineering, and medicine. In: Antarctic sea ice variability in the Southern Ocean-climate system: proceedings of a workshop. The National Academies Press, Washington, DC. https://doi.org/10.17226/24696

Nelson WA (2012) Phylum rhodophyta: red algae. In: Gordon DP (ed) New Zealand inventory of biodiversity. Canterbury University Press, Christchurch

Ocaranza-Barrera P, González-Wevar C, Guillemin ML, Rosenfeld S, Mansilla A (2018) Molecular divergence between Iridaea cordata (Turner) Bory de Saint-Vincent from the Antarctic Peninsula and the Magellan Region. J App Phycol. 31:939–949. https://doi.org/10.1007/s10811-018-1656-2

Orsi AH, Whitworth T, Nowlin WD Jr (1995) On the meridional extent and fronts of the Antarctic Circumpolar Current. Deep-Sea Res 42:641–673. https://doi.org/10.1016/0967-0637(95)00021W

Papenfuss GF (1964) Catalogue and bibliography of Antarctic and Subantarctic benthic marine algae. Antar Res Ser 1:1–76

Parmesan C (2006) Ecological and evolutionary responses to recent climate change. Annu Rev Ecol Evol Syst 37:637–669

Pellizzari F, Santos-Silva MC, Medeiros A, Oliveira MC, Yokoya NS, Pupo D, Rosa L (2017) Diversity and spatial distribution of seaweeds in the South Shetland Islands, Antarctica: an updated database for environmental monitoring under climate change scenarios. Polar Biol. https://doi.org/10.1007/s00300-017-2092-5

Pibernat RA, Ellis-Evans C, Hinghofer-Szalkay HG (eds) (2007) Life in extreme environments. Springer, The Netherlands. https://doi.org/10.1007/978-1-4020-6285-8

Pierrat B, Sausede T, Brayard A, David B (2013) Comparative biogeography of echinoids, bivalves and gastropods from the Southern Ocean. J Biogeogr 40:1374–1385

Pritchard HD, Ligtenberg SR, Fricker HA, Vaughan DG, van den Broeke MR, Padman L (2012) Antarctic ice-sheet loss driven by basal melting of ice shelves. Nature 484:502–505. https://doi.org/10.1038/nature10968

Quartino ML, Deregibus D, Campana GL, Latorre GEJ, Momo FR (2013) Evidence of macroalgal colonization on newly ice-free areas following glacial retreat in Potter Cove (South Shetland Islands), Antarctica. PLoS One 8(3):e58223. https://doi.org/10.1371/journal.pone.0058223

Ramírez ME, Santelices B (1991) Catálogo de las algas marinas bentónicas de la costa temperada del Pacífico de Sudamérica. Facultad de Ciencias Biológicas, Pontificia Universidad Católica de Chile

Ricker RW (1987) Taxonomy and biogeography of Macquarie Island seaweeds. British Museum (Natural History), London

Sanches PF, Pellizzari FM, Horta PH (2016) Multivariate analyses of Antarctic and sub- Antarctic seaweed distribution patterns: an evaluation of the role of the Antarctic Circumpolar Current. J Sea Res 110:29–38. https://doi.org/10.1016/j.seares.2016.02.002

Sangil C, Sansón M, Afonso-Carillo J, Herrera R, Rodríguez A, Martín-García L, Díaz-Villa T (2012) Changes in sub- tidal assemblages in a scenario of warming: proliferations of ephemeral benthic algae in the Canary Islands (eastern Atlantic Ocean). Mar Environ Res 77:120–128

Sjøtun K, Husa V, Asplin L, Sandvik AD (2015) Climatic and environmental factors influencing occurrence and distribution of macroalgae — a fjord gradient revisited. Mar Ecol Prog Ser 532:73–88. https://doi.org/10.3354/meps11341

Smith KL, Baldwin RJ, Kaufmann RS, Sturz A (2003) Ecosystem studies at Deception Island, Antarctica: an overview. Deep-Sea Res II 50:1595–1609. https://doi.org/10.1016/S0967-0645(03)00081-X

Sorte CJB, Williams SL, Carlton JT (2010) Marine range shifts and species introductions: comparative spread rates and community impacts. Glob Ecol Biogeogr 19:303–316

Stammerjohn SE, Martinson DG, Smith RC, Yuan X, Rind D (2008) Trends in Antarctic annual sea ice retreat and advance and their relation to El Niño–Southern Oscillation and Southern Annular Mode variability. J Geophys Res 113. https://doi.org/10.1029/2007jc004269

Taton A, Grubisic S, Balthasart P, Hodgson DA, Laybourn-Parry J, Wilmotte A (2006) Biogeographical distribution and ecological ranges of benthic cyanobacteria in East Antarctic lakes. FEMS Microbiol Ecol 57(2):272–289. https://doi.org/10.1111/j.1574-6941.2006.00110.x

Thomas CD (2010) Climate change and range boundaries. Divers Distrib 16:488–495

Wells E, Brewin P, Brickle P (2011) Intertidal and subtidal benthic seaweed diversity of South Georgia. Norfolk, UK

Wernberg T, Thomsen MS, Tuya F, Kendrick GA, Staehr PA, Toohey BD (2010) Decreasing resilience of kelp beds along a latitudinal temperature gradient: potential implications for a warmer future. Ecol Lett 13:685–694
Wernberg T, Russell BD, Thomsen MS, Gurgel CFD, Bradshaw CJA, Poloczanska ES, Connell SD (2011a) Seaweed communities in retreat from ocean warming. Curr Biol 21(21):1828–1832. https://doi.org/10.1016/j.cub.2011.09.028
Wernberg T, Russell BD, Moore PJ, Ling SD, Smale DA, Campbell A, Coleman MA, Steinberg PD, Kendrick GA, Connell SD (2011b) Impacts of climate change in a global hotspot for temperate marine biodiversity and ocean warming. J Exp Mar Biol Ecol 400:7–16
Wernberg T, Bennett S, Babcock RC, Bettignies T, De Cure K, Depczynski M, Wilson S (2016) Climate-driven regime shift of a temperate marine ecosystem. Science 353:169–172
Wiencke C, Amsler CD (2012) Seaweeds and their communities in polar regions. In: Wiencke C, Bischof K (eds) Seaweed biology: novel insights into ecophysiology, ecology and utilization. Springer-Verlag, Berlin, pp. 265–291
Wiencke C, Clayton MN (2002) Antarctic seaweeds. In: Wagele JW (ed) Synopses of the Antarctic benthos. Lichtensein, Germany
Wiencke C, Tom Dieck IT (1990) Temperature requirements for growth and survival of macroalgae from Antarctica and southern Chile. Mar Ecol Prog Ser 24:157–170
Wiencke C, Clayton MN, Gómez I, Iken K, Luder UH, Amsler CD, Karsten U, Hanelt D, Bischof K, Dunton K (2007) Life strategy, ecophysiology and ecology of seaweeds in polar waters. Rev Environ Sci Biotechnol 6: 95–126. https://doi.org/10.1007/s11157-006-9106-z
Wiencke C, Amsler CD, Clayton MN (2014) Chapter 5.1. Macroalgae. In: De Broyer C, Koubbi P, Griffiths HJ, Raymond B, Cd UA (eds) Biogeographic Atlas of the Southern Ocean. Scientific Committee on Antarctic Research, Cambridge, pp 66–73
Wulff A, Iken K, Quartino ML, Al-Handal A, Wiencke C, Clayton MN (2009) Biodiversity, biogeography and zonation of marine benthic micro–and macroalgae in the Arctic and Antarctic. Bot Mar 52:491–507. https://doi.org/10.1515/bot.2009.072
Zacher K, Rautenberger R, Hanelt D, Wulff A, Wiencke C (2009) The abiotic environment of polar marine benthic algae. Bot Mar 52:483–490. https://doi.org/10.1515/bot.2009

Chapter 6
Comparative Phylogeography of Antarctic Seaweeds: Genetic Consequences of Historical Climatic Variations

Marie-Laure Guillemin, Claudio González-Wevar, Leyla Cárdenas, Hélène Dubrasquet, Ignacio Garrido, Alejandro Montecinos, Paula Ocaranza-Barrera, and Kamilla Flores Robles

Abstract In the Southern Ocean, rapid climatic fluctuations during the Quaternary are thought to have induced range contractions and bottlenecks, which drastically impacted marine communities. For photosynthetic macroalgae that are restricted to very shallow waters, survival in deepwater refugia is not possible. Comparing pattern of distribution of genetic diversity using sequences of mitochondrial and chloroplast markers in distinct species of green, brown and red macroalgae, we sought to detect common responses to the effect of these glacial cycles. All the Antarctic macroalgae were characterized by very low genetic diversity, absence of genetic structure and significant signatures of recent population expansion. The eight studied species seem to have barely survived glacial events in situ, in a unique refugium from which they recolonized their current distribution area. We propose that polynyas or areas showing long-term geothermal activity along Antarctic continental

M.-L. Guillemin (✉)
Instituto de Ciencias Ambientales y Evolutivas, Universidad Austral de Chile, Campus Isla Teja, Valdivia, Chile

CNRS, UMI 3614 Evolutionary Biology and Ecology of Algae, Sorbonne Universités, Roscoff, France

Research Center Dynamics of High Latitude Marine Ecosystems (IDEAL), Valdivia, Chile
e-mail: marie-laure.guillemin@uach.cl

C. González-Wevar
Instituto de Ciencias Marinas y Limnológicas, Facultad de Ciencias, Universidad Austral de Chile, Campus Isla Teja, Valdivia, Chile

Research Center Dynamics of High Latitude Marine Ecosystems (IDEAL), Valdivia, Chile
e-mail: claudio.gonzalez@uach.cl

L. Cárdenas
Instituto de Ciencias Ambientales y Evolutivas, Universidad Austral de Chile, Campus Isla Teja, Valdivia, Chile

Research Center Dynamics of High Latitude Marine Ecosystems (IDEAL), Valdivia, Chile

I. Gómez, P. Huovinen (eds.), *Antarctic Seaweeds*,
https://doi.org/10.1007/978-3-030-39448-6_6

margins or peri-Antarctic islands could be good candidate as glacial refugium, but more variable genetic markers will be needed to precisely pinpoint its location. Common haplotypes, scattered over hundreds or even thousands of kilometres of coastline, point out to long-distance dispersal of fronds drifting on the strong oceanic currents in the region as the main mechanism of postglacial expansion.

Keywords Glacial refugia · Genetic diversity · Last glacial maximum · Population bottleneck · Quaternary · Southern ocean

6.1 Historical Isolation of Antarctic Marine Macroalgae

Antarctic marine macroalgae diversity, described as less diverse than other areas of the Southern Ocean, is still characterized by high levels of endemism reaching up to 35% (Wiencke and Clayton 2002; Wiencke et al. 2014; also see chapter by Oliveira et al. 2020 in this volume). As recorded for much of the Antarctic flora and fauna (Clarke et al. 2005; Allcock and Strugnell 2012), the presence of various endemic Antarctic macroalgal species has been linked to the major tectonic, oceanographic and climatic changes that have affected the region during the last 50 million years (Ma) (Dell 1972; Crame 1999, 2018; Mackensen 2004; Aronson et al. 2007; Moon et al. 2017; Halanych and Mahon 2018). Indeed, the fragmentation of the continental landmasses and the initiation of the Antarctic Circumpolar Current (ACC) (Fig. 6.1) gradually isolated the Antarctic waters from the rest of the Southern Ocean (Crame 1999; Scher and Martin 2006; Aronson et al. 2007; Dalziel et al. 2013; Sijp et al. 2014; Scher et al. 2015). Currently, the Antarctic Polar Front (APF) (Fig. 6.1) is considered as an effective barrier for many near-shore marine benthic

H. Dubrasquet · K. Flores Robles
Instituto de Ciencias Ambientales y Evolutivas, Universidad Austral de Chile, Campus Isla Teja, Valdivia, Chile

I. Garrido
Research Center Dynamics of High Latitude Marine Ecosystems (IDEAL), Valdivia, Chile

Instituto de Ciencias Marinas y Limnológicas, Facultad de Ciencias, Universidad Austral de Chile, Campus Isla Teja, Valdivia, Chile

Département de Biologie et Quebec Océan– Ocean Institute, Université Laval, Quebec City, QC, Canada

A. Montecinos
Instituto de Ciencias Ambientales y Evolutivas, Universidad Austral de Chile, Campus Isla Teja, Valdivia, Chile

CNRS, UMI 3614 Evolutionary Biology and Ecology of Algae, Sorbonne Universités, Roscoff, France

P. Ocaranza-Barrera
Laboratorio de Ecosistemas Marinos Antárticos y Subantárticos, Universidad de Magallanes (LEMAS), Punta Arenas, Chile

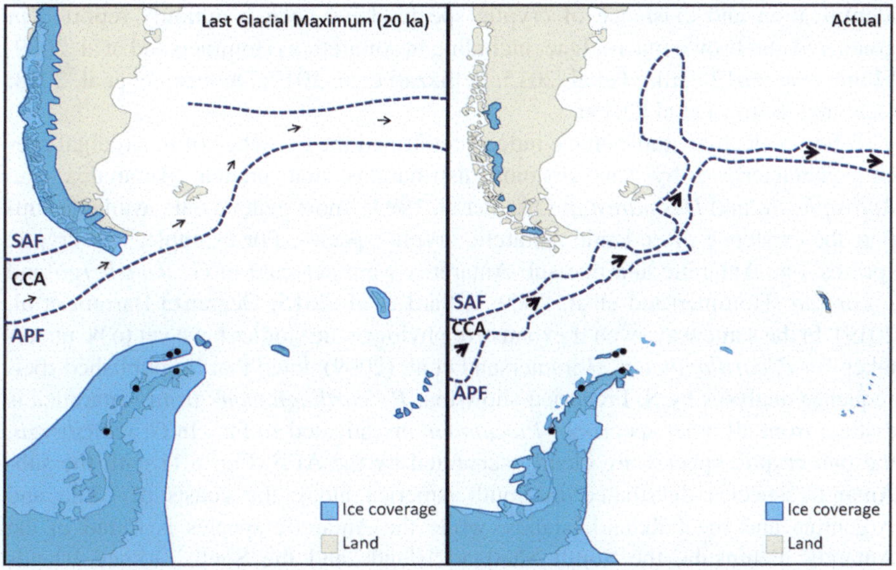

Fig. 6.1 Reconstruction of the Antarctic and sub-Antarctic ice coverage during the Last Glacial Maximum (LGM; ~20,000 ka) and in the present day. Putative LGM ice extension is based on various recent glaciological studies published for the tip of South America (McCulloch et al. 2000), the Antarctic Peninsula and the South Shetland Islands (Simms et al. 2011; Larter et al. 2014; O'Cofaigh et al. 2014) and South Georgia (White et al. 2018). Current ice extension is drawn based on ®GoogleEarth satelital images. Arrow: Antarctic Circumpolar Current (ACC); thickness represent the strength of the ACC (Roberts et al. 2017). Dotted blue lines: Antarctic Polar Front (APF; mean position of the Polar Front is represented) and sub-Antarctic Front (SAF) (Roberts et al. 2017). The five sampling areas where Antarctic seaweeds were sampled are noted with back points (more details are given in the text and in Figs. 6.3 and 6.4)

taxa (but see chapter by Oliveira et al. 2020 in this volume), especially between Antarctic and sub-Antarctic provinces (Clarke et al. 2005; González-Wevar et al. 2012, 2017; Poulin et al. 2014, Billard et al. 2015; Crame 2018; Halanych and Mahon 2018). Non-endemic Antarctic seaweeds have classically included (1) cosmopolitan species reported on most Antarctic and sub-Antarctic/temperate coasts of the Southern Ocean (e.g. *Plocamium cartilagineum*), (2) cold-water species with disjoint amphiequatorial distribution (e.g. *Acrosiphonia arcta* and *Desmarestia viridis/confervoides*) and (3) more broadly distributed species also occurring in peri- and sub-Antarctic areas of the Southern Ocean as South Georgia, South Sandwich Islands or even the southern tip of South America (e.g. *Iridaea cordata*, *Gigartina skottsbergii* and *Adenocystis utricularis*) (Wiencke and Clayton 2002; Wiencke et al. 2014).

Nevertheless, reconstruction of these Southern Ocean macroalgae species geographic distribution maps relies solely on classical taxonomy. Macroalgal anatomy and high degree of phenotypic plasticity could lead to incorrect taxonomic classification of specimens and molecular tools are now recognized as essential for species determination in these organisms (Saunders 2005, 2008). Inaccurate morphological identification and existence of cryptic species have been commonly reported in green, red and brown macroalgae, including in Antarctica (Hommersand et al. 2009; Moniz et al. 2012; Billard et al. 2015; Pellizzari et al. 2017; Dubrasquet et al. 2018; Ocaranza-Barrera et al. 2019).

If few molecular studies have indeed confirmed the existence of macroalgal species characterized by vast disjoint distributions that include Antarctica (i.e. *Acrosiphonia* and *Desmarestia*; Olsen et al. 1993), most genetic data available support the existence of endemic Antarctic cryptic species. For example, two cryptic species, one Antarctic and one sub-Antarctic, were detected in *G. skottsbergii* and *I. cordata* (Hommersand et al. 2009; Billard et al. 2015; Ocaranza-Barrera et al. 2019). In the same way, even if exhaustive phylogenetic studies have yet to be undertaken for *P. cartilagineum*, Hommersand et al. (2009) noted that 'unpublished rbcL sequence analyses by S. Fredericq show that *P. "cartilagineum"* from Antarctica is distinct from all other species of *Plocamium* investigated so far'. In *G. skottsbergii*, the two cryptic species are clearly separated by the APF (Fig. 6.1) with the sub-Antarctic species distributed in South America along the coasts of Chile and Argentina and the Falkland Islands, while the Antarctic species is found in the Antarctic Peninsula, the South Shetland Islands and the South Orkney Islands (Billard et al. 2015). Divergence time estimations between Antarctic and sub-Antarctic cryptic macroalgal species indicate a split between two lineages occurring at the end of the Miocene, some 10–5 Ma ago (Hommersand et al. 2009; Billard et al. 2015; Ocaranza-Barrera et al. 2019), long after the physical fragmentation of the Southern Ocean continental landmasses or even to the formation of the ACC (Crame 1999; Mackensen 2004). However, the time of divergence between Antarctic and sub-Antarctic lineages of macroalgae predates the Quaternary glacial cycles (~2 Ma).

The existence of macroalgal species (both nominal species and cryptic species detected only when using molecular tools), entirely restricted to the Antarctic shelf, or at most including some adjacent islands of the maritime Antarctic (i.e., South Shetland Islands, South Orkney Islands and Balleny Islands) and offshore peri-Antarctic islands located south of the APF (i.e., South Georgia and Sandwich Islands), evidenced the long evolutionary history of these organisms within the Antarctic waters. In contrast to the hypothesis that most of the Antarctic biota could be recent colonists, arrived after the deglaciation of the Last Glacial Maximum (LGM) ice sheets (Clarke and Crame 1992), various Antarctic marine species, including macroalgae, have withstood Quaternary glacial cycles in situ (Convey et al. 2008, 2009; Allcock and Strugnell 2012).

6.2 Antarctic Marine Macroalgae: Surviving Quaternary Glacial Cycles in Situ

The onset of glaciations in Antarctica began during the Eocene-Oligocene boundary, and after a slightly warmer stage (27–15 Ma) that seemed to have reduced the extent of Antarctic ice, gradual cooling during the Miocene-Pliocene transition led to the re-establishment of major ice sheets covering Antarctica (Kennett 1977; Zachos et al. 2001). Later on, the Quaternary (2 Ma to 20 Ka) was characterized by the alternation between glacial and interglacial periods that greatly affected the seasonality and intensity of sea ice formation in the region (Clarke and Crame 1992; Barker and Thomas 2004; Gersonde et al. 2005; Kemp et al. 2010). During the LGM, ice sheet expansions were accompanied by eustatic drop in sea level (~120 m lower than at present), as well as important latitudinal changes in the position of the ACC (Fig. 6.1; McCulloch et al. 2000; Simms et al. 2011; Larter et al. 2014; O'Cofaigh et al. 2014; Roberts et al. 2017; White et al. 2018).

The development of large ice sheets generated major landscape and seascape changes in Antarctica (Zachos et al. 2001; Thatje et al. 2005; Davies et al. 2012). For instance, glacial reconstructions of the Antarctic Peninsula during the LGM (20 Ka) showed stable grounding ice sheets advancing up to the continental shelf edge (Larter et al. 2014; O'Cofaigh et al. 2014) (Fig. 6.1). Major glacial events in Antarctica are thought to have led to mass extinctions, with ice scouring pushing near-shore marine communities down the steep shelf slopes surrounding the Antarctic continent (Clarke and Crame 1992; Thatje et al. 2005). Antarctic benthic communities, especially sessile organisms with narrow depth distribution range, as macroalgae, would have been remarkably vulnerable to continental sea ice advances and retreats (Thatje et al. 2005; Dambach et al. 2012). Nevertheless, it seems that grounded ice did not completely cover the Antarctic continental shelf, not even during the glacial maxima (Klages et al. 2017), and some isolated areas could have acted as in situ glacial refugia. A wealth of evidences support this view, including genetic data obtained for various terrestrial and marine organisms (Convey et al. 2008, 2009; Allcock and Strugnell 2012), fossil evidences (Hiller et al. 1988) and glaciological and oceanographic reconstructions (Paillard and Parrenin 2004; Bentley et al. 2014; Larter et al. 2014; O'Cofaigh et al. 2014; Klages et al. 2017).

Several mechanisms could have sustained Antarctic shelf areas uncovered by ice during the Quaternary glacial events. First, areas of open water (polynyas) formed along Antarctic continental margins could have acted as in situ marine refugia (Thatje et al. 2008). The possible existence of much more saline waters surrounding Antarctica during glacial periods could have produced more stable and widespread polynyas than previously postulated (Paillard and Parrenin 2004; Thatje et al. 2008). Alternatively, in situ refugia could also be associated to volcanoes or areas of geothermal activities (Fraser et al. 2014). Such geothermal refugia could be very important in areas where clusters of long-lived volcanoes are recorded such as the tip of the Antarctic Peninsula, the South Shetland Islands or the Ross Sea (Fraser et al. 2012, 2014) (see Chaps. 3 and 4 in this volume). Moreover, geological evidence supports the diachrony of ice sheet extensions around Antarctica (Anderson et al. 2002) and near-shore marine organisms could ultimately have survived, hopping from one open ice-free continental shelf area to another during glacial periods (Thatje et al. 2008; Allcock and Strugnell 2012; Hughes et al. 2013).

At last, Antarctic near-shore macroalgae could also have endured the Quaternary glacial events south of the APF but beyond the Antarctic continental shelf margins. The coasts of peri-Antarctic islands have been postulated as potential glacial refugia for some species since their distribution generally includes islands of the Scotia Arc such as South Georgia and the South Sandwich Islands and Balleny Islands (Wiencke and Clayton 2002; Wiencke et al. 2014). South Georgia represents the northern limit of distribution for many species of the Antarctic benthos (Barnes et al. 2006), and this area was reported as glacial refugium for some Antarctic marine invertebrates (e.g. the Antarctic limpet *Nacella concinna*: González-Wevar et al. 2013).

6.3 Persistence in Multiple Isolated Glacial Refugia Versus a Single Antarctic Refugium

Quaternary climatic oscillations dramatically affected species geographic range and demography, especially at high latitude (Hewitt 2000, 2004; Maggs et al. 2008; Provan and Bennett 2008; Allcock and Strugnell 2012; Fraser et al. 2012). The expansion-contraction model (Provan and Bennett 2008) proposed an alternation between the contraction of species distribution ranges during glacial advances and subsequent rapid expansions during interglacial periods. These expansion/contraction cycles had a strong impact on the distribution of genetic variation of high latitude organisms and can be detected, nowadays, using molecular tools and phylogeographic analyses (Hewitt 2000, 2004; Provan and Bennett 2008; Allcock and Strugnell 2012). Depending on the number and localization of glacial refugia, different evolutionary scenarios can arise (Fig. 6.2).

For species that have survived glacial events in a unique refugium from which they recolonized their current distribution area, a very low level of genetic diversity and a high spatial genetic homogeneity are expected (see scenario 1 in Fig. 6.2). In this case, the effect of genetic drift during glacial bottlenecks could be amplified by gene surfing at expanding frontiers leading to extreme erosion of genetic diversity and even complete fixation of pioneer alleles over huge areas (Excoffier and Ray 2008; Excoffier et al. 2009; Hallatschek and Nelson 2010). Supporting this idea, complete genetic fixation has been recorded over distance of hundreds or even thousands of kilometres for some temperate Southern Hemisphere macroalgae, with the same mitochondrial haplotype covering all regions of the species distribution previously scoured by ice during glacial maxima (e.g. along the southern coast of Chile: Fraser et al. 2009; Montecinos et al. 2012).

In contrast, population fragmentation in multiple disjoint glacial refugia (see scenario 2 in Fig. 6.2) could lead to repeated events of divergence and diversification (Clarke and Crame 1992; Allcock and Strugnell 2012). Indeed, small populations could rapidly diverge due to strong genetic drift and, in addition, divergence could be enhanced due to selection if environmental differences exist between refugia. Divergence during isolation could be sufficient to generate reproductive barrier and lead to speciation (Wilson et al. 2009; Allcock and Strugnell 2012; Lecointre et al. 2013). This phenomenon, classically known as the climate 'diversity pump' (Haffer 1969), has been proposed to be at the origin of recent evolutionary radiation events

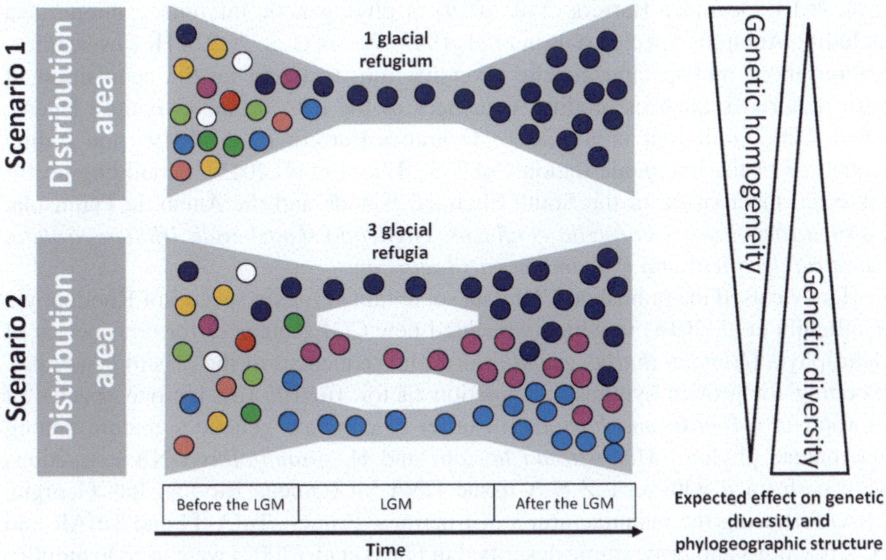

Fig. 6.2 Putative effects of genetic bottleneck due to intense ice scouring during glacial periods (i.e. the LGM) and postglacial range expansion following ice retreat yielding to the species current distribution. Two distinct scenarios have been considered: (1) the existence of only one in situ glacial refugium and (2) the existence of multiple (here three) in situ glacial refugia. Species distribution area is drawn in grey; each circle represents one individual; distinct colours represent genetic variants (e.g. distinct haplotypes). Effects of bottleneck and postglacial expansion on the current genetic diversity and phylogeographic structure potentially observed nowadays in species affected by the glacial cycles are given on the right

in Antarctica (Clarke and Crame 1989; Near et al. 2012; Crame 2018). Such process is expected to be more effective for species having limited dispersal capabilities such as those with non-pelagic development or very short free-living larval stage (Pearse et al. 2009). Indeed, highly differentiated genetic lineages/sister species have been detected in various Antarctic marine organisms characterized by low dispersal capacity and related to a scenario of isolation in multiple refugia during glacial events (Wilson et al. 2009; Allcock and Strugnell 2012; Hemery et al. 2012). In macroalgae, population fragmentation in multiple refugia has been commonly reported in the North East Atlantic where genetic distinctiveness characterizes patchy populations located at low latitude and inhabiting former refugial areas (Neiva et al. 2016).

6.4 Antarctic Macroalgae Genetic Diversity: COI and TufA Sequences Data Sets

Little is known about Antarctic macroalgae genetic diversity and most of the studies have used molecular markers primarily for species identification (Hommersand et al. 2009; Moniz et al. 2012; Mystikou et al. 2014; Pellizzari et al. 2017; Dubrasquet et al. 2018; Ocaranza-Barrera et al. 2019) or phylogenetic inferences among taxa including Antarctic species (Olsen et al. 1993; Peters et al. 2000). However, information about intraspecific genetic diversity, mostly based on the acquisition of mitochondrial sequences (coding sequences of the cytochrome c oxidase I gene, noted COI, Guillemin et al. 2018; Ocaranza-Barrera et al. 2019; non-coding sequences of the intergenic region Cox2–3, Billard et al. 2015), is building up for some red macroalgae in the South Shetland Islands and the Antarctic Peninsula: *Curdiea racovitzae, Georgiella confluens, Gigartina skottsbergii, Iridaea cordata, Palmaria decipiens* and *Plocamium cartilagineum*.

Here we used the published COI data set obtained for six species of Rhodophyta (Guillemin et al. 2018) to which we added new COI sequences for one species of Ochrophyta (*Himantothallus grandifolius*) and sequences of the plastid gene tufA, encoding for protein synthesis elongation factor Tu (EF-Tu), for one species of Clorophyta (*Monostroma hariotii*) in order to compare genetic signature among macroalgae phylum. *Monostroma hariotii* and H. *grandifolius* DNA extractions were performed with an E.Z.N.A tissue DNA kit (Omega Bio-tek, Inc. Georgia, USA) following the manufacturer's instructions. Primers TufAgf4 and TufAR and PCR reactions and programme described in Famà et al. (2002) were used to amplify tufA fragments in *M. hariotii*, while a fragment of the COI gene was amplified in *H. grandifolius* using the primers GazF2 and GazR2 and methodology described in Lane et al. (2007). PCR products were purified using the commercial kit UltraCleanTM (MO BIO Laboratories, Carlsbad, USA) and sequenced in both directions at the AUSTRAL-omics Core-Facility (Universidad Austral de Chile, Chile).

Sequences of all macroalgae except *H. grandifolius* were obtained from five sampling localities: two located in the South Shetland Islands (in Greenwich and King George Island), two along the Northern part of the West Antarctic Peninsula (WAP) near the Chilean O'Higgins Antarctic base and in Paradise Bay and one along the central part of the WAP (Marguerite Bay) (Figs. 6.1, 6.3, and 6.4). *Himantothallus grandifolius* COI data set includes 17 new sequences obtained from Greenwich Island, King George Island, O'Higgins and Paradise Bay (GENBANK N° MK503231–MK503247) and two sequences available in GENBANK (HE866784, King George Island, Yang et al. 2014; GQ368262, Terre Adelie, Silberfeld et al. 2010). *Monostroma hariotii* tufA sequences correspond to GENBANK N° MK507414–MK507450. For each species-specific data set, we calculated five diversity indices, Tajima's D (Tajima 1989) and Fu's Fs (Fu 1997) statistics and estimated the observed distributions of pairwise differences between sequences using Arlequin v 3.5 (Excoffier and Lisher 2010). For Tajima's D and Fu's Fs, significant departure from mutation drift equilibrium was tested using 1000 bootstrap replicates in Arlequin (Excoffier and Lisher 2010). Observed distributions of pairwise differences between sequences were compared to estimated values under a model of sudden demographic expansion through a generalized least squares approach and goodness of fit was tested using 1000 permutations in Arlequin (Excoffier and Lisher 2010). Haplotype networks were reconstructed using NETWORK v 4.510 (Bandelt et al. 1999).

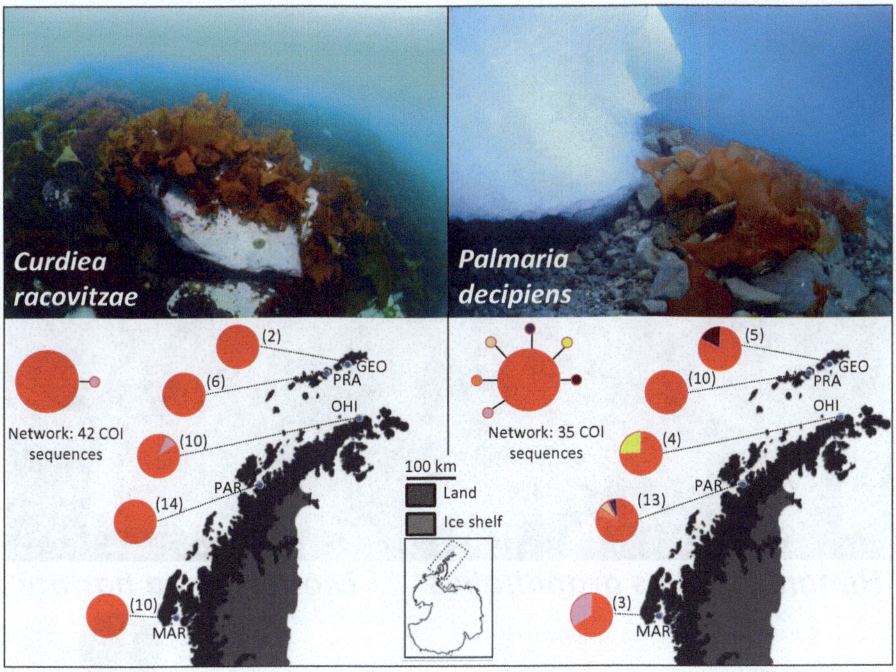

Fig. 6.3 Haplotype networks and pie charts showing the geographical distribution of haplotypes for the mitochondrial genetic marker COI (632 bp) in *Curdiea racovitzae* and *Palmaria decipiens*. In the networks, each circle represents a haplotype and its size is proportional to the frequency in which the haplotype was encountered, and black lines correspond to one mutational step. The five localities correspond, from north to south, to King George Island (GEO), Greenwich Island (PRA), O'Higgins (OHI), Paradise Bay (PAR) and Marguerite Bay (MAR). The number of sequenced individuals is given between brackets. From the six species of red algae studied (Guillemin et al. 2018), only results obtained for *C. racovitzae* and *P. decipiens* are illustrated since these two species represent the less and the most diverse Rhodophyta sampled along the coasts of the Antarctic Peninsula and the South Shetland Islands, respectively. Underwater photographs of specimen's characteristic of both species were taken by I. Garrido; please note the impact of ice scour on nearby sea bed *P. decipiens* populations. (King George Island)

6.5 Brown, Red and Green Macroalgae: Sharing a Common Pattern of Glacial Impact and Postglacial Populations Recovery?

All the analysed macroalgae are non-buoyant and commonly found in the Antarctic waters. Nevertheless, they present some noticeable ecological differences. While *I. cordata*, *G. skottsbergii*, *C. racovitzae* and *P. decipiens* are mostly found in the intertidal down to the shallow subtidal, *G. confluens* and *P. cartilagineum* are generally found deeper in the subtidal, as understory of large brown macroalgae. *Monostroma hariotii* presents very thin, delicate thalli, is very common in both the intertidal and subtidal zones and is considered a pioneer species able to colonize

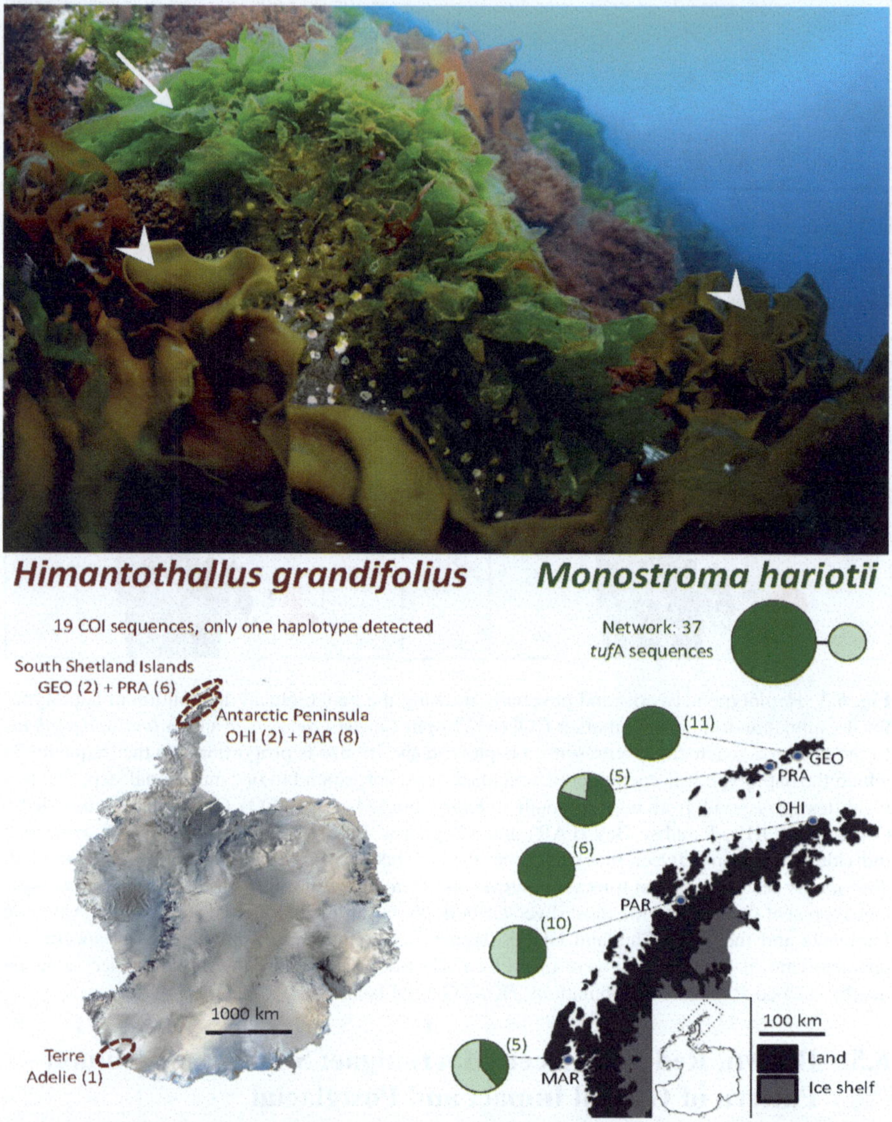

Fig. 6.4 Haplotype network and pie charts showing the geographical distribution of haplotypes for the plastid genetic marker tufA (772 bp) in *Monostroma hariotii* and distribution of the unique haplotype detected in the 19 COI sequences (mitochondrial marker, 619 bp, include HE866784 from King George Island and GQ368262 from Terre Adelie already published in GENBANK) in *Himantothallus grandifolius*. In the network, each circle represents a haplotype and its size is proportional to the frequency in which the haplotype was encountered; black line corresponds to one mutational step. The five localities correspond, from north to south, to King George Island (GEO), Greenwich Island (PRA), O'Higgins (OHI), Paradise Bay (PAR) and Marguerite Bay (MAR). The number of sequenced individuals is given between brackets. An underwater photograph of a mixed high subtidal bed of *M. hariotii* (arrow) and *H. grandifolius* (arrow heads) is given on the top. (King George Island) (Photo by Ignacio Garrido)

areas with freshwater influence from glacier run-off, presenting heavy ice scour or recently available for colonization after glacier retreat (Quartino et al. 2013; Wiencke et al. 2014). With thalli more than ten meters long, *H. grandifolius* is the largest seaweed reported in Antarctica. The species, in combination with other brown macroalgae of the genus *Desmarestia*, form dense beds of canopy-forming algae, dominating the subtidal rocky shores.

6.5.1 Signature of a Drastic Impact of the Last Glacial Maximum

Regardless of their ecological differences or classification into distinct taxonomic divisions, molecular data show that all eight macroalgae have been strongly impacted by glacial events (Table 6.1, Figs. 6.3 and 6.4). Extremely low genetic diversity was observed at the intraspecific level with a number of haplotypes varying between one in *H. grandifolius* up to only seven in the case of *I. cordata* and *P. decipiens* (Table 6.1). These results could be related to historical changes in macroalgal population size in the study area. The complete lack of diversity observed for *H. grandifolius* does not allow to further test this hypothesis; however, the seven other species presented additional results that support this idea: a starlike topology was observed in haplotype networks (Figs. 6.3 and 6.4), mismatch distributions were unimodal (hypothesis of sudden expansion could not be rejected when calculated for the sum of squared deviation, SSD, or Harpending's raggedness index, Rag, except for *M. hariotii* Rag for which a $p < 0.05$ was obtained; Table 6.1; see also Guillemin et al. 2008 for more details on red algae species) and all but the green alga *M. hariotii* presented negative values of Tajima's D and Fu's Fs statistics (Table 6.1).

None to very few mutations have been accumulated in the populations since expansion from their glacial refugium, strengthening the idea that the demographic expansion occurred recently. We proposed (as in Billard et al. 2015; Guillemin et al. 2018; estimations based on the calculation of the parameter Tau from models of sudden expansion and previously published mutation rate per sequence per generation for mitochondrial genes in macroalgae) that the demographic bottleneck corresponds to population contraction of macroalgae in refugium during the LGM and that recolonization of the coast begun some 18,000 years ago, at most, a date congruent with time of deglaciation in the area (Simms et al. 2011; O'Cofaigh et al. 2014). Similarly, other studies have related the recent expansion detected in the southern most populations of sub-Antarctic macroalgae to population recolonizing areas previously covered by ice after the LGM (Fraser et al. 2009, 2012, 2013; Montecinos et al. 2012; Billard et al. 2015).

The level of genetic diversity estimated for Antarctic macroalgae is within the lowest reported for Antarctic marine species with broad distribution, with values of haplotype diversity (H) ranging between 0.000 and 0.398 and nucleotide diversity

Table 6.1 Genetic diversity indices and neutrality test in six red algae, one brown alga and one green alga sampled along the Antarctic Peninsula and South Shetland Islands. Sequences used for analyses of all six species of Rhodophyta and *Himantothallus grandifolius* correspond to the mitochondrial gene COI, while the plastid gene *tuf*A was used for analyses of *Monostroma hariotii*

Species	*N*	*k*	*S*	*H*	Π	π ($\times 10^2$)	Tajima's D	Fu's FS	Mismatch distribution
Rhodophyta									
Curdiea racovitzae	42	2	1	0.048	0.048	0.008	−1.120*	−1.491 ns	Unimodal[a]
Georgiella confluens	20	4	4	0.363	0.489	0.077	−1.638*	−1.613 ns	Unimodal[a]
Gigartina skottsbergii	28	2	1	0.071	0.071	0.011	−1.151 ns	−1.155 ns	Unimodal[a]
Iridaea cordata	90	7	5	0.398	0.623	0.102	−0.797 ns	−2.882 ns	Unimodal[a]
Palmaria decipiens	35	7	6	0.318	0.343	0.054	−2.103***	−7.041***	Unimodal[a]
Plocamium cartilagineum	64	4	4	0.122	0.155	0.025	−1.759***	−3.466**	Unimodal[a]
Ochrophyta									
Himantothallus grandifolius	19[b]	1	0	0.000	0.000	0.000			
Clorophyta									
Monostroma hariotii	37[c]	2	1	0.378	0.378	0.049	0.846[ns]	1.223[ns]	Unimodal[d]

N, number of sampled specimens; *k*, number of haplotypes; *S*, polymorphic sites; *H*, haplotype diversity; Π, average number of nucleotide difference; π, nucleotide diversity; *ns*, non-significant. All results obtained for Rhodophyta were taken from Guillemin et al. (2018). COI sequences are of 632 base pair (bp)

*$p < 0.05$

**$p < 0.01$

***$p < 0.001$

[a]Hypothesis of sudden expansion could not be rejected when calculated for the sum of squared deviation or Harpending's raggedness index (i.e. 0.000 < SSD < 0.007; 0.179 < Rag < 0.821; all $p > 0.05$). See Guillemin et al. (2018) for more detail

[b]Include two sequences available in GENBANK: HE866784 from King George Island (Yang et al. 2014) and GQ368262 from Terre Adelie (Silberfeld et al. 2010). All COI sequences are of 619 bp

[c]Sequences for the plastid gene *tuf*A were all of 772 bp, except for the individual GGVMLG0575, which was 52 bp shorter. Since the 3′ part of the *tuf*A gene was not variable in our whole data set, GGVMLG0575 was completed to 772 bp as in the other individuals before analyses

[d]Goodness-of-fit tests for a model of sudden expansion calculated in ARLEQUIN v.3.5 for the sum of squared deviation (SSD = 0.005, p = 0.0001) and for Harpending's raggedness index (Rag = 0.202, p = 0.459)

($\pi \times 10^2$) ranging between 0.000 and 0.102 (lowest values in *H. grandifolius* and highest values in *I. cordata*; Table 6.1). Rather low level of diversity and complete lack of genetic structure have also been reported in some widespread Antarctic marine invertebrates as *Sterechinus neumayeri* (H = 0.257; $\pi \times 10^2$ = 0.036; Díaz et al. 2018), *Chorismus antarcticus* (H = 0.639; $\pi \times 10^2$ = 0.209; Raupach et al.

2010), *Parborlasia corrugatus* ($H = 0.762$; $\pi \times 10^2 = 0.118$; Thornhill et al. 2008), *Euphausia superba* ($H = 0.856$; $\pi \times 10^2 = 1.394$; Bortolotto et al. 2011), *Nematocarcinus lanceopes* ($H = 0.902$; $\pi \times 10^2 = 0.568$; Raupach et al. 2010) and *Nymphon australe* ($H = 0.918$; $\pi \times 10^2 = 0.657$; Soler-Membrives et al. 2017). The lowest levels of genetic diversity have been observed in benthic marine invertebrates with distribution restricted to shallow waters of the Antarctic shelf and have been related to the extreme effect of ice scouring in these organisms, not able to migrate to deeper waters during glacial maxima (Raupach et al. 2010; Díaz et al. 2018). It is then not surprising to observe genetic diversity congruent with extreme events of mortality due to LGM ice impact in photosynthetic macroalgae, these organisms being characterized by one of the most restricted bathymetry in the marine realm.

Indeed, Antarctic macroalgae are anchored to the sea bottom and cannot survive under ice, without direct sunlight, for long periods of time. Glacial refugia for these organisms should then be limited in depth (generally dense beds of Antarctic macroalgae are reported growing down to 40–50 m) and to areas where the sea ice melt during summer. However, periods of ice-free summer as short as 2 months allow macroalgal bed growth (Norkko et al. 2004) and living specimens of various macroalgae species were retrieved in Terre Adelie under ice sheets in an area that did not melt during summer for 2 years in a row (S. Hourdez pers. com., specimens kept in the Museum of Natural History, Paris, France). This time period could, however, be considered as a maximum before severe population decline. Indeed, none of the fleshy macroalgae were considered as healthy (but half the encrusting coralline algae were still pigmented) after 3 years of permanent sea ice coverage in the Commonwealth Bay, a polynya affected by the grounding of a huge drifting iceberg in 2010 (Clark et al. 2015). Moreover, Antarctic macroalgae are extremely shade adapted and have been observed to grow at depth of 90 m (Wiencke and Clayton 2002; Wiencke et al. 2007). In other regions of the world, the existence of kelp beds reaching depth of 200 m was even predicted using oceanographic and ecophysiological models (Graham et al. 2007; see also Spalding et al. 2019 for reports of macroalgae growing down to 140 m). Additionally, long-term survival of Antarctic macroalgae in areas highly impacted by ice could be facilitated by the synthesis and accumulation of protective compounds (Wiencke et al. 2007). Some species, such as the green algae *Acrosiphonia arcta* and *Prasiola* spp., have developed remarkable adaptations to cope with ice disturbance characteristic of Antarctic habitats. Indeed, both synthesize cryoprotectants that prevent damage by ice to membranes or enzymes and *Prasiola* has been reported to photosynthesize at temperatures as low as −15 °C (Jacob et al. 1991; Raymond and Fritsen 2001; Wiencke et al. 2007). Antarctic macroalgae could then be tougher than expected and may have survived in small areas highly impacted by ice during the LGM.

6.5.2 One Refugium to Rule Them all

Classically, locations of glacial refugia have been inferred by comparing genetic diversity between populations located within areas potentially free of ice during the LGM and the ones located in recently colonized areas, previously covered by ice (Hewitt 2000, 2004; Maggs et al. 2008; Provan and Bennett 2008). Since population effective size is more stable in glacial refugia than in recently colonized areas, a much higher genetic diversity is expected to characterize the former than the later (Hewitt 2000, 2004; Maggs et al. 2008; Provan and Bennett 2008). If the particularly low genetic variability observed in our species lead to the idea of one unique refugium in Antarctica, it frustratingly does not allow pinpointing its location. Indeed, whatever the species under study, none of our sampled populations shows a specifically high genetic diversity (Figs. 6.3 and 6.4; see also Guillemin et al. 2018 for more details on distribution of genetic diversity of *I. cordata*, *G. confluens*, G. *skottsbergii* and *P. cartilagineum*).

Various areas could be proposed as in situ refugia along the coast of Antarctica for our seaweeds and these include localities close to active volcanoes (as Deception Island, Penguin Island and Bridgeman Island in the South Shetland Islands or the coast of the Western Ross Sea; Fraser et al. 2014) or localities where the shelf seafloor was free of grounded ice during glacial maxima (i.e. Alexander Island, Eastern Amundsen Sea Embayment, Western Ross Sea, George V Land and Prydz Bay; Klages et al. 2017). Moreover, all eight macroalgal distributions include offshore peri-Antarctic Islands located south of the APF, such as the South Orkney Islands, South Sandwich Islands and South Georgia (Wiencke and Clayton 2002; Wiencke et al. 2014; Griffiths and Waller 2016). We then cannot rule out a recolonization of the Antarctic coasts from these more northern areas. Supporting this idea, a study based on molecular markers allowed the recent discovery of one cryptic species of the brown alga *Adenocystis utricularis* endemic to South Georgia (Fraser et al. 2013). The authors related this result to a persistence of *Adenocystis* in local refugium along the island coasts during the LGM (Fraser et al. 2013). Antarctic samples were not included in the study of Fraser et al. (2013) and the level of genetic divergence between peri-Antarctic and Antarctic *Adenocystis* remains to be estimated in order to test for a possible recolonization of the Antarctic shelf from a South Georgia glacial refugium. However, ice coverage during the LGM strongly impacted South Georgia (White et al. 2018) and coastal glacial refugia could have been as scarce in the offshore peri-Antarctic Islands as at the margins of the Antarctic shelf itself (Fig. 6.1). Indeed, the complete eradication of populations of another brown alga, the intertidal/shallow subtidal *Durvillaea antarctica*, from the coasts of most peri-Antarctic islands (in particular in Macquarie Island, Falkland Islands, South Georgia, Marion Island and Kerguelen Islands) during the LGM has been established (Fraser et al. 2009, 2012).

Unfortunately, in order to locate glacial refugia in the region, we will need more variable genetic markers and a much better sampling of the coasts of the Antarctic shelf (i.e. Alexander Island, Western Ross Sea, George V Land and Prydz Bay; see

Griffiths and Waller 2016), adjacent islands of the maritime Antarctic and offshore peri-Antarctic Islands south of the APF where populations of macroalgae have been reported (in particular, South Orkney Islands, South Sandwich Islands and South Georgia, Bouvet Island, Heard Island and Balleny Islands). Sampling localities that could have been less affected by ice scour during the Last Glacial Maximum should be a priority.

The proposed scenario of one unique in situ refugium in Antarctica contrasts with history of glacial perturbations in Arctic macroalgae. Indeed, based on patterns of genetic diversity and distribution of private haplotypes, the existence of various marine glacial refugia has been proposed in the Northern Hemisphere for these organisms (Maggs et al. 2008; Provan and Bennett 2008; Neiva et al. 2016). Less severe cooling in the Arctic than the Antarctic could explain, in part, these differences between hemispheres (Pointing et al. 2015). Moreover, Arctic macroalgae distribution during glacial cycles could have shifted along continuous coastlines spanning a huge latitudinal gradient (Fraser et al. 2012; Neiva et al. 2016), while no such opportunity was available for Antarctic species, bounded by the deep water of the Southern Ocean, leading to a more thorough extinction along the Antarctic coastline during the LGM.

6.5.3 Postglacial Recolonization: Widespread Haplotypes Drifting Around Antarctica?

Apart from their very low genetic diversity, all eight algae under study present a common striking characteristic: one or a few haplotypes were encountered in all localities even when situated hundreds or even thousands of kilometres apart (Figs. 6.3 and 6.4). For all six red algae and for the green alga *M. hariotii*, the most common haplotype was observed from King George Island in the South Shetland Islands down to Marguerite Bay, a sampling point located along the central part of the WAP (Figs. 6.3 and 6.4). These common haplotypes are spread over more than 450 km of coast and within two distinct biogeographic subregions (the South Shetland Islands and the WAP; Linse et al. 2006; Spalding et al. 2007; Terauds et al. 2012). Shared haplotypes have been observed even between the WAP and South Orkney Island, reaching distances greater than 1600 km, in the red alga *G. skottsbergii* (same Cox2–3 haplotype from Marguerite Bay to the South Orkney Islands; Billard et al. 2015). The same pattern is observed in *I. cordata* for which the same COI haplotype has also been sequenced along the WAP and South Orkney Island (Guillemin M-L. unpublished data, no genetic differences detected between the most common haplotype encountered along the WAP and three individuals from South Orkney Island sequenced for 632 pb). For *H. grandifolius*, the distribution of the unique COI haplotype encountered in our study zone could be extended to Terre Adelie, in a sampling site located almost on the other side of the Antarctic continent (sequence from Silberfeld et al. 2010). *Himantothallus grandifolius* is then the first register of a non-buoyant species displaying a true Antarctic circumpolar distribution supported by genetic data. Complete circum-Antarctic distribution has been described for other seaweeds, including common species of brown (*Desmarestia menziesii*), green (*M. hariotii* and *Urospora penicilliformis*) and red (*I. cordata*, P. *cartilagineum*, *P. decipiens*, *Phyllophora antarctica* and *Phycodrys antarctica*) macroalgae (Wiencke et al. 2014). However, for these species, all sequences available to date have been obtained for specimens located in the South Shetland Islands and/or the WAP (Olsen et al. 1993; Peters et al. 2000; Hommersand et al. 2009; Dubrasquet et al. 2018; Ocaranza-Barrera et al. 2019). One could thus wonder if other common Antarctic macroalgae species will present the same pattern as *H. grandifolius* and samples should be gathered from the Western Ross Sea, Terre Adelie and Prydz Bay and sequenced in order to test this hypothesis.

In contrast to what was reported for pelagic species or benthic species exhibiting extensive dispersal by planktonic stage (Lange et al. 2002; Raupach et al. 2010; Sromek et al. 2015; Soler-Membrives et al. 2017; Caccavo et al. 2018), we believe that the presence of widespread haplotypes in our data sets clearly reflects Antarctic macroalgae past glacial demographic histories but not necessarily contemporary high gene flow. Benthic macroalgae free-living stages (i.e. for all species except buoyant ones, Macaya et al. 2016, also see chapter by Macaya et al. 2020 in this volume) are generally extremely restricted to short-lived gametes and spores that quickly sink a few meters, at most, from thalli of origin (Valero et al. 2011). However, in these organisms, large-scale spread of haplotypes was commonly related to strong genetic drift and allelic surfing at species' leading edges during postglacial range expansions or after transoceanic colonization (Maggs et al. 2008; Provan and Bennett 2008; Montecinos et al. 2012; Fraser et al. 2013; Guillemin et al. 2014; Neiva et al. 2016). Non-buoyant macroalgal species, for which both population genetic and phylogeographic data sets are available, show well this discrepancy with, on the one hand, the presence of haplotypes distributed over thousands of kilometres in area of recent colonization and, on the other hand, current reproduction and recruitment occurring at a very local scale (in *Agarophyton* [former *Gracilaria*]: Guillemin et al. 2008, 2014, in *Fucus*: Neiva et al. 2016; Jueterbock et al. 2018). Exceptional events of thalli rafting on strong oceanic currents and colonizing new areas when depleted of abundant algal coverage (Waters et al. 2013) have been advanced to explain these patterns (Montecinos et al. 2012; Fraser et al. 2013; Guillemin et al. 2014; Neiva et al. 2016). In order to test for actual gene flow between studied populations of macroalgae, more variable genetic markers, such as microsatellites or SNPs (i.e. single nucleotide polymorphism), will be needed. Indeed, highly variable microsatellite loci revealed a clear regional substructure of populations in *Phaeocystis antarctica*, a brown micro algal planktonic species for which no clear phylogeographic patterns were described before around Antarctica using nuclear and plastid sequences (Lange et al. 2002; Gäbler-Schwarz et al. 2015).

Even if none of the study models possess floating structures, large amounts of all eight species can easily be found cast ashore after storm events, including huge adult specimens of *H. grandifolius* (authors pers. obs.). Passive transport of detached fronds could therefore play a major role in Antarctic non-buoyant macroalgae long-distance dispersal, a mechanism most probably facilitated by the strong oceanic currents in the region (Loeb 2007). Classically, the circumpolar current (ACC, see Fig. 6.1) was reported as a structuring force in the Southern Ocean and has been proposed to be the main connecting current between populations of marine organisms located around Antarctica (e.g. Fraser et al. 2009; Soler-Membrives et al. 2017). For the red algae *G. skottsbergii* and *I. cordata*, fronds rafting on the ACC could easily explain the spread of the most common haplotypes between localities of the South Shetland Islands, the Antarctic Peninsula and even South Orkney Island (Billard et al. 2015; Guillemin et al. 2018; Ocaranza-Barrera et al. 2019). Indeed, the ACC flows northeastward from the Bellingshausen Sea through the South Shetland Islands and the South Orkney Islands, before entering the Scotia Sea (Loeb 2007). However, since it deflects from the Antarctic coasts during most of its journey around the continent (from the Weddell Sea to Prydz Bay and the Western Ross Sea, see Loeb 2007), rafting on the ACC does not easily explain the presence of the same COI haplotype in Terre Adelie and the Antarctic Peninsula for *H. grandifolius*. It is highly probable that fronds transport between these two distant regions was promoted by the Antarctic Costal Current (ACoC), a current that flows in a counterclockwise motion and located near the coast of the Antarctic shelf (Loeb 2007). The importance of transport by the ACoC was also proposed to be at the origin of the widespread haplotypes observed in marine species with distribution restricted to the shallow waters of the shelf, as in the sea urchin *S. neumayeri* (Díaz et al. 2018). Remarkably, a study following the track of icebergs around Antarctica shows that a large majority do travel counterclockwise around the continental shelf and that the ACoC can very easily connect Terre Adelie with the tip of the Antarctic Peninsula (Stuart and Long 2011).

6.6 Concluding Remarks

Macroalgae are very important primary producers and ecosystem engineers in Antarctic coastal ecosystems, serving as food for herbivores and detritivores and providing habitat for many associated organisms (see Chaps. 15 and 16 of this book). Even if these species have developed remarkable adaptations to cope with the high level of disturbances characteristic of Antarctic swallow depth (Wiencke et al. 2007, 2014), they could be highly vulnerable to actual threats intensifying in the region as global warming, pollution and introduction of invasive species. As seawater temperature rise around Antarctica, a trend particularly noticeable along the Antarctic Peninsula, extent and duration of sea ice coverage decline and glaciers are in retreat (Cook et al. 2016). At first sight, this scenario seems highly favourable for Antarctic macroalgae that can rapidly colonize these newly deglaciated areas (Quartino et al. 2013; also see chapter by Quartino et al. 2020 in this volume). Indeed, in the Arctic, the long-term study of Kortsch et al. (2012) shows that decrease in sea ice cover can produce rapid shift in marine communities from benthic habitats dominated by calcareous algae and sea anemones to dense beds of fleshy, habitat-forming macroalgae. However, climatic changes are also associated with higher probability of ice sheet collapse and iceberg calving. Scouring by these drifting icebergs, which can even get stuck and ground in polynyas, may have a highly detrimental effect on macroalgae beds (Clark et al. 2015). Climate changes have also been related to possible alteration in large-scale ocean circulation and gyre and eddy kinetic in the Southern Ocean, leading to breaches in the dispersal barriers (APF and ACC) established since the Eocene/Oligocene in the region (Barker and Thomas 2004; Fraser et al. 2018). When coupled with the intensification of maritime transport (linked to both touristic and scientific activities) between Antarctic and sub-Antarctic provinces, these oceanic changes could lead to an increase in propagule pressure and introduction of sub-Antarctic macroalgae species in the Antarctic waters. To the date, only one green alga (i.e. *Ulva intestinalis*) has been reported as an established introduced alien species in the South Shetland Islands, probably arriving as fouling on the hull of a visiting vessel (Clayton et al. 1997). However, macroalgae highly successful in recolonising sub-Antarctic coasts after glacial retreat have recently been reported to recurrently reach the Antarctic coasts as cast ashore living thalli (i.e. *Durvillaea antarctica*, Fraser et al. 2018; also see chapter by Macaya et al. 2020 in this volume). The capacity of Antarctic macroalgae to withstand competition from sub-Antarctic invaders has not been studied yet. Nonetheless, Antarctic macroalgae are organisms highly adapted to the local stenothermal environment. Studies report temperature optima between −2 °C and 10 °C for most Antarctic species with limitation or even failure in growth or gametogenesis at temperature higher than 5 °C in *G. confluens*, *G. skottsbergii*, *P. cartilagineum* and *Desmarestia antarctica* (Wiencke et al. 2007; Wiencke et al. 2014). In the actual context of rapid temperature increase, populations of sub-Antarctic or even more temperate colonizers may then more easily settle along the Antarctic coasts.

Antarctic macroalgae were clearly able to cope with changes related to the cooling and then freezing of the Southern Ocean waters and have survived Quaternary glacial perturbations in situ. However, our study shows that maximal glacial events have led to mass extinction in all eight model species, whatever their taxonomic divisions. Will these species be able to keep pace with the current rapid environmental fluctuations? Genetic adaptation could be key in enhancing resistance and resilience to climatic changes and a high level of genetic diversity has classically been related to high population fitness, resilience and future adaptability (Reed and Frankham 2003; Jump and Peñuelas 2005). Due to strong bottleneck during the LGM, standing genetic diversity in Antarctic macroalgae populations seems to be extremely depleted and an adaptive response of Antarctic algae populations to future changes could be limited. Nevertheless, recent studies on the brown seaweed *Sargassum muticum* show a complete absence of genetic diversity in all populations of the invaded range when tested using mitochondrial sequences or even nuclear microsatellites and only a very low level of diversity when using more than 8000 SNPs (Cheang et al. 2010; Le Cam et al. 2019). Rapid and successful invasion of Europe and the USA was clearly not impaired by the extremely low genetic diversity in *S. muticum*. In order to predict the possible outcome of future climatic changes on macroalgae populations, it is now essential to use more variable genetic markers in order to detect a subtle genetic structure along the Antarctic coasts and point out restricted regional or local genotypes that could be of interest (see results using SNPs in Le Cam et al. 2019) and to experimentally test for existence and extent of phenotypic plasticity/acclimatization potential (a mechanism that can potentially buffer negative effects of climate change long enough to allow for population adaptation; Jump and Peñuelas 2005) and potential for an in situ adaptive response to climate change of these organisms.

Acknowledgments Funding by Instituto Antártico Chileno (INACH) T_16-11 and RG_15-16 and the Centro FONDAP IDEAL no. 15150003 is gratefully acknowledged. At King George Island, sampling of *Monostroma hariotii* and *Himantothallus grandifolius* was done with the help of the Project Anillo ART1101 from Conicyt-Chile. The authors thank P. Brunning, J. L. Kappes, T. Heran, Y. Henriquez and L. Vallejos for their help in the field. The authors would also like to thank the Chilean Navy (especially the captain and crew of the ships Almirante Oscar Viel and Lautaro), the staff from the Chilean Army in the O'Higgins base and the Air Force of Chile (FACh) for the logistic support of our fieldwork in Subantarctica and Antarctica.

References

Allcock AL, Strugnell JM (2012) Southern Ocean diversity: new paradigms from molecular ecology. Trends Ecol Evol 27:520–528

Anderson JB, Shipp SS, Lowe AL, Wellner JS, Mosola AB (2002) The Antarctic ice sheet during the last glacial maximum and its subsequent retreat history: a review. Quat Sci Rev 22:49–70

Aronson RB, Thatje S, Clarke A, Peck LS, Blake DB, Wilga CD, Seibel BA (2007) Climate change and invasibility of the Antarctic benthos. Annu Rev Ecol Evol Syst 38:129–154

Bandelt HJ, Forster P, Röhl A (1999) Median-Joining networks for inferring intraspecific phylogenies. Mol Biol Evol 16:37–48

Barker P, Thomas E (2004) Origin, signature and palaeoclimatic influence of the Antarctic Circumpolar Current. Earth Sci Rev 66:143–162

Barnes DK, Hodgson DA, Convey P, Allen CS, Clarke A (2006) Incursion and excursion of Antarctic biota: past, present and future. Glob Ecol Biogeogr 15:121–142

Bentley MJ, Ó Cofaigh CO, Anderson JB, Conway H, Davies B, Graham AG et al (2014) A community-based geological reconstruction of Antarctic ice sheet deglaciation since the Last Glacial Maximum. Quat Sci Rev 100:1–9

Billard E, Reyes J, Mansilla A, Faugeron S, Guillemin ML (2015) Deep genetic divergence between austral populations of the red alga *Gigartina skottsbergii* reveals a cryptic species endemic to the Antarctic continent. Polar Biol 38:2021–2034

Bortolotto E, Bucklin A, Mezzavilla M, Zane L, Patarnello T (2011) Gone with the currents: lack of genetic differentiation at the circum-continental scale in the Antarctic krill *Euphausia superba*. BMC Genet 12:32

Caccavo JA, Papetti C, Wetjen M, Knust R, Ashford JR, Zane L (2018) Along-shelf connectivity and circumpolar gene flow in Antarctic silverfish (*Pleuragramma antarctica*). Sci Rep 8:17856

Cheang CC, Chu KH, Fujita D, Yoshida G, Hiraoka M, Critchley A, Choi HG, Duan D, Serisawa Y, Ang PO Jr (2010) Low genetic variability of *Sargassum muticum* (Phaeophyceae) revealed by a global analysis of native and introduced populations. J Phycol 46:1063–1074

Clark GF, Marzinelli EM, Fogwill CJ, Turney CS, Johnston EL (2015) Effects of sea-ice cover on marine benthic communities: a natural experiment in Commonwealth Bay, East Antarctica. Polar Biol 38:1213–1222

Clarke A, Crame JA (1989) The origin of the Southern Ocean marine fauna. In: Crame JA (ed) Origins and evolution of the Antarctic biota, vol 47. The Geol Soc Spec Publ, London, pp 253–268

Clarke A, Crame JA (1992) The Southern Ocean benthic fauna and climate change: a historical perspective. Phil Trans R Soc Lond B 338:299–309

Clarke A, Barnes DKA, Hodgson DA (2005) How isolated is Antarctica? Trends Ecol Evol 20:1–3

Clayton MN, Wiencke C, Klöser H (1997) New records and sub-Antarctic marine benthic macroalgae from Antarctica. Polar Biol 17:141–149

Convey P, Gibson JA, Hillenbrand CD, Hodgson DA, Pugh PJ, Smellie JL, Stevens MI (2008) Antarctic terrestrial life–challenging the history of the frozen continent? Biol Rev 83:103–117

Convey P, Stevens M, Hodgson D, Smellie J, Hillenbrand C, Barnes D, Clarke A, Pugh P, Linse K, Cary S (2009) Exploring biological constraints on the glacial history of Antarctica. Quat Sci Rev 28:3035–3048

Cook AJ, Holland PR, Meredith MP, Murray T, Luckman A, Vaughan DG (2016) Ocean forcing of glacier retreat in the western Antarctic Peninsula. Science 353:283–286

Crame JA (1999) An evolutionary perspective on marine faunal connections between southernmost South America and Antarctica. Sci Mar 63:1–14

Crame JA (2018) Key stages in the evolution of the Antarctic marine fauna. J Biogeogr 45:986–994

Dalziel IWD, Lawver LA, Pearce JA, Barker PF, Hastie AR, Barfod DN, Schenke HW, Davis MB (2013) A potential barrier to deep Antarctic circumpolar flow until the late Miocene? Geology 41:947–950

Dambach J, Thatje S, Rödder D, Basher Z, Raupach MJ (2012) Effects of late-Cenozoic glaciation on habitat availability in Antarctic benthic shrimps (Crustacea: Decapoda: Caridea). PLoS One 7:e46283

Davies B, Hambrey M, Smellie J, Carrivick J, Glasser N (2012) Antarctic Peninsula ice sheet evolution during the Cenozoic Era. Quat Sci Rev 31:30–66

Dell RK (1972) Antarctic benthos. Adv Mar Sci 10:1–216

Díaz A, Gerard K, González-Wevar C, Maturana C, Féral JP, David B, Saucède T, Poulin E (2018) Genetic structure and demographic inference of the regular sea urchin *Sterechinus neumayeri* (Meissner, 1900) in the Southern Ocean: the role of the last glaciation. PLoS One 13:e0197611

Dubrasquet H, Reyes J, Sanchez RP, Valdivia N, Guillemin ML (2018) Molecular-assisted revision of red macroalgal diversity and distribution along the Western Antarctic Peninsula and South Shetland Islands. Cryptogamie Algol 39:409–430

Excoffier L, Lisher H (2010) Arlequin suite ver 3.5: a new series of programs to perform population genetics analyses under Linux and Windows. Mol Ecol Resour 10:56

Excoffier L, Ray N (2008) Surfing during population expansions promotes genetic revolutions and structuration. Trends Ecol Evol 23:347–351

Excoffier L, Foll M, Petit RJ (2009) Genetic consequences of range expansions. Annu Rev Ecol Evol Syst 40:481–501

Famà P, Wysor B, Kooistra WH, Zuccarello GC (2002) Molecular phylogeny of the genus *Caulerpa* (Caulerpales, Chlorophyta) inferred from chloroplast tufA gene. J Phycol 38:1040–1050

Fraser CI, Nikula R, Spencer HG, Waters JM (2009) Kelp genes reveal effects of Subantarctic sea ice during the Last Glacial Maximum. Proc Natl Acad Sci 106:3249–3253

Fraser CI, Nikula R, Ruzzante DE, Waters JM (2012) Poleward bound: biological impacts of Southern Hemisphere glaciation. Trends Ecol Evol 27:462–471

Fraser CI, Zuccarello GC, Spencer HG, Salvatore LC, Garcia GR, Waters JM (2013) Genetic affinities between trans-oceanic populations of non-buoyant macroalgae in the high latitudes of the Southern Hemisphere. PLoS One 8:e69138

Fraser CI, Terauds A, Smellie J, Convey P, Chown SL (2014) Geothermal activity helps life survive glacial cycles. Proc Natl Acad Sci 111:5634–5639

Fraser CI, Morrison AK, Hogg AM, Macaya EC, van Sebille E, Ryan PG, Padovan A, Jack C, Valdivia N, Waters JM (2018) Antarctica's ecological isolation will be broken by storm-driven dispersal and warming. Nat Clim Chang 8:704

Fu YX (1997) Statistical tests of neutrality of mutations against population growth, hitchhiking and background selection. Genetics 147:915–925
Gäbler-Schwarz S, Medlin LK, Leese F (2015) A puzzle with many pieces: the genetic structure and diversity of *Phaeocystis antarctica* Karsten (Prymnesiophyta). Eu. J Phycol 50:112–124
Gersonde R, Crosta X, Abelmann A, Armand L (2005) Sea-surface temperature and sea ice distribution of the Southern Ocean at the EPILOG Last Glacial Maximum—a circum-Antarctic view based on siliceous microfossil records. Quat Sci Rev 24:869–896
González-Wevar CA, Díaz A, Gerard K, Poulin E, Cañete JI (2012) Divergence time estimations and contrasting patterns of genetic diversity between Antarctic and southern South America benthic invertebrates. Rev Chil Hist Nat 85:445–456
González-Wevar CA, Saucède T, Morley SA, Chown SL, Poulin E (2013) Extinction and recolonization of maritime Antarctica in the limpet *Nacella concinna* (Strebel, 1908) during the last glacial cycle: toward a model of Quaternary biogeography in shallow Antarctic invertebrates. Mol Ecol 22:5221–5236
González-Wevar CA, Hüne M, Segovia NI, Nakano T, Spencer HG, Chown SL, Saucède T, Johnstone G, Mansilla A, Poulin E (2017) Following the Antarctic Circumpolar Current: patterns and processes in the biogeography of the limpet *Nacella* (Mollusca: Patellogastropoda) across the Southern Ocean. J Biogeogr 44:861–874
Graham MH, Kinlan BP, Druehl LD, Garske LE, Banks S (2007) Deep-water kelp refugia as potential hotspots of tropical marine diversity and productivity. Proc Natl Acad Sci 104:16576–16580
Griffiths HJ, Waller CL (2016) The first comprehensive description of the biodiversity and biogeography of Antarctic and Sub-Antarctic intertidal communities. J Biogeogr 43:1143–1155
Guillemin ML, Faugeron S, Destombe C, Viard F, Correa JA, Valero M (2008) Genetic variation in wild and cultivated populations of the haploid–diploid red alga *Gracilaria chilensis*: how farming practices favor asexual reproduction and heterozygosity. Evolution 62:1500–1519
Guillemin ML, Valero M, Faugeron S, Nelson W, Destombe C (2014) Tracing the trans-Pacific evolutionary history of a domesticated seaweed (*Gracilaria chilensis*) with archaeological and genetic data. PLoS One 9:e114039
Guillemin ML, Dubrasquet H, Reyes J, Valero M (2018) Comparative phylogeography of six red algae along the Antarctic Peninsula: extreme genetic depletion linked to historical bottlenecks and recent expansion. Polar Biol 41:827–837
Haffer J (1969) Speciation in Amazonian forest birds. Science 165:131–137
Halanych KM, Mahon AR (2018) Challenging dogma concerning biogeographic patterns of Antarctica and the Southern Ocean. Ann Rev Ecol Evol S 49:355–378
Hallatschek O, Nelson DR (2010) Life at the front of an expanding population. Evolution 64:193–206
Hemery LG, Eléaume M, Roussel V, Améziane GC, Steinke D, Cruaud C, Couloux A, Wilson NG (2012) Comprehensive sampling reveals circumpolarity and sympatry in seven mitochondrial lineages of the Southern Ocean crinoid species *Promachocrinus kerguelensis* (Echinodermata). Mol Ecol 21:2502–2518
Hewitt GM (2000) The genetic legacy of the Quaternary ice ages. Nature 405:907
Hewitt GM (2004) Genetic consequences of climatic oscillations in the Quaternary. Phil Trans R Soc Lond B 359:183–195
Hiller A, Wand U, Kämpf H, Stackebrandt W (1988) Occupation of the Antarctic Continent by petrels during the past 35000 years: inferences from a ^{14}C study of stomach oil deposits. Polar Biol 9:69–77
Hommersand MH, Moe RL, Amsler CD, Fredericq S (2009) Notes on the systematics and biogeographical relationships of Antarctic and sub-Antarctic Rhodophyta with descriptions of four new genera and five new species. Bot Marina 52:509–534
Hughes PD, Gibbard PL, Ehlers J (2013) Timing of glaciation during the last glacial cycle: evaluating the concept of a global 'Last Glacial Maximum'(LGM). Earth Sci Rev 125:171–198
Jacob A, Kirst GO, Wiencke C, Lehmann H (1991) Physiological responses of the Antarctic green alga *Prasiola crispa* ssp. *antarctica* to salinity stress. J Plant Physiol 139:57–62

Jueterbock A, Coyer JA, Olsen JL, Hoarau G (2018) Decadal stability in genetic variation and structure in the intertidal seaweed *Fucus serratus* (Heterokontophyta: Fucaceae). BMC Evol Biol 18:94

Jump AS, Peñuelas J (2005) Running to stand still: adaptation and the response of plants to rapid climate change. Ecol Lett 8:1010–1020

Kemp A, Grigorov I, Pearce R, Naveira Garabato A (2010) Migration of the Antarctic Polar Front through the mid-Pleistocene transition: evidence and climatic implications. Quat Sci Rev 29:1993–2009

Kennett JP (1977) Cenozoic evolution of Antarctic glaciation, the Circum-Antarctic ocean, and their impact on global paleoceanography. J Geophys Res 82:3843–3860

Klages JP, Kuhn G, Hillenbrand CD, Smith JA, Graham AG, Nitsche FO, Frederichs T, Jernas PE, Gohl K, Wacker L (2017) Limited grounding-line advance onto the West Antarctic continental shelf in the easternmost Amundsen Sea Embayment during the last glacial period. PLoS One 12:e0181593

Kortsch S, Primicerio R, Beuchel F, Renaud PE, Rodrigues J, Lønne OJ, Gulliksen B (2012) Climate-driven regime shifts in Arctic marine benthos. Proc Natl Acad Sci 109:14052–14057

Lane CE, Lindstrom SC, Saunders GW (2007) A molecular assessment of northeast Pacific *Alaria* species (Laminariales, Phaeophyceae) with reference to the utility of DNA barcoding. Mol Phylogenet Evol 44:634–648

Lange M, Chen YQ, Medlin LK (2002) Molecular genetic delineation of *Phaeocystis* species (Prymnesiophyceae) using coding and non-coding regions of nuclear and plastid genomes. Eur J Phycol 37:77–92

Larter RD, Anderson JB, Graham AGC, Gohl K, Hillenbrand CD, Jakobsson M et al (2014) Reconstruction of changes in the Amundsen Sea and Bellingshausen sea sector of the West Antarctic ice sheet since the last glacial maximum. Quat Sci Rev 100:55–86

Le Cam S, Daguin-Thiébaut C, Bouchemousse S, Engelen AH, Mieszkowska N, Viard F (2019) A genome-wide investigation of the worldwide invader *Sargassum muticum* shows high success albeit (almost) no genetic diversity. Evol Appl 00:1–15 DOI: 10.1111/eva.12837

Lecointre G, Améziane N, Boisselier MC, Bonillo C, Busson F, Causse R et al (2013) Is the species flock concept operational? The Antarctic shelf case. PLoS One 8:e68787

Linse K, Griffiths HJ, Barnes DK, Clarke A (2006) Biodiversity and biogeography of Antarctic and sub-Antarctic mollusca. Deep Sea Res Pt II 53:985–1008

Loeb V (2007) Environmental variability and the Antarctic marine ecosystem. In: Vasseur DA, McCann KS, Vasseur DA (eds) The impact of environmental variability on ecological systems. Springer, Dordrecht, pp 197–225

Macaya EC, López B, Tala F, Tellier F, Thiel M (2016) Float and raft: role of buoyant seaweeds in the phylogeography and genetic structure of non-buoyant associated flora. In: Hu ZM, Fraser C (eds) Seaweed phylogeography. Springer, Dordrecht, pp 97–130

Macaya EC, Tala F, Hinojosa I, Rothäusler E (2020) Detached seaweeds as important dispersal agents across the Southern Ocean. In: Gómez I, Huovinen P (eds) Antarctic seaweeds: diversity, adaptation and ecosystem services. Springer, Cham, Switzerland, pp 59–75

Mackensen A (2004) Changing Southern Ocean paleocirculation and effects on global climate. Antarct Sci 16:369–386

Maggs CA, Castilho R, Foltz D, Henzler C, Jolly MT, Kelly J et al (2008) Evaluating signatures of glacial refugia for North Atlantic benthic marine taxa. Ecology 89:S108–S122

McCulloch RD, Bentley MJ, Purves RS, Hulton NRJ, Sugden DE, Clapperton CM (2000) Climatic inferences from glacial and palaeoecological evidence at the last glacial termination, southern South America. J Quat Sci 15:409–417

Moniz MB, Rindi F, Novis PM, Broady PA, Guiry MD (2012) Molecular phylogeny of Antarctic *Prasiola* (Prasiolales, Trebouxiophyceae) reveals extensive cryptic diversity. J Phycol 48:940–955

Montecinos A, Broitman BR, Faugeron S, Haye PA, Tellier F, Guillemin ML (2012) Species replacement along a linear coastal habitat: phylogeography and speciation in the red alga *Mazzaella laminarioides* along the south east Pacific. BMC Evol Biol 12:01–97

Moon KL, Chown SL, Fraser CI (2017) Reconsidering connectivity in the sub-Antarctic. Biol Rev 92:2164–2181

Mystikou A, Peters AF, Asensi AO, Fletcher KI, Brickle P, van West P, Convey P, Küpper FC (2014) Seaweed biodiversity in the south-western Antarctic Peninsula: surveying macroalgal community composition in the Adelaide Island/Marguerite Bay region over a 35-year time span. Polar Biol 37:1607–1619

Near TJ, Dornburg A, Kuhn KL, Eastman JT, Pennington JN, Patarnello T, Zane L, Fernández DA, Jones CD (2012) Ancient climate change, antifreeze, and the evolutionary diversification of Antarctic fishes. Proc Natl Acad Sci 109:3434–3439

Neiva J, Serrão EA, Assis J, Pearson GA, Coyer JA, Olsen JL, Hoarau G, Valero M (2016) Climate oscillations, range shifts and phylogeographic patterns of North Atlantic Fucaceae. In: Hu ZM, Fraser C (eds) Seaweed phylogeography. Springer, Dordrecht, pp 279–308

Norkko A, Thrush SF, Cummings VJ, Funnell GA, Schwarz AM, Andrew NL, Hawes I (2004) Ecological role of *Phyllophora antarctica* drift accumulations in coastal soft-sediment communities of McMurdo Sound, Antarctica. Polar Biol 27:482–494

O'Cofaigh C, Davies BJ, Livingstone SJ, Smith JA, Johnson JS, Hocking EP et al (2014) Reconstruction of ice-sheet changes in the Antarctic Peninsula since the Last Glacial Maximum. Quat Sci Rev 100:87–110

Ocaranza-Barrera P, González-Wevar CA, Guillemin ML, Rosenfeld S, Mansilla A (2019) Molecular divergence between *Iridaea cordata* (Turner) Bory de Saint-Vincent from the Antarctic Peninsula and the Magellan Region. J Appl Phycol 31:939–949

Oliveira MC, Pellizzari F, Medeiros AS, Yokoya NS (2020) Diversity of Antarctic seaweeds. In: Gómez I, Huovinen P (eds) Antarctic seaweeds: diversity, adaptation and ecosystem services. Springer, Cham, Switzerland, pp 23-39

Paillard D, Parrenin F (2004) The Antarctic ice sheet and the triggering of deglaciations. Earth Planet Sci Lett 227:263–271

Pearse JS, Mooi R, Lockhart SJ, Brandt A (2009) Brooding and species diversity in the Southern Ocean: selection for brooders or speciation within brooding clades? In: Krupnik I, Lang MA, Miller SE (eds) Smithsonian at the poles: contributions to international polar year science. Smithsonian Institution Scholarly Press, Washington, DC, pp 181–196

Pellizzari F, Silva MC, Silva EM, Medeiros A, Oliveira MC, Yokoya NS, Pupo D, Rosa LH, Colepicolo P (2017) Diversity and spatial distribution of seaweeds in the South Shetland Islands, Antarctica: an updated database for environmental monitoring under climate change scenarios. Polar Biol 40:1671–1685

Peters AF, Ramírez ME, Rülke A (2000) The phylogenetic position of the subantarctic marine macroalga *Desmarestia chordalis* (Phaeophyceae) inferred from nuclear ribosomal ITS sequences. Polar Biol 23:95–99

Pointing SB, Buedel B, Convey P, Gillman L, Koerner C, Leuzinger S, Vincent WF (2015) Biogeography of photoautotrophs in the high polar biome. Front Plant Sci 6:692

Poulin E, González-Wevar C, Díaz A, Gérard K, Hüne M (2014) Divergence between Antarctic and South American marine invertebrates: what molecular biology tells us about Scotia Arc geodynamics and the intensification of the Antarctic Circumpolar Current. Glob Planet Chang 123:392–399

Provan J, Bennett KD (2008) Phylogeographic insights into cryptic glacial refugia. Trends Ecol Evol 23:564–571

Quartino ML, Deregibus D, Campana GL, Latorre GEJ, Momo FR (2013) Evidence of macroalgal colonization on newly ice-free areas following glacial retreat in Potter Cove (South Shetland Islands), Antarctica. PLoS One 8:e58223

Quartino ML, Saravia LA, Campana GL, Deregibus D, Matula CV, Boraso AL, Momo FR (2020) Production and biomass of seaweeds in newly ice-free areas: implications for coastal processes in a changing Antarctic environment. In: Gómez I, Huovinen P (eds) Antarctic seaweeds: diversity, adaptation and ecosystem services. Springer, Cham, Switzerland, pp 155–168

Raupach MJ, Thatje S, Dambach J, Rehm P, Misof B, Leese F (2010) Genetic homogeneity and circum-Antarctic distribution of two benthic shrimp species of the Southern Ocean, *Chorismus antarcticus* and *Nematocarcinus lanceopes*. Mar Biol 157:1783–1797

Raymond JA, Fritsen CH (2001) Semi purification and ice recrystallization inhibition activity of ice active substances associated with Antarctic photosynthetic organisms. Cryobiology 43:63–70

Reed DH, Frankham R (2003) Correlation between fitness and genetic diversity. Conserv Biol 17:230–237

Roberts J, McCave IN, McClymont EL, Kender S, Hillenbrand CD, Matano R, Hodell DA, Peck VL (2017) Deglacial changes in flow and frontal structure through the Drake Passage. Earth Planet Sci Lett 474:397–408

Saunders GW (2005) Applying DNA barcoding to red macroalgae: a preliminary appraisal holds promise for future applications. Phil Trans R Soc Lond B 360:1879–1888

Saunders GW (2008) A DNA barcode examination of the red algal family Dumontiaceae in Canadian waters reveals substantial cryptic species diversity. 1. The foliose Dilsea–Neodilsea complex and Weeksia. Botany 86:773–789

Scher HD, Martin EE (2006) Timing and climatic consequences of the opening of Drake Passage. Science 312:428–430

Scher HD, Whittaker JM, Williams SE, Latimer JC, Kordesch WE, Delaney ML (2015) Onset of Antarctic Circumpolar Current 30 million years ago as Tasmanian Gateway aligned with westerlies. Nature 523:580

Sijp WP, von der Heydt AS, Dijkstra HA, Flögel S, Douglas P, Bijl PK (2014) The role of ocean gateways on cooling climate on long time scales. Glob Planet Chang 119:1–22

Silberfeld T, Leigh JW, Verbruggen H, Cruaud C, De Reviers B, Rousseau F (2010) A multi-locus time-calibrated phylogeny of the brown algae (Heterokonta, Ochrophyta, Phaeophyceae): investigating the evolutionary nature of the "brown algal crown radiation". Mol Phylogenet Evol 56:659–674

Simms AR, Milliken KT, Anderson JB, Wellner JS (2011) The marine record of deglaciation of the South Shetland Islands, Antarctica since the Last Glacial Maximum. Quat Sci Rev 30:1583–1601

Soler-Membrives A, Linse K, Miller KJ, Arango CP (2017) Genetic signature of Last Glacial Maximum regional refugia in a circum-Antarctic sea spider. R Soc Open Sci 4:170615

Spalding MD, Fox HE, Allen GR, Davidson N, Ferdaña ZA, Finlayson MAX et al (2007) Marine ecoregions of the world: a bioregionalization of coastal and shelf areas. Bioscience 57:573–583

Spalding HL, Amado-Filho GM, Bahia RG, Ballantine DL, Fredericq S, Leichter JJ, Nelson WA, Slattery M, Tsuda RT (2019) Macroalgae. In: Loya Y, Puglise KA, Bridge TCL (eds) Mesophotic coral ecosystems. Springer, Cham, pp 507–536

Sromek L, Lasota R, Wolowicz M (2015) Impact of glaciations on genetic diversity of pelagic mollusks: Antarctic *Limacina antarctica* and Arctic *Limacina helicina*. Mar Ecol Prog Ser 525:143–152

Stuart KM, Long DG (2011) Tracking large tabular icebergs using the SeaWinds Ku-band microwave scatterometer. Deep Sea Res Pt II 58:1285–1300

Tajima F (1989) Statistical method for testing the neutral mutation hypothesis by DNA polymorphism. Genetics 123:585–595

Terauds A, Chown SL, Morgan F, Peat H, Watts DJ, Keys H, Convey P, Bergstrom DM (2012) Conservation biogeography of the Antarctic. Divers Distrib 18:726–741

Thatje S, Hillenbrand CD, Larter R (2005) On the origin of Antarctic marine benthic community structure. Trends Ecol Evol 20:534–540

Thatje S, Hillenbrand CD, Mackensen A, Larter R (2008) Life hung by a thread: endurance of Antarctic fauna in glacial periods. Ecology 89:682–692

Thornhill DJ, Mahon AR, Norenburg JL, Halanych KM (2008) Open-ocean barriers to dispersal: a test case with the Antarctic Polar Front and the ribbon worm *Parborlasia corrugatus* (Nemertea: Lineidae). Mol Ecol 17:5104–5117

Valero M, Destombe C, Mauger S, Ribout C, Engel CR, Daguin-Thiebaut C, Tellier F (2011) Using genetic tools for sustainable management of kelps: a literature review and the example of *Laminaria digitata*. Cah Biol Mar 52:467

Van Oppen MJH, Olsen JL, Stam WT, van Denhoek C, Wiencke C (1993) Arctic-Antarctic disjunctions in the benthic seaweeds *Acrosiphonia arcta* (Chlorophyta) and *Desmarestia viridis/willii* (Phaeophyta) are of recent origin. Mar Biol 115:381–386

Waters JM, Fraser CI, Hewitt GM (2013) Founder takes all: density-dependent processes structure biodiversity. Trends Ecol Evol 28:78–85

White DA, Bennike O, Melles M, Berg S, Binnie SA (2018) Was South Georgia covered by an ice cap during the Last Glacial Maximum? In: Siegert MJ, SSR J, White DA (eds) Exploration of subsurface Antarctica: uncovering past changes and modern processes. The Geological Society of London, pp 49–59

Wiencke C, Clayton MN (ed) (2002) Antarctic seaweeds. ARG Gantner Verlag, KG Ruggell, p 239

Wiencke C, Clayton MN, Gómez I, Iken K, Lüder UH, Amsler CD, Karsten U, Hanelt D, Bischof K, Dunton K (2007) Life strategy, ecophysiology and ecology of seaweeds in polar waters. Rev Environ Sci Biotechnol 6:95–126

Wiencke C, Amsler CD, Clayton MN (2014) Chapter 5.1. Macroalgae. In: De Broyer C, Koubbi P, Griffiths HJ, Raymond B, Dudekem DAC et al (eds) Biogeographic atlas of the Southern Ocean. Scientific Committee on Antarctic Research, Cambridge, pp 66–73

Wilson NG, Schrödl M, Halanych KM (2009) Ocean barriers and glaciation: evidence for explosive radiation of mitochondrial lineages in the Antarctic sea slug *Doris kerguelenensis* (Mollusca, Nudibranchia). Mol Ecol 18:965–984

Yang EC, Peters AF, Kawai H, Stern R, Hanyuda T, Bárbara I, Müller DG, Strittmatter M, van Reine WFP, Küpper FC (2014) Ligulate *Desmarestia* (Desmarestiales, Phaeophyceae) revisited: *D. japonica* sp. nov. and *D. dudresnayi* differ from *D. ligulata*. J Phycol 50:149–166

Zachos J, Pagani M, Sloan L, Thomas E, Billups K (2001) Trends, rhythms, and aberrations in global climate 65 Ma to present. Science 292:686–693

Part III
Physiology, Productivity and Environmental Reponses

Chapter 7
Underwater Light Environment of Antarctic Seaweeds

Pirjo Huovinen and Iván Gómez

Abstract Antarctic seaweeds are highly shade-adapted organisms, which can photosynthesize under very dim light. This remarkable characteristic allows them colonizing over 30 m depths and surviving extended dark periods during the polar winter. On the other hand, they are well equipped to cope with high light stress, which points to a trade-off between shade adaptation and efficient UV stress tolerance. Optical properties of water determine both the underwater light climate for photosynthesis and the risk of seaweeds for UV exposure in their habitats. Thus, understanding the natural (spatial, temporal) and anthropogenic-driven changes in spectral transparency of water and factors governing it is fundamental in evaluating the state of seaweeds under current and future environmental scenarios. In the present chapter the aspects related to the optical properties determining the underwater habitat of Antarctic seaweeds are summarized, along with the potential changes in water optics as a result of climate change, ozone depletion and other environmental and emerging threats, and their interactions.

Keywords Antarctic costal waters · Light climate · Ozone depletion · Photosynthesis · Pollution · UV radiation · Water column optics

7.1 Introduction

Aquatic organisms in the polar regions have adapted to survive wide seasonal changes in the light field of their habitats (McMinn and Martin 2013). The polar winter accompanied by ice cover leads to extended periods of dim light conditions or even darkness (Berge et al. 2015), while in spring-summer, water column receives increasing levels of solar radiation, which is furthermore enhanced after icebreak

P. Huovinen (✉) · I. Gómez
Instituto de Ciencias Marinas y Limnológicas, Facutad de Ciencias, Universidad Austral de Chile, Valdivia, Chile

Research Center Dynamics of High Latitude Marine Ecosystems (IDEAL), Valdivia, Chile
e-mail: pirjo.huovinen@uach.cl; igomezo@uach.cl

I. Gómez, P. Huovinen (eds.), *Antarctic Seaweeds*,
https://doi.org/10.1007/978-3-030-39448-6_7

and also during episodes of ozone depletion leading to higher risk of UV exposure during austral spring (Frederick et al. 1989; Neale et al. 1998). In order to withstand such widely changing conditions, Antarctic seaweeds are characterized by remarkable photobiological adaptations to grow and photosynthesize at irradiances as low as 10 μmol m^{-2} s^{-1}, allowing them surviving extended dark periods during winter and colonizing over 30 m depths. On the other hand, they present efficient mechanisms to tolerate high UV stress (Gómez et al. 2009; Karsten et al. 2009; Gómez and Huovinen 2015).

Underwater light climate is governed by the optical properties that define the depth limits for photosynthesis and hence the vertical distribution of seaweeds as well as the risk for exposure to detrimental UV levels in Antarctic coastal waters (see Chap. 11 by Gómez and Huovinen). Thus, understanding the natural (temporal, spatial) and anthropogenic-driven changes in water optics and factors controlling it is essential, not only in evaluating aquatic primary production and UV risk in aquatic habitats (Holm-Hansen et al. 1993; Cullen and Neale 1994) but also for predictions of the status of these ecosystems under current and future scenarios of global climate change (Vincent and Belzile 2003). The knowledge on light attenuation and the potential impact, e.g., of glacier-derived freshwater input, are also needed in order to explain the spatial variation in primary production and carbon fluxes with far-reaching implications for the benthic communities with a potential to modify the biogeochemical gradients in the whole Antarctic coastal system (see review by Sulzberger et al., 2019, and Chap. 8 by Quartino et al.). In the present chapter, we summarize the aspects related to the optical characteristics determining the underwater habitat of Antarctic seaweeds with implications for their photobiology, along with the potential changes in water optics as a result of interactions with climate change factors and other environmental and emerging threats.

7.2 Optics of Antarctic Coastal Waters

7.2.1 Light in Aquatic Environment

In natural waters, sunlight may be either absorbed or scattered. Principles of underwater optics have been described and reviewed by Hargreaves (2003), Kirk (2011), and Mobley (2015). In summary, the main components responsible for light absorption have been described: (1) water itself, (2) colored dissolved organic matter (CDOM), (3) living photosynthetic organisms (phytoplankton), and (4) nonliving organic and inorganic particulate material. Pure water absorbs and thus attenuates light increasingly in red (>550 nm) and infrared (IR) bands, while absorption by CDOM (mainly yellow dissolved humic substances with diverse intrinsic properties, e.g., molecular weight, origin, age; see Hessen and Tranvik, 1998) increases strongly towards blue and UV bands. Phytoplankton attenuates photosynthetically active radiation (PAR) through their light-absorbing photosynthetic pigments and also by causing scattering of light as a result of their particulate form. Nonliving particulate matter attenuates light strongly through scattering, whereas their light absorption (increasing towards blue and UV wavelengths) is weaker (for details, see Hargreaves 2003, Kirk 2011 and Mobley 2015).

Underwater light climate varies strongly in different types of water bodies according to their inherent optical properties (IOPs), i.e., those describing aquatic medium (such as absorption coefficient and volume scattering function) independent of the light field (thus measurable in a water sample). Together with IOPs, environmental conditions (e.g., radiance, solar angle, waves) partially influence apparent optical properties (AOPs, such as reflectance and diffuse attenuation coefficient), which can be used to describe water optics through in situ measurements (see Kirk 2011; Mobley 2015). In an optically homogeneous water column, the downward irradiance (E_d) diminishes in an exponential manner with depth (z); thus diffuse attenuation coefficient for downward irradiance (K_d) can be calculated according to Kirk (2011) as:

$$K_d = -\mathrm{dln}\, E_d\left(z,\lambda\right) / \mathrm{dz}$$

Consequently, a depth where 1% or 10% of the subsurface irradiance is available can be estimated from K_d as:

$$z_{z1} = 4.6 / K_d \; and \; z_{z10} = 2.3 / K_d$$

The upper water column down to a depth where 1% of PAR remains is generally defined as the euphotic zone, which can range from the upmost water layer to hundred meters. For example, in saline-alkaline lakes and turbid estuaries, $z_{1\%}$ can reach less than half and one meter, respectively (Oduor and Schagerl 2007; Coljin et al. 1987), while in some oceanic areas (e.g., Sargasso Sea in the Atlantic and the coast of Hawaii in the Pacific), light can penetrate more than hundred of meters (Tyler 1975; Bienfang et al. 1984). Penetration of UV wavelengths in the water column can also range widely, from few centimeters (e.g., in small humic lakes; Scully and Lean 1994; Huovinen et al. 2003) to dozens of meters in highly transparent oligotrophic waters (e.g., $z_{1\%}$ around 40 m in Sargasso Sea; Smith and Baker 1979).

7.2.2 Light Climate in Antarctic Waters

An extensive study around the Antarctic Peninsula in mid-1980s by Mitchell and Holm-Hansen (1991) revealed euphotic zone of approximately 75 m in the offshore waters in Drake Passage, while in coastal areas PAR attenuated to 1% within the upper 15 m. The clear open water areas were characterized by negligible sediment load and terrigenous material, which can be explained by the scarcity of terrestrial vegetation (the ice-free landscape is largely dominated by tundra-type vegetation like moss and lichen; Convey 2010) and hence a low contribution of allochtonous organic matter from the catchment. In contrast, shallow near-shore waters around the South Shetland Islands with glaciers showed high turbidity due to meltwater runoff and related increase of organic matter decreasing transparency (Mitchell and Holm-Hansen 1991). In fact, long-term measurements from Potter Cove (King George Island, South Shetland Islands) indicate that increased light attenuation due to strong sedimentation in areas impacted by glacier retreat is modifying the growth conditions for seaweeds and their upper distribution limits (Deregibus et al. 2016; see also Chap. 9 by Deregibus et al.). Also in Fildes Bay, another well-studied area in King George Island, biomass of subtidal seaweeds displays a glacier-related gradient (Valdivia et al. 2015). In Bransfield and Gerlache Straits, seasonal variation in light penetration was observed: in January, light was strongly attenuated ($z_{1\%}$ around 4.6 m based on K_d (488 nm)) in coastal areas due to massive phytoplankton blooms, which disappeared by February–March leading to deeper (over 30 m) light penetration, reducing also the spatial (onshore-offshore) gradient in water transparency (Mitchell and Holm-Hansen 1991). Strong spatial heterogeneity was also observed by Figueroa (2002) with euphotic zone ranging 9–58 m in open waters of Gerlache and Bransfield Straits. On the other hand, in Drake Passage where no blooms occurred, the waters remained clear ($z_{1\%}$ 46–76 m based on K_d (488 nm)) from December to March (Mitchell and Holm-Hansen 1991). In coastal waters of Fildes Bay, $z_{1\%}$ PAR ranging 19–34 m has been reported (Huovinen et al. 2016) (Figs. 7.1 and 7.2). In Potter Cove, higher water transparency in November–December ($z_{1\%}$ PAR over 25 m) has been shown to decrease in January–February, light attenuating already in the upper meters in areas influenced by meltwater (Klöser et al. 1993). Here, in winter and early spring, light attenuation is mainly regulated by phytoplankton, while towards summer, particulate matter from increasing meltwater contributes strongly (up to 47%) (Schloss and Ferreyra 2001).

During the open water season, relatively high UV transparency has been measured in Fildes Bay (Huovinen et al. 2016), although spatial variation between sites and with depth is observed (Fig. 7.1). UV ($z_{10\%}$) penetration has been shown to range here from approximately 3 to 6 m for UV-B_{305nm} to 8 to 19 m for UV-A_{395nm} (Huovinen et al. 2016). Based on theoretical estimations (tropospheric ultraviolet and visible (TUV) model) for summer solstice and mean K_d of the bay area, sunlight levels at $z_{10\%}$ could reach 0.13 W m^{-2} for UV-B (at 5 m), 4.4 W m^{-2} for UV-A (at 10 m), and 170 µmol m^{-2} s^{-1} for PAR (at 15 m) (Fig. 7.2). In Potter Cove, penetration depth ($z_{10\%}$) of up to 8 m has been reported for UV-B radiation. Subsurface (10 cm) levels of up to 1.5 and 26 W m^{-2} of UV-B (280–320 nm) and UV-A (320–400 nm), respectively, were measured. Corresponding levels at 2 m depth were decreased to 0.26 and 8.6 W m^{-2}. PAR levels at corresponding depths were 1178 and 515 µmol m^{-2} s^{-1} (Quartino et al. 2005; Zacher et al. 2007a). It should be noted that these authors considered 320 nm as the limit between UV-B and UV-A bands, resulting in markedly higher UV-B levels than when limiting the band to 315 nm (see Huovinen et al. 2006). Maximal penetration depths ($z_{10\%}$) for UV-B (305 nm) and UV-A (340 nm) of 8 m and 11 m, respectively, have been reported for Bransfield and Gerlache Straits and Palmer Station (Helbling et al. 1995; Figueroa 2002). In Bellinghausen

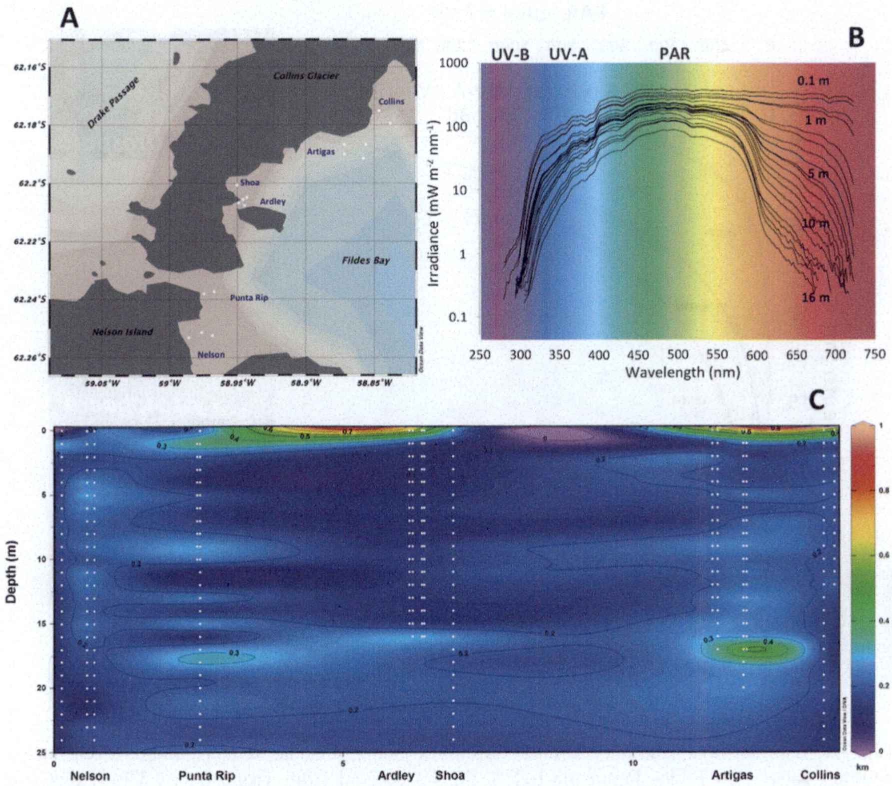

Fig. 7.1 Fildes Bay (King George Island, Maritime Antarctica (**a**). Variation of spectral irradiance with depth in Fildes Bay based on measurement with the hyperspectral radiometer RAMSES-ACC2-UV–vis (Trios Optical Sensors, Oldenburg, Germany) (**b**). Spatial (with depth and distance) variation of K_d at 395 nm (m^{-1}) determined at 1 m intervals from measurements with the radiometer PUV-2500 (Biospherical Instruments Inc., USA) in six areas of Fildes Bay and visualized with Ocean Data View software (Schlitzer R., Ocean Data View, odv. Awi.de, 2015) (Adpated from Huovinen et al. 2016) (**c**)

Sea, even higher penetration depths for UV-B ($z_{10\%}$ 310 nm around 12 m) and UV-A (340 nm around 17 m) have been measured. Here, UV-B radiation could be detected down to 60–70 m (Smith et al. 1992).

Potential for DNA damage at depths ranging 3.5–16 m (based on $z_{10\%}$ DNA weighted irradiance) have been reported for Antarctic waters (Huot et al. 2000; Buma et al. 2001). Under ozone hole the effective UV-B penetration depth was found to increase by 7 m (Smith et al. 1992). For the estimations on net UV impact on aquatic organisms, knowledge on spectral attenuation is needed as it varies with depth (see Figs. 7.1b and 7.2), and shifts in underwater irradiance spectra have implications for repair and other processes that depend on spectral light composition (Neale 2000; Williamson et al. 2001). For example, a higher ratio (mean around 1.0) between $z_{10\%}$ 305 and 340 nm has been reported in Antarctic waters under

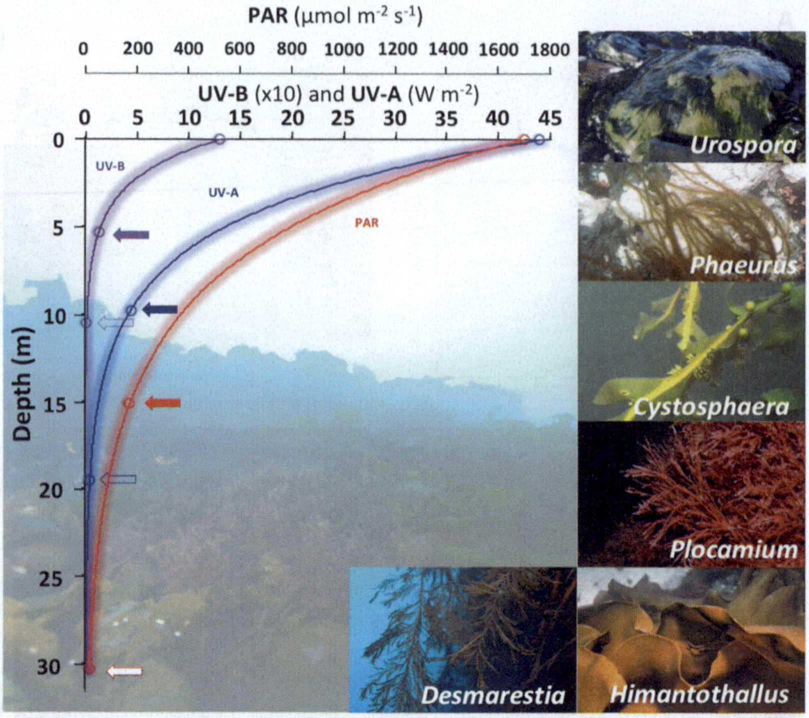

Fig. 7.2 Estimation of irradiance levels at different water depths, based on irradiance levels during summer solstice in Fildes Peninsula (62°S, 50°W) derived from Tropospheric Ultraviolet and Visible (TUV 5.3; Madronich and Flocke (1999)) model (https://www2.acom.ucar.edu/modeling/tropospheric-ultraviolet-and-visible-tuv-radiation-model) and on K_d values (mean of 305 and 313 nm for UV-B; mean of 320, 340, 380, and 395 nm for UVA) of the bay area (data from Huovinen et al. 2016). Penetration depths $z_{1\%}$ (solid arrows) and $z_{10\%}$ (open arrows) are indicated. Representative seaweeds for different depths are shown. (Photos by Ignacio Garrido except *Urospora* by Iván Gómez)

ozone hole conditions than in other oceanic areas (without ozone depletion) (mean 0.54) (Tedetti and Sempéré 2006).

In polar regions, seasonal ice cover plays an important role in governing the underwater light climate (Lesser et al. 2004; Fritsen et al. 2011; Taskjelle et al. 2016). In McMurdo Sound (Ross Sea), 0.16% of incident irradiance (mainly blue-green band) was transmitted through a 2-m-thick ice (Schwarz et al. 2003) and 0.05% (0.2–0.6 μmol m^{-2} s^{-1}) reached bottom (23 m) (Robinson et al. 1995). Under ice cover, water was highly transparent with K_d for PAR 0.09 m^{-1} (Schwarz et al. 2003). In scenarios of global climate change, the importance of sea-ice duration has been highlighted and is regarded, together with light penetration and sedimentation, a major driver structuring the shallow Antarctic benthos (Clark et al. 2017). Thinner ice cover and its shorter duration results in enhanced exposure to solar radiation for extended periods, which can bring consequences for photosynthesis (Runcie and

Riddle 2007), zonation patterns of seaweeds (Campana et al. 2009; Clark et al. 2017) and increase UV damage (Fountoulakis et al. 2014). Timing of ice breakup may be crucial for the establishment of canopy-forming seaweed communities in spring (Johnston et al. 2007). When large brown algae are established, their canopies can markedly reduce light levels reaching their understory species, and it can be also altered by tidal fluctuations (Huovinen and Gómez 2011). In Antarctic coastal waters, the presence of large brown algae growing at depths as shallow as 5 m can affect considerably the incident irradiance at deeper locations where seaweeds coexist with abundant populations of understory species, especially red algae (Klöser et al. 1996; Gómez et al. 2019; see also Chap. 11 by Gómez and Huovinen).

7.3 Adaptations of Antarctic Seaweeds to Extreme Light Conditions

7.3.1 Photosynthetic Shade Adaptation of Antarctic Seaweeds

The strong seasonality and turbidity of coastal waters (Zacher et al. 2009) imply severe constraints for photosynthesis of Antarctic seaweeds (see Wiencke et al. 2009). They can overall be characterized by very low requirements of light and constitutively high efficiency of photosynthesis, allowing them colonizing deep habitats with low light availability and coping with periods of darkness (reviewed by Gómez et al. 2009). For example, some Antarctic crustose red algae (corallines) in the Ross Sea live permanently under ice cover and remain in darkness for several months in winter (Schwarz et al. 2005). In fact, shade adaptation is considered a metabolic prerequisite that allows survival under wide gradient of light (Weykam et al. 1996; Huovinen and Gómez 2013; Gómez and Huovinen 2015).

Photosynthetic adaptations to cope with low light availability allow Antarctic seaweeds supplying their light requirements over a broad depth distribution (Gómez et al. 1997). The growth and photosynthesis especially of the endemic brown seaweeds has been shown not to be limited even at deep locations close to 30–40 m (Drew 1977; Weykam et al. 1996; Schwarz et al. 2003). Although some species from King George Islands collected at shallow waters displayed higher saturated net photosynthesis (P_{max}) than species from deeper sites, their photosynthetic efficiency (α) and light demands for photosynthesis (E_k) did not vary with depth (Weykam et al. 1996). These photosynthetic characteristics have been confirmed more recently with chlorophyll fluorescence techniques (Huovinen and Gómez 2013). Also photosynthetic efficiency (α_{ETR}) has been found to be high along the depth gradient (0–30 m) (Gómez and Huovinen 2015) in spite of relatively high water transparency (PAR at 20 m depth around 50 μmol m^{-2} s^{-1}) (Huovinen et al. 2016; Fig. 7.2). Despite their wide vertical distribution (from 5 m downwards), lower distribution limit (close to 30 m) of Antarctic seaweeds such as *Desmarestia mensiezii*, *D. anceps*, *Palmaria decipiens*, and *Gigartina skottsbergii* coincides with the depth of compensation irradiance level (Klöser et al. 1996; Gómez et al. 1997; Deregibus et al. 2016). Also, although it can be found at depths close to 5–10 m, *Himantothallus grandifolious* starts to dominate only at depth below 30 m when substrate and competitive balance with other large Desmarestiales, e.g., *D. anceps*, are favorable (Zielinski 1990; Klöser et al. 1996). Thus, the vertical distribution patterns of Antarctic seaweeds cannot be explained by light limitation alone, but rather in combination with other biotic (e.g., herbivory, competition) and abiotic (e.g., substrate characteristics, ice-induced perturbations, water movement) factors (Klöser et al. 1996; Iken et al. 1998; Amsler et al. 2011; see Chap. 13 by Valdivia and Chap. 17 by Amsler et al.).

7.3.2 Tolerance of Antarctic Seaweeds to High PAR and UV

Antarctic coastal waters are often characterized by high water transparency with light (PAR) penetration ($z_{1\%}$) ranging 19–34 m (Huovinen et al. 2016; Figs. 7.1 and 7.2). In coastal waters of King George Island, PAR has been found to penetrate down to 40 m depth (Klöser et al. 1993), and levels around 50 μmol m^{-2} s^{-1} have been measured at 30 m (Gómez et al. 1997). This contrasts with the light conditions in some Arctic near-shore waters, e.g., in the Beaufort Sea with muddy bottoms and estuarine characteristics where PAR penetration is markedly lower ($z_{1\%}$ 3–11 m; in offshore waters >4.6 m) (Dunton et al. 2009). After Antarctic icebreak in spring-summer, seaweeds experience sudden increase of underwater light; however, water optics may present strong variation due to turbidity, e.g., from glacier melting or freshwater runoff (Klöser et al. 1993; Deregibus et al. 2016; see Chap. 9 by Deregibus et al.). Under these conditions, seaweeds exhibit a suite of physiological mechanisms to cope with high levels of solar radiation. One of them is dynamic photoinhibition, downregulation of photosynthesis under high solar radiation, dissipating excess absorbed energy as heat in photosystem II (PSII) (Adams III et al. 2006). This protective mechanism has been reported in Antarctic seaweeds exposed to natural solar irradiation, showing decreased photosynthetic activity around solar noon followed by recovery towards evening (Hanelt et al. 1994). Thus, the capacity to withstand high solar radiation allows these shade-adapted organisms thriving also at environments where light levels exceed their requirements for saturation of photosynthesis, such as in intertidal and shallow waters (Huovinen and Gómez 2013; Gómez et al. 2019).

Together with high PAR levels, seaweeds can also be exposed to harmful levels of UV radiation in their habitats (Huovinen and Gómez 2013, Huovinen et al. 2016; Figs. 7.1 and 7.2). Deleterious effects of UV-B radiation on aquatic organisms, including seaweeds, are widely recognized (reviewed by Holm-Hansen et al. 1993; Vincent and Neale 2000; Karsten et al. 2009; Burritt and Lamare 2016). UV-B radiation can directly affect cellular components (e.g., nucleus, chloroplast), their ultrastructure and processes, as well as target important biomolecules, such as DNA (leading, e.g., to formation of cyclobutane dimers (CPDs) and 6–4 photoproducts that interfere with replication) (Mitchell and Karentz 1993), proteins (e.g., D1 protein in PSII), and photosynthetic pigments (Gerber and Häder 1992). It can also interfere with uptake of nutrients (Döhler et al. 1991) and metabolism of fatty and amino acids (Goes et al. 1994, 1995). It may cause oxidative stress by inducing production of reactive oxygen species (ROS, such as singlet-oxygen, hydroxyl, and superoxide radicals) that are damaging to biomolecules and can cause, e.g., peroxidation of lipids (Bischof and Rautenberger 2012). ROS may also be formed in aquatic environment when UV interacts with DOM (Kieber et al. 2003) (see also Sect. 7.4.1). Disturbance or damage to important cellular components and biomolecules can impair biochemical and physiological processes, such as photosynthesis (through effect on pigments, enzymes, photosynthetic apparatus, etc.), growth, and reproduction (reviewed in Bischof et al. 2006).

The magnitude of final harmful impact depends on the balance between produced damage and the efficiency of protective, e.g., UV-shielding compounds such as mycosporine-like amino acids (MAAs) in red algae and phlorotannins in brown algae (see Chap. 18 by Gómez and Huovinen) and repair (e.g., light-induced repair of DNA damage involving photolyase enzyme) mechanisms to mitigate it (Karentz et al. 1991; Mitchell and Karentz 1993; Vincent and Roy 1993; Nuñez-Pons et al. 2018). Antioxidants (e.g., carotenoids, phlorotannins, enzymes such as superoxide dismutase and glutathione peroxidase) furthermore serve as defense mechanism through ROS scavenging (Bischof and Rautenberger 2012). Also, downregulation of PSII (dynamic photoinhibition) has been proposed as a protective mechanism, not only against high PAR, but also against UV radiation, contributing to the physiological tolerance of seaweeds (Bischof et al. 2006).

Early life stages of seaweeds with a thinner wall are more vulnerable to UV damage than multicellular adult stages. Marked DNA damage and physiological stress has been reported in propagules of Antarctic seaweeds upon UV exposure (Roleda et al. 2006; Zacher et al. 2007b), although they also show capacity for recovery (Navarro et al. 2019) and UV-screening compounds (Roleda et al. 2006; see also Chap. 10 by Navarro et al.). Furthermore, adult thalli of intertidal species display high tolerance to light stress as they have to cope with high levels of solar radiation, especially during low tidal levels (Cruces et al. 2013). Here, dynamic photoinhibition seems to play a key role as a protective mechanism (Hanelt et al. 1994). Interestingly, high photosynthetic tolerance has also been reported for subtidal species that are not exposed to high UV levels in their habitat (deeper than 20 m) (Huovinen and Gómez 2013). Especially large endemic brown algae have shown marked UV stress tolerance over their broad vertical distribution (5–30 m), which has been related to efficient morph-functional mechanisms and constitutively high levels of phlorotannins (Gómez and Huovinen 2015; Flores-Molina et al. 2016; Gómez et al. 2019; see also Chaps. 11 and 18 by Gómez and Huovinen). Overall, UV sensitivity of seaweeds at their different life stages is considered one of the ecologically important factors defining their vertical zonation and distribution limits (Bischof et al. 1998; Wiencke et al. 2000, 2006; see Chap. 10 by Navarro et al.). In contrast to short-term laboratory experiments, organisms in their natural habitats are influenced by complex interactions of multiples environmental factors, which may mitigate or potentiate the adverse impact of UV radiation (see Sect. 7.4). For example, a global field study examining UV impact on marine benthic community level found that any effect disappeared after few months during succession (Wahl et al. 2004). However, recent reports based on meta-analyses point to overall negative UV impact at all trophic levels (summarized by Williamson et al. 2019).

7.4 Consequences for Light Field Under Current and Future Threats

7.4.1 Ozone Depletion

The deleterious effects of solar radiation are mainly associated with the UV wavebands, i.e., UV-C (100–280 nm), UV-B (280–315 nm), and UV-A (315–400 nm) radiation, with greater effectiveness for biological damage occurring towards shorter wavelengths (Setlow 1974; Cullen et al. 1992). The most damaging UV-C waveband is absorbed by oxygen and ozone in the atmosphere thus not reaching the earth's surface. On the other hand, UV-A waveband is not absorbed by atmospheric ozone, while UV-B waveband is partially absorbed and its levels are affected by changes in the thickness of the ozone layer (Frederick et al. 1989; Madronich et al. 1998). Based on radiation amplification factors (RAF), it has been estimated that 1% ozone depletion leads to approximately 2–3% increase in DNA-damaging UV-B dose, i.e., weighted with DNA action spectra (Madronich 1994). More than four decades ago, Molina and Rowland (1974) discovered the destructing effect of chlorofluorocarbons (CFCs) for ozone layer. A decade later, Farman et al. (1985) reported for the first time the "ozone hole" over the Antarctic. Although the biological risk from UV radiation also in aquatic environments had already been recognized (Smith and Baker 1979; Calkins and Thordardottir 1980), stratospheric ozone depletion and consequently enhanced levels solar UV-B radiation started to gain importance as one of the major anthropogenic-driven threats for Antarctic biota thereafter (Holm-Hansen et al. 1993; Weiler and Penhale 1994). After several decades of alarming trends, recent reports indicate that the ozone layer over the Antarctic is showing recovery (Solomon et al. 2016, 2017; summarized by Bais et al. 2018, 2019). Since 2000, the observed recovery seems to be in agreement with a decreasing amount of ozone-depleting substances. Although global trend over all latitudes is masked due to strong variability (Chipperfield et al. 2017), recovering trend has been reported for some areas, including Antarctica (Kuttippurath and Nair 2017). However, the complex interactions of stratospheric ozone with climate change effects in the atmosphere and ocean lead to uncertainties in estimating the timescale for recovery, and changes (either increase or decrease) in UV levels are estimated to vary in different regions (summarized by Bais et al. 2018, 2019). Together with greenhouse gases, ozone depletion has led to latitudinal shifts of climate (by moving the Southern Annular Mode (SAM) towards a more positive

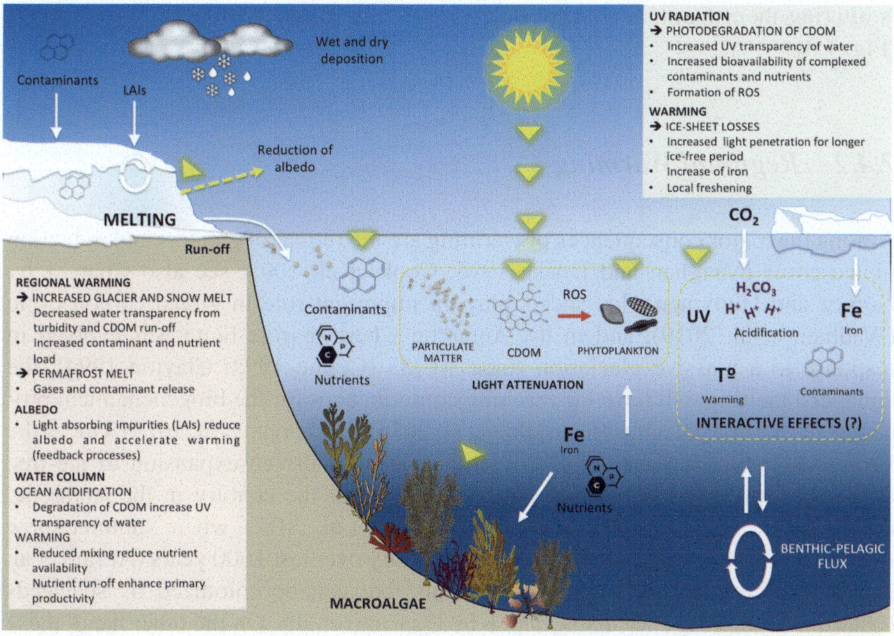

Fig. 7.3 Overview of the major components related to underwater optics and processes occurring under changing environmental context in Antarctic coastal waters, with implications for photobiology of Antarctic seaweeds. See text for details and references

phase), which results in the strengthing of the westerly winds over Antarctic, with impact on aquatic ecosystems, e.g., their mixed layer depth (Bais et al. 2019; Williamson et al. 2019). Because climate change also modifies surface reflectivity (through changes in snow and ice cover), clouds, and aerosols, it together with ozone layer plays key role in defining the UV exposure of Antarctic ecosystems in future (Bais et al. 2019).

Under episodes of ozone depletion during austral spring, band ratios can be modified resulting in enhanced UV-B exposure in proportion to UV-A/PAR, i.e., affecting the balance between UV damage and photorepair (Mitchell and Karentz 1993), thus leading to a higher risk for Antarctic ecosystems. In addition to direct adverse effects on organisms (see Sect. 7.3.2 in the present chap), enhanced UV levels imply elevated potential for photochemical reactions between UV radiation and CDOM, leading to photodegradation of CDOM and hence increasing water transparency (Morris and Hargreaves 1997), bioavailability of complexed contaminants (e.g., heavy metals), and nutrients (reviewed by Zepp 2003), formation of biologically damaging products (e.g., singlet oxygen) (reviewed by Kieber et al. 2003), and stimulating bacterial growth (Kieber et al. 1989). Overall, in recent years increasing attention has been paid to the role of UV radiation in a variety of biogeochemical processes (e.g., carbon cycling, enhanced photodegradation) and how global change is altering them (summarized by Sulzberger et al. 2019; Williamson et al. 2019) (Fig. 7.3).

7.4.2 Regional Warming

Among the major consequences of warming are the retreat of glaciers and reduction of ice cover (Vaughan and Doake 1996; Cook et al. 2005; see also Chap. 1 by Gómez and Huovinen). Ice sheets play an important role in global carbon cycle (Wadham et al. 2019), and in the Antarctic where marine biota was evolved in response to a massive glaciation since 30 Ma (Crame 1992; Clayton 1994), the enhanced melting will have far-reaching consequences for the biogeochemical processes in vast coastal areas (Constable et al. 2014; Sulzberger et al. 2019). Under the strongest forcing scenario of IPCC, climate change-driven expansion of ice-free areas is estimated to be around 25% by the end of the century in the Antarctic, mainly in the WAP (Amesbury et al. 2017; Lee et al. 2017), where summer snow melting has currently reached its highest intensity over last 1000 years (Abram et al. 2013). New ice-free areas provide habitats that can be colonized by seaweeds (Quartino et al. 2013; see also Chap. 8 by Quartino et al.). On the other hand, these areas are often characterized by high turbidity (low light penetration) as a result of sediment runoff (see Fig. 7.3), which can modify the zonation patterns of some species towards shallower waters (Deregibus et al. 2016) and cause reduction of seaweed productivity (Jerosch et al. 2019; see Chap. 9 by Deregibus et al.). In the Arctic, the growth of kelp *Laminaria solidungula* has been found to depend directly on the water transparency, which is strongly governed by resuspension of sediments especially during increased frequency of storm events (Dunton et al. 2009). Increased turbidity can furthermore interfere with disinfection of pathogens and other microbiomes by UV radiation in surface waters with consequences hitherto not well understood (Williamson et al. 2017, 2019; see Chap. 14 by Gaitan and Schmid).

In polar regions, high permafrost temperatures have been registered, with potential impact on global climate through emissions of CO_2 and release of methane. In the Antarctic, temperature increased by 0.37 °C in zones with continuos permafrost during 2007–2016 (IPCC 2019). Increased input of organic matter from catchment as a result of glacier melting can also lead to enhanced nutrient (e.g., nitrates, iron) transfer to coastal areas (Hodson et al. 2017; Ducklow et al. 2018), potentially increasing phytoplankton biomass and thus decreasing light penetration. Increased nutrient availability can also modify the responses of seaweeds to environmental stress, e.g., mitigating adverse effects of UV radiation and metals as observed in sub-Antarctic kelps (Huovinen et al. 2010). Elevated temperature is also known to improve the efficiency of repair processes, leading to higher UV tolerance of Antarctic seaweeds (Rautenberger et al. 2015). Increase of Fe from runoff may cause oxidative stress in Antarctic seaweeds (González et al. 2017).

7.4.3 Feedback with Other Emergent Threats

Growing evidence on the presence of anthropogenic pollution in the Antarctic, especially in the WAP region, is changing our vision of a pristine environment (reviewed by Bargagli 2008). In spite of natural barriers (Antarctic Circumpolar Current and atmospheric circulation) and remoteness, these ecosystems receive pollution, e.g., persistent organic pollutants (POPs), from other regions through long-range atmospheric transport (LRAT), in addition to local sources (Bengtson Nash 2011; Vecchiato et al. 2015; Khairy et al. 2016). In fact, growing human activities (e.g., waste incineration, sewage effluents, fuel combustion) within this region are leaving their marks in the environment (reviewed by Bargagli 2008). Emerging contaminants, such as microplastics, have also already been reported in the Antarctic environment (Waller et al. 2017; Lacerda et al. 2019). In this context, more accessible areas such as the Antarctic Peninsula are increasingly threatened by human impact (e.g., from scientific stations and visitors), which is evidenced by chemical contamination including heavy metals (de Moreno et al. 1997; Farías et al. 2002; Amaro et al. 2015; Padeiro et al. 2016; Chu et al. 2019) and polycyclic aromatic hydrocarbons (PAHs) (Na et al. 2011; Préndez et al. 2011) in the sediments, water, and snow in these ecosystems. The fate of environmental contaminants can be drastically modified under changing climate and ozone depletion scenarios implying multiple and complex consequences (Wrona et al. 2006; Schiedek et al. 2007; Noyes et al. 2009; Grannas et al. 2013; Galic et al. 2017; Sulzberger et al. 2019) (Fig. 7.3). For instance, increased melting of cryospheric environments can release chemical contaminants that have been stored in ice and snow for prolonged periods (Wania and Westgate 2008; Grannas et al. 2013; Hauptmann et al. 2017). The risk of enhanced levels and availability of contaminants like POPs and mercury due to climate change is receiving increased attention (IPCC 2019; Sulzberger et al. 2019). Certain light-absorbing impurities (LAIs) deposited on snow and ice surfaces can induce complex feedback processes that further accelerate melting (Lutz et al. 2014; Tedesco et al. 2016; Huovinen et al. 2018), potentially leading to unpredicted pollution levels in the surrounding environments. The fate of LAIs reaching aquatic environment and their potential for underwater light attenuation is still largely unknown. On the other hand, interaction of certain environmental contaminants (such as PAHs) with UV radiation can lead directly or via ROS to their photomodification (photooxidation, photolysis) resulting in photodegradation or formation of photoproducts with different characteristics and effects (including lower toxicity) as the original compound. They can also achieve higher toxicity (phototoxicity) through photosensitization reactions when they interact with UV radiation (reviewed by Björn and Huovinen 2015; Sulzberger et al. 2019) (Fig. 7.3). Therefore, even when PAH levels in the environment are low, such as those reported for King George Island (Na et al. 2011; Préndez et al. 2011), their potential for UV-induced or UV-enhanced toxicity under ozone depletion implies an unpredicted risk. UV radiation can also contribute to generation of microplastics (summarized by Andrady and Pandey 2019) and interact with other emerging contaminants (e.g., pharmaceutical and personal care products such as sunscreens) (see Björn and Huovinen 2015; Bais et al. 2019; Williamson et al. 2019).

Since 1980s, the oceans as carbon sink have removed 20–30% of CO_2 from anthropogenic emissions, leading to increased ocean acidification (summarized in Gattuso and Hansson 2011). Such conditions are corrosive to marine organisms, causing a decline (3.9% during 1998–2014 in the Southern Ocean) in calcification rates in skeleton- and shell-forming species (IPCC 2019). Ocean systems at high latitudes are especially vulnerable to experience changes of pH due to their lower buffering capacity (McNeil and Matear 2008; Jewett and Romanou 2017). Decreased pH can increase UV transparency of water through photodegradation of CDOM (Schindler et al. 1996; Yan et al. 1996) (Fig. 7.3). In systems affected by freshening due to local glacier melt, acidification can be exacerbated increasing calcium carbonate corrosivity (Evans et al. 2014). To date, studies carried out in calcareous and non-calcareous seaweeds indicate that responses to acidification depend on the taxonomic status, biogeographic location, prevailing metabolism of species (e.g., algae with carbon concentrating mechanisms versus those that rely on CO_2 transport), the type of experimental approach and time of responses, etc. (Hurd et al. 2009; Roleda and Hurd 2012; Hofmann and Bischof 2014). Overall, interaction of acidification, climate change, and UV radiation has been found to impact negatively some calcifying seaweed species (e.g., by reducing structural UV protection) (Russell et al. 2011) or affecting photosynthesis (Gao and Häder 2017). In the Antarctic seaweeds *Desmarestia anceps* and *D. menziesii*, changes in protein and lipid contents were observed when the algae were exposed to combined treatments of pH and temperatures (Schram et al. 2017); however, these conditions did not affect markedly other physiological parameters such as photosynthesis and phlorotannin content (Schoenrock et al. 2015).

7.5 Concluding Remarks

Global change together with increasing anthropogenic impact lead to complex scenarios for Antarctic coastal waters: enhanced turbidity from glacier and ice melt impedes the light penetration, but earlier ice breakup in late winter-spring increases underwater light levels during extended open-water period, whereas UV radiation and acidification can lead to degradation of organic matter and hence increase light penetration. Under favorable light conditions in the water column, increased nutrient load from terrestrial and glacier runoff can stimulate phytoplankton growth causing blooms with negative impact on water clarity and light availability for benthic productivity. Therefore, complex implications for seaweeds are to be expected, ranging from beneficial to harmful, depending on local processes as well as the structure and function of the seaweed communities. How other feedbacks related, e.g., with transient and persistent contaminants, ocean acidification and local freshening will impact the underwater light climate in the Antarctic and their consequences for the biota are difficult to predict. Overall, the scarcity of long-term data that connects the changes in climate and habitat with ecosystem responses has been identified among the major gaps when predicting the climate change impact on Antarctic cryosphere and its associated marine realm (Fountain et al. 2012). Similarly, susceptibility of seaweeds to the predicted changes in the Antarctic habitat and their photobiological adaptations need to be addressed from molecular to systemic scales. In fact, the novel molecular approaches are likely to improve our understanding of the mechanisms that cold-adapted organisms display to cope with the projected environmental variability (Lyon and Mock 2014).

Acknowledgments The funding provided by CONICYT (through Projects Anillo ART1101, FONDECYT 1161129 and FONDAP 15150003) and by Instituto Antártico Chileno (INACH; Grant T-20-09) to carry out our research in the Antarctic is greatly acknowledged. We are also grateful to the members of our research group in Universidad Austral de Chile and the staff of the Instituto Antártico Chileno for their invaluable cooperation.

References

Abram NJ, Mulvaney R, Wolff EW, Triest J, Kipfstuhl S et al (2013) Acceleration of snow melt in an Antarctic Peninsula ice core during the twentieth century. Nat Geosci 6:404–411. https://doi.org/10.1038/NGEO1787

Adams WW III, Zarter CR, Mueh KE, Amiard V, Demmig-Adams B (2006) Energy dissipation and photoinhibition: a continuum of photoprotection. In: Demmig-Adams B, Adams WW III, Mattoo AK (eds) Photoprotection, photoinhibition, gene regulation, and environment. Springer, The Netherlands, pp 49–64

Amaro E, Padeiro A, Mão de Ferro A, Mota AM, Leppe M et al (2015) Assessing trace element contamination in Fildes Peninsula (King George Island) and Ardley Island. Ant Mar Poll Bull 15:523–527

Amesbury MJ, Roland TP, Royles J, Hodgson DA, Convey P et al (2017) Widespread biological response to rapid warming on the Antarctic Peninsula. Curr Biol 27:1616–1622. https://doi.org/10.1016/j.cub.2017.04.034

Amsler CD, Iken K, McClintock JB, Beaker BJ (2011) Defenses of polar macroalgae against herbivores and biofoulers. In: Wiencke C (ed) Biology of polar benthic algae. Walter de Gruyter GmBH & Co, KG, Berlin, New York, pp 101–120

Andrady AL, Pandey KK, Heikkilä (2019) Interactive effects of solar UV radiation and climate change on material damage. Photochem Photobiol Sci 18:804–825. https://doi.org/10.1039/c8pp90065e

Bais AF, Lucas RM, Bornman JF, Williamson CE, Sulzberger B et al (2018) Environmental effects of ozone depletion, UV radiation and interactions with climate change: UNEP environmental effects assessment panel, update 2017. Photochem Photobiol Sci 17:127–179. https://doi.org/10.1039/c7pp90043k

Bais AF, Bernhard G, McKenzie RL, Aucamp PJ, Young AJ et al (2019) Ozone-climate interactions and effects on solar ultraviolet radiation. Photochem Photobiol Sci 18:602–640. https://doi.org/10.1039/c8pp90059k

Bargagli R (2008) Environmental contamination in Antarctic ecosystems. Sci Total Environ 400:212–226

Bengtson Nash S (2011) Persistent organic pollutants in Antarctica: current and future research priorities. J Environ Monit 13:497–504. https://doi.org/10.1039/c0em00230e

Berge J, Renaud P, Darnis G, Cottier F, Last K et al (2015) In the dark: a review of ecosystem processes during the Arctic polar night. Prog Oceanogr 139:258–271. https://doi.org/10.1016/j.pocean.2015.08.005

Bienfang PK, Szyper JOP, Okamoto MY, Noda EK (1984) Temporal and spatial variability of phytoplankton in a subtropical ecosystem. Limnol Oceanogr 29:527–539

Bischof K, Rautenberger R (2012) Seaweed responses to environmental stress: reactive oxygen and antioxidative strategies. In: Wiencke C, Bischof K (eds) Seaweed biology: novel insights into ecophysiology, ecology and utilization. Springer, Berlin, Heidelberg, pp 109–132

Bischof K, Hanelt D, Wiencke C (1998) UV-radiation can affect depth-zonation of Antarctic macroalgae. Mar Biol 131:597–605

Bischof K, Gómez I, Molis M, Hanelt U, Karsten D et al (2006) Ultraviolet radiation shapes seaweed communities. Rev Environ Sci Biotechnol 5:141–166

Björn LO, Huovinen P (2015) Phototoxicity. In: Björn LO (ed) Photobiology: the science of light and life, 3rd edn. Springer, New York, pp 335–345

Buma AGJ, de Boer MK, Boelen P (2001) Depth distributions of DNA damage in Antarctic marine phyto-and bacterioplankton exposed to summertime ultraviolet radiation. J Phycol 37:200–208

Burritt DJ, Lamare MD (2016) The cellular responses of marine algae and invertebrates to ultraviolet radiation, alone and in combination with other common abiotic stressors. In: Solan M, Whiteley NM (eds) Stressors in the marine environment. Oxford University Press, Oxford, pp 117–134

Calkins J, Thordardottir T (1980) The ecological significance of solar UV radiation on aquatic organisms. Nature 283:563–566

Campana GL, Zacher K, Fricke A, Molis M, Wulff A et al (2009) Drivers of colonization and succession in polar benthic macro-and microalgal communities. Bot Mar 52:655–667

Chipperfield MP, Bekki S, Dhomse S, Harris NRP, Hassler B et al (2017) Detecting recovery of the stratospheric ozone layer. Nature 549:211–218. https://doi.org/10.1038/nature23681

Chu WL, Dang NL, Kok YY, Yap KSI, Phang SM, Convey P (2019) Heavy metal pollution in Antarctica and its potential impact on algae. Pol Sci 20:75–83. https://doi.org/10.1016/j.polar.2018.10.004

Clark GF, Stark JS, Palmer AS, Riddle MJ, Johnston EL (2017) The roles of sea-ice, light and sedimentation in structuring shallow Antarctic benthic communities. PLoS One 12(1):e0168391. https://doi.org/10.1371/journal.pone.0168391

Clayton MN (1994) Evolution of the Antarctic benthic algal flora. J Phycol 30:897–904

Coljin F, Admiraal W, Baretta JW, Ruardij P (1987) Primary production in a turbid estuary, the Ems-Dollard: field and model studies. Cont Shelf Res 7:1405–1409

Constable AJ, Melbourne-Thomas J, Corney SP, Arrigo KR, Barbraud C et al (2014) Climate change and Southern Ocean ecosystems. I: how changes in physical habitats directly affect marine biota. Glob Chang Biol 20:3004–3025. https://doi.org/10.1111/gcb.12623

Convey P (2010) Terrestrial biodiversity in Antarctica–recent advances and future challenges. Polar Sci 4(2):135–147

Cook AJ, Fox AJ, Vaughan DG, Ferrigno JG (2005) Retreating glacier fronts on the Antarctic Peninsula over the past half-century. Science 308:541–544

Crame JA (1992) Evolutionary history of the polar regions. Hist Biol 6:37–60

Cruces E, Huovinen P, Gómez I (2013) Interactive effects of UV radiation and enhanced temperature on photosynthesis, phlorotannin induction and antioxidant activities of two sub-Antarctic brown algae. Mar Biol 160:1–13

Cullen JJ, Neale PJ (1994) Ultraviolet radiation, ozone depletion, and marine photosynthesis. Photosynth Res 39:303–320

Cullen JJ, Neale PJ, Lesser MP (1992) Biological weighting function for the inhibition of phytoplankton photosynthesis by ultraviolet radiation. Science 258:646–650

de Moreno JEA, Gerpe MS, Moreno VJ, Vodopivez C (1997) Heavy metals in Antarctic organisms. Polar Biol 17:131–140

Deregibus D, Quartino ML, Campana GL, Momo FR, Wiencke C, Zacher K (2016) Photosynthetic light requirements and vertical distribution of macroalgae in newly ice-free areas in Potter Cove, South Shetland Islands, Antarctica. Polar Biol 39:153–166

Döhler G, Hegmeier E, Grigoleit E, Krause KD (1991) Impact of solar UV radiation on uptake of ^{15}N-ammonia and ^{15}N-nitrate by marine diatoms and natural phytoplankton. Biochem Physiol Pflanzen 187:293–303

Drew EA (1977) Physiology of photosynthesis and respiration in some Antarctic marine algae. Brit Ant Surv Bull 46:59–76

Ducklow HW, Stukel MR, Eveleth R, Doney SC, Jickells T et al (2018) Spring–summer net community production, new production, particle export and related water column biogeochemical processes in the marginal sea ice zone of the Western Antarctic Peninsula 2012–2014. Phil Trans R Soc A 376:20170163

Dunton KH, Schonberg SV, Funk DW (2009) Interannual and spatial variability in light attenuation: evidence from three decades of growth in the arctic kelp, *Laminaria solidungula*. In: Proceedings of Smithsonian at the Poles Symposium, Smithsonian Institution, Washington, DC, 3–4 May 2007. Smithsonian Institute Scholarly Press, Washington, DC, pp 271–284

Evans W, Mathis JT, Cross JN (2014) Calcium carbonate corrosivity in an Alaskan inland sea. Biogeosciences 11:365–379

Farías S, Pérez Arisnabarreta S, Vodopivez C, Smichowski P (2002) Levels of essential and potentially toxic metals in Antarctic macro algae. Spectrochim Acta B 57:2133–2140

Farman JC, Gardiner BG, Shanklin JD (1985) Large losses of total ozone in Antarctica reveal seasonal ClOx/NOx interaction. Nature 315:207–210

Figueroa FL (2002) Bio-optical characteristics of Gerlache and Bransfield Strait waters during an Antarctic summer cruise. Deep-Sea Res II Top Stud Oceanogr 49:675–691

Flores-Molina MR, Rautenberger R, Muñoz P, Huovinen P, Gómez I (2016) Stress tolerance of the endemic Antarctic brown alga *Desmarestia anceps* to UV radiation and temperature is mediated by high concentrations of phlorotannins. Photochem Photobiol 92:455–466

Fountain AG, Campbell JL, Schuur EAG, Stammerjohn SE, Williams MW, Ducklow HW (2012) The disappearing cryosphere: impacts and ecosystem responses to rapid cryosphere loss. Bioscience 62:405–415

Fountoulakis I, Bais AF, Tourpali K, Fragkos K, Misios S (2014) Projected changes in solar UV radiation in the Arctic and sub-Arctic Oceans: effects from changes in reflectivity, ice transmittance, clouds, and ozone. J Geophys Res Atmosp 119:8073–8090

Frederick JE, Snell HE, Haywood EK (1989) Solar ultraviolet radiation at the earth's surface. Photochem Photobiol 50:443–450

Fritsen CH, Wirthlin ED, Momberg DK, Lewis MJ, Ackley SF (2011) Bio-optical properties of Antarctic pack ice in the early austral spring. Deep-Sea Res II 58:1052–1061. https://doi.org/10.1016/j.dsr2.2010.10.028

Galic N, Grimm V, Forbes VE (2017) Impaired ecosystem process despite little effects on populations: modeling combined effects of warming and toxicants. Glob Chang Biol 23:2973–2989. https://doi.org/10.1111/gcb.13581

Gao K, Häder DP (2017) Effects of ocean acidification and UV radiation on marine photosynthetic carbon fixation. In: Kumar M, Ralph P (eds) Systems biology of marine ecosystems. Springer, Cham, pp 235–250

Gattuso JP, Hansson L (2011) Ocean acidification: background and history. In: Gattuso JP, Hansson L (eds) Ocean acidification. Oxford University Press, Oxford, pp 1–27

Gerber S, Häder DP (1992) UV effects on photosynthesis, proteins and pigmentation in the flagellate *Euglena gracilaris*: biochemical and spectroscopic observations. Biochem Syst Ecol 20:485–492

Goes JI, Handa N, Taguchi S, Hama T (1994) Effect of UV-B radiation on the fatty acid composition of the marine phytoplankter *Tetraselmis* sp.: relationship to cellular pigments. Mar Ecol Prog Ser 114:259–274

Goes JI, Handa N, Taguchi S, Hama T, Saito H (1995) Impact of UV radiation on the production pattern and composition of dissolved free and combined amino acids in marine phytoplankton. J Plakton Res 17:1337–1362

Gómez I, Huovinen P (2015) Lack of physiological depth patterns in conspecifics of endemic Antarctic brown algae: a trade-off between UV stress tolerance and shade adaptation? PLoS One 10(8):e0134440. https://doi.org/10.1371/journal.pone.0134440

Gómez I, Weykam G, Klöser H, Wiencke C (1997) Photosynthetic light requirements, daily carbon balance and zonation of sublittoral macroalgae from King George Island (Antarctica). Mar Ecol Prog Ser 148:281–293

Gómez I, Wulff A, Roleda MY, Huovinen P, Karsten U et al (2009) Light and temperature demands of marine benthic micro-algae and seaweeds in the polar regions. Bot Mar 52:593–608

Gómez I, Navarro NP, Huovinen P (2019) Bio-optical and physiological patterns in Antarctic seaweeds: a functional trait based approach to characterize vertical zonation. Prog Oceanogr 174:17–27. https://doi.org/10.1016/j.pocean.2018.03.013

González PM, Deregibus D, Malanga G, Campana GL, Zacher K et al (2017) Oxidative balance in macroalgae from Antarctic waters. Possible role of Fe. J Exp Mar Biol Ecol 486:379–386. https://doi.org/10.1016/j.jembe.2016.10.018

Grannas AM, Bogdal C, Hageman KJ, Halsall C, Harner T et al (2013) The role of the global cryosphere in the fate of organic contaminants. Atmos Chem Phys 13:3271–3305. https://doi.org/10.5194/acp-13-3271-2013

Hanelt D, Jaramillo J, Nultsch W, Senger S, Westermeier R (1994) Photoinhibition as a regulative mechanisms of photosynthesis in marine algae of Antarctica. Ser Cient INACh 44:67–77

Hargreaves BR (2003) Water column optics and penetration of UVR. In: Helbling EW, Zagarese HE (eds) UV effects in aquatic organisms and ecosystems. Royal Society of Chemistry, Cambridge, pp 59–105

Hauptmann AL, Sicheritz-Ponten T, Cameron KA, Baelum J, Plichta DR et al (2017) Contamination of the Arctic reflected in microbial metagenomes from Greenland ice sheet. Environ Res Lett 12:074019

Helbling EW, Marguet ER, Villafañe VE, Holm-Hansen O (1995) Bacterioplankton viability in Antarctic waters as affected by solar ultraviolet radiation. Mar Ecol Prog Ser 126:293–298

Hessen DO, Tranvik LJ (eds) (1998) Aquatic humic substances: ecology and biogeochemistry. Springer, Berlin Heidelberg

Hodson A, Nowak A, Sabacka M, Jungblut A, Navarro F et al (2017) Climatically sensitive transfer of iron to maritime Antarctic ecosystems by surface runoff. Nat Commun 8:14499

Hofmann LC, Bischof K (2014) Ocean acidification effects on calcifying macroalgae. Mar Ecol Prog Ser 22:261–279

Holm-Hansen O, Lubin D, Helbling WE (1993) Ultraviolet radiation and its effects on organisms in aquatic environments. In: Young AR, Björn LO, Moan J, Nultsch W (eds) Environmental UV photobiology. Plenum Press, New York, pp 379–425

Huot Y, Jeffrey WH, Davis RF, Cullen JJ (2000) Damage to DNA in bacterioplankton: a model of damage by ultraviolet radiation and its repair as influenced by vertical mixing. Photochem Photobiol 72:62–74

Huovinen P, Gómez I (2011) Spectral attenuation of solar radiation in Patagonian fjord and coastal waters and implications for algal photobiology. Cont Shelf Res 31:254–259

Huovinen P, Gómez I (2013) Photosynthetic characteristics and UV stress tolerance of Antarctic seaweeds along the depth gradient. Polar Biol 36:1319–1332. https://doi.org/10.1007/s00300-013-1351-3

Huovinen PS, Penttilä H, Soimasuo MR (2003) Spectral attenuation of solar ultraviolet radiation in humic lakes in Central Finland. Chemosphere 51:205–214

Huovinen P, Gómez I, Lovengreen C (2006) A five-year study of solar ultraviolet radiation in Southern Chile (39° S): potential impact on physiology of coastal marine algae? Photochem Photobiol 82:515–522

Huovinen P, Leal P, Gómez I (2010) Interacting effects of copper, nitrogen and ultraviolet radiation on the physiology of three south Pacific kelps. Mar Freshw Res 61:330–341

Huovinen P, Ramírez J, Gómez I (2016) Underwater optics in sub-Antarctic and Antarctic coastal ecosystems. PLoS One 11(5):e0154887. https://doi.org/10.1371/journal.pone.0154887

Huovinen P, Ramírez J, Gómez I (2018) Remote sensing of albedo-reducing snow algae and impurities in the Maritime Antarctica. ISPRS J Photogram Rem Sens 146:507–517

Hurd CL, Hepburn CD, Currie KI, Raven JA, Hunter KA (2009) Testing the effects of ocean acidification on algal metabolism: considerations for experimental designs. J Phycol 45:1236–1251

Iken K, Quartino ML, Barrera-Oro E, Palermo J, Wiencke C, Brey T (1998) Trophic relations between macroalgae and herbivores. Ber Polarforsch 299:258–262

IPCC (2019) Special report on the ocean and cryosphere in a changing climate. https://www.ipcc.ch/srocc/download-report/

Jerosch K, Scharf FK, Deregibus D, Campana GL, Zacher K et al (2019) Ensemble modeling of Antarctic macroalgal habitats exposed to glacial melt in a Polar Fjord. Front Ecol Evol 7:207. https://doi.org/10.3389/fevo.2019.00207

Jewett L, Romanou A (2017) Ocean acidification and other ocean changes. In: Wuebbles DJ, Fahey DW, Hibbard KA, Dokken DJ, Stewart BC et al (eds) Climate science special report: fourth national climate assessment, vol I. U.S. Global Change Research Program, Washington, DC, pp 364–392

Johnston EL, Connell SD, Irving AD, Pile AJ, Gillanders BM (2007) Antarctic patterns of shallow subtidal habitat and inhabitants in Wilke's Land. Polar Biol 30:781–788

Karentz D, McEuen FS, Land MC, Dunlap WC (1991) Survey of mycosporine-like amino acid compounds in Antarctic marine organisms: potential protection from ultraviolet exposure. Mar Biol 108:157–166

Karsten U, Wulff A, Roleda MY, Müller R, Steinhoff FS et al (2009) Physiological responses of polar benthic algae to ultraviolet radiation. Rev Bot Mar 52:639–654. https://doi.org/10.1515/BOT.2009.077

Khairy MA, Luek JL, Dickhut R, Lohmann R (2016) Levels, sources and chemical fate of persistent organic pollutants in the atmosphere and snow along the western Antarctic Peninsula. Environ Pollut 216:304–313. https://doi.org/10.1016/j.envpol.2016.05.092

Kieber DJ, McDaniel J, Mopper J (1989) Photochemical source of biological substrates in seawater: implications for carbon cycling. Nature 341:637–639

Kieber DJ, Peake BM, Scully NM (2003) Reactive oxygen species in aquatic ecosystems. In: Helbling EW, Zagarese HE (eds) UV effects in aquatic organisms and ecosystems. Royal Society of Chemistry, Cambridge, pp 251–290

Kirk JTO (ed) (2011) Light and photosynthesis in aquatic ecosystems, 3rd edn. Cambridge University Press, Cambridge

Klöser H, Ferreyra G, Schloss I, Mercuri G, Laturnus F, Curtosi A (1993) Seasonal variation of algal growth conditions in sheltered Antarctic bays: the example of Potter Cove (King George Island, south Shetlands). J Mar Syst 4:289–301

Klöser H, Quartino ML, Wiencke C (1996) The distribution of macroalgae and macroalgal communities in gradients of physical conditions in Potter Cove, King George Island, Antarctica. Hydrobiologia 333:1–17

Kuttippurath J, Nair PJ (2017) The signs of Antarctic ozone hole recovery. Sci Rep 7:585. https://doi.org/10.1038/s41598-017-00722-7

Lacerda ALF, Rodrigues LS, van Sebille E, Rodrigues FL, Ribeiro L et al (2019) Plastics in sea surface waters around the Antarctic Peninsula. Sci Rep 9:3977. https://doi.org/10.1038/s41598-019-40311-4

Lee JR, Raymond B, Bracegirdle TJ, Chadès I, Fuller RA et al (2017) Climate change drives expansion of Antarctic ice-free habitat. Nature 547:49. https://doi.org/10.1038/nature22996

Lesser MP, Lamare MD, Barker MF (2004) Transmission of ultraviolet radiation through the Antarctic annual sea ice and its biological effects on sea urchin embryos. Limnol Oceanogr 49:1957–1963

Lutz S, Anesio AM, Villar J, Benning LG (2014) Variations of algal communities cause darkening of a Greenland glacier. FEMS Microbiol Ecol 89:402–414. https://doi.org/10.1111/1574-6941.12351

Lyon BR, Mock T (2014) Polar microalgae: new approaches towards understanding adaptations to an extreme and changing environment. Biology 3:56–80
Madronich (1994) Increases in biologically damaging UV-B radiation due to stratospheric ozone reductions: a brief review. Arch Hydrobiol Beih Ergebn Limnol 43:17–30
Madronich S, McKenzie RL, Björn LO, Caldwell MM (1998) Changes in biologically active ultraviolet radiation reaching the earth's surface. J Photochem Photobiol B Biol 46:5–19
Madronich S, Flocke S (1999) The role of solar radiation in atmospheric chemistry. In: Boule P (ed) Environmental photochemistry. Springer-Verlag, Berlin, pp 1–26
McMinn A, Martin A (2013) Dark survival in a warming world. Proc R Soc B 280:20122909. https://doi.org/10.1098/rspb.2012.2909
McNeil BI, Matear RJ (2008) Southern Ocean acidification: a tipping point at 450-ppm atmospheric CO_2. PNAS 105(48):18860–18864
Mitchell BG, Holm-Hansen O (1991) Bio-optical properties of Antarctic Peninsula waters: differentiation from temperate ocean models. Deep-Sea Res 38:1009–1028
Mitchell DL, Karentz D (1993) The induction and repair of DNA photodamage in the environment. In: Young AR, Björn LO, Moan J, Nultsch W (eds) Environmental UV photobiology. Plenum Press, New York, pp 345–377
Mobley CD (2015) Underwater light. In: Björn LO (ed) Photobiology: the science of light and life, 3rd edn. Springer, New York Springer, pp 77–84
Molina MJ, Rowland RS (1974) Stratospheric sink for chlorofluoromethanes: chlorine atom catalyzed destruction of ozone. Nature 249:810–812
Morris DP, Hargreaves BP (1997) The role of photochemical degradation of dissolved organic carbon in regulating the UV transparency of three lakes on the Pocono Plateau. Limnol Oceanogr 42:239–249
Na G, Liu C, Wang Z, Ge L, Ma X, Yao Z (2011) Distribution and characteristic of PAHs in snow of Fildes Peninsula. J Environ Sci 23:1445–1451
Navarro NP, Huovinen P, Gómez I (2019) Photosynthetic characteristics of geographically disjunct seaweeds: a case study on the early life stages of Antarctic and Subantarctic species. Prog Oceanogr 174:28–36. https://doi.org/10.1016/j.pocean.2018.11.001
Neale PJ (2000) Spectral weighting functions for quantifying effects of UV radiation in marine ecosystems. In: de Mora S, Demers S, Vernet M (eds) Effects of UV radiation in the marine environment, vol 10. Cambridge University Press, Environmental Chemical Series, pp 72–100
Neale PJ, Davis RF, Cullen JJ (1998) Interactive effects of ozone depletion and vertical mixing on photosynthesis of Antarctic phytoplankton. Nature 392:585–589
Noyes PD, McElwee MK, Miller HD, Clark BW, Van Tiem LA et al (2009) The toxicology of climate change: environmental contaminants in a warming world. Rev Env Int 35:971–986
Nuñez-Pons L, Avila C, Romano G, Verde C, Giordano D (2018) UV-protective compounds in marine organisms from the Southern Ocean. Rev Mar Drugs 16:336. https://doi.org/10.3390/md16090336
Oduor S, Schagerl M (2007) Phytoplankton primary productivity characteristics in response to photosynthetically active radiation in three Kenyan Rift Valley saline-alkaline lakes. J Plankton Res 29:1041–1050. https://doi.org/10.1093/plankt/fbm078
Padeiro A, Amaro E, dos Santos MMC, Araujo MF, Gomes SS et al (2016) Trace element contamination and availability in the Fildes Peninsula, King George Island, Antarctica. Environ Sci: Processes Impacts 18:648–657
Préndez M, Barra C, Toledo C, Richter B (2011) Alkanes and polycyclic aromatic hydrocarbons in marine surficial sediment near Antarctic stations at Fildes Peninsula. King George Island Antarctic Sci 23:578–588
Quartino M, Zaixso H, Boraso de Zaixso A (2005) Biological and environmental characterization of marine macroalgal assemblages in Potter Cove, South Shetland Islands, Antarctica. Bot Mar 48:187–197
Quartino ML, Deregibus D, Campana GL, Latorre GEJ, Momo FR (2013) Evidence of macroalgal colonization on newly ice-free areas following glacial retreat in Potter Cove (South Shetland Islands), Antarctica. PLoS One 8(3):e58223. https://doi.org/10.1371/journal.pone.0058223

Rautenberger R, Huovinen P, Gómez I (2015) Effects of increased seawater temperature on UV tolerance of Antarctic marine macroalgae. Mar Biol 162:1087–1097

Robinson D, Arrigo K, Iturriaga R, Sullivan C (1995) Microalgal light-harvesting in extreme low-light environments in McMurdo Sound, Antarctica. J Phycol 31:508–520

Roleda MY, Hurd CL (2012) Seaweed responses to ocean acidification. In: Wiencke C, Bischof K (eds) Seaweed biology: insights into ecophysiology, ecology and utilization. Springer, Berlin, pp 407–431

Roleda MY, Wiencke C, Lüder UH (2006) Impact of ultraviolet radiation on cell structure, UV-absorbing compounds, photosynthesis, DNA damage and germination in zoospores of Arctic *Saccorhiza dermatodea*. J Exp Bot 57:3847–4856

Runcie JW, Riddle MJ (2007) Photosynthesis of marine macroalgae in ice-covered and ice free environments in East Antarctica. Eur J Phycol 41:223–233

Russell BD, Passarelli CA, Connell SD (2011) Forecasted CO_2 modifies the influence of light in shaping subtidal habitat. J Phycol 47:744–752

Schiedek D, Sundelin B, Readman JW, Macdonald RW (2007) Interactions between climate change and contaminants. Rev Mar Poll Bull 54:1845–1856

Schindler DW, Curtis PJ, Parker BP, Stainton MP (1996) Consequences of climate warming and lake acidification for UV-B penetration in North American boreal lakes. Nature 379:705–708

Schloss IR, Ferreyra GA (2001) Primary production, light and vertical mixing in Potter Cove, a shallow bay in the maritime Antarctic. Polar Biol 25:41–48

Schram JB, Schoenrock KM, McClintock JB, Amsler CD, Angus RA (2017) Ocean warming and acidification alter Antarctic macroalgal biochemical composition but not amphipod grazer feeding preferences. Mar Ecol Prog Ser 581:45–56

Schwarz AM, Hawes I, Andrew N, Norkko A, Cummings V, Thrush S (2003) Macroalgal photosynthesis near the southern global limit for growth; Cape Evans, Ross Sea, Antarctica. Polar Biol 26:789–799

Schwarz A-M, Hawes I, Andrew N, Mercer S, Cummings V, Thrush S (2005) Primary production potential of non-geniculate coralline algae at Cape Evans, Ross Sea, Antarctica. Mar Ecol Prog Ser 294:131–140

Schoenrock KM, Schram JB, Amsler CA, McClintock JB, Angus RA (2015) Climate change impacts on overstory *Desmarestia* spp. from the western Antarctic Peninsula. Mar Biol 162:377–389

Scully NM, Lean DRS (1994) The attenuation of ultraviolet radiation in temperate lakes. Arch Hydobiol Ergebn Limnol 43:135–144

Setlow RB (1974) The wavelengths in sunlight effective in producing skin cancer: a theoretical analysis. Proc Natl Acad Sci U S A 71:3363–3366

Smith RC, Baker KS (1979) Penetration of UV-B and biologically effective dose-rates in natural waters. Photochem Photobiol 29:311–323

Smith RC, Prézelin BB, Baker KS, Bidigare RR, Boucher NP et al (1992) Ozone depletion: ultraviolet radiation and phytoplankton biology in Antarctic waters. Science 255:952–959

Solomon S, Ivy DJ, Kinnison D, Mills MJ, Neely RR, Schmidt A (2016) Emergence of healing in the Antarctic ozone layer. Science 353(6296):269–274. https://doi.org/10.1126/science.aae0061

Solomon S, Ivy D, Gupta M, Bandoro J, Santer B et al (2017) Mirrored changes in Antarctic ozone and stratospheric temperature in the late 20th versus early 21st centuries. J Geophys Res Atmos 122:8940–8950. https://doi.org/10.1002/2017JD026719

Sulzberger B, Austin AT, Cory RM, Zepp RG, Paul ND (2019) Solar UV radiation in a changing world: roles of cryosphere-land-water-atmosphere interfaces in global biogeochemical cycles. Photochem Photobiol Sci 18:747–774. https://doi.org/10.1039/c8pp90063a

Taskjelle T, Hudson SR, Granskog MA, Nicolaus M, Lei R et al (2016) Spectral albedo and transmittance of thin young Arctic sea ice. J Geophys Res Oceans 121:540–553. https://doi.org/10.1002/2015JC011254

Tedesco M, Doherty S, Fettweis X, Alexander P, Jeyaratnam J, Stroeve J (2016) The darkening of the Greenland ice sheets: trends, drivers, and projections (1981–2100). Cryosphere 10:477–496

Tedetti M, Sempéré R (2006) Penetration of ultraviolet radiation in the marine environment. A review. Photochem Photobiol 82:389–397
Tyler JE (1975) The in situ quantum efficiency of natural phytoplankton populations. Limnol Oceanogr 20:976–980
Valdivia N, Díaz MJ, Garrido I, Gómez I (2015) Consistent richness-biomass relationship across environmental gradients in a marine macroalgal-dominated subtidal community on the Western Antarctic Peninsula. PLoS One 10(9):e0138582. https://doi.org/10.1371/journal.pone.0138582
Vaughan DG, Doake CSM (1996) Recent atmospheric warming and retreat of ice shelves on the Antarctic Peninsula. Nature 379:328–331
Vecchiato M, Argiriadis E, Zambon S, Barbante C, Toscano G et al (2015) Persistent Organic Pollutants (POPs) in Antarctica: occurrence in continental and coastal surface snow. Microchem J 119:75–82. https://doi.org/10.1016/j.microc.2014.10.010
Vincent WF, Belzile C (2003) Biological UV exposure in the polar oceans: Arctic-Antarctic comparisons. In: Huiskes AHL, Gieskes WWC, Rozema J, Schorno RML, van der Vies SM, Wolff WJ (eds) Antarctic biology in a global context. Backhuys Publishers, Leiden, The Netherlands, pp 176–181
Vincent WF, Neale PJ (2000) Mechanisms of UV damage to aquatic organisms. In: de Mora S, Demers S, Vernet M (eds) The effects of UV radiation in the marine environment, vol. 10. Cambridge University Press, Cambridge, pp 149–176
Vincent WF, Roy S (1993) Solar ultraviolet-B radiation and aquatic primary production: damage, protection, and recovery. Environ Rev 1:1–12
Wadham JL, Hawkings JR, Tarasov L, Gregoire LJ, Spencer RGM et al (2019) Ice sheets matter for the global carbon cycle. Nat Commun 10:3567. https://doi.org/10.1038/s41467-019-11394-4
Wahl M, Molis M, Davis A, Dobretsov S, Dürr ST et al (2004) UV effects that come and go: a global comparison of marine benthic community level impacts. Glob Chang Biol 10:1962–1972. https://doi.org/10.1111/j.1365-2486.2004.00872.x
Waller CL, Griffiths HJ, Waluda CM, Thorpe SE, Loaiza I et al (2017) Microplastics in the Antarctic marine system: an emerging area of research. Review. Sci Total Environ 598:220–227. https://doi.org/10.1016/j.scitotenv.2017.03.283
Wania F, Westgate JN (2008) On the mechanisms of mountain cold-trapping of organic chemicals. Environ. Sci Technol 42:9092–9098
Weiler CS, Penhale PA (eds) (1994) Ultraviolet radiation in Antarctica: measurements and biological effects, vol 62. American Geophysical Union, Antarctic Research Series, Washington DC
Weykam G, Gómez I, Wiencke C, Iken K, Klöser H (1996) Photosynthetic characteristics and C: N ratios of macroalgae from King George Island (Antarctica). J Exp Mar Biol Ecol 204:1–22
Wiencke C, Gómez I, Pakker H, Flores-Moya A, Altamirano M et al (2000) Impact of UV-radiation on viability, photosynthetic characteristics and DNA of brown algal zoospores: implications for depth zonation. Mar Ecol Prog Ser 197:217–229
Wiencke C, Roleda MY, Gruber A, Clayton MN, Bischof K (2006) Susceptibility of zoospores to UV radiation determines upper depth distribution limit of Arctic kelps: evidence through field experiments. J Ecol 94:455–463
Wiencke C, Gómez I, Dunton K (2009) Phenology and seasonal physiological performance of polar seaweeds. Bot Mar 52:585–592
Williamson CE, Neale PJ, Grad G, de Lange HJ, Hargreaves BR (2001) Beneficial and detrimental effects of UV on aquatic organisms: implications of spectral variation. Ecol Appl 11:1843–1857
Williamson CE, Madronich S, Lal A, Zepp RG, Lucas RM et al (2017) Climate change-induced increases in precipitation are reducing the potential for solar ultraviolet radiation to inactivate pathogens in surface waters. Sci Rep 7:13033
Williamson CE, Neale PJ, Hylander S, Rose KC, Figueroa FL et al (2019) The interactive effects of stratospheric ozone depletion, UV radiation, and climate change on aquatic ecosystems. Photochem Photobiol Sci 18:717–746. https://doi.org/10.1039/c8pp90062k
Wrona FJ, Prowse TD, Reist JD, Hobbie JE, Levesque LMJ et al (2006) Effects of ultraviolet radiation and contaminant-related stressors on Arctic freshwater ecosystems. Ambio 35:388–401

Yan ND, Keller W, Scully NM, Lean DRS, Dillon PJ (1996) Increased UV-B penetration in a lake owing to drought-induced acidification. Nature 381:141–143

Zacher KD, Hanelt D, Wiencke C, Wulff A (2007a) Grazing and UV radiation effects on an Antarctic intertidal microalgal assemblage: a long-term field study. Polar Biol 30:1203–1212

Zacher K, Roleda MY, Hanelt D, Wiencke C (2007b) UV effects on photosynthesis and DNA in propagules of three Antarctic seaweeds (*Adenocystis utricularis*, *Monostroma hariotii* and *Porphyra endiviifolium*). Planta 225:1505–1516

Zacher K, Rautenberger R, Hanelt D, Wulff A, Wiencke C (2009) The abiotic environment of polar marine benthic algae. Bot Mar 52:483–490

Zepp RG (2003) Solar UVR and aquatic carbon, nitrogen, sulfur and metals cycles. In: Helbling EW, Zagarese HE (eds) UV effects in aquatic organisms and ecosystems. Royal Society of Chemistry, Cambridge, pp 137–184

Zielinski K (1990) Bottom macroalgae of the Admiralty Bay (King George Island, Antarctica). Pol Polar Res 11:95–131

Chapter 8
Production and Biomass of Seaweeds in Newly Ice-Free Areas: Implications for Coastal Processes in a Changing Antarctic Environment

María L. Quartino, Leonardo A. Saravia, Gabriela L. Campana, Dolores Deregibus, Carolina V. Matula, Alicia L. Boraso, and Fernando R. Momo

Abstract The Antarctic rocky coasts are mainly colonized by extensive seaweed communities, which play key roles as food resource, habitat, and refuge for many benthic and pelagic organisms. Due to climate warming, Antarctic marine ecosystems are being affected by glacier retreat opening new habitats, e.g., newly ice-free areas that can be colonized by macroalgae. As a consequence, primary production and fate of macroalgae are changing in these new polar environments. In these ecosystems, the carbon production, especially from large brown algae, is an important food source to the benthic invertebrate communities mainly when other resources are scarce. Thus, in new areas colonized by seaweeds, the trophic structure and biogeochemical fluxes can vary considerably. Moreover, when seaweeds die or are removed by water movement, ice scouring, or storms, they are detached, fragmented, and degraded, incorporating and releasing particulate and dissolved organic matter to the coastal food webs, i.e., they support a large fraction of the secondary production of the benthos. The present chapter is a review of the knowledge on seaweed biomass and production in the coastal Antarctic ecosystem opening a discussion on the role of these organisms as main energy sources in, e.g., small fjords and glacier-influenced sites, impacted by recent climatic change.

Keywords Carbon flux · Glacier retreat · Ice-free areas · Potter Cove · Seaweed production

M. L. Quartino (✉)
Departamento de Biología Costera, Instituto Antártico Argentino (IAA), Buenos Aires, Argentina

Museo Argentino de Ciencias Naturales B Rivadavia (MACN), Buenos Aires, Argentina
e-mail: lquartino@dna.gov.ar

I. Gómez, P. Huovinen (eds.), *Antarctic Seaweeds*,
https://doi.org/10.1007/978-3-030-39448-6_8

8.1 Introduction: Seaweeds in Coastal Marine Ecosystems

Worldwide, macroalgae are essential components of coastal waters providing habitat, nursery and refuge for fish and numerous mobile and sessile invertebrates (Hurd et al. 2014) and can potentially also play a role in providing refuge from climate change driven stressors, e.g., ocean acidification (Hurd 2015; Krause-Jensen et al. 2016; Wahl et al. 2018; see also Chap. 11 by Gómez and Huovinen). Furthermore, large brown algae, e.g., species of order Laminariales (known as "kelps"), are regarded as ecosystem engineers that could significantly modify the habitat characteristics such as water velocity, light penetration, and physic-chemical properties of seawater (Jones et al. 1994; Dawson et al. 2010), while creating an understory condition favorable for species adapted to, e.g., low light intensity (Steneck et al. 2002).

The macroalgal biomass is an important component of the functioning of the marine ecosystem; a large portion of seaweeds is not consumed by herbivores but returns to the environment as decaying organic matter (Cebrián 2004). These large

L. A. Saravia
Instituto de Ciencias, Universidad de General Sarmiento (UNGS),
Los Polvorines, Buenos Aires, Argentina
e-mail: lsaravia@campus.ungs.edu.ar

G. L. Campana
Departamento de Biología Costera, Instituto Antártico Argentino (IAA),
Buenos Aires, Argentina

Departamento de Ciencias Básicas, Universidad Nacional de Luján (UNLu),
Luján, Buenos Aires, Argentina
e-mail: gcampana@dna.gov.ar

D. Deregibus
Departamento de Biología Costera, Instituto Antártico Argentino (IAA),
Buenos Aires, Argentina

Consejo Nacional de Investigaciones Científicas y Técnicas, (CONICET),
Buenos Aires, Argentina
e-mail: dderegibus@dna.gov.ar

C. V. Matula
Departamento de Biología Costera, Instituto Antártico Argentino (IAA),
Buenos Aires, Argentina
e-mail: ucv@mrecic.gov.ar

A. L. Boraso
Instituto de Desarrollo Costero, Universidad Nacional de la Patagonia San Juan Bosco
(UNPSJB), Comodoro Rivadavia, Argentina

F. R. Momo
Instituto de Ciencias, Universidad de General Sarmiento (UNGS),
Los Polvorines, Buenos Aires, Argentina

Departamento de Ciencias Básicas, Universidad Nacional de Luján (UNLu),
Luján, Buenos Aires, Argentina
e-mail: fmomo@campus.ungs.edu.ar

quantities of seaweed biomass can follow different paths: they can be detached from the hard bottom and drift by the water movements (Baring et al. 2018). Also the whole thalli can drift and slowly be fragmented and be thrown by the waves to the coastline. On the beach they can be accumulated and, once degraded, enter again to the system by the tides sinking to the bottom of the seabed where they decompose (Krause-Jensen and Duarte 2016). Thus, this algal debris can supply important amounts of organic matter to benthic ecosystems and provide food, directly for detritivores or indirectly, by stimulating bacterial metabolism (Rossi et al. 2013).

Primary production is the formation of organic matter (carbon) in a photosynthetic organism per time unit (Israel 1995). In general, ecological studies of primary production refer to "net primary production" (NPP), which is that portion of gross primary production from photosynthesis that remains after consumption via respiration. This may represent different processes depending on the methodology used to measure it. In most studies, net primary production represents the dry mass of plant matter produced per unit area per unit time (Reed et al. 2008).

Seaweed biomass can be measured using destructive and nondestructive sampling. The first one corresponds to the collection of algal biomass in terms of grams by unit area (g m^{-2}). The nondestructive sampling implies the use of underwater images or video, which mainly permit to gain information on algal cover. Coverage data must be converted into biomass, usually through linear regression to estimate finally biomass (Allison 2004). Several improvements have been proposed to this method, taking into account the species-specific and nonlinear nature of the relationship between cover and biomass due to the different algal architectures. Thus, a better prediction of biomass would be obtained by fitting a multivariate generalized additive model (GAM) considering all species present in each quadrat (Pedersen et al. 2019). Various studies have focused on quantifying the macrophyte production to assess its contribution to the marine ecosystem worldwide. For example, macroalgae and seagrasses store about 0.4% and 16% of their net primary production in the sediments, respectively (Duarte and Cebrián 1996), and some of the excess organic matter can be exported to adjacent water column (Barrón and Duarte 2009; Wada and Hama 2013). Duarte and Cebrián (1996) reported that the export of organic matter accounts, on average, for 25% of the net primary production of macroalgae. Thus, macroalgae can be regarded as the most extensive and productive primary producers in coastal zones around the world. It was estimated that they cover about 3.5 million km^2 and can account for a global net primary production of about 1.5 Tg Cyr^{-1} (Krause-Jensen and Duarte 2016).

8.2 Macroalgae and Carbon Fluxes in Antarctic Coastal Areas

In contrast to the global algal production, there is not enough information on macroalgal production in Antarctica. In Antarctic marine ecosystems, macroalgae are one of the major primary producers that grow and develop associated to rocky substrates of different slope and size (Klöser et al. 1996; Wiencke and Amsler 2012). In general, Antarctic benthic macroalgae are distributed from the intertidal down to the lower subtidal zone at 40 m depths and present a clear zonation: an intertidal fringe is usually abraded by winter ice floes, it is mainly colonized by opportunistic species, and it is devoid of large perennial macroalgae. The upper subtidal is dominated by the brown seaweeds *Desmarestia menziesii* J.Agardh and *Ascoseira mirabilis* Skottsberg under strong turbulence. The central subtidal (above 15 m) is mainly dominated by *Desmarestia anceps* Montagne under moderate turbulence, and it is gradually replaced by *Himantothallus grandifolius* (Gepp and Gepp) Zinova. This latter species becomes dominant at deeper waters, where turbulence is usually negligible (Klöser et al. 1996; Quartino et al. 2005; Wulff et al. 2009; see Chap 11 by Gómez and Huovinen).

Antarctic macroalgae are thought to deliver huge amounts of organic matter to the Antarctic coastal food webs (Fischer and Wiencke 1992; Gómez et al. 2009, Marina et al. 2018). Furthermore, they are a food source for numerous mobile consumers, including invertebrates and some nototheniid fish (Iken et al. 1998; Amsler et al. 2005; Barrera Oro et al. 2019). Sublittoral rocky shores dominated by macroalgae have been regarded as important areas hosting high values of abundance and biomass of organisms (Amsler et al. 1995; Brouwer et al. 1995; Quartino and Boraso de Zaixso 2008; Wulff et al. 2009).

One of the main macroalgal carbon contributions to the ecosystem is probably the great amounts of biomass scattered on the coasts, an accumulation of allochthonous organic debris known as "wrack" that is commonly observed in the intertidal zone. In the maritime Antarctic zone, this type of organic matter has been recognized as an important source of carbon for diverse consumers (Zenteno et al. 2019). Grange and Smith (2013) reported drifting macroalgae as very abundant in some fjord basins compared to the open shelf. These authors observed substantial abundances of drifting algae ranging from ~8 to 130 cm m^{-2}.

Once on the coast, the stranded algae can also be used by seals and sea elephants as resting places. Westermeier et al. (1992) recorded a mean value of 22 kg m^{-2} corresponding to the highest abundance of drifting macroalgae along the coastline of Maxwell Bay. In Admiralty Bay, the amount of macroalgae deposited on the beach was estimated to be 104 kg m^{-2} dry weight (Rakusa-Suszczewski 1995). The organic matter of the decomposed algae mixed with excrement of seabirds (guano) reenters the marine ecosystem by the waves, winds, and tide action. Leaching of nutrients to the environment and subsequently to the coastal waters, are the final stages of macroalgal decomposition, after which the dissolved nutrients fertilize the shore through the tidal beach runoff with nitrate, nitrite, ammonia, and phosphorous (Nedzarek and Rakusa-Suszczewski 2004). Thus, the volumes of algal debris can reflect the high productivity normally assigned to Antarctic seaweeds populating different regions across the Antarctic Peninsula.

8.3 Macroalgal Biomass Studies in Antarctica

In the decade 1980–1990s and early the 2000s, several studies described the biomass (standing crop) of macroalgae at the Antarctic benthos, particularly around the West Antarctic Peninsula (Table 8.1). These early works implied frequently the use of "destructive sampling," which allowed a more accurate assessment of the identity and biomass of each taxa (DeLaca and Lipps 1976; Miller and Pearse 1991; Amsler et al. 1995; Brouwer et al. 1995; Quartino et al. 2001, 2005). These studies showed that in the northern portion of the Western Antarctic Peninsula and adjacent islands, seaweeds dominate shallow benthic communities on hard substrates, covering over 80% of the seafloor, with standing biomass levels in the range of 5–10 kg m^{-2} wet weight (Wiencke and Amsler 2012, Gómez et al. 2009), a number comparable to temperate kelp forests (Mann 1972; Duggins 1988).

During the following decades, many benthic community studies were conducted using underwater photo and/or video (nondestructive sampling) (Quartino et al. 2013; Sahade et al. 2015; Valdivia et al. 2015; Lagger et al. 2018; Jerosch et al. 2019). These methodologies facilitate fast sampling and are a great advantage in

Table 8.1 Maximum biomass (g m^{-2}, wet or dry weight) determined for four species of macroalgae reported by different surveys (depths between brackets when provided by the authors)

Locality	*Desmarestia anceps*	*Desmarestia menziesii*	*Himantothallus grandifolius*	*Iridaea cordata*	Reference
Wet weight					
(1)	–	2050	–	–	Richardson (1979)
(2)	5660 (11 m)	1850 (11 m)	1250 (11 m)	–	Brouwer et al. (1995)
(3)	–			3440 (4–5 m)	Cormaci et al. (1996)
(4) a	6044 (20 m)	6737 (5 m)	10,336 (20 m)	2554.60 (0 m)	Quartino and Boraso de Zaixso (2008)
(4) b	3900 (5 m)	3470 (3 m)		923.8 (3 m)	C. Matula, personal communication
	4198 (10 m)	3975 (5 m)		575 (5 m)	
Dry weight					
(5)	–	800 (4 m)	600 (12 m)	–	DeLaca and Lipps (1976)
(2)	1000 (11 m)	460 (11 m)	240 (11 m)	–	Brouwer et al. (1995)
	3300 (5 m)				
(4) a	671 (20 m)	749 (5 m)	1152 (20 m)	255.36 (0 m)	Quartino and Boraso de Zaixso (2008)
(4) b	821 (5 m)	730.5 (3 m)		194.5 (3 m)	C. Matula, personal communication
	883.8 (10 m)	836.85 (5 m)		121 (5 m)	

Localities: (1) Borge Bay; South Orkney Islands (60° 43′S, 45° 36′W); (2) Signy Island, South Orkney Islands (60° 42′S 45° 36′W); (3) Ross Sea (74° 30′S, 165° 30′E); (4) Potter Cove, South Shetland Islands (62° 14′S, 58° 38′W), a, sampling done in 1994–1996, and b, sampling done in 2015–2016; (5) Ansvers Island (62° 46′S, 64° 04′W) this extreme and rough environment. Values recorded by these methods are mainly percent coverage of macroalgae by unit area. Although cover data are valuable, it is always convenient to transform it into biomass in order to calculate and quantify the algal production as mentioned previously.

8.4 The Ecosystem of Potter Cove: An Outstanding Case Study

Potter Cove (62° 14′S, 58° 40′W), a small fjord at the Isla 25 de Mayo/King Gorge Island, South Shetland Islands, has become an ideal place to carry out biological, geological, and oceanographic studies. This cove is divided into two characteristic areas, the mouth (or outer part) and an inner part separated by a shallow sill (Klöser et al. 1996). Both areas are biologically distinctive: the rocky shores of the outer part of Potter Cove host a high biomass of macroalgae (Klöser et al. 1996; Quartino et al. 2005), whereas the inner cove has one of the largest concentrations of benthic filter feeders found in Antarctica (Tatián et al. 2004).

In fact, the first surveys on macroalgae had identified an important benthic algal community associated with the presence of hard bottom substrate and light availability (Quartino et al. 2001, 2005). In most of the studied sites, abundance of large Desmarestiales was observed from 5 down to 30 m depth, in a continuous vertical zonation. This high algal biomass was detected on rocky coastal areas close to the mouth of the cove, where the more transparent, oceanic water prevailed (Quartino et al. 2001, 2005).

Macroalgal production of the most abundant species decreased during summer months (Quartino and Boraso de Zaixso 2008) following the seasonal strategy exhibited normally by "season anticipators," which start their growth under short-day conditions in late winter/spring, often under the sea-ice, reaching their maximum biomass in summer months (Wiencke et al. 2009; Wiencke et al. 2014). Quartino and Boraso de Zaixso (2008) calculated an average standing stock of 792, 84 ton and an algal production of 1401 ton during summer season for the whole Potter Cove. This study provided the first estimation of seaweed production in Antarctica, combining field biomass data with experimental growth rates calculated by Wiencke (1990a, b) and Gómez and Wiencke (1997). Additionally, an estimation of their decomposition was included to describe the macroalgal mass balance in this Antarctic fjord (Fig. 8.1). Due to the usually low pelagic primary productivity in Potter Cove (Schloss and Ferreyra 2002), the study suggested that macroalgae were the main carbon source, supporting a large fraction of the secondary production of the benthic system (Tatián et al. 2004; Sahade et al. 2004; Quartino et al. 2008).

The continuous monitoring of this marine ecosystem has permitted to survey the changes in the physiognomy of the cove along the last three decades. One of the most singular observations was the progressive melting of the Fourcade Glacier (Rückamp et al. 2011), which surrounds Potter Cove. The retreat of this glacier has

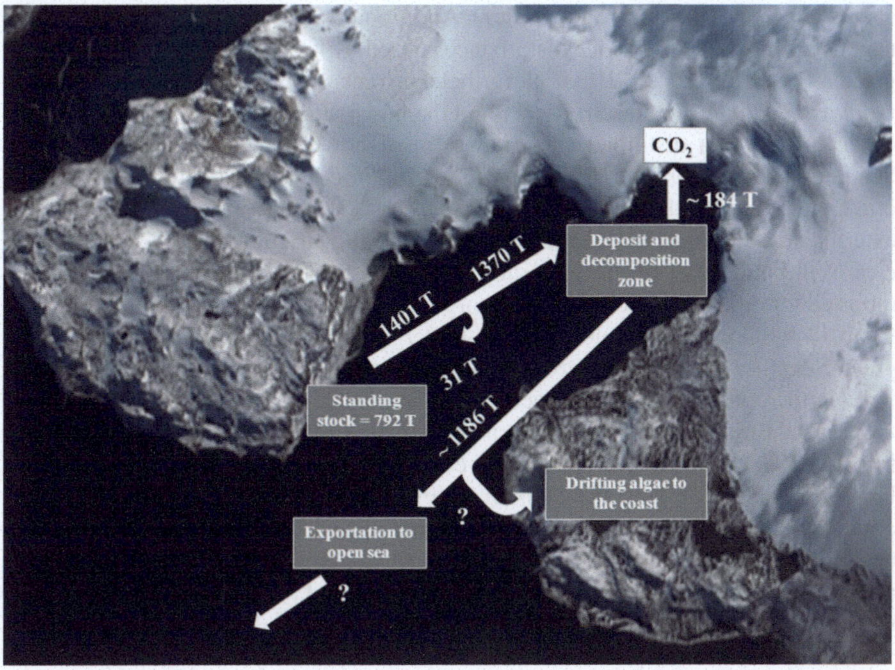

Fig. 8.1 Pathways of macroalgae production in Potter Cove. The average macroalgal standing stock (792 ton) was calculated for the summer season considering the entire cove area (Quartino and Boraso de Zaixso 2008). The production (1401 ton) and the flux of biomass to the ecosystem (1370 ton) were calculated using only the dominant macroalgal species (Quartino et al. 2008). Detached macroalgae are expected to enter the inner cove via the prevailing cyclonic water circulation (Schloss and Ferreyra 2002). The decomposed fraction was calculated using a rate of 0.0016 day^{-1}, estimated by Brouwer (1996) for *Desmarestia anceps*

been uncovering new hard bottom ice-free areas available for benthic colonization, especially macroalgae (Quartino et al. 2013). These newly ice-free areas are extremely disturbed, presenting a considerable alteration of the water column (due to the increase of sediment input and salinity changes particularly during the summer melting season) and on the ice disturbance patterns (Eraso and Domínguez 2007; Quartino et al. 2013). In fact, the photographs and video records revealed a well-developed macroalgal community even in close proximity to the retreating glacier where the sediment load was high. In these sites, the increase in the sediment runoff reduces the light penetration and can constitute a constraint for photosynthesis (Schloss et al. 2012; Wiencke and Amsler 2012; Quartino et al. 2013). Particularly, this can result in a change in the vertical distribution of those species adapted to dark conditions, which probably will move to shallower depths adjusting their light requirements in this new environment with higher sediment load (Deregibus et al. 2016; see also Chap. 9 by Deregibus et al.).

Thus, under these optical conditions, development and vertical distribution of the macroalgae can be strongly affected by the high turbidity (Deregibus et al. 2016; see also Chap. 7 by Huovinen and Gómez). Additionally macroalgal colonization was negatively impacted by the ice disturbance caused by ice blocks originated from the glacier (Quartino et al. 2013). Nevertheless some species such as the red alga *Palmaria decipiens* (Reinsch) Ricker and the green alga *Monostroma hariotii* Gain thrive in the most disturbed sites. It was also shown that *Gigartina skottsbergii* Setchell and Gardner and *Himantothallus grandifolius* were adapted to harsh abiotic conditions, but the red alga *G. skottsbergii* seemed to be more sensitive to the external stress (González et al. 2017). The vertical distribution of the subtidal species in the inner part of the cove did not fit with the typical macroalgal zonation reported for other sites around King George Island (Huovinen and Gómez 2013 and Chap 11 by Gómez and Huovinen); in Potter Cove algal species usually found at greater depth were observed growing at shallow waters. The main results showed that (1) the complexity of the macroalgal community was positively correlated to the elapsed time from the ice retreat: newly ice-free areas closer to the outer side of the cove presented mature macroalgal communities dominated by perennial and large brown algae of the genus *Desmarestia*, (2) algal development depended on the optical conditions and the sediment input with some species being limited by light availability, (3) macroalgal colonization was negatively affected by ice disturbance, and (4) the colonization was determined by the size and type of substrate and by the slope of the seafloor (Fig. 8.2).

In this new warming scenario, ice melting is probably the primary cause of changes in the macroalgal communities in the inner Cove, and it could be mediated by different associated phenomena (Fig. 8.3). Glacier retreat originates newly ice-free areas providing hard substrate available for benthic colonization, which can be positively related to increased diversity, richness, and macroalgal cover at different spatial scales (Valdivia et al. 2014). In addition, glacier melting increases ice scouring, sediment runoff in the water column and enhances turbidity having a negative effect on these ecological parameters because algae do not have enough light available for photosynthesis. Consequently, macroalgae may shift their vertical distribution increasing overlapping and competition, resulting in a negative effect (Fig. 8.3). Additionally a complementary PAR model was performed with the variables affecting the annual PAR availability in this climate change context (see Chap 9 by Deregibus et al.).

8.5 A Dynamic Growth Model for Antarctic Macroalgae Under a Fast-Changing Environment

Dynamic growth models have been developed mainly for farmed macroalgae (Hadley et al. 2015; Zhang et al. 2016; Broch and Slagstad 2012; Duarte and Ferreira 1997) and also to determine the growth of species in eutrophicated sites and under algal blooms at mid latitudes (Ren et al. 2014; Perrot et al. 2014). Despite their recognized great importance as primary producers (Hurd et al. 2014), there are no

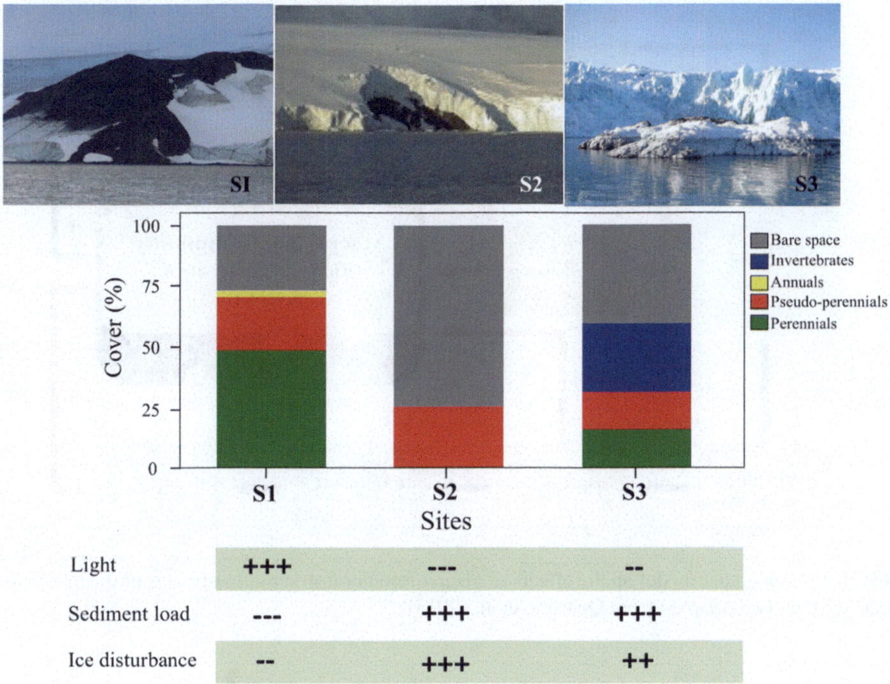

Fig. 8.2 Three different situations observed in the newly ice-free areas of Potter Cove. S1 corresponded to sites unaffected by sedimentation and dominated by large brown algae. S2 corresponded to a site dominated by the red algae *Palmaria decipiens* in close proximity to a retreating glacier with high sediment inflow and high ice disturbance. S3 represented a newly ice-free area in the inner side of the cove close to the glacier and with influence of sediment runoff from small creeks. In this site the coexistence of macroalgae and macrofauna was observed. Macroalgae were classified according to their life history as annuals, pseudo perennials and perennials. Symbols + and – correspond to the grade of intensity. (Adapted from Quartino et al. (2013))

dynamic models for Antarctic macroalgae. In this line, we recently described a new model adapted to the particular conditions of the Antarctic environment (Guillaumot et al. 2018). The model describes the biomass dynamics of algal assemblages without considering a particular species (Fig. 8.4) and can also be applied to dominant species that do not experience interspecific competition. The outputs of this approach can be integrated in carbon flux models of Antarctic coastal ecosystems. The assumptions of the model are that the growth is not limited by nutrient availability (Drew and Hastings 1992; Ducklow et al. 2007; Wiencke and Amsler 2012) and that substrate availability, light, and ice scouring should be regarded as the main environmental variables (Quartino et al. 2013). Considering that the levels of irradiance that reach the bottom is mediated by depth, suspended sediments, and incident radiation, the model takes into account the high variability in incident light and turbidity, two factors with high seasonal fluctuations in Antarctic. The estimations use a daily time sequence, assuming that the numbers of days available for growth are

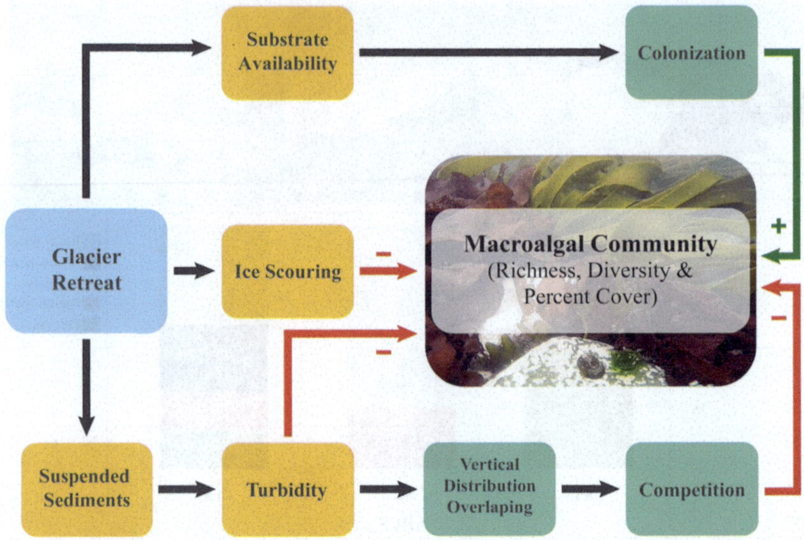

Fig. 8.3 Conceptual model on the effects of glacier retreat on structural attributes of the macroalgal community. (Adapted from Quartino et al. (2013))

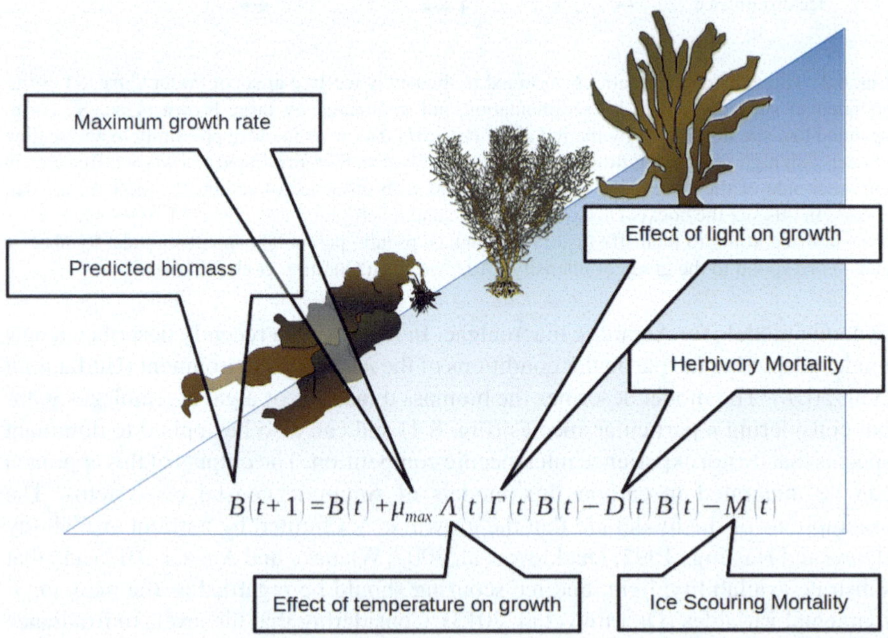

Fig. 8.4 Overview of the dynamic growth model of Antarctic macroalgae and the main variables entered in the algorithm. (Adapted from Gómez and Huovinen (2015))

determined by the period when the zone considered is free of ice. Then the model equations would be:

$$B(t+1) = B(t) + \mu_{\max}\Lambda(t)\Gamma(t)B(t) - D(t)B(t) - M(t)$$

where $B(t)$ is the biomass; μ_{max} is the maximum intrinsic growth rate; $\Lambda(t)$ and Γ represent the effects of light and temperature on growth, respectively; and $D(t)$ and $M(t)$ are the biomass dependent mortality rates associated with herbivory and ice scouring respectively.

Each of these terms is also dependent on other environmental characteristics defined as:

$$\Lambda = \frac{I}{I_0} e^{\left(1-\frac{I}{I_0}\right)}$$

where I_0 is the optimum irradiance and I is the light available at the seabed that will have more dependencies:

$$I = I_s e^{-K_d Z}$$

where I_s is the irradiance at the sea surface, K_d is the extinction coefficient (Kirk 1994), and Z is the depth. The incident light I_s must be recorded in situ and averaged over several years to represent the mean conditions, while K_d is also affected by the turbidity produced by glacier melting (Deregibus et al. 2016; see also Chap. 9 by Deregibus et al.).

The dependence of growth on temperature is modeled as a function of an optimal temperature range (Duarte et al. 2003):

$$\Gamma = \frac{2.0(1+\beta)X_T}{X_T^2 + 2.0\beta X_T + 1.0}$$

where T_{max} is the upper temperature above which growth ceases and β is an adjustment parameter. X_T is defined as:

$$X_T = \frac{T - T_{opt}}{T_{opt} - T_{max}}$$

where T_{opt} is the optimal temperature for growth and T is the average daily water temperature which must be supplied from local measurements or estimated. Even although this kind of mechanistic model requires experimental and field data to be calibrated and validated, it can be a useful tool to provide predictions of macroalgal biomass in different scenarios of climate change.

8.6 Seaweed Production in Present and Future Warming Scenarios

In recent decades, the ice shelves of the WAP have changed rapidly and consequently a significant retreat was observed on both sides of the Peninsula (Cook et al. 2005). In response to relatively small environmental change, some ecosystems can undergo abrupt transformation (Clarke et al. 2013). For example, decreases in the spatial and temporal extent of sea ice will affect the underwater light environment during winter and early spring altering photosynthesis of organisms adapted to specific light regimes (See Chap. 9 by Deregibus et al.). Moreover, a shorter permanence of sea ice and enhanced sediment inflow from the glaciers will modify the environment reducing light availability for algal photosynthesis and can also affect survival and germination rates directly (Arakawa and Matsuike 1992; Chapman and Fletcher 2002; Eriksson and Johansson 2005). A significant change in the annual light budget is likely to have major consequences for benthic ecosystems in shallow waters: the timing of annual sea ice breakup affects the composition of benthic communities, primarily due to a change in the available light on an annual basis (Clarke et al. 2013; see also Chap. 7 by Huovinen and Gómez). Thus, global warming is expected to cause an earlier sea-ice melting, which will likely induce tipping points for many areas at shallow depths, probably causing ecosystems to shift from predominantly heterotrophic to predominantly autotrophic, the later state dominated by macroalgal production (Clarke et al. 2013; see also Chap. 15 by Momo et al. and Chap. 16 by Ortiz et al.).

Linking macroalgal decomposition and the trophic network in the future coastal scenarios, the colonization of macroalgae in the new areas probably will change the structure and composition of the whole ecosystem. As it was mentioned before, this situation is being detected in Potter Cove, where the increasingly higher colonization of macroalgae in newly ice-free areas is providing carbon sources (food) for the second fraction of the trophic web. In this system sediment retains the nitrogen and carbon released from the macroalgae degradation process, whereas a relatively small amount of macroalgal-derived nutrients is released back into the overlying water, with a negative feedback on phytoplankton production in the water column (Braeckman et al. 2019). These authors suggested that this "sink effect" or "food bank" would also explain the prolonged food availability in the euphotic Antarctic sedimentary benthos in the form of microphytobenthic biomass, at least as long as the overlying waters are not too turbid or ice covered (Mincks et al. 2005).

A recent study in Potter Cove using an ensemble model predicted a reduction of macroalgal summer production under increasing sedimentation (Jerosch et al. 2019). Thus, the projection of macroalgal distribution in a near future scenario shows a dramatic reduction of summerly macroalgal productivity inside the cove, while the glacier continues melting and increasing the discharge of sediments. However, this process could be mitigated in part by an increased colonization and productivity of macroalgae in shallow newly ice-free hard bottom areas inside the cove (Quartino et al. 2013; Deregibus et al. 2016; Campana et al. 2018).

8.7 Future Prospects

Even though macroalgae are the most productive marine macrophytes worldwide, they have been excluded from "blue carbon" assessments as they typically grow in habitats that are not considered to accumulate large stocks of organic carbon (Krause-Jensen and Duarte 2016). Recently, more studies have highlighted the potential of marine vegetation as a sink for anthropogenic C emissions (known as "Blue Carbon") suggesting that marine macroalgae can sequestrate anthropogenic CO_2 (Chung et al. 2011; Queirós et al. 2019). So far, there are also some few reports of the presence of macroalgal carbon in marine sediments, suggesting that the presence of macroalgal carbon may be widespread, extending from shallow to deep-sea sediments and from polar to tropical regions. It has been also reported that this type of carbon can be found across a broad range of depths into the sediment, from surface and subsurface layers down to deeper than a hundred metres into the sediment (Krause-Jensen et al. 2018).

Despite the evidence of macroalgae as carbon sinks, the rates and magnitude of this process has not been estimated precisely. Some studies have delivered a first-order estimate of the contribution of macroalgae to carbon sequestration from burial in coastal sediments and its export to the deep sea (Duarte and Cebrián 1996; Krause-Jensen and Duarte 2016). This last step of macroalgal C sequestration is particularly important because the exchange of carbon with the atmosphere is precluded over extended time periods, even after being remineralized (Krause-Jensen and Duarte 2016).

In all, meltwater influences coastal areas where macroalgae are the dominant species covering most of the rocky bottoms of many coastal areas along the WAP. However, there is still little knowledge on the macroalgal carbon production in Antarctic shallow waters and its fate under changing environmental conditions. Therefore, it would be necessary to implement monitoring programs to measure the expansion of these organisms in other locations of the WAP to estimate at large scale the fraction of macroalgae production sink in the sediment, the budget of macroalgal production, their decomposition, and further export to other marine ecosystems.

Acknowledgments Special thanks to Silvia Rodríguez for her valuable collaboration and technical assistance at the Coastal Biology Department of Instituto Antártico Argentino. We also thank the support of the diving and logistic teams at Carlini Station. This study was performed with the financial support of Instituto Antártico Argentino/Dirección Nacional del Antártico, ANPCyT-DNA (PICTO 2010-116, PICT 2017-2691). This chapter also presents an outcome of the international Research Network IMCONet funded by the Marie Curie Action IRSES IMCONet (FP7 IRSES, Action No. 318718) and of CoastCarb (Funding ID 872690, H2020-MSCA-RISE-2019 - Research and Innovation Staff Exchange).

References

Allison G (2004) The influence of species diversity and stress intensity on community resistance and resilience. Ecol Monogr 74(1):117–134

Amsler CD, Rowley RJ, Laur DR et al (1995) Vertical distribution of Antarctic peninsular macroalgae: cover, biomass and species composition. Phycologia 34(5):424–430

Amsler CD, Iken K, McClintock JB et al (2005) Comprehensive evaluation of the palatability and chemical defenses of subtidal macroalgae from the Antarctic Peninsula. Mar Ecol Prog Ser 294:141–159

Arakawa H, Matsuike K (1992) Influence on insertion of zoospores, germination, survival, and maturation of gametophytes of brown algae exerted by sediments. Nippon Suis Gakkai 58:619–625

Baring RJ, Fairweather PG, Lester RE (2018) Nearshore drift dynamics of natural versus artificial seagrass wrack. Estuar Coast Shelf Sci 202:164–171

Barrera-Oro E, Moreira E, Seefeldt MA et al (2019) The importance of macroalgae and associated amphipods in the selective benthic feeding of sister rockcod species *Notothenia rossii* and *N. coriiceps* (Nototheniidae) in West Antarctica. Polar Biol 42(2):317–334

Barrón C, Duarte CM (2009) Dissolved organic matter release in a *Posidonia oceanica* meadow. Mar Ecol Prog Ser 374:75–84

Braeckman U, Pasotti F, Vazquez S, Zacher K, Hoffmann R, Elvert M, Marchant H, Buchner C, Quartino ML et al (2019) Degradation of macroalgal detritus in shallow coastal Antarctic sediments. Limnol Oceanogr 64:1423–1441

Broch OJ, Slagstad D (2012) Modelling seasonal growth and composition of the kelp. J Appl Phycol 24:759–776

Brouwer PE (1996) Decomposition in situ of the sublittoral Antarctic macroalga *Desmarestia anceps* Montagne. Polar Biol 16(2):129–137

Brouwer PE, Geilen EFM, Gremmen NJ et al (1995) Biomass, cover and zonation pattern of sublittoral macroalgae at Signy Island, South Orkney Islands, Antarctica. Bot Mar 38:259–270

Campana GL, Zacher K, Deregibus D et al (2018) Succession of Antarctic benthic algae (Potter Cove, South Shetland Islands): structural patterns and glacial impact over a four-year period. Polar Biol 41(2):377–396

Cebrián J (2004) Role of first-order consumers in ecosystem carbon flow. Ecol Lett 7(3):232–240

Chapman AS, Fletcher RL (2002) Differential effects of sediments on survival and growth of *Fucus serratus* embryos (Fucales, Phaeophyceae). J Phycol 38:894–903

Chung IK, Beardal J, Mehta S et al (2011) Using marine macroalgae for carbon sequestration: a critical appraisal. J Appl Phycol 23(5):877–886

Clarke GF, Stark JS, Johnston EL, Runcie JW, Goldsworthy PM, Raymond B, Riddle MJ (2013) Light-driven tipping points in polar ecosystems. Glob Chang Biol 19(12):3749–3761

Cook AJ, Fox AJ, Vaughan DG et al (2005) Retreating glacier fronts on the Antarctic Peninsula over the past half-century. Science 308(5721):541–544

Cormaci M, Furnari G, Scammacca B, Alongi G (1996) Summer biomass of a population of Iridaea cordata *Iridaea cordata* (Gigartinaceae, Rhodophyta) from Antarctica. In: Lindstrom SC, Chapman DJ (eds) Proceedings of the 15th International Seaweed Symposium. Hydrobiologia 326/327:267–272

Dawson TP, Rounsevell MD, Kluvánková-Oravská T et al (2010) Dynamic properties of complex adaptive ecosystems: implications for the sustainability of service provision. Biodivers Conserv 19(10):2843–2853

DeLaca TE, Lipps JH (1976) Shallow-water marine associations, Antarctic Peninsula. Antarct J 11:12–20

Deregibus D, Quartino ML, Campana GL et al (2016) Photosynthetic light requirements and vertical distribution of macroalgae in newly ice-free areas in Potter Cove, South Shetland Islands, Antarctica. Polar Biol 39(1):153–166

Drew EA, Hastings RM (1992) A year-round ecophysiological study of *Himantothallus grandifolius* (Desmarestiales, Phaeophyta) at Signy Island, Antarctica. Phycologia 31:262–277

Duarte CM, Cebrián J (1996) The fate of marine autotrophic production. Limnol Oceanogr 41(8):1758–1766

Duarte P, Ferreira JG (1997) A model for the simulation of macroalgal population dynamics and productivity. Ecol Model 98:199–214

Duarte P, Meneses R, Hawkins AJS, Zhu M, Fang J, Grant J (2003) Mathematical modelling to assess the carrying capacity for multi-species culture within coastal waters. Ecol Model 168(1–2):109–143

Ducklow HW, Hansell DA, Morgan JA (2007) Dissolved organic carbon and nitrogen in the Western Black Sea. Mar Chem 105(1–2):140–150

Duggins DO (1988) The effects of kelp forests on nearshore environments: biomass, detritus, and altered flow. In: Van Blaricom GR, Estes JA (eds) The community ecology of sea otters. Ecological studies (analysis and synthesis), vol 65. Springer, Berlin, pp 192–201

Eraso A, Domınguez MA (2007) Physicochemical characteristics of the subglacier discharge in Potter Cove, King George Island, Antarctica. In: Tyk A, Stefaniak K (eds) Karst and cryokarst, vol 45. Studies of the Faculty of Earth Sciences, University of Silesia, Sosnowiec, pp 111–122

Eriksson BK, Johansson G (2005) Effects of sedimentation on macroalgae species-specific responses are related to reproductive traits. Oecologia 143:438–448

Fischer G, Wiencke C (1992) Stable carbon isotope composition, depth distribution and fate of macroalgae from the Antarctic Peninsula region. Polar Biol 12:341–348

Gómez I, Huovinen P (2015) Lack of physiological depth patterns in conspecifics of endemic Antarctic brown algae: a trade-off between UV stress tolerance and shade adaptation?. PLoS One 10(8):e0134440

Gómez I, Wiencke C (1997) Seasonal growth and photosynthetic performance of Antarctic macroalga *Desmarestia menziesii* (Phaeophyceae) cultivated under fluctuating Antarctic daylengths. Bot Acta 110:25–31

Gómez I, Wulff A, Roleda MY, Huovinen P, Karsten U, Quartino ML, Dunton K, Wiencke C (2009) Light and temperature demands of marine benthic micro-algae and seaweeds in the polar regions. Bot Mar 52:593–608

González PM, Deregibus D, Malanga G et al (2017) Oxidative balance in macroalgae from Antarctic waters. Possible role of Fe. J Exp Mar Biol Ecol 486:379–386

Grange LJ, Smith CR (2013) Megafaunal communities in rapidly warming fjords along the West Antarctic Peninsula: hotspots of abundance and beta diversity. PLoS One 8(12):e77917

Guillaumot C, Aguera A, Danis B, Deregibus D, Quartino ML, Saravia LA (2018) Growth model of Antarctic macroalgae in a fast-changing environment. Abstract Proceedings of Scar Open Science Conference, Davos, Switzerland, p 863

Hadley S, Wild-Allen K, Johnson C et al (2015) Modeling macroalgae growth and nutrient dynamics for integrated multi-trophic aquaculture. J Appl Phycol 27(2):901–916

Huovinen P, Gómez I (2013) Photosynthetic characteristics and UV stress tolerance of Antarctic seaweeds along the depth gradient. Polar Biol 36:1319–1332

Hurd CL (2015) Slow-flow habitats as refugia for coastal calcifiers from ocean acidification. J Phycol 51(4):599–605

Hurd CL, Harrison PJ, Bischof K, Lobban CS (eds) (2014) Seaweed ecology and physiology. Cambridge University Press, Cambridge

Iken K, Quartino ML, Barrera-Oro E, Palermo J, Wiencke C, Brey T (1998) Trophic relations between marcoalgae and herbivores. Rep Polar Res 299:201–206

Israel A (1995) Determinacion de la produccion primaria en macroalgas marinas. In: Alveal K, Ferrario ME, Oliveira EC, Sar E (eds) Manual de métodos ficológicos. Universidad de Concepción, Concepción, pp 795–823

Jerosch K, Scharf FK, Deregibus D et al (2019) Ensemble modelling of Antarctic macroalgal habitats exposed to glacial melt in a polar fjord. Front Ecol Evol 7(207):1–16

Jones CG, Lawton JH, Shachak M (1994) Organisms as ecosystem engineers. Oikos 69:373–386

Kirk JT (ed) (1994) Light and photosynthesis in aquatic ecosystems. Cambridge University Press, Cambridge

Klöser H, Quartino ML, Wiencke C (1996) Distribution of macroalgae and macroalgal communities in gradients of physical conditions in Potter Cove, King George Island, Antarctica. Hydrobiology 333:1–17

Krause-Jensen D, Duarte CM (2016) Substantial role of macroalgae in marine carbon sequestration. Nat Geosci 9(10):737–743

Krause-Jensen D, Marbà N, Sanz-Martin M et al (2016) Long photoperiods sustain high pH in Arctic kelp forests. Sci Adv 2(12):e1501938

Krause-Jensen D, Lavery P, Serrano O, Marbà N, Masque P, Duarte CM (2018) Sequestration of macroalgal carbon: the elephant in the blue carbon room. Biol Lett 14:20180236

Lagger C, Nime M, Torre L et al (2018) Climate change, glacier retreat and a new ice-free island offer new insights on Antarctic benthic responses. Ecography 40:1–12

Mann KH (1972) Ecological energetics of the seaweeds zone in a marine bay on the Atlantic coast of Canada. II productivity of seaweeds. Mar Biol 14:199–209

Marina TI, Salinas V, Cordone G et al (2018) The food web of Potter Cove (Antarctica): complexity, structure and function. Estuar Coast Shelf Sci 200:141–151

Miller KA, Pearse JS (1991) Ecological studies of seaweeds in McMurdo Sound, Antarctica. Am Zool 31:35–48

Mincks SL, Smith CR, DeMaster DJ (2005) Persistence of labile organic matter and microbial biomass in Antarctic shelf sediments: evidence of a sediment food bank. Mar Ecol Prog Ser 300:3–19

Nedzarek A, Rakusa-Suszczewski S (2004) Decomposition of macroalgae and the release of nutrient Admiralty Bay, King George, Antarctica. Polar Biosci 17:26–35

Pedersen EJ, Miller DL, Simpson GL, Ross N (2019) Hierarchical generalized additive models in ecology: an introduction with mgcv. Peer J 7:e6876

Perrot T, Rossi N, Ménesguen A et al (2014) Modelling green macroalgal blooms on the coasts of Brittany, France to enhance water quality management. J Mar Syst 132:38–53

Quartino ML, Boraso de Zaixso AL (2008) Summer macroalgal biomass in Potter Cove, South Shetland Islands, Antarctica: its production and flux to the ecosystem. Polar Biol 31:281–294

Quartino ML, Klöser H, Schloss IR et al (2001) Biomass and associations of benthic marine macroalgae from the inner Potter Cove (King George Island, Antarctica) related to depth and substrate. Polar Biol 24(5):349–355

Quartino ML, Zaixso HE, Boraso de Zaixso AL (2005) Biological and environmental characterization of marine macroalgal assemblages in Potter Cove, South Shetland Islands, Antarctica. Bot Mar 48:187–197

Quartino ML, Boraso de Zaixso AL, Momo FR (2008) Macroalgal production and the energy cycle of Potter Cove. In: Wiencke C, Ferreyra GA, Abele D, Marenssi S (eds) The Antarctic ecosystem of Potter Cove, King-George Island (Isla 25 de Mayo). Synopsis of research performed 1999–2006 at the Dallmann Laboratory and Jubany Station, p 68–74

Quartino ML, Deregibus D, Campana GL et al (2013) Evidence of macroalgal colonization on newly ice-free areas following glacial retreat in Potter Cove (South Shetland Islands), Antarctica. PLoS One 8:e58223

Queirós AM, Stephens N, Widdicombe S et al (2019) Connected macroalgal-sediment systems: blue carbon and food webs in the deep coastal ocean. Ecol Monogr 89(3):e01366

Rakusa-Suszczewski S (1995) The hydrography of Admiralty Bay and its inlets, coves and lagoons (King George Island, Antarctica). Pol Polar Res 16:61–70

Reed DC, Rassweiler A, Arkema KK (2008) Biomass rather than growth rate determines variation in net primary production by giant kelp. Ecology 89(9):2493–2505

Ren JS, Barr NG, Scheuer K et al (2014) A dynamic growth model of macroalgae: application in an estuary recovering from treated wastewater and earthquake-driven eutrophication. Estuar Coast Shelf Sci 148:59–69

Richardson MG (1979) The distribution of Antarctic marine macro-algae related to depth and substrate. Br Antarct Surv Bull 49:1–13

Rossi F, Gribsholt B, Gazeau F et al (2013) Complex effects of ecosystem engineer loss on benthic ecosystem response to detrital macroalgae. PLoS One 8(6):e66650

Rückamp M, Braun M, Suckro S et al (2011) Observed glacial changes on the King George Island ice cap, Antarctica, in the last decade. Glob Planet Change 79:99–109

Sahade R, Tatián M, Esnal GB (2004) Reproductive ecology of the ascidian Cnemidocarpa verrucosa *Cnemidocarpa verrucosa* at Potter Cove, South Shetland Islands, Antarctica. Mar Ecol Prog Ser 272:131–140

Sahade R, Lagger C, Torre L et al (2015) Climate change and glacier retreat drive shifts in an Antarctic benthic ecosystem. Sci Adv 1:e1500050

Schloss IR, Ferreyra GA (2002) Primary production, light and vertical mixing in Potter Cove, a shallow bay in the maritime Antarctic. Polar Biol 25(1):41–48

Schloss IR, Abele D, Moreau S et al (2012) Response of phytoplankton dynamics to 19-year (1991–2009) climate trends in Potter Cove (Antarctica). J Mar Syst 92:53–66

Steneck RS, Graham MH, Bourque BJ et al (2002) Kelp forest ecosystems: biodiversity, stability, resilience and future. Environ Conserv 29(4):436–459

Tatián M, Sahade M, Esnal GB (2004) Diet components in the food of Antarctic ascidians living at low levels of primary production. Antarct Sci 16:123–128

Valdivia N, Díaz M, Holtheuer J, Garrido I, Huovinen P, Gómez I (2014) Up, down, and all around: scale-dependent spatial variation in rocky coastal communities of Fildes Peninsula, King George Island, Antarctica. PLoS One 9(6):e100714

Valdivia N, Díaz MJ, Garrido I, Gómez I (2015) Consistent richness-biomass relationship across environmental stress gradients in a marine macroalgal-dominated subtidal community on the Western Antarctic Peninsula. PLoS One 10(9):e0138582.

Wada S, Hama T (2013) The contribution of macroalgae to the coastal dissolved organic matter pool. Estuar Coast Shelf Sci 129:77–85

Wahl M, Schneider Covachã S, Saderne V et al (2018) Macroalgae may mitigate ocean acidification effects on mussel calcification by increasing pH and its fluctuations. Limnol Oceanogr 63(1):3–21

Westermeier R, Gómez I, Rivera PJ, Müller DG (1992) Macroalgas marinas antárticas: distribución, abundancia y necromasa en Isla Rey Jorge, Shetland del Sur, Antártica. Ser Cient INACH 42:21–34

Wiencke C (1990a) Seasonality of brown macroalgae from Antarctica a long-term culture study under fluctuating Antarctic daylengths. Polar Biol 10:589–600

Wiencke C (1990b) Seasonality of red and green macroalgae from Antarctica—long-term culture study under fluctuating Antarctic daylengths. Polar Biol 10:601–607

Wiencke C, Amsler CD (2012) Seaweeds and their communities in polar regions. In: Seaweed biology, novel insights into ecophysiology, ecology and utilization. Springer, Berlin, Heidelberg, pp 265–291

Wiencke C, Gómez I, Dunton K (2009) Phenology and seasonal physiological performance of polar seaweeds. Bot Mar 52:585–592

Wiencke C, Amsler CD, Clayton MN (2014) Chapter 5.1. Macroalgae. In: De Broyer C, Koubbi P, Griffiths HJ, Ramond B, Udekem d'Acoz C, Van de Putte A, Danis B, David B, Grant S, Gutt J et al (eds) Biogeographic atlas of the Southern Ocean. Scientific Committee on Antarctic Research, Cambridge, pp 66–73

Wulff A, Iken K, Quartino ML et al (2009) Biodiversity, biogeography and zonation of marine benthic micro- and macroalgae in the Arctic and Antarctic. Bot Mar 52(6):491–507

Zenteno L, Cárdenas L, Valdivia N et al (2019) Unraveling the multiple bottom-up supplies of an Antarctic nearshore benthic community. Prog Oceanogr 174:55–63

Zhang J, Wu W, Ren JS, Lin F (2016) A model for the growth of mariculture kelp *Saccharina japonica* in Sanggou Bay, China. Aquac Environ Interact 8:273–283

Chapter 9
Carbon Balance Under a Changing Light Environment

Dolores Deregibus, Katharina Zacher, Inka Bartsch, Gabriela L. Campana, Fernando R. Momo, Christian Wiencke, Iván Gómez, and María L. Quartino

Abstract The natural environment of Antarctic seaweeds is characterized by changing seasonal light conditions. The ability to adapt to this light regime is one of the most important prerequisites for their ecological success. Thus, the persistence of seaweeds depends on their capacity to maintain a positive carbon balance (CB)

D. Deregibus (✉)
Departamento de Biología Costera, Instituto Antártico Argentino, Buenos Aires, Argentina

Consejo Nacional de Investigaciones Científicas y Técnicas (CONICET), Buenos Aires, Argentina
e-mail: dderegibus@dna.gov.ar

K. Zacher · I. Bartsch · C. Wiencke
Alfred-Wegener-Institute, Helmholtz Centre for Polar and Marine Research, Bremerhaven, Germany
e-mail: katharina.zacher@awi.de; inka.bartsch@awi.de; Christian.wiencke@awi.de

G. L. Campana
Departamento de Biología Costera, Instituto Antártico Argentino, Buenos Aires, Argentina

Departamento de Ciencias Básicas, Universidad Nacional de Luján (UNLu), Luján, Buenos Aires, Argentina
e-mail: gcampana@dna.gov.ar

F. R. Momo
Instituto de Ciencias, Universidad de General Sarmiento, Los Polvorines, Buenos Aires, Argentina

Departamento de Ciencias Básicas, Universidad Nacional de Luján (UNLu), Luján, Buenos Aires, Argentina
e-mail: fmomo@campus.ungs.edu.ar

I. Gómez
Instituto de Ciencias Marinas y Limnológicas, Facultad de Ciencias, Universidad Austral de Chile, Valdivia, Chile

Research Center Dynamics of High Latitude Marine Ecosystems (IDEAL), Valdivia, Chile
e-mail: igomezo@uach.cl

M. L. Quartino
Departamento de Biología Costera, Instituto Antártico Argentino, Buenos Aires, Argentina

Museo Argentino de Ciencias Naturales "B. Rivadavia", Buenos Aires, Argentina
e-mail: lquartino@dna.gov.ar

I. Gómez, P. Huovinen (eds.), *Antarctic Seaweeds*,
https://doi.org/10.1007/978-3-030-39448-6_9

for buildup of biomass over the course of the year. A positive CB in Antarctica occurs only during the ice-free period in spring and summer, when photosynthetically active radiation (PAR, 400–700 nm) penetrates deeply into the water column. The accumulated carbon compounds during this period are stored and remobilized to support metabolism for the rest of the year.

Over the last decades climate warming has induced a severe glacial retreat in Antarctica and has opened newly ice-free areas. Increased sediment runoff, and reduced light penetration due to melting during the warmer months, may lead to a negative CB with changes in the vertical distribution of seaweeds. Furthermore, warmer winters and springs result in earlier sea-ice melt, causing an abrupt increase in light, compensating the reduction in PAR in summer or increasing the annual light budget. Studies performed in Potter Cove, Isla 25 de Mayo/King George Island, reveal that algae growing in newly ice-free areas did not acclimate to the changing light conditions. Lower or even negative CB values in areas close to the glacier runoff seem to be primarily dependent on the incoming PAR that finally determines the lower distribution limit of seaweeds. The present chapter discusses how carbon balance respond to the changing Antarctic light environment and its potential implications for the fate of benthic algal communities.

Keywords Metabolic carbon balance · Glacier runoff · Light requirements · Photosynthesis · Photosynthetic acclimation · Turbidity

9.1 Introduction

9.1.1 Climate Change in the Antarctic Coastal Zone

Rapid regional warming will continue to be one of the major forcing elements in Antarctica during this century (Hendry et al. 2018; IPCC 2014, 2019). The Western Antarctic Peninsula (WAP) is a hotspot of climate warming evidenced by a rapid increase in air temperature and, as a consequence, strong sea-ice decline and fast glacial retreat (Cook et al. 2005; Meredith and King 2005; Stammerjohn et al. 2008; Turner et al. 2009). This region is also highly impacted by ice disturbance, and the rates of iceberg scour have been shown to increase with consequences still not well understood for the benthic life (Barnes and Souster 2011; Barnes et al. 2014). Due to a high density of research stations, accessibility, and mild climate, the WAP coastal areas are natural laboratories for studying how ecosystems respond to rapid climate change (Constable et al. 2014; Deregibus et al. 2017; Lagger et al. 2018). In fact, a high diversity of benthic assemblages concentrate in the upper 70 m zone, which is subject to diverse physical perturbations.

A notorious phenomenon in coastal zones of the WAP is the formation of "newly ice-free areas" as a consequence of glacier retreat (Cook et al. 2005; Rückamp et al. 2011; Quartino et al. 2013). In these areas new substrates become available for benthic colonization (Constable et al. 2014; Lagger et al. 2017, 2018). In parallel, the enhanced glacier melting leads to increased turbidity and decreases salinity in the water column (Barnes and Conlan 2007; Campana et al. 2009; Schloss et al. 2012; Bers et al. 2013; Grange and Smith 2013). Moreover, ice-scouring events can be intensified (Barnes and Souster 2011; Barnes et al. 2018). These climate-driven rapid shifts and concomitant new stressors (e.g., sedimentation, ice disturbance) can have complex consequences for the functioning of the coastal ecosystems (Gutt et al. 2015). Thus, various hypotheses have arisen inspiring numerous studies on their effects on coastal ecosystems (Smale and Barnes 2008; Gutt et al. 2011; Schofield et al. 2010; Torre et al. 2012; Ducklow et al. 2013; Barnes et al. 2014; Moon et al. 2015; Sahade et al. 2015; Valdivia et al. 2015). The question whether rising air temperatures could cause changes in the light environment in Antarctic coastal zones and their effect on primary producers is central to understand the future benthic primary production in these coastal ecosystems (Zacher et al. 2009; Clark et al. 2013, 2017). In this context, given that photosynthesis in coastal areas is the most important biological process affected by light and due to its vital importance in neritic areas, seaweeds can be considered as sentinel taxa to evaluate these changes.

9.1.2 *Antarctic Seaweeds and the Changing Light Environment*

The availability of light for photosynthesis and growth is the major factor governing depth distribution of seaweeds (Lüning 1990). At the southern distribution limit of seaweeds in the Antarctic, the polar night lasts for about 4 months. Sea-ice cover extends the period of hibernal darkness even more (reviewed in Zacher et al. 2009). Besides phytoplankton blooms, circulation and wind have major importance in limiting light penetration into the water column (Schloss et al. 2012). Thus, Antarctic seaweeds are well adapted to cope with extended periods of darkness showing low light requirements for photosynthesis (E_k) and high photosynthetic efficiencies (α). In addition, a number of species have also the ability to cope with high light conditions in summer (Wiencke 1990; Gómez et al. 1997, 2009; Wiencke and Amsler 2012; see also Chap. 7 by Gómez and Huovinen).

Seasonality strongly determines the fluctuations in abiotic factors, which affect the physiological and ecological performance of seaweeds (Wiencke and Amsler 2012; Marcías et al. 2017). In Antarctica, adaptation to the seasonality of the light regime is a fundamental prerequisite for the ecological success of seaweeds (Wiencke et al. 2011). There are two different growth strategies, i.e., season anticipators and season responders *sensu* Kain (1989). The season anticipators start growing in late winter/spring. In contrast, the season responders start growth and reproduction later in spring and summer when the light conditions are favorable (Wiencke and Amsler 2012; see also Chap. 10 by Navarro et al.). However, both groups are highly dependent on the light availability in spring and summer when primary production mostly occurs as the solar angle is high and fast ice is absent (Miller and Pearse 1991; Wiencke et al. 2006; Runcie and Riddle 2012; see also Chap. 7 by Huovinen and Gómez).

Changes in light availability do not only have natural causes; besides the high seasonal variability in Antarctica, changes in external variables that influence seaweeds are due to anthropogenic reasons (e.g., higher regional air temperatures due to the CO_2 emissions) (Vaughan et al. 2003; Turner et al. 2013). Especially in coastal areas, climate change may significantly shrink the annual light budget available for benthic primary producers due to an enhancement of sedimentation and a decrease in the duration of the fast ice season (Clark et al. 2013; Quartino et al. 2013). Algae growing in newly ice-free areas are subjected to a reduction of light penetration, which constitutes a constraint for photosynthesis (Schloss et al. 2012; Wiencke and Amsler 2012; Deregibus et al. 2016; González et al. 2017; see also Chap. 8 by Quartino et al.). Furthermore, glacier melting also causes accumulation of sediment on the seafloor, which could affect the attachment of benthic algae (Johnston et al. 2007). On the other hand, beneficial effects of sediment input could be considered during spring and summer by attenuating high PAR, which may inhibit the recruitment of macroalgal species (Graham 1996; Hanelt 1996; Hanelt et al. 1997) or may protect against harmful UV radiation (Roleda et al. 2009). However, previous studies have shown a positive relationship between the degree of light penetration and the complexity of the macroalgal community (in terms of diversity and the presence of large perennial species) in the newly ice-free areas (Quartino et al. 2013). Thus, the time of occurrence of the thawing is extremely relevant for seaweeds as favorable light conditions for algal growth are constrained to only a few months of the year during the bright light season (Wiencke and Amsler 2012; see also Chap. 7 by Huovinen and Gómez). Here, the maintenance of positive carbon balance is probably one of the most significant physiological adjustments of seaweeds to cope with these changing light conditions (Gómez et al. 1997; Deregibus et al. 2016).

9.1.3 Carbon Balance: Concepts and Methodological Challenges

The production of organic matter via photosynthesis using carbon dioxide, water, and sunlight is known as primary production, while primary productivity is the rate at which energy is converted to organic substances by photosynthetic organisms (Hurd et al. 2014). The rates of primary productivity refer to the efficiency at which solar energy is used to fix inorganic carbon and create biomass, and hence it is an essential parameter reflecting the ecosystem function (Falkowski and Raven 1997).

The carbon balance (CB) equalizes the assimilated carbon (C) during photosynthesis in relation to the C that is lost due to respiration (Wiencke and Amsler 2012). Photosynthesis versus irradiance (P-E) curves describes how the photosynthetic rate (e.g., net O_2 evolution) varies with increasing irradiance. Integrating data on daily changes of in situ irradiance with the P-E-derived parameters, E_k, the photosynthetic capacity (P_{max}), and dark respiration (Deregibus et al. 2016), it is possible to calculate the daily metabolic CB as an indicator of the physiological ability to live at a certain depth (Hanelt and Figueroa 2012). In coastal areas not affected by glacial melt, PAR can penetrate into the water column as deep as 30 m (1% depth) during late winter-spring, still allowing a positive CB (Gómez et al. 1997; see also Chap. 7 by Huovinen and Gómez).

The lower distribution limit of algae is determined by their capacity to maintain a positive CB to build up biomass (Hanelt and Figueroa 2012). During this period a positive CB replenishes the energy budget to be used to survive the long periods of darkness during the rest of the year (Wiencke et al. 2011; Deregibus et al. 2016). Several studies indicate that the daily exposure time to light is more important than the intensity of light for macroalgal productivity in coastal areas (Dennison and Alberte 1985; Matta and Chapman 1991; Gómez et al. 1997). Direct associations between increases in turbidity and decreases in macroalgal productivity have been reported from a variety of systems worldwide (Airoldi 2003; Anthony et al. 2004; Spurkland and Iken 2011; Pritchard et al. 2013). Similarly, changes in the productivity of seaweeds in relation to sea ice variations have been reported in Antarctic (Clark et al. 2013) and Arctic (Krause-Jensen and Duarte 2014) assemblages. Apparently, direct and indirect effects of climate change on light availability will have dramatic effects on the annual macroalgal CB, which could result in changes in benthic primary productivity (Runcie and Riddle 2012; Bartsch et al. 2016; Deregibus et al. 2016; Gómez et al. 2019; see also Chap. 8 by Quartino et al.).

Various mathematical models have related irradiance to photosynthesis in order to estimate primary productivity (Jassby and Platt 1976; Nelson and Siegrist 1987; Henley 1993; Jones et al. 2014). However, the photosynthetic parameters α, E_k, and P_{max} derived from P-E curves may differ depending on the selected fitting model (Smith 1936; Steele 1962; Webb et al. 1974; Jassby and Platt 1976; Cullen 1990; Frenette et al. 1993; Henley 1993). Thus, productivity estimates using distinct models may not be comparable and even lead to erroneous conclusions (Frenette et al. 1993; Deregibus et al. 2016). Therefore, the use of best mathematical fit will improve considerably the quality of our productivity estimations (Jassby and Platt 1976; Nelson and Siegrist 1987).

The primary productivity of seaweeds can be estimated in several ways, e.g., through changes in net weight, through growth, in situ productivity measurements, or measurements of dissolved oxygen production (Gómez et al. 2009; Runcie et al. 2009; Runcie and Riddle 2011; Hanelt and Figueroa 2012). These different methodologies show constraints in their ability to accurately describe the primary productivity in marine environments (Runcie and Riddle 2012). In the WAP, Gómez et al. (1997) did an initial effort to calculate the seaweed primary productivity through the calculation of the daily metabolic CB using the amount of in situ daily hours when algae are light saturated. Their findings could explain well the vertical zonation of seaweeds according to the light regime at the WAP (Gómez et al. 1997).

9.2 Carbon Balance: A Case Study in Potter Cove

Over the last years, numerous studies have been performed in Potter Cove, Isla 25 de Mayo/King George Island, which is a distinctive example of an area with a special focus on studies related to the impacts of climate-forced glacier retreat across an entire shallow water ecosystem.

Adjacent to Carlini Station, Potter Cove is surrounded by the Fourcade Glacier. Over the last decades, this glacier has been retracting and melt water inflow in the cove increased (Eraso and Dominguez 2007; Rückamp et al. 2011) (Fig. 9.1). Newly ice-free areas in costal shallow areas were formed and are still appearing (Quartino et al. 2013; Lagger et al. 2017, per obs). In this cove, increase in sedimentation has negatively affected species and changed benthic communities (Torre et al. 2012; Quartino et al. 2013; Pasotti et al. 2014; Sahade et al. 2015; Campana et al. 2018; Meredith et al. 2018; Jerosch et al. 2019; see Chap. 8 by Quartino et al.). As a consequence of glacier melting, a spatial gradient developed along the cove, providing the opportunity to analyze the primary productivity under a natural abiotic gradient (Deregibus et al. 2016). For the first time, the primary productivity of seaweeds growing in these new areas in a gradient of glacial influence was estimated. High "glacier influence" was defined as a decrease in light penetration in the water col-

Fig. 9.1. Small creek originated by glacier melting during summer months causing a discharge of turbid fresh water filled with terrigenous sediments into the inner zone of Potter Cove. (Adapted from Deregibus (2017) with permission (photos by Dolores Deregibus))

umn (increased turbidity) due to sediment input. Improvements on the calculations were proposed through a novel and more accurate approximation of seaweed CB calculations with the inclusion of in situ continuous-light measurements (Deregibus et al. 2016).

9.2.1 Light Availability

The major new achievement and challenge was to install and perform continuous underwater PAR measurements along a depth gradient (5, 10, 20, and 30 m) in several areas along the turbidity gradient in summer 2010, for the first time in an Antarctic environment. Since then, daily continuous PAR measurements are obtained at 10 m depth and are utilized in combination with the photosynthetic parameters to estimate the CB of seaweeds over the seasons and along turbidity gradients (Deregibus 2017).

The light loggers (for details see Deregibus et al. 2016) were installed in an upright position, on a specially constructed concrete base and are periodically secured by SCUBA divers at the respective sites (Fig. 9.2). The PAR data could then be used to plot irradiance vs time with a time interval of 15 min (Fig. 9.3).

As an example, in Fig. 9.3, we present the mean PAR records of one selected week in January 2010. At 5 m depth, seaweeds were exposed to PAR for 12 h (mean daily irradiance of 22.2 μmol photons m^{-2} s^{-1}). At 10 m depth, PAR values showed an abrupt decrease indicating a very low daily light budget below 5 m in summer in this area. Also, the daily exposure time to sunlight PAR decreased with increasing depths shortening the photoperiod (Fig. 9.3).

Despite high solar angle, the question rises how the investigated seaweeds may survive and grow under these conditions in Potter Cove. It became evident that it is insufficient to consider not only summer periods to assess productivity but also annual irradiance (Fig. 9.4), and carbon budgets need to be calculated. Accordingly,

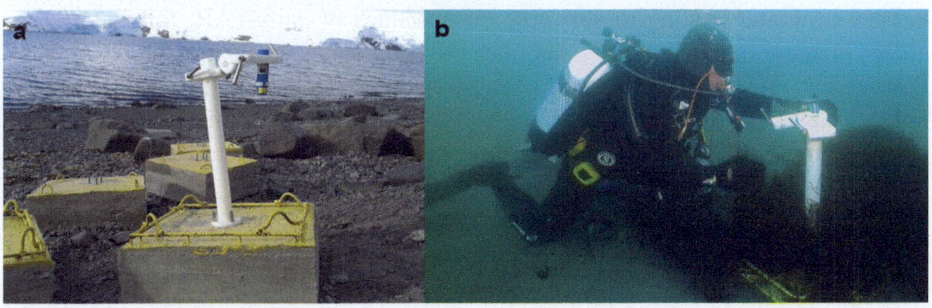

Fig. 9.2 Underwater PAR measurements: (**a**) Odyssey Logger Sensor for continuous PAR measurements. Setup (concrete base and adjustable arm) that holds the sensor. (**b**) Diver exchanging the sensor. (Adapted from Deregibus (2017) (photos by Dolores Deregibus))

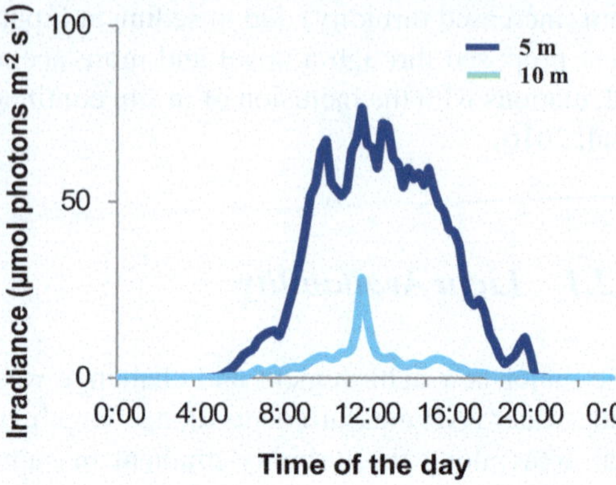

Fig. 9.3 Daily course of irradiance in newly ice-free area in Potter Cove at 5 and 10 m depth between Jan 15 and 22, 2010 (Data were averaged over seven continuous days). (Adapted from Deregibus et al. (2016))

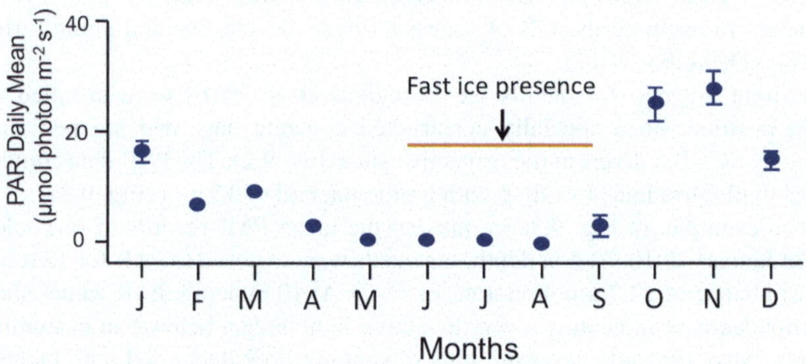

Fig. 9.4. Annual course of incident irradiance measured at Potter Cove, Isla 25 de Mayo/King George Island. Daily mean of PAR was integrated over single months in a newly ice-free area during 2015

whole-year continuous PAR measurements are being performed to estimate the light climate over an entire year – and between years in Potter Cove (e.g. Fig. 9.4).

During 2015, which serves as an example here, there was an evident presence of fast ice in winter leading to very low mean values of daily PAR, followed by an abrupt increase of the PAR intensity once the pack-ice broke up in spring (Fig. 9.4). Whole-year underwater in situ irradiances show that the most productive season may start as early as October with highest mean daily PAR values. This is in contrast to our expectations as during the warmer summer months (December to March), the light intensity was lower than in spring, mainly due to the input of sediment of terrestrial origin into the water column. On an interannual scale, these values may differ considerably (Deregibus unpublished). Similar results were found for the Arctic in Kongsfjorden (Spitsbergen), a fjord system highly affected by melting glaciers (Bartsch et al. 2016; Pavlov et al. 2019). Overall, our long-term, year-round PAR measurements can be further used to estimate the annual primary productivity of Antarctic seaweeds in coastal areas subjected to extreme changes in light availability.

9.2.2 Photosynthetic Acclimation

The endemic brown alga *Himantothallus grandifolius* (A. Gepp and E. S. Gepp) Zinova and the red alga *Palmaria decipiens* (Reinsch) Ricker, presented here as examples, were sampled at 5 and 10 m in the vicinity of the light logger. Photosynthetic oxygen evolution and dark respiration were measured under laboratory conditions, and photosynthesis versus irradiance curves (P-E curves) were obtained and used to calculate photosynthetic parameters with the hyperbolic tangent function of Jassby and Platt (1976). This function, which was proved to be the best fitting model to the experimental data, is expressed as:

$$P = P_{max} * \tanh(\alpha E / P_{max}) + R$$

where P is the photosynthetic rate, P_{max} is the maximum photosynthetic rate, tanh is the hyperbolic tangent, α is the initial slope of the curve at low irradiance, E is the incident irradiance, and R is the dark respiration rate (Fig. 9.5).

The hyperbolic tangent function (Jassby and Platt 1976) was found to be the best fit for our data compared to other commonly used models (e.g., Webb et al. 1974). Jassby and Platt (1976) found that their hyperbolic tangent equation proved to be the best overall fit to 200 types of datasets in a comparison of various equations (Jassby and Platt 1976; Jones et al. 2014).

In Potter Cove, photosynthetic parameters such as P_{max}, α, and E_k of seaweeds growing in newly ice-free areas with variable glacial influence and at different depths were generally quite similar. This indicates a low acclimation potential of photosynthesis to different irradiance regimes (Deregibus et al. 2016). The absence of photoacclimation in Antarctic seaweeds living under different? light availability has been reported previously (Gómez et al. 2009). Apparently, this is a characteristic that forms part of suite of photobiological features conferring the extreme shade adaptation of these organisms, developed primarily to cope with dim light in autumn-winter (Gómez and Huovinen 2015). In fact, similar values of P_{max} and α were reported at different depths for *H. grandifolius* (Drew and Hastings 1992;

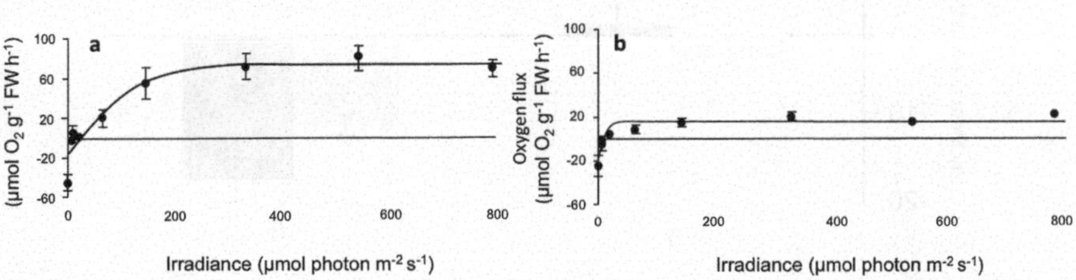

Fig. 9.5. Photosynthesis-irradiance (P-E) curves of *Palmaria decipiens* (**a**) and *Himantothallus grandifolius* (**b**) in a newly ice-free area in Potter Cove, which represents a typical P-E curve for a shade-adapted species. Generally, the oxygen production under light saturation is lower in *H. grandifolius* than in *P. decipiens*. Both species did not show any sign of photoinhibition even under highest irradiances of 800 µmol photons m^{-2} s^{-1}. (Adapted from Deregibus et al. (2016))

Gómez et al. 1997), and in *P. decipiens*, *Gigartina skottsbergii* Setchell and Gardner *Trematocarpus antarcticus* (Hariot) Frederic and Moe, and *Desmarestia anceps* Montagne (Gómez et al. 1997). Interestingly, other studies, carried out in the Ross Sea, Antarctica (Schwarz et al. 2003), in Greenland (Kühl et al. 2001), and in the Arctic (Krüger 2016) using chlorophyll fluorescence, suggest a greater acclimation potential of seaweeds to low light at deeper depths as reflected in their lower E_c (photosynthetic compensation point), E_k, and $rETR_{max}$ values. Furthermore, a recent study reports different photosynthetic response to low light between algae growing in Antarctica and in the Subantarctic (Navarro et al. 2019).

9.2.3 Daily Carbon Balance of Seaweeds

Calculations of daily net CB (mg C g^{-1} FW day^{-1}) integrated the photosynthetic parameters derived from the P-E curves and the incident underwater irradiances according to the following formula:

$$P = P_{max} * \tanh\left(\alpha\left(\text{average}\frac{\frac{E_1}{E_2}}{P_{max}}\right)\right) + R$$

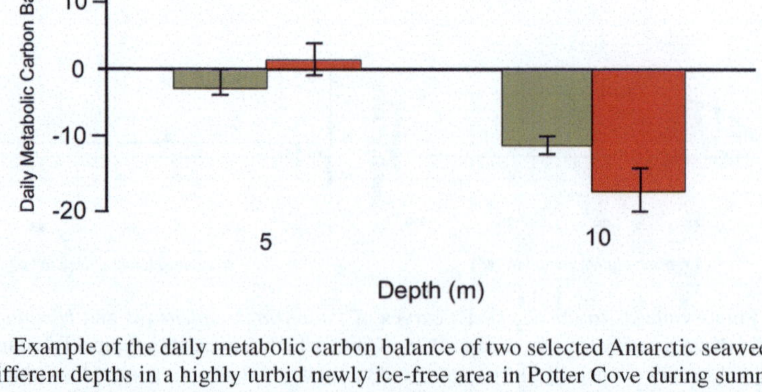

Fig. 9.6 Example of the daily metabolic carbon balance of two selected Antarctic seaweeds growing at different depths in a highly turbid newly ice-free area in Potter Cove during summer 2010. Values correspond to an overall net gain or loss of C during 24 h. (Adapted after Deregibus et al. (2016))

where "average (E_1:E_2)" is the average of two incident irradiances between time 1 and time 2. This formula improved the calculation of the net oxygen production along an entire day (Fig. 9.6) and shows that in an area with high turbidity due to sedimentation, the CB was only positive for *P. decipiens* at 5 m depth with a mean value of 0.2 (±0.4) mg C g^{-1} FW day^{-1}, but negative at 10 m. In the case *H. grandifolius*, it was negative for both depths (Fig. 9.6). Negative values were mainly due to the low light availability and a reduction of the time to which algae were exposed to light (Deregibus et al. 2016). The results agree with the in situ distribution of both species: they grow only to a maximum depth of 10 m in this highly disturbed area (Deregibus et al. 2016; Deregibus 2017).

9.3 New Scenarios and Their Implications for Algal Photosynthesis

Carbon balance is the most conclusive parameter to understand and explain the zonation patterns in Antarctic seaweeds (Gómez et al. 2009). This parameter is directly related to light availability as it is lower or negative in more turbid areas and/or at deeper depths compared to shallower depths and areas with low turbidity. Isla 25 de Mayo/King George Island is characterized by a marked increase in the atmospheric temperature in recent decades (Ferron et al. 2004; Schloss et al. 2012; Bers et al. 2013). Falk and Sala (2016) indicated an anticipation of the thawing period towards spring and an extension towards autumn in this area. The latter, and the fact that turbidity has increased not only in summer but also in spring in Potter Cove over the last years (Schloss et al. 2012; Deregibus et al. 2016), and that in some cases negative CB have also been measured in spring (Deregibus 2017) raises additional questions. For example: What would happen if the thawing season starts earlier due to increased air temperatures, leading also to negative CB in spring? Would the storage compounds produced in summer be enough to support the energy requirements the rest of the year?

Conversely, it should also be considered that warmer winters and springs lead to earlier fast ice melting (Schloss et al. 2012; Deregibus et al. 2017), which causes an abrupt increase of light, probably compensating the reduction of PAR in summer or even significantly increasing the annual light budget for seaweeds (Johnston et al. 2007; Clark et al. 2013). This last assumption leads to additional questions: Will the total CB values become really lower under a future warming? Or will there be compensation due to a higher light availability in late winter and early spring (less or no fast ice cover)?

If the annual photon doses do not sustain growth and reproduction at a certain area, seaweeds will not survive (Runcie and Riddle 2012). Furthermore, light not only serves as a source of energy for seaweeds, but it is also an environmental signal inducing changes in processes which are dependent on daylength as trigger signal

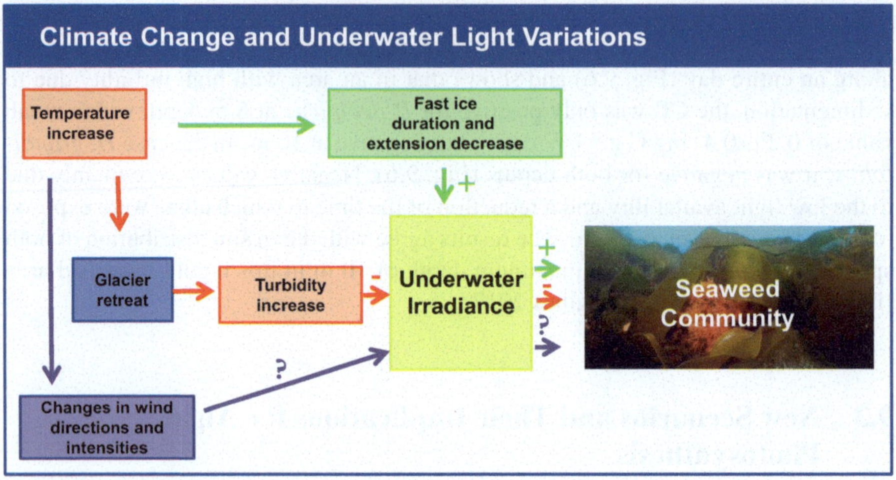

Fig. 9.7 Conceptual model on the impact of direct and indirect climate change factors on the underwater irradiance in newly ice-free areas at the Western Antarctic Peninsula. (Adapted from Quartino et al. (2013) and Deregibus (2017))

(Hanelt and Figueroa 2012). The seasonal development of seaweeds must be tuned to the strong seasonality of the light conditions (Wiencke et al. 2011). This is very important as variations in PAR would not only affect the primary energy source but could also alter the information to modulate the morphogenetic development and the other signals for growth and reproduction. It is key and maybe relevant to elucidate the growth seasons that are being more/most? affected by global warming during the year, as higher temperatures in winter impact differently on the light availability than in spring and summer (Wiencke and Amsler 2012).

The conceptual model in Fig 9.7 describes the ways through which these environmental changes could be affecting the light availability for the benthic autotrophic organisms. Environmental conditions directly affect, and are reflected, in the underwater annual PAR budget. Light changes are mediated by a series of factors: firstly, glacier melting increases the amount of sediments in the water column and enhances turbidity. This factor has most impact on benthic primary producers as it directly and negatively reduces light availability, which is a constraint for photosynthesis. Secondly, the decrease in fast ice duration may increase the open water period and thereby the availability of underwater light in winter and spring. Thirdly, wind, depending on its intensity and direction, has a significant impact on sediment resuspension and distribution processes of the sediments limiting light penetration. This model is complementary to the general conceptual model shown in Chap. 8 by Quartino et al., on the environmental factors affecting the coastal macroalgal community.

9.4 Concluding Remarks and Future Prospects

So far, glacial retreat has opened new space (hard substratum) in the inner Potter Cove and seaweeds have colonized and persisted in newly ice-free areas (Quartino et al. 2013; Campana et al. 2018). Although newly ice-free areas are highly affected by the glacial influence, several seaweed species grow under high sedimentation showing their exceptional ability to survive and successfully reproduce under such conditions (Becker et al. 2011; Quartino et al. 2013; Deregibus et al. 2016).

The realization that seaweeds are the most productive marine macrophytes and identified as very relevant contributors to global blue carbon sequestration (Hill et al. 2011; Krause-Jensen and Duarte 2016) has consigned them in a key role at a global scale. Considering that the spatial distribution of the seaweed community has expanded to the inner side of Potter Cove, it is likely that the increase in seaweed biomass leads to an enhanced production in this area (Quartino et al. 2013), with cascading effects to the rest of the food web (Marina et al. 2018). This raises the question of how changes in seaweed productivity could affect the rest of the coastal ecosystem in Potter Cove. In this context, given that seaweeds in Potter Cove are in a constant tradeoff between extending their distribution into newly ice-free areas and being affected by climate change, an interest to reveal whether this expansion will give way to a persistence and maturation of these communities in these new areas exists.

Due to climate change in polar ecosystems, shifts from predominantly heterotrophic to autotrophic states of shallow polar seabeds have been predicted (Bartsch et al. 2016; Clark et al. 2013, 2017). Kortsch et al. (2012) and Scherrer et al. (2018) reported marked community shifts with abrupt and persistent increase in macroalgal cover in the Arctic following the extension in the ice-free period. Krause-Jensen et al. (2012) also reported substantial increases in the productivity and maximal depth distribution of kelps in Greenland. Furthermore, in the Arctic it was also stated that longer ice-free seasons have extended the growth season of seaweeds favoring an increased diversity of algae and macrozoobenthos (Paar et al. 2016). In this context, it is also expected that further glacier retreat will continue to favor seaweed colonization in new coastal areas potentially resulting in higher productivity and carbon sequestration (Clark et al. 2013; Krause-Jensen and Duarte 2016).

The future projection in case study systems such as Potter Cove would be to continue with the long-term monitoring of multiple abiotic factors and primary productivity calculations. Knowledge on the minimum light requirements for seaweeds and on their ecophysiological characteristics are needed to better understand and predict macroalgal survival and possible changes in primary productivity, distribution, and depth zonation in areas affected by glacial melting due to climate change. Continuous PAR measurements are currently being performed in the newly ice-free areas of Potter Cove, which is a unique data set in terms of time span (more than five years) and continuity in a gradient of glacial influence in Antarctica.

Finally, as a way to understanding the regional variability and its impacts on the biota, these monitoring programs require international collaboration including integrated actions among different Antarctic stations (Deregibus et al. 2017).

Acknowledgments The fieldwork described in this chapter has been performed at Carlini Station-Dallmann Laboratory within the framework of the scientific collaboration existing between Instituto Antártico Argentino/Dirección Nacional del Antártico and the Alfred-Wegener-Institute, Helmholtz Centre for Polar and Marine Research (AWI). The research was supported by Grants from DNA-IAA (PICTA 7/2008–2011) and ANPCyT-DNA (PICTO 0116/2012–2015, PICT 2017-2691). These studies were also supported by MINCYT-BMBF Program (AL/17/06-01DN18024) and Conicyt-Chile (Center FONDAP IDEAL 15150003). We are especially grateful to the scientific, logistic, and diving groups of Carlini Station. We thank Carolina Matula for her contribution to improve the annual light figure. We gratefully acknowledge financial support by the AWI. We thank Facultad de Ciencias Exactas y Naturales, Universidad de Buenos Aires, for allowing the reuse of material published in Deregibus 2017 Thesis. The present research also presents an outcome of the EU project IMCONet (FP7 IRSES, Action No. 319718) and EU project CoastCarb, Funding ID 872690, H2020-MSCA-RISE-2019 – Research and Innovation Staff Exchange.

References

Airoldi L (2003) The effects of sedimentation on rocky coast assemblages. In: Atkinson RJA, Gibson RN (eds) Oceanogr Mar Biol Ann Rev. CRC Press, London, pp 161–236

Anthony KRN, Ridd PV, Orpin AR, Larcombe P, Lough J (2004) Temporal variation in light availability in coastal benthic habitats: effects of clouds, turbidity, and tides. Limnol Oceanogr 49:2201–2211. https://doi.org/10.4319/lo.2004.49.6.2201

Barnes DKA, Conlan KE (2007) Disturbance, colonization and development of Antarctic benthic communities. Philos Trans R Soc Lond Ser B Biol Sci 362(1477):11e38. https://doi.org/10.1098/rstb.2006.1951

Barnes DKA, Souster T (2011) Reduced survival of Antarctic benthos linked to climate-induced iceberg scouring. Nat Clim Chang 1:1–4. https://doi.org/10.1038/nclimate1232

Barnes DKA, Fenton M, Cordingley A (2014) Climate-linked iceberg activity massively reduces spatial competition in Antarctic shallow waters. Curr Biol 24(12):R553–R554. https://doi.org/10.1016/j.cub.2014.04.040

Barnes DKA, Fleming A, Sands CJ, Quartino ML, Deregibus D (2018) Icebergs, blue carbon and Antarctic climate feedbacks. Philos Trans A Math Phys Eng Sci 376(2122):20170176. https://doi.org/10.1098/rsta.2017.0176

Bartsch I, Paar M, Fredriksen S, Schwanitz M, Daniel C, Hop H, Wiencke C (2016) Changes in kelp forest biomass and depth distribution in Kongsfjorden, Svalbard, between 1996–1998 and 2012–2014 reflect Arctic warming. Polar Biol 39:2021–2036. https://doi.org/10.1007/s00300-015-1870-1

Becker S, Quartino ML, Campana GL, Bucolo P (2011) The biology of an Antarctic rhodophyte, *Palmaria decipiens*: recent advances. Antarct Sci 23(5):419–430. https://doi.org/10.1017/S0954102011000575

Bers AV, Momo F, Schloss IR, Abele D (2013) Analysis of trends and sudden changes in long-term environmental data from King George Island (Antarctica): relationships between global climatic oscillations and local system response. Clim Chang 116(34):789–803

Campana GL, Zacher K, Fricke A, Molis M, Wulff A, Quartino ML, Wiencke C (2009) Drivers of colonization and succession in polar benthic macro- and microalgal communities. Bot Mar 52:655–667. https://doi.org/10.1515/BOT.2009.076

Campana GL, Zacher K, Deregibus D, Momo F, Wiencke C, Quartino ML (2018) Succession of Antarctic benthic algae (Potter Cove, South Shetland Islands): structural patterns and glacial impact over a four-year period. Polar Biol 41(2):377–396

Clark GF, Stark JS, Johnston EL (2013) Light-driven tipping points in polar ecosystems. Glob Chang Biol 12:3749–3761. https://doi.org/10.1111/gcb.12337

Clark GF, Stark JS, Palmer AS, Riddle MJ, Johnston EL (2017) The roles of sea-ice, light and sedimentation in structuring shallow Antarctic benthic communities. PLoS One 12(1):e0168391. https://doi.org/10.1371/journal.pone.0168391

Constable AJ, Melbourne -Thomas J, Corney SP, Arrigo KR, Barbraud C, Barnes DK, Bindoff NL et al (2014) Climate change and Southern Ocean ecosystems. I: How changes in physical habitats directly affect marine biota. Glob Chang Biol 20:3004–3025. https://doi.org/10.1111/gcb.12623

Cook AJ, Fox AJ, Vaughan DG, Ferrigno JG (2005) Retreating glacier fronts on the Antarctic Peninsula over the past half-century. Science 308(5721):541–544

Cullen JJ (1990) On models of growth and photosynthesis in phytoplankton. Deep Sea Res Part A 37:667–683. https://doi.org/10.1016/0198-0149(90)90097-F

Dennison WC, Alberte RS (1985) Role of daily light period in the depth distribution of *Zostera manna* (eelgrass). Mar Ecol Prog Ser 25:51–61

Deregibus D (2017) Efecto del retroceso glaciario inducido por el cambio climático sobre la comunidad de macroalgas en nuevas áreas libres de hielo en un ecosistema costero antártico (Caleta Potter, I.25 de Mayo, I. Shetland del Sur). PhD thesis, Universidad de Buenos Aires. http://hdl.handle.net/20.500.12110/tesis_n6241_Deregibus

Deregibus D, Quartino ML, Campana GL, Momo F, Wiencke C, Zacher K (2016) Photosynthetic light requirements and vertical distribution of macroalgae in newly ice-free areas in Potter Cove, South Shetland Islands, Antarctica. Polar Biol 39:153–166. https://doi.org/10.1007/s00300-015-1679-y

Deregibus D, Quartino ML, Zacher K, Campana GL, Barnes D (2017) Understanding the link between sea ice, ice scour and Antarctic benthic biodiversity; the need for cross station and nation collaboration. Polar Rec 53:143–152. https://doi.org/10.1017/S0032247416000875

Drew EA, Hastings RM (1992) A year-round ecophysiological study of *Himantothallus grandifolius* (Desmarestiales, Phaeophyta) at Signy Island, Antarctica. Phycologia 31:262–277

Ducklow HW, Fraser WR, Meredith MP, Stammerjohn SE, Doney SC, Martinson DG, Sévrine F et al (2013) West Antarctic Peninsula: an ice-dependent coastal marine ecosystem in transition. Oceanography 26(3):190–203. https://doi.org/10.5670/oceanog.2013.62

Falk U, Sala H (2016) Winter melt conditions of the inland ice cap on King George Island, Antarctic Peninsula. Erdkunde 341–363. https://doi.org/10.3112/erdkunde.2015.04.04

Falkowski PG, Raven JA (eds) (1997) Aquatic photosynthesis. Blackwell Scientific, Oxford

Ferron FA, Simões JC, Aquino FE, Setzer AW (2004) Air temperature time series for King George Island, Antarctica. Pesqui Antarct Bras 4:155–169

Frenette J, Demers S, Legendre L, Dodson J (1993) Lack of agreement among models for estimating photosynthetic parameters. Limnol Oceanogr 38(3):679–678. https://doi.org/10.4319/lo.1993.38.3.0679

Gómez I, Huovinen P (2015) Lack of physiological depth patterns in conspecifics of endemic Antarctic brown algae: a trade-off between UV stress tolerance and shade adaptation? PLoS One 10(8):e0134440

Gómez I, Weykam G, Klöser H, Wiencke C (1997) Photosynthetic light requirements, metabolic carbon balance and zonation of sublittoral macroalgae from King George Island (Antarctica). Mar Ecol Prog Ser 148:281–293

Gómez I, Wulff A, Roleda MY, Huovinen P, Karsten U, Quartino ML, Dunton K, Wiencke C (2009) Light and temperature demands of marine benthic microalgae and seaweeds in polar regions. Bot Mar 52:593–608. https://doi.org/10.1515/BOT.2009.073

Gómez I, Navarro NP, Huovinen P (2019) Bio-optical and physiological patterns in Antarctic seaweeds: a functional trait based approach to characterize vertical zonation. Prog Oceanogr 174:17–27. https://doi.org/10.1016/j.pocean.2018.03.013

González PM, Deregibus D, Malanga G et al (2017) Oxidative balance in macroalgae from Antarctic waters. Possible role of Fe. J Exp Mar Biol Ecol 486:379–386

Graham MH (1996) Effect of high irradiance on recruitment of the giant kelp *Macrocystis* (Phaeophyta) in shallow water. J Phycol 32:903–906. https://doi.org/10.1111/j.0022-3646.1996.00903.x

Grange LJ, Smith CR (2013) Megafaunal communities in rapidly warming fjords along the West Antarctic Peninsula: hotspots of abundance and beta diversity. PLoS One 8(12):e77917

Gutt J, Barratt I, Domack E, d'Udekem d'Acozd C, Dimmlere W, Grémare A, Heilmayer O, Isla E et al (2011) Biodiversity change after climate-induced ice-shelf collapse in the Antarctic. Deep Sea Res II 58(1):74–83

Gutt J, Bertler N, Bracegirdle TJ, Buschmann A, Comiso J, Hosie G, Isla E, Schloss IR et al (2015) The Southern Ocean ecosystem under multiple climate change stresses – an integrated circumpolar assessment. Glob Chang Biol 21:1434–1453. https://doi.org/10.1111/gcb.12794

Hanelt D (1996) Photoinhibition of photosynthesis in marine macroalgae. Sci Mar 60:243–248

Hanelt D, Figueroa FL (2012) Physiological and photomorphogenic effects of light on marine macrophytes. In: Wiencke C, Bischof K (eds) Seaweed biology: novel insights into ecophysiology, ecology and utilization. Springer, Berlin, pp 3–23

Hanelt D, Melchersmann B, Wiencke C, Nultsch W (1997) Effects of high light stress on photosynthesis of polar macroalgae in relation to depth distribution. Mar Ecol Prog Ser 149:255–266

Hendry KR, Meredith MP, Ducklow HW (2018) The marine system of the West Antarctic Peninsula: status and strategy for progress. Phil Trans R Soc A 376:20170179. https://doi.org/10.1098/rsta.2017.0179

Henley WJ (1993) Measurements and interpretation of photosynthetic light-response curves in algae in the context of photoinhibition and diel changes. J Phycol 29:729–739. https://doi.org/10.1111/j.0022-3646.1993.00729.x

Hill R, Bellgrove A, Macreadie PI, Petrou K, Beardall J, Steven A, Ralph PJ (2011) Can macroalgae contribute to blue carbon? An Australian perspective. Limnol Oceanogr 60(5):1689–1706. https://doi.org/10.1002/lno.10128

Hurd CL, Harrison PJ, Bischof K, Lobban CS (eds) (2014) Seaweed ecology and physiology. Cambridge University Press, Cambridge

IPCC (Intergovernmental Panel on Climate Change) (2014) Climate change 2014: impacts, adaptation, and vulnerability. Summaries, frequently asked questions, and cross-chapter boxes. A Contribution of Working Group II to the Fifth Assessment Report of the Intergovernmental Panel on Climate Change. World Meteorological Organization, Geneva. 190 pp

IPCC (Intergovernmental Panel on Climate Change) (2019) Special report on the ocean and cryosphere in a changing climate. https://www.ipcc.ch/srocc/download-report/

Jassby AD, Platt T (1976) Mathematical formulation of the relationship photosynthesis and light for phytoplankton. Limnol Oceanogr 21:540–547. https://doi.org/10.4319/lo.1976.21.4.0540

Jerosch K, Scharf FK, Deregibus D, Campana GL, Zacher K, Pehlke H, Falk U, Hass HC, Quartino ML, Abele D (2019) Ensemble modeling of Antarctic macroalgal habitats exposed to glacial melt in a polar fjord. Front Ecol Evol 7:207. https://doi.org/10.3389/fevo.2019.00207

Johnston EL, Connell SD, Irving AD, Pile AJ, Gillanders BM (2007) Antarctic patterns of shallow subtidal habitat and inhabitants in Wilke's Land. Polar Biol 30:781–788

Jones CT, Craig SE, Barnett AB, MacIntyre HL, Cullen JJ (2014) Curvature in models of the photosynthesis-irradiance response. J Phycol 50(2):341–355. https://doi.org/10.1111/jpy.12164

Kain JM (1989) The seasons in the subtidal. Br Phycol J 24:203–215

Kortsch S, Primicerio R, Beuchel F, Renaud PE, Rodrigues J, Lønne OJ, Gulliksen B (2012) Climate-driven regime shifts in Arctic marine benthos. PNAS 109(35):14052–14057. https://doi.org/10.1073/pnas.1207509109

Krause-Jensen D, Duarte CM (2014) Expansion of vegetated coastal ecosystems in the future Arctic. Front Mar Sci 1:77. https://doi.org/10.3389/fmars.2014.00077

Krause-Jensen D, Duarte CM (2016) Substantial role of macroalgae in marine carbon sequestration. Nat Geosci 9(10):737

Krause-Jensen D, Marbà N, Olesen B, Sejr MK, Christensen PB, Rodrigues J, Renaud PE et al (2012) Seasonal sea ice cover as principal driver of spatial and temporal variation in depth extension and annual production of kelp in Greenland. Glob Chang Biol 18:2981–2994

Krüger M (2016) Photosynthese Lichtkurven ausgewählter makroalgenarten des Kongsfjords (Spitsbergen, Norwegen) als Grundlage für Abschätzungnezur Produktivitäät des arktischen Kelpwaldes. Diploma Thesis, Technical University Bergakademie Freiberg, 194 pp

Kühl M, Glud R, Borum J, Roberts R, Rysgaard S (2001) Photosynthetic performance of surface-associated algae below sea ice as measured with a pulse-amplitude-modulated (PAM) fluorometer and O_2 microsensors. Mar Ecol Prog Ser 223:1–14. https://doi.org/10.3354/meps223001

Lagger C, Servetto N, Torre L, Sahade R (2017) Benthic colonization in newly ice-free soft-bottom areas in an Antarctic fjord. PLoS One 12:e0186756

Lagger C, Nime M, Torre L, Servetto N, Tatián M, Sahade R (2018) Climate change, glacier retreat and a new ice-free island offer new insights on Antarctic benthic responses. Ecography 40:1–12

Lüning K (ed) (1990) Seaweeds, their environment biogeography and ecophysiology. Wiley, New York. 527 pp

Marcías ML, Deregibus D, Saravia L, Campana GL, Quartino ML (2017) Life between tides: spatial and temporal variations of an intertidal macroalgal community at Potter Peninsula, South Shetland Islands, Antarctica. Estuar Coast Shelf Sci 187:193–203

Marina TI, Salinas V, Cordone G, Campana GL, Moreira E, Deregibus D, Torre L, al e (2018) The food web of Potter Cove (Antarctica): complexity, structure and function. Estuar Coast Shelf Sci 200:141–151

Matta JL, Chapman DJ (1991) Photosynthetic responses and daily carbon balance of *Colpomenia peregrina*: seasonal variations and differences between intertidal and subtidal populations. Mar Biol 108:303–313. https://doi.org/10.1007/BF01344345

Meredith MP, King JC (2005) Rapid climate change in the ocean west of the Antarctic Peninsula during the second half of the 20th century. Geophys Res Lett 32:L19604

Meredith MP, Falk U, Bers AV, Mackensen A, Schloss IR, Ruiz Barlett E, Jerosch K, Silva Busso A, Abele D (2018) Anatomy of a glacial meltwater discharge event in an Antarctic cove. Philos Trans R Soc A 376:20170163. https://doi.org/10.1098/rsta.2017.0163

Miller KA, Pearse JS (1991) Ecological studies of seaweeds in McMurdo Sound, Antarctica. Am Zool 31:35–48. https://doi.org/10.1093/icb/31.1.35

Moon HW, Hussin WMRW, Kim HC, Ahn I-Y (2015) The impacts of climate change on Antarctic nearshore megaepifaunal benthic assemblages in a glacial fjord on King George Island: responses and implications. Ecol Indic 57:280–292. https://doi.org/10.1016/j.ecolind.2015.04.031

Navarro NP, Huovinen P, Gómez I (2019) Photosynthetic characteristics of geographically disjunct seaweeds: a case study on the early life stages of Antarctic and Subantarctic species. Prog Oceanogr 174:28–36. https://doi.org/10.1016/j.pocean.2018.11.001

Nelson S, Siegrist AW (1987) Comparison of mathematical expressions describing light-saturation curves for photosynthesis by tropical marine macroalgae. Bull Mar Sci 41:617–622

Paar M, Voronkov A, Hop H, Brey T, Bartsch I, Schwanitz M et al (2016) Temporal shift in biomass and production of macrozoobenthos in the macroalgal belt at Hansneset, Kongsfjorden, after 15 years. Polar Biol 39:2065. https://doi.org/10.1007/s00300-015-1760-6

Pasotti FE, Manini D, Giovannelli D, Wolfl A-K, Monien D, Verleyen E, Braeckman U, Abele D, Vanreusel A (2014) Antarctic shallow water benthos under glacier retreat forcing. Mar Ecol 36:716–733. https://doi.org/10.1111/maec.12179

Pavlov AK, Leu E, Hanelt D, Bartsch I, Karsten U, Hudson SR, Gallet J-C, Cottier F et al (2019) The underwater light climate in Kongsfjorden and its ecological implications. In: Hop H, Wiencke C (eds) The ecosystem of Kongsfjorden, Svalbard, Advances in Polar Ecology, vol 2. Springer, Cham, pp 137–172

Pritchard DW, Hurd CL, Beardall J, Hepburn CD (2013) Survival in low light: photosynthesis and growth of a red alga in relation to measured *in situ* irradiance. J Phycol 49(5):867–879. https://doi.org/10.1111/jpy.12093

Quartino ML, Deregibus D, Campana GL, Latorre GEJ, Momo FR (2013) Evidence of macroalgal colonization on newly ice-free areas following glacial retreat in Potter Cove (South Shetland Islands), Antarctica. PLoS One 8(3):e58223

Roleda MY, Campana GL, Wiencke C, Hanelt D, Quartino ML, Wulff A (2009) Sensitivity of Antarctic *Urospora penicilliformis* (Ulotrichales, Chlorophyta) to ultraviolet radiation is life-stage dependent. J Phycol 45:600–609. https://doi.org/10.1111/j.1529-8817.2009.00691.x

Rückamp M, Braun M, Suckro S, Blindow N (2011) Observed glacial changes on the King George Island ice cap, Antarctica, in the last decade. Glob Planet Change 79:99–109. https://doi.org/10.1016/j.gloplacha.2011.06.009

Runcie KW, Riddle MJ (2011) Distinguishing downregulation from other non-photochemical quenching of an Antarctic benthic macroalga using *in situ* fluorometry. Eur J Phycol 46:171–180

Runcie JW, Riddle MJ (2012) Estimating primary productivity of marine macroalgae in East Antarctica using in situ fluorometry. Eur J Phycol 47(4):449–460

Runcie JW, Paulo D, Santos R, Sharon Y, Beer S, Silva J (2009) Photosynthetic responses of *Halophila stipulacea* to a light gradient: I – In situ energy partitioning of non photochemical quenching. Aquat Biol 7:143–152

Sahade R, Lagger C, Torre L, Momo FR, Monien P, Schloss I et al (2015) Climate change and glacier retreat drive shifts in an Antarctic benthic ecosystem. Sci Adv 1:e1500050. https://doi.org/10.1126/sciadv.1500050

Scherrer KJN, Kortsch S, Varpe Ø, Weyhenmeyer GA, Gulliksen B, Primicerio R (2018) Mechanistic model identifies increasing light availability due to sea ice reductions as cause for increasing macroalgae cover in the Arctic. Limnol Oceanogr 64(1):330–341. https://doi.org/10.1002/lno.11043

Schloss IR, Abele D, Moreau S, Norkko A, Cummings V, Thrush S (2012) Response of phytoplankton dynamics to 19 year (1991–2009) climate trends in Potter Cove (Antarctica). J Mar Syst 92:53–66. https://doi.org/10.1016/j.Jmarsys.2011.10.006

Schofield O, Ducklow HW, Martinson DG, Meredith MP, Moline MA, Fraser WR (2010) How do polar marine ecosystems respond to rapid climate change? Science 328:1520–1523. https://doi.org/10.1126/science.1185779

Schwarz AM, Hawes I, Andrew N, Norkko A, Cummings V, Thrush S (2003) Macroalgal photosynthesis near the southern global limit for growth; Cape Evans, Ross Sea, Antarctica. Polar Biol 26:789–799. https://doi.org/10.1007/s00300-003-0556-2

Smale DA, Barnes DKA (2008) Likely responses of the Antarctic benthos to climate-related changes in physical disturbance during the 21st century, based primarily on evidence from the West Antarctic Peninsula region. Ecography 31:289–305. https://doi.org/10.1111/j.0906-7590.2008.05456.x

Smith EL (1936) Photosynthesis in relation to light and carbon dioxide. Proc Natl Acad Sci 22:504–511

Spurkland T, Iken K (2011) Salinity and irradiance effects on growth and maximum photosynthetic quantum yield in subarctic *Saccharina latissima* (Laminariales, Laminariaceae). Bot Mar 54:355–365. https://doi.org/10.1515/BOT.2011.042

Stammerjohn SE, Martinson DG, Smith RC, Ianuzzi RA (2008) Sea ice in the western Antarctic Peninsula region: spatio-temporal variability from ecological and climate change perspectives. Deep-Sea Res II 55:2041–2058. https://doi.org/10.1016/j.dsr2.2008.04.026

Steele J (1962) Environmental control of photosynthesis in the sea. Limnol Oceanogr 7:137–150. https://doi.org/10.4319/lo.1962.7.2.0137

Torre L, Servetto N, Eöry ML, Momo F, Tatián M, Abele D, Sahade R (2012) Respiratory responses of three Antarctic ascidians and a sea pen to increased sediment concentrations. Polar Biol 35:1743–1748. https://doi.org/10.1007/s00300-012-1208-1

Turner J, Bindschadler RA, Convey P, Di Prisco G, Fahrbach E, Gutt J, Hodgson DA et al (2009) Antarctic climate change and the environment. SCAR, Cambridge

Turner J, Barrand NE, Bracegirdle TJ, Convey P, Hodgson DA, Jarvis M et al (2013) Antarctic climate change and the environment - an update. Polar Rec 50(3):237–259. https://doi.org/10.1017/S0032247413000296

Valdivia N, Díaz MJ, Garrido I, Gómez I (2015) Consistent richness-biomass relationship across environmental gradients in a marine macroalgal-dominated subtidal community on the western Antarctic Peninsula. PLoS One 10(9):e0138582

Vaughan DG, Marshall GJ, Connolley WM, Parkinson C, Mulvaney R, Hodgson DA et al (2003) Recent rapid regional climate warming on the Antarctic Peninsula. Clim Chang 60:243–274. https://doi.org/10.1023/A:1026021217991

Webb WL, Newton M, Starr D (1974) Carbon dioxide exchange of *Alnus rubra*: a mathematical model. Oecologia 17:281–291. https://doi.org/10.1007/BF00345747

Wiencke C (1990) Seasonality of brown macroalgae from Antarctica-a long-term culture study under fluctuating Antarctic daylengths. Polar Biol 10:589–600. https://doi.org/10.1007/BF00239370

Wiencke C, Amsler CD (2012) Seaweeds and their communities in polar regions. In: Wiencke C, Bischof K (eds) Seaweed biology: novel insights into ecophysiology, ecology and utilization. Ecological studies, vol 219. Springer, Heidelberg, pp 265–292. https://doi.org/10.1007/978-3-642-28451-9

Wiencke C, Clayton MN, Gómez I, Iken K, Lüder UH, Amsler CD et al (2006) Life strategy, ecophysiology and ecology of seaweeds in polar waters. Rev Environ Sci Biotechnol 6:95–126

Wiencke C, Gómez I, Dunton K (2011) Phenology and seasonal physiological performance of polar seaweeds. In: Wiencke C (ed) Biology of polar benthic algae. De Gruyter, Berlin, pp 181–194. https://doi.org/10.1007/978-1-4020-6285-8_13

Zacher K, Rautenberger R, Hanelt D, Wulff A, Wiencke C (2009) The abiotic environment of polar marine benthic algae. Bot Mar 52:483–490. https://doi.org/10.1515/BOT.2009.082

Chapter 10
Life History Strategies, Photosynthesis, and Stress Tolerance in Propagules of Antarctic Seaweeds

Nelso Navarro, Pirjo Huovinen, and Iván Gómez

Abstract Reproduction is one of the most important processes to maintain seaweed populations. In general, growth and reproduction of seaweeds depend on environmental cues, such as change in temperature, light, and nutrients. However, the fact that Antarctic waters show a small variation in temperature and nutrient levels over the year, these biological processes depend mainly on variables related to light conditions, especially daylength. This seems to be more obvious in the eulittoral and shallow sublittoral species, because the reproduction and growth coincides with the spring season. However, in species inhabiting the deeper sublittoral zone, reproduction seems to be controlled by a free-running endogenous clock synchronized by the seasonal variation of daylength or by photoperiodisms. Whatever the case, the Antarctic environment imposes physiological constraints to reproductive output, settlement and development of propagules, recruitment, and growth of seaweeds. Early life stages (e.g., spores, gametes, propagules, and plantlets) are extremely shade-adapted and susceptible to environmental stress, such as exposure to UV radiation; however, they are thermally well adapted, at least for short periods of time, allowing them to develop in a highly variable environment. In this chapter, we review the main reproduction strategies that Antarctic seaweeds display to cope with the extreme environment. Additionally, we review recent studies on stress tolerance of early developmental stages from selected species. In scenarios of the changing Antarctic environment due to warming, UV radiation, freshening, and

N. Navarro (✉)
Laboratorio de Ecofisiología y Biotecnología de Algas (LEBA), Universidad de Magallanes, Punta Arenas, Chile

Instituto de Ciencias Marinas y Limnológicas, Facultad de Ciencias, Universidad Austral de Chile, Valdivia, Chile
e-mail: nelso.navarro@umag.cl

P. Huovinen · I. Gómez
Instituto de Ciencias Marinas y Limnológicas, Facultad de Ciencias, Universidad Austral de Chile, Valdivia, Chile

Research Center Dynamics of High Latitude Marine Ecosystems IDEAL, Valdivia, Chile
e-mail: pirjo.huovinen@uach.cl; igomezo@uach.cl

I. Gómez, P. Huovinen (eds.), *Antarctic Seaweeds*,
https://doi.org/10.1007/978-3-030-39448-6_10

other emergent stressors, the knowledge on adaptive life strategies of early developmental phases can allow us better predicting the fate of seaweed communities.

Keywords Life history stages · Photosynthetic light requirements · Reproduction · Seaweed propagules · Seasonal development

10.1 Seasonal Strategies and Life History Cycles

In their environment, seaweeds are exposed to a complex suite of abiotic variables, whose interaction may affect reproduction synergistically or antagonistically. In the case of Antarctic seaweeds, the life strategy of the individual species is regulated by the strong seasonal variation in light conditions (Wiencke et al. 2009). Two different growth (and reproduction) strategies have been identified: the season anticipator and season responder strategy (sensu Kain 1989; Wiencke and Clayton 2002). These strategies have been corroborated through long-term laboratory culture experiments in which temperature and nutrient levels were kept constant and only light and daylength were modified in order to simulate the seasonally fluctuating Antarctic irradiances (Wiencke 1990a, b; Dummermuth and Wiencke 2003). Moreover, other phenological events such as seasonal induction of propagules, their release, and the growth of early developmental stages have been examined in the field (Roleda et al. 2007, 2008; Zacher et al. 2009; Navarro et al. 2016).

10.1.1 Season Anticipators

This group of algae grows and reproduces in winter under short-day and low-light conditions (Fig. 10.1). Thus, physiological and reproduction processes seem to be controlled by a free-running endogenous annual rhythm synchronized by the seasonal changes of daylength or by photoperiodisms and not by environmental conditions (such as levels of light or temperature) as demonstrated by Lüning and tom Dieck (1989), tom Dieck (1989), and Lüning and Kadel (1993) in several species from temperate regions. Likewise, the growth of Antarctic season anticipators has been related to increasing daylength during the late winter and early spring (Wiencke et al. 2007, 2009). Many endemic Antarctic seaweeds with sublittoral distribution are regarded as season anticipators, e.g., the brown seaweeds *Himantothallus grandifolius* (A.Gepp and E.S.Gepp) Zinova, *Desmarestia anceps* Montagne, *D. antarctica* Moe *et* Silva, *Phaeurus antarcticus* Skottsberg (Wiencke 1990a), *Ascoseira mirabilis* Skottsberg (Gómez et al. 1995, 1996; Wiencke 1990a), and *D. menziesii* J. Agardh (Gómez and Wiencke 1996) and the red seaweeds *Palmaria decipiens*

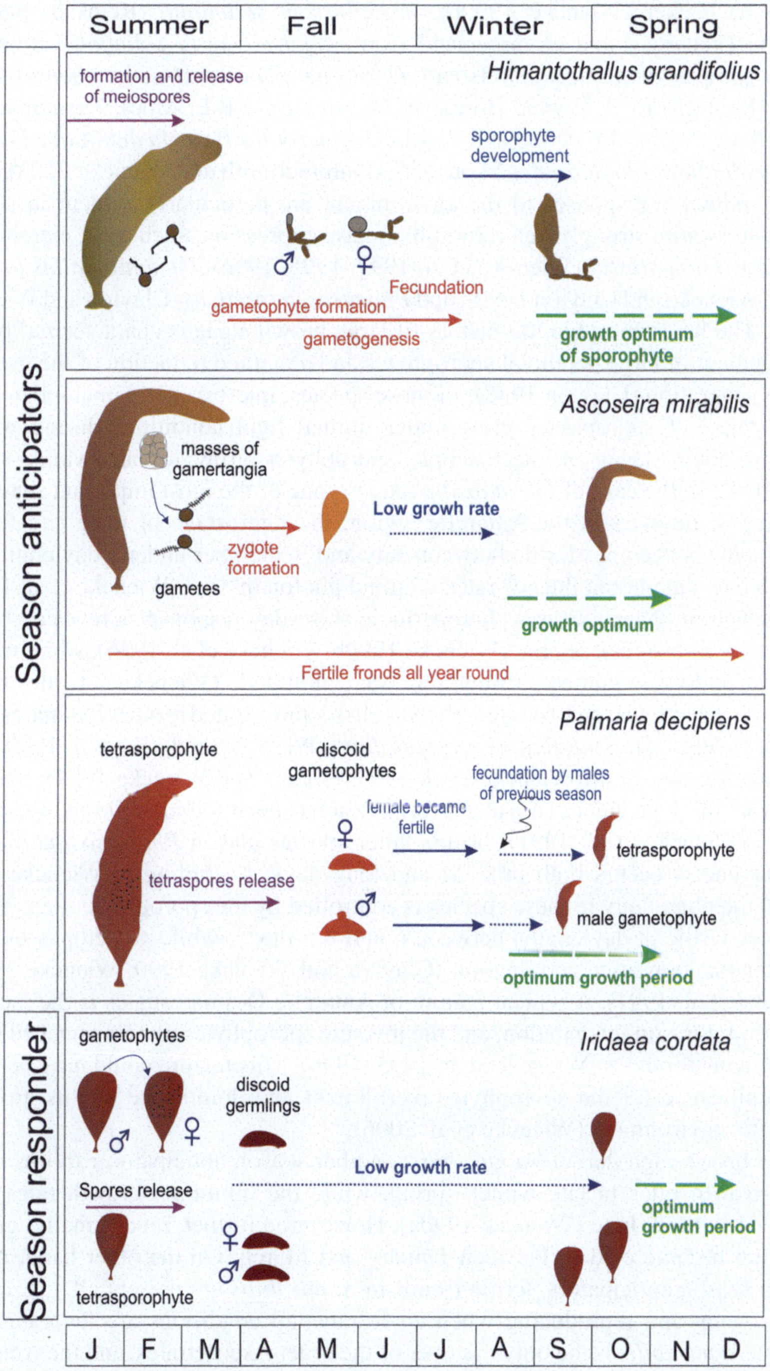

Fig. 10.1 Life history development of some conspicuous Antarctic seaweeds. The green line shows the period when growth starts. In the case of season anticipators, growth take place in late winter onwards, while in responders, growth occur during spring onwards

(Reinsch) Ricker (Wiencke 1990b), *Paraglossum salicifolia* (Reinsch) Schowe M. Lin, Fredericq, and Hommersand (formerly *Delesseria salicifolia* Reinsch), *Gymnogongrus antarcticus* Skottsberg; *G. turquetii* Hariot, *Hymenocladiopsis prolifera* (Reinsch) M. J. Wynne (formerly *H. crustigena* R.L. Moe), *Trematocarpus antarcticus* (Hariot) Fredericq and R. Moe (formerly *Kallymenia antarctica* Hariot), and *Phyllophora ahnfeltioides* Skottsberg (Dummermuth and Wiencke 2003).

Reproductive responses to the environment are particularly evident in season anticipators with strongly heteromorphic phase expression, such as in members of the genus *Desmarestia* (Wiencke et al. 1991, 1995, 1996), *Himantothallus grandifolius* (Wiencke and Clayton 1990), and *Phaeurus antarcticus* (Clayton and Wiencke 1990). The heteromorphic life history of large brown algae is characterized by the development of large perennial sporophytes and a marked reduction of the gametophytic generation (Clayton 1988). In these species, microscopic gametophytes and early stages of sporophytes grow under limited light conditions during winter, whereas adult stages of macroscopic sporophytes grow in late winter–spring (Fig. 10.1). In the case of *Desmarestia anceps*, one of the most important seaweeds in terms of biomass in the Antarctic region, the microstage of male and female gametophytes becomes fertile between July and September under a daylength of 5 and 7 h day^{-1} at photon fluence rates <3 μmol photon m^{-2} s^{-1} (Wiencke et al. 1996). The induction of fertility is a photoperiodic short-day response as revealed by the effect of a night-break regime (Wiencke 1990b; Wiencke et al. 1996), while in continuous darkness gamete formation was inhibited (Wiencke et al. 1996). Gametogenesis under short daylengths was also demonstrated in other Desmarestiales members, e.g., *Himantothallus grandifolius* (Wiencke and Clayton 1990) and *Desmarestia menziesii* (Wiencke et al. 1995; Gómez and Wiencke 1997), whereas no daylength dependence of gamete formation has been found in *Desmarestia antarctica* (Wiencke et al. 1991). In this latter species and in *Phaeurus antarcticus*, gametogenesis occurs both in short and long days. According to Wiencke et al. (2009) the phenology in these species is controlled by the sporophytic stage, which becomes fertile at daylengths between 6 and 8 h day^{-1}, while gametophytes form gametangia soon after germination (Clayton and Wiencke 1990; Wiencke 1990a; Wiencke et al. 1991). A typical feature of Antarctic Desmarestiales is the fact that they exhibit in situ fecundation, and the juvenile sporophytes remain attached to the female gametophytes (Wiencke et al. 1995, 1996). This feature could have ecological significance for the sporophytes recruitment and dominance of this group in Antarctic environment (Wiencke et al. 2006).

The brown alga *Ascoseira mirabilis,* another season anticipator, exhibits maximum growth rates in late winter–spring, while the minimum growth rates were recorded in May–June (Wiencke 1990a). However, a further, much smaller growth optimum became evident between January and March. On the other hand, unlike typical season anticipators, fertile fronds in *A. mirabilis* are present all year round, and growing and reproducing when environmental conditions are favorable (see below). *A. mirabilis* is the only member of the order Ascoseirales, and the Antarctic environmental constraints might have exerted an evolutionary pressure to develop a unique life history and reproductive biology when compared with other

Phaeophyceae (Roleda et al. 2007). The species is monoecious with sexual (isogamia) reproduction. There is one, free-living diploid generation, and zygotes develop into new individuals (Wiencke and Clayton 2002). Conceptacles are scattered all over the blades and the extrusion of gametangial masses through the ostioles precedes the release of heterokont gametes (Müller et al. 1990). Zygote formation follows immediately after fusion of gametes.

For red seaweeds, the phenology of six season anticipator species from the Antarctic, *Paraglossum salicifolium*, *Gymnogongrus antarcticus*, *Gymnogongrus turquetii*, *Hymenocladiopsis prolifera*, *Trematocarpus antarcticus*, and *Phyllophora ahnfeltioides* were investigated by Dummermuth and Wiencke (2003) in a two-year culture study under fluctuating daylengths simulating the Antarctic conditions. The period of highest growth rate in these species was registered between September and November (late winter–spring) and the formation of new blades occurred from January/February onwards. Before the summer solstice, growth ceased. However, in *Hymenocladiopsis prolifera*, the seasonal growth peak was observed in August when the light conditions increased from 3 to 25 μmol photon $m^{-2}\,s^{-1}$. This suggests that the phenology of season anticipators could not only be controlled by daylength but also by photon fluence rates. Thus, in species distributed along wide ranges of depth (e.g., 5–30 m), the seasonal growth peak could be later in the season at deeper water depths and earlier in shallower waters. Reproductive fronds were not observed, except in *Trematocarpus antarcticus,* which completed its life cycle with carpospore formation between June and August, but with the first cystocarps found in March (Dummermuth and Wiencke 2003). These results agree with the findings reported by Lamb and Zimmermann (1977), who reported cystocarps in thalli of *T. antarcticus* in January. Similarly, in *Gymnogongrus antarcticus,* cystocarps are formed in the summer (Skottsberg 1953; Cormaci et al. 1992). Cystocarps and tetrasporangia of *Paraglossum salicifolium* have been observed in late winter (Wynne 1982). Likewise, spermatangia, cystocarps, and tetrasporangia in *Delesseria sanguinea*, a comparable species in the same family, are formed during winter (Kornmann and Sahling 1977). In the case of *G. turquetii* (Kylin and Skottsberg 1919; Skottsberg 1953) and *Phyllophora ahnfeltioides* (Kylin and Skottsberg 1919), cystocarpic fronds have been reported between May and June (autumn).

The red seaweed *Palmaria decipiens*, one of the dominating species in terms of biomass, is considered as season anticipator (Wiencke 1990b; Weykam and Wiencke 1996) and displays a heteromorphic life history perfectly adapted to the Antarctic conditions. In this species female gametophytes represent the microscopic phase, while the male gametophyte develops into a macro-thallus similar in morphology to the tetrasporophytes (Fig. 10.1). The male and tetrasporophytic blades are formed in winter (Wiencke 1990b; Weykam et al. 1997) and the optimum growth period and high rates of net photosynthesis and photosynthetic efficiency coincide with increasing light intensities in spring (Wiencke 1990b; Weykam and Wiencke 1996). Tetrasporophytes become fertile in February and tetraspores develop in May into semiglobular to discoid gametophytes. The females become fertile only once from May to June. After fertilization, the female gametophyte is overgrown by the developing sporophyte, which matures and releases tetraspores in the next summer.

Interestingly, it takes about a year until male gametophytes become fertile; thus fertilization of females is only possible by mature males of the previous season, indicating a life span of the species of several years (Wiencke 1990b).

10.1.2 Season Responders

In these organisms growth and reproduction coincide with favorable light conditions in spring and summer. Thus, these species react directly to the primary factors in their environment (such as light availability) and show an opportunistic life strategy (Wiencke 1990a, b). Most of the season responder species are distributed in the eulittoral and upper sublittoral zone, and they can have temperate or cold-temperate affinities (Wiencke et al. 2007; Navarro et al. 2019; see Chap. 12 by Campana et al.). Well-known members of this group are the red seaweeds *Iridaea cordata* (Turner) Bory (Weykam et al. 1997) and *Gigartina skottsbergii* Setchell *et* N.L. Gardner (Wiencke 1990b), the brown alga *Adenocystis utricularis* (Bory) Skottsberg (Wiencke 1990a), and the green seaweeds *Ulva hookeriana* (Kützing) H. S. Hayden, Blomster, Maggs, P. C. Silva, Stanhope, and Walland (formerly *Enteromorpha bulbosa* (Suhr) Montagne and *Acrosiphonia arcta* (Dillwyn) J. Agardh (Wiencke 1990b).

The pseudoperennial *Gigartina skottsbergii* and *Iridaea cordata* have a triphasic life history with isomorphic haploid male and female gametophytes and a diploid tetrasporophyte. They occur normally in eulittoral pools and in the upper sublittoral, but also can be found down to 30 m (Wiencke and Clayton 2002; Navarro et al. 2016). Both species show the maximum growth rate during the spring-summer season (e.g., December), while the minimum growth rates were recorded from May to July (Wiencke 1990b). Mature tetrasporophytes and gametophytes of *Iridaea cordata* were observed during spring-summer (Roleda et al. 2008; Navarro et al. 2016). Tetraspores and carpospores of this species germinate normally forming a discoid germling from which new plantlets arise from July onwards. The plantlets show a growth optimum between September and November and large blades are formed in summer (Wiencke 1990b). Regrowth from the perennial basal parts of the blades is possible (Wiencke and Clayton 2002), which could explain its dominance at the eulittoral (Marcías et al. 2017). In the case of *G. skottsbergii*, Wiencke (1990b) reported the induction of sporangium formation in tetrasporophytes in the laboratory by the end of September, when irradiances were between 27 and 46 μmol photon m^{-2} s^{-1}, but spores were not released before June. In contrast, in the field, reproductive fronds with viable propagules have been collected in October (Roleda et al. 2008) and January (Navarro et al. 2016). This discrepancy in reproductive periods might be related to differences related to the experimental setup of laboratory cultures by Wiencke (1990b). As suggested for the season anticipator *Hymenocladiopsis prolifera*, photon fluence rates might also control the seasonal phenology of this species. Thus, these algae apparently have the capacity to reproduce during a prolonged time span under changing environmental conditions.

Adenocystis utricularis has a heteromorphic life cycle with a sporophytic macrothallus and gametophytic microthallus. Spores from the macrothalli develop into microscopic filamentous, dioecious gametophytes (Wiencke and Clayton 2002), the dominant life phase under winter conditions in laboratory culture (Wiencke 1990a). Macrothalli start to develop asexually on crustose parts of the microthalli from June onwards. Between October and December, growth rates of macrothalli are optimal. Reproductive macrothalli are present in January–February, while release of spores occurs in February, after which the thalli disintegrate.

As suggested for other species with heteromorphic phase expression, the microthallus is probably an important over-wintering stage (Wiencke and Clayton 2002). However, eventually all developmental stages can be present at the same time depending where they occur. De Reviers and Délépine (1981) reported that macrothalli are present throughout the year with juveniles being most abundant in October in the eulittoral zone, while in the sublittoral zone, small macrothalli are present only from November to June. Laboratory experiments where the photon fluence rates varied from 2 to 46 μmol photon m^{-2} s^{-1} confirmed this field observation (Wiencke 1990a). Thus, in the eulittoral zone, *A. utricularis* occurs as an aseasonal annual, while in upper sublittoral zone, the species probably occurs as a seasonal annual due to less available light. The species has been reported to occur down to 20 m (Wiencke and Clayton 2002), and at these depths the alga possibly is biannual as suggested by Wiencke (1990a).

10.2 Photosynthetic Light Requirements of Early Stages

Irrespective of the life history strategy, the Antarctic environment imposes physiological constraints to the reproductive output (propagules), settlement, recruitment, and growth of seaweeds. However, seaweeds have adapted their biological processes by evolving different functional mechanisms to cope with the Antarctic light and temperature conditions. In general, Antarctic seaweeds are very low light adapted, adult phases being able to photosynthesize at irradiances as low as 10 μmol photon m^{-2} s^{-1}, while propagules can photosynthesize at even lower irradiances (Gómez et al. 2009). In Table 10.1 the saturating irradiances of photosynthesis (E_k values) of different Antarctic seaweed propagules are summarized, and in Fig. 10.2 mean values of E_k for propagules from eulittoral and sublittoral algae are plotted. With the exception of some eulittoral species, most of the studied Antarctic seaweeds exhibit E_k values lower than 60 (μmol photon m^{-2} s^{-1}). Although differences in saturation irradiance between eulittoral (65 ± 20 μmol photon m^{-2} s^{-1}) and sublittoral (40 ± 15 μmol photon m^{-2} s^{-1}) exist, propagules are able to adapt to different light conditions (quantity and quality) mainly during the winter–spring transition. This is particularly evident in propagules of species that colonize a wide range of vertical distribution (e.g., *Desmarestia anceps*). After sea ice breakup in King George Island (South Shetland Islands), light can penetrate down to 30 m

Table 10.1 Overview of light requirement for saturation (E_k), photosynthetic efficiency (α_{ETR}), inhibition by PAR and PAR + UV radiation, and subsequent recovery in different life history stages of Antarctic seaweeds collected from different depths. Propagule size and date of collection are also indicated. Irradiance values represent conditions during incubation in the laboratory. PAR and PAR + UV inhibition were calculated after 4 h of exposition according to information provided in the papers)

Species	Depth	Date of collection	Cell diameter (μm)	Irradiance (μmol m^{-2} s^{-1})	E_k	α_{ETR}	F_v/F_m	PAR inhibition	PAR + UV inhibition	Recovery	Reference
Urospora penicilliformis zoospores, gametes	Upper eulittoral (0 m)	Oct 2004	6	22	87	0.16	~0.50	No	~37% (PA; 4 h); ~39% (PAB; 4 h)	100% (24 h)	Roleda et al. (2009)
Urospora penicilliformis gametophytes	Upper eulittoral (0 m)	Oct 2004		22	252	0.18	~0.51	No	~20% (PA; 4 h); ~23% (PAB; 4 h)	94% (1 h)	Roleda et al. (2009)
Pyropia endiviifolia carpospores	Eulittoral (0 m)	Jan–Feb 2015	12 ± 1	13	101 ± 19	0.14 ± 0.01	0.35 ± 0.01	No	~16% (PAB, 4 h)	100% (4 h)	Navarro et al. (2016, 2019)
Pyropia endiviifolia monospores	Upper eulittoral (0 m)	Jan–Mar 2005	15 ± 2	23	33	0.12	0.49 ± 0.04	46% (4 h)	~56% (PA; 4 h); ~60% (PAB; 4 h);	100% (48 h)	Zacher et al. (2007)
Adenocystis utricularis zoospores	Eulittoral (0–1 m)	Jan–Feb 2015	6 ± 1	13	59 ± 6	0.45 ± 0.01	0.64 ± 0.02	No	~2% (PAB, 4 h)	100% (4 h)	Navarro et al. (2016)
Adenocystis utricularis zoospores	Eulittoral (0–1 m)	Jan–Mar 2005	4	23	64	0.14	0.46 ± 0.11	No	~20% (PA; 4 h); ~57% (PAB; 4 h);	100% (48 h)	Zacher et al. (2007)

Monostroma hariotii gametes	Eulittoral (0–1 m)	Jan–Feb 2015	5 ± 0.5	13	295 ± 84	0.09 ± 0.01	0.26 ± 0.01	23% (4 h)	~28% (PAB; 4 h)	100% (4 h)	Navarro et al. (2016, 2019)
Monostroma hariotii gametes	Eulittoral (0 m)	Jan–Mar 2005	7	23	83	0.065	0.29 ± 0.04	50% (4 h)	~66% (PA; 4 h); ~71% (PAB; 4 h);	100% (48 h)	Zacher et al. (2007)
Iridaea cordata tetraspores	Eulittoral (0–1 m)	Jan–Feb 2015	18 ± 2	13	46 ± 6	0.17 ± 0.03	0.39 ± 0.06	No	~20% (PAB; 4 h)	100% (4 h)	Navarro et al. (2016, 2019)
Iridaea cordata tetraspores	Upper sublittoral (0 m)	Oct 2004	20 ± 2	22	57	0.12	0.47 ± 0.04	53% (4 h)	~78% (PA; 4 h); ~82% (PAB; 4 h)	100% (2 d)	Zacher et al. (2009)
Gigartina skottsbergii carpospores	Sublittoral (5–8 m)	Jan–Feb 2015	25 ± 2	13	27 ± 9	0.27 ± 0.05	0.42 ± 0.32	52% (4 h)	~80% (PAB, 4 h)	35–60% (4 h)	Navarro et al. (2016)
Gigartina skottsbergii carpospores	Sublittoral (3 m)	Oct 2004	27 ± 2	22	54 ± 2	0.14	0.40 ± 0.03	73%	~87% (PA; 4 h);89% (PAB; 4 h)	100% (2 d)	Roleda et al. (2008)
Gigartina skottsbergii tetraspores	Sublittoral (3 m)	Oct 2004	23 ± 1.5	22	44 ± 21	0.14	0.31 ± 0.07	69% (4 h)	~86% (PA; 4 h);90% (PAB; 4 h)	100% (2 d)	Roleda et al. (2008)
Ascoseira mirabilis gametangia	Deep sublittoral (17–30 m)	Jan–Feb 2015	22 ± 4	13	17 ± 5	0.21 ± 0.03	0.48 ± 0.02	16% (4 h)	~30% (PAB; 4 h)	85% (4 h)	Navarro et al. (2016)

(continued)

Table 10.1 (continued)

Species	Depth	Date of collection	Cell diameter (µm)	Irradiance (µmol m^{-2} s^{-1})	E_k	α_{ETR}	F_v/F_m	PAR inhibition	PAR + UV inhibition	Recovery	Reference
Ascoseira mirabilis gametangia	Upper sublittoral (2 m)	Oct 2004	8–10	22	52	0.09	0.40 ± 0.06	85% (4 h)	~91% (PA; 4 h); 95% (PAB; 4 h)	100% (2 d)	Roleda et al. (2007)
Ascoseira mirabilis conceptacles	Shallow sublittoral (1 m)	Jan 2014		13	52 ± 14	0.20 ± 0.04	~0.6		~0% (6 h)	100% (14 h)	Huovinen and Gómez (2015)
Cystosphaera jacquinotii receptacles	Deep sublittoral	Jan–Feb 2015		13	13 ± 6	0.28 ± 0.01			~15–21% (2 h)	~ 90% (4 h)	Huovinen and Gómez (2015)

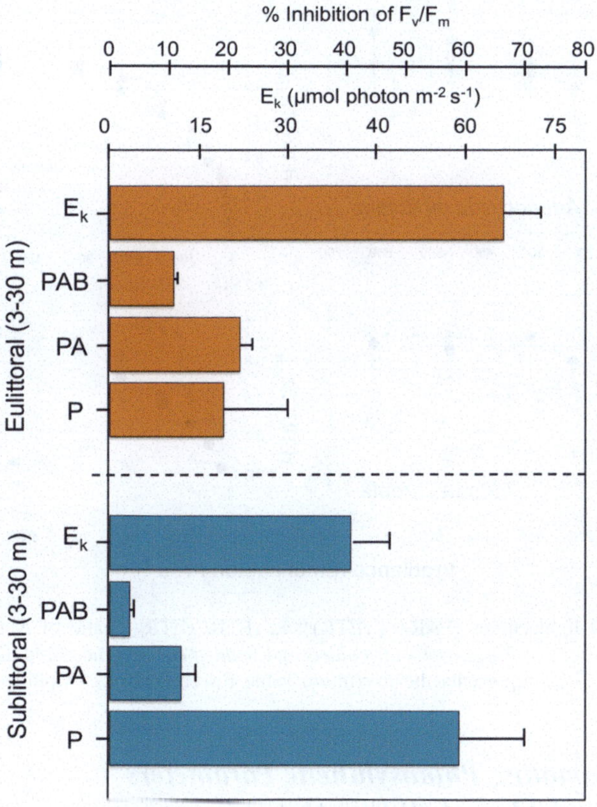

Fig. 10.2 Summary of light requirement for saturation (E_k) and inhibition by PAR and PAR + UV radiation in propagules of eulittoral and sublittoral Antarctic seaweeds. PAR and PAR + UV inhibition were calculated after 4-h exposure according to the information provided in each of the studies consulted. For references, see Table 10.1

depth reaching an average of photon fluence rates of 70 µmol photon m^{-2} s^{-1} (Gómez et al. 1997). This level of irradiance can also be strongly attenuated in terms of spectral characteristics under the canopy of large brown algae (Huovinen et al. 2016; Gómez et al. 2019). Below the canopy the spectrum is enriched in green and in far red light, probably affecting photosynthesis as well as the photomorphogenetic development of the understory species (Salles et al. 1996) (see Chap. 7 by Huovinen and Gómez).

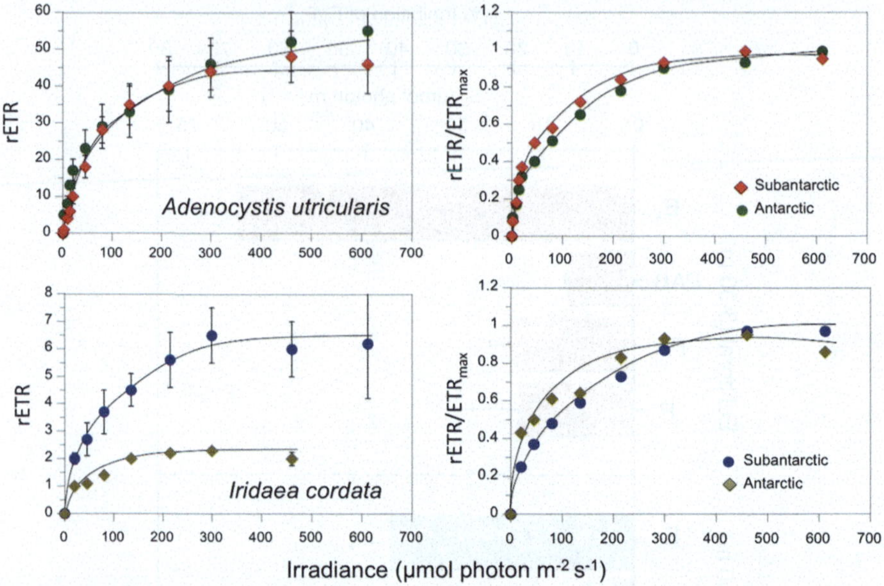

Fig. 10.3 Rapid light curves (PAR vs rETR) and rETR/ $rETR_{max}$ ratio of Antarctic and sub-Antarctic populations of *Adenocystis utricularis* and *Iridaea cordata* measured using chlorophyll fluorescence. $rETR_{max}$ represents the maximum value for each curve. (Modified from Navarro et al. 2019)

10.2.1 Estimating Photosynthetic Parameters from Chlorophyll Fluorescence

Photosynthetic characteristics of propagules of Antarctic seaweeds are normally determined estimating photosynthetic parameters (ETR_{max}, alpha and saturation irradiance, E_k) calculated from P-E curves (summarized in Gómez et al. 2009). When P-E curves are based on chlorophyll fluorescence measurements, the electron transport rates (ETR) are commonly used as a parameter (Fig. 10.3). Considering the limitations of the fluorescence method, as well as various factors that can affect light requirements, e.g., form, season, size, number of cells, chlorophyll concentration, etc., the P-E curve-derived light requirements for photosynthesis (E_k) represent the best measures to estimate shade adaptation in adult (Huovinen and Gómez 2013) and early stages (Zacher et al. 2007; Roleda et al. 2008, 2009; Navarro et al. 2016, 2019). However, two important aspects for calculation of electron transport rates are sometimes not considered: (1) the proportional rates of chlorophyll *a* at each photosystem (FII factor), which is different between red, green, and brown algae (Grzymski et al. 1997), and (2) the amount of light absorbed by the algal samples (absorptance). Due to the difficulty of measuring absorptance in a propagule suspension, the use of the relative ETR has been proposed (Beer et al. 2001). This parameter provides useful information for the description of relative changes in photosynthetic activity if the experimental approach uses standardized measurements.

Thus, the comparison of photosynthetic parameters of propagules using the ETR-based P-E curves must be made with caution. Alternatively, the rETR can be normalized to $rETR_{max}$ (with $rETR_{max}$ as the maximum value for each curve) expressing the rETR curve between 0 and 1 (relative units), which allows comparing propagules from species with very different ETR values. For example, Navarro et al. (2019) showed differences in photosynthetic performance of propagules from congeneric and conspecific Antarctic/sub-Antarctic seaweeds using the rETR and $rETR/rETR_{max}$ ratio curves. While the rETR curve demonstrates that tetraspores of *I. cordata* from Antarctic populations exhibit very low rETR values when compared to the sub-Antarctic population, the $rETR/rETR_{max}$ ratio allows detecting differences in the shape of the curves (Fig. 10.3).

10.3 Effects of Environmental Factors on the Biology of Propagules

10.3.1 High Solar Radiation

Environmental shifts will affect recruitment, and consequently, the whole fate of the seaweed population and their maintenance through time. Once Antarctic seaweed spores or gametes are released, they face a completely different physical environment than what existed when they were housed in the parental reproductive structures (Amsler et al. 1992; Zacher 2014). During the last decade, various studies have examined the effects of stress factors (e.g., temperature and UV radiation) on microscopic stages, e.g., propagules and plantlets, of some selected Antarctic seaweeds (Gómez et al. 2009; Roleda et al. 2009; Zacher et al. 2009; Navarro et al. 2016). There is a consensus that propagules are the most susceptible life stage of seaweeds in terms of their responses to environmental perturbations. However, the effect of a given factor on the propagule physiology is highly variable, depending on a suite of environmental and endogenous factors, which can act synergistically or antagonistically.

It is well known that UV wavelengths cause direct and indirect effects on algal cells (e.g., Karsten et al. 2009). The direct effects are normally mediated by absorption of UV by important biomolecules, in particular the DNA, enzymes, and membrane components (Vass 1997). In the case of propagules, which attain small size, translucent cytosol and an incipient development of the cell wall, UV radiation can easily reach the DNA where diverse injuries are produced, e.g., formation of cyclobutane pyrimidine dimers (CPDs) (Wiencke et al. 2000). This results in the inability of RNA and DNA polymerases to recognize the damaged sectors, causing the interruption of gene transcription and DNA replication (Britt 1995).

Consequently, modifications in the metabolism, cellular division, and germination of unicellular propagules can occur (Huovinen et al. 2000).

DNA damage has been reported in propagules of eulittoral Antarctic seaweeds *Adenocystis utricularis*, *Monostroma hariotii*, and *Iridaea cordata* after exposure to different doses of UV radiation (Roleda et al. 2007, 2008; Zacher et al. 2007, 2009). In general, the amount of CPDs increases with increasing UV-B dose; however, lesions can be effectively repaired after 48 h under photoreactivation processes (Zacher et al. 2007, 2009). Besides, contrasting patterns have been detected in species from different depths: propagules of *Pyropia endiviifolia* from upper eulittoral did not exhibit CPDs under different UV-B doses (Zacher et al. 2007), while propagules of *Gigartina skottsbergii* and *Ascoseira mirabilis* from the deep sublittoral were more affected and not able to repair their damaged DNA completely after 8-h UV-B exposure (0.4 Wm^{-2}) (Roleda et al. 2007, 2008). Interestingly, in the case of *Gigartina skottsbergii,* the accumulation of DNA damage was related to the ploidy level of the propagules: DNA damage was lower in diploid carpospores (2*n*) compared to haploid tetraspores (*n*) suggesting that diploid carpospores are more tolerant to UV radiation in terms of UV-B-induced DNA damage (Roleda et al. 2008). These authors suggested that higher UV-stress tolerance of diploid carpospores than haploid tetraspores could be related to the genetic buffering hypothesis, which says that diploid organisms are more vigorous and tolerant to stress than haploid ones, i.e., the two copies of every gen confer them advantages to withstand the effects of deleterious recessive mutations (Raper and Flexer 1970; Gerstein et al. 2010). However, diverse studies have stated many important genetic advantages of haploidy such as lower mutation load and more rapid spread of beneficial alleles and of diploidy, e.g., protection from somatic mutation and heterozygote advantage (Otto and Gerstein 2008). In fact, in spite of the higher DNA damage, tetraspores of *G. skottsbergii* exhibited a higher DNA damage repair rate than carpospores when the UVR was excluded. It must be noted that DNA damage in spores exposed to high UV-B dose was not repaired completely after 2 days of post-cultivation, and the remaining DNA damage was lower in carpospores than in tetraspores (Roleda et al. 2008).

UV radiation affects also photochemical processes, especially inhibiting the energy transfer within the PSII reaction center by blocking the electron flow. UV-B radiation affects the D1/D2 protein complex (Richter et al. 1990) mainly by fragmenting the D1 protein (Vass 1997; Bischof et al. 2006) through UV-active chromophores on both the donor and acceptor side of this protein (Bouchard et al. 2006). On the oxidizing side, the oxygen evolving system (water splitting complex) is another sensitive target of UV-B (Renger et al. 1986). Moreover, it has been suggested that UV-B can affect the antenna complex through the functional shutdown of the photosystem, resulting in a failure in the transfer of energy to the reaction center (Renger et al. 1986; Lorenz et al. 1997; Bischof et al. 2006). In propagules of Antarctic seaweeds, UV radiation has also been pointed out as responsible for the decrease in photosynthetic activity, measured as decreases in optimum quantum yield-F_v/F_m. For example, Navarro et al. (2016) reported that propagules of species from the eulittoral (e.g., *Iridaea cordata*, *Pyropia endiviifolia*, *Adenocystis utricularis*) showed <20% inhibition in F_v/F_m from UV (1.5 and 0.26 Wm^{-2} of UV-A and UV-B, respectively) after 4 h of exposure, while propagules of the red alga *G. skottsbergii* collected in the sublittoral were more sensitive exceeding 30% inhibition in F_v/F_m in the same condition. It is important to emphasize, however, that photochemical reactions of propagules from Antarctic seaweeds can also be strongly photoinhibited by PAR (Fig. 10.2). For example, 1-h exposure under 22 μmol photons m^{-2} s^{-1} of PAR decreased F_v/F_m in propagules of the sublittoral *G. skottsbergii* (53–58%) (Roleda et al. 2008) and *A. mirabilis* (62%) (Roleda et al. 2007) and in the eulittoral *M. hariotii* (62%), *P. endiviifolia* (81%) (Zacher et al. 2007), and *I. cordata* (~25%) (Zacher et al. 2009). Increasing exposure time further reduced F_v/F_m in all these species, with exception of *M. hariotii* (Zacher et al. 2007). In contrast, in the case of propagules of the eulittoral *Adenocystis utricularis* (Zacher et al. 2007) and *Urospora penicilliformis* (Roleda et al. 2009), the photosynthetic activity was not affected by PAR. PAR supplemented with UV-A (~4.3 Wm^{-2}) decreased photosynthetic efficiency significantly compared to only PAR treatment in all mentioned species during 1-h exposure. However, additional UV-B (~0.35 Wm^{-2}) revealed a further decrease of F_v/F_m only in sublittoral *Ascoseira mirabilis* (25%) and *G. skottsbergii* (3–7%) (Roleda et al. 2007, 2008). Although UV radiation further decreased photosynthetic efficiency in these species, all propagules recovered completely after 48 h (Table 10.1).

Additionally, the UV susceptibility has been related to propagule size as cell path length affects various bio-optical processes such as scattering and spectral extinction (Swanson and Druehl 2000; Roleda et al. 2008; Navarro et al. 2016). However, at the cellular level, UV tolerance does not seem to respond to complex biochemical and bio-optical processes. For example, tetraspores of *I. cordata* from Antarctica exhibit a smaller size but very high UV tolerance compared to tetraspores of the same species from sub-Antarctic region (Navarro et al. 2019). UV tolerance can also be related to the presence and/or the capacity to induce formation of UV-absorbing compounds, what could result in a more effective UV photoprotection, still in small propagules (Roleda et al. 2008). To our knowledge, only few studies have described absorption of UV in Antarctic seaweed propagules under UV stress. Higher concentration of palythine (λ_{max} = 320 nm) than shinorine (λ_{max} = 334 nm) has been reported in freshly released tetraspores of *G. skottsbergii* (Roleda et al. 2008) and *I. cordata* (Zacher et al. 2009). However, contrasting patterns in MAA content were observed after 8 h under PAR or PAR + UV treatments, while the total content of MAAs in tetraspores of *G. skottsbergii* was not significantly different between control (freshly released spores) and treatment. In contrast, MAA concentration in spores of *I. cordata* decreased in treated compared to freshly released spores. Based on these findings, it could be suggested that (1) freshly released propagules could have a basal level of UV-absorbing substances due to the higher in situ incident solar radiation in the field and (2) the level of UV-absorbing substances can acclimate depending on environmental conditions. In the first case, the synthesis of UV-absorbing substances would take place when the spores are still protected by the thick tissue of the parental thalli (tetrasporangial tissue in the case of tetraspores of *G. skottsbergii* and *I. cordata*). For *I. cordata*, Karsten et al. (2000) reported a higher amount of MAAs in tetrasporangial tissue than in vegetative parts of the thalli. Similarly, Huovinen and Gómez (2015) reported that reproductive tissue of *Ascoseira mirabilis* and *Cystosphaera jacquinotii* contain higher amounts of soluble phlorotannins, a type of UV-absorbing phenols found in brown algae. The presence of these compounds in reproductive tissues could ensure the maturation, survival, and germination of released propagules when they are exposed to UV radiation in the water column. Although photoprotection was only partial in laboratory experiments, propagules of *I. cordata* and *G. skottsbergii* tetraspores exposed to UV-B radiation showed the higher total MAAs in comparison with those incubated under only PAR (Roleda et al. 2008; Zacher et al. 2009).

10.3.2 Temperature

Antarctic seaweed propagules are adapted to low temperature. Cold adaptation was confirmed by the high photosynthetic efficiency (in terms of maximum quantum yield of fluorescence – F_v/F_m) at 0 °C in six Antarctic distributed species (Navarro et al. 2016). This low temperature requirement for photosynthesis is certainly the result of the long Antarctic cold-water history of at least 14 Ma (Crame 1993). However, it is well known that photosynthesis increases progressively with increasing temperature and then rapidly declines near upper critical temperature (Davison 1991). In the case of Antarctic species, the optimum temperature for photosynthesis is between 10 and 20 °C (Eggert and Wiencke 2000; Eggert 2012), lower than that reported for algae from other geographic regions (reviewed in Gómez et al. 2009). In propagules of eulittoral species such as *Adenocystis utricularis*, *Monostroma hariotii*, and *Pyropia endiviifolia* and shallow sublittoral *Ascoseira mirabilis*, the highest photosynthetic efficiency was observed at 25 °C (Navarro et al. 2016). This suggests that propagules of these species are thermally well adapted (eurythermal species), allowing them to develop in a highly variable environment or in different biogeographic regions. For example, *A. utricularis* and *M. hariotii* are widely distributed in sub-Antarctic and temperate coasts of South America (Huovinen and Gómez 2012, see Chap. 2 by Oliveira et al. and Chap. 4 Macaya et al.). In contrast, the high photosynthetic efficiency exhibited by propagules of Antarctic endemic *Ascoseira mirabilis* at 25 °C could be explained by the upper vertical distribution of the parental sporophytes or could be a conserved trait related to the fact that the species is probably a relic of Mesozoic (Gondwana) marine flora, which was highly diverse when the average water temperatures were close to 12 °C (Clayton 1994).

Temperature is a factor modifying the susceptibility/tolerance to UV radiation. The influence of this factor apparently depends on the position of parental thalli on the shore. In this context, a recent study provided evidence that propagules of Antarctic seaweeds are relatively tolerant to enhanced temperature, which can furthermore modulate UV tolerance at least under laboratory conditions (Navarro et al.

2016). These authors observed that the exposure of propagules to a combination of UV radiation and temperature stress inhibits the photosynthetic capacity of propagules of six Antarctic seaweed species from the eulittoral (*Pyropia endiviifolia*, *Iridaea cordata*, *Adenocystis utricularis*, and *Monostroma hariotii*) and the sublittoral (*Ascoseira mirabilis,* and *Gigartina skottsbergii*), the former group being more tolerant to UV and enhanced temperature than the sublittoral group. Additionally, propagules of eulittoral species *P. endiviifolia*, *I. cordata,* and *A. utricularis* exhibit negative UV effects at 2 °C compared to 7 and 12 °C, suggesting that enhanced temperature improves UV tolerance. On the contrary, this positive interaction was not observed in propagules of the shallow sublittoral *A. mirabilis*, where an increase in temperature exacerbates the reduction of photosynthetic efficiency (Navarro et al. 2016). It is known that various processes related to photoprotection, e.g., D1 protein turnover, enzyme repair mechanisms, and dissipative quenching, operate more efficiently at higher temperatures (Wünschmann and Brand 1992; Becker et al. 2010). Thus, the lower inhibition of photosynthesis observed at 12 °C compared to 2 and 7 °C can be regarded as an efficient acclimation of photosynthesis in these cells. Even though photosynthesis was inhibited by UV radiation, propagules from eulittoral species recover completely after 4 h under dim visible light, whereas sublittoral ones do not. A fast turnover of D1 protein may be responsible for the fast reversible photoinhibition of photosynthesis in eulittoral macroalgae as suggested for *Urospora penicilliformis* propagules (Roleda et al. 2009). However, the recovery is not influenced by a temperature increase in the studied species (Navarro et al. 2019).

Antarctic propagules can retain their capacity to tolerate elevated temperatures, which is evident when they are compared with their sub-Antarctic counterparts. For example, F_v/F_m measured in *I. cordata* tetraspores from Antarctica was not inhibited by UV radiation at 2 °C or 8 °C, while propagules from sub-Antarctic populations exhibited a decrease after a 4-h exposure, mainly at 2 °C in PAR (30%) and PAR + UV (67%). Considering only the effects of temperature, F_v/F_m decreased by 14% in tetraspores from sub-Antarctic population exposed at 2 °C when compared to the control (8 °C). Surprisingly, photosynthetic activity in tetraspores from Antarctic increased by 2% relative to control. These results suggest that low temperatures may exacerbate UV stress to photosynthesis in spores from the sub-Antarctic population, whereas Antarctic spores would be adapted to low temperature and UV. The results also confirm previous evidence obtained in adult thalli of *Ulva* spp. from Antarctic and sub-Antarctic region by Rautenberger and Bischof (2006). At 10 °C the inhibition of *Ulva hookeriana* (known as *Enteromorpha bulbosa* (Suhr) Montagne) from Antarctica was comparable to its sub-Antarctic counterpart *Ulva clathrata* (10% of control). However, at 0 °C, inhibition was of 50% in the sub-Antarctic *Ulva clathrata* and 37% in *U. hookeriana* (Rautenberger and Bischof 2006). Overall, the results indicate that in cold-adapted species, stress tolerance can be efficient, which allow many shallow sublittoral, and especially eulittoral species, to thrive under extremely changing thermal conditions.

10.3.3 Other Environmental Stressors

In the Antarctic environment, seaweeds are also facing fluctuations of other environmental factors such as salinity, influenced by local meltwater influx and calving glaciers as well as desiccation when algae are exposed to air during low tides. Although the effects of salinity on seaweeds are relatively well known (reviewed in Kirst 1990 and Karsten 2012), few studies have been conducted on Antarctic seaweeds (e.g., Jacob et al. 1991, 1992a, b; Karsten et al. 1991a, b). In general, it has been reported that seaweeds respond to external salinity changes with osmotic acclimation processes involving the control of internal organic (e.g., proline, sucrose, β-dimethylsulphoniopropionate) and inorganic (K^+, Na^+, Mg^{2+}, Cl^-, SO_4^{2-}, and PO_4^{3-}) ions (Karsten et al. 1991a, b; Kirst 1990). Antarctic seaweeds inhabiting the eulittoral and supralittoral zone can be characterized as euryhaline organisms, which can survive salinities between 7 and 102 PSU with a low rate of mortality. Most taxa grow, photosynthesize, and respire optimally under normal seawater conditions with rather broad tolerances between 7 and 68 PSU. Hitherto, there is no information of the effect of salinity on Antarctic seaweed propagules. On the other hand, emergent stressors in Antarctic environment, e.g., ocean acidification (Hurd et al. 2009) and marine pollution (Goutte et al. 2013), can pose risks to adult and early phases of Antarctic seaweeds. Ocean acidification can affect the physiology of seaweeds; however, practically no data exist on their effects on early phases of macroalgae. In the giant kelp *Macrocystis pyrifera*, pH between 7.59 and 7.60 reduced meiospore germination, which was ameliorated when CO_2 was added (Roleda et al. 2011). Hitherto there is no information on the effects of these compounds on the biology of Antarctic propagules.

It has been suggested that metals may inhibit reproduction in brown algae by interfering with the ability of sperm to find eggs, perhaps via interference of the pheromone attractant (Maier 1993). However, the effect of trace metals is expected to be detrimental to propagules (spores, gametes, and zygotes) due to poor development of the protective cell wall. Moreover, cell walls of brown seaweeds composed of alginate and fucoidan can bind cations and have a high affinity for copper (Lignell et al. 1982), affecting the settlement and germination of propagules. For example, in *Lessonia*, copper drastically affected spore release by mature sporophytes as well as spore settlement. The highest copper concentration applied interrupted the development of the spores totally after settlement (Contreras et al. 2007). In all, the importance of studying the effects of metals and other pollutants (hydrocarbons, pesticides, other persistent pollutants, and so on) on Antarctic algae propagules lies in the recent increase of contaminant concentration in Antarctic due to human activities (Bargagli 2008). On the other hand, although the harmful effects of metals, e.g., copper toxicity, have been analyzed in brown species (reviewed by Coelho et al. 2000; Contreras et al. 2007), the effects of these new, emergent stressors on the biology of Antarctic seaweeds have to be examined in a context of the combined action of multiple factors (see Chap. 7 by Huovinen and Gómez).

10.4 Concluding Remarks: Biology of Propagules under Climate Change

Despite having a crucial importance in the biology of seaweeds, propagules have not been sufficiently studied in relation to their physiological requisites to respond to climate change. The importance of understanding the effects of global climate change on reproductive stages lies in the fact that early stages of development are essential for recruitment, especially for those species that rely their dominance entirely on reproductive abilities.

The predicted increase of temperature and the prevalence of episodes of depleted ozone around the Antarctic Peninsula region and adjacent islands will impose physiological constraints to reproductive output, settlement, and recruitment of different species of seaweeds. Increase in seawater temperature could also influence the phenology and the formation of propagules (spores and/or gametes) and consequently, the timing and formation of juvenile thalli, especially in species inhabiting the eulittoral and shallow sublittoral zone (Zacher et al. 2007; Campana et al. 2009; see Chap. 12 by Campana et al.). Furthermore, as a consequence of temperature increase, glaciers can retreat opening new free space for recruitment of benthic organisms, including macroalgae (Quartino et al. 2013; see Chap. 8 by Quartino et al.). In these new open areas, however, alteration in light, salinity, sedimentation, and disturbance processes can occur, limiting settlement of established communities and even favoring the arrival of cold-temperate species (see also Chap. 9 by Deregibus et al.). Increased turbidity can have, however, contrasting implications for the biology of reproductive cells, which can become favored by a minimized impact of UV radiation, but decreasing available irradiance for photosynthesis.

Undoubtedly, Antarctic seaweeds have developed life strategies to colonize and form a complex structure in the coastal ecosystems. In scenarios of climate change and warming in the Antarctic, dispersal and colonization of Antarctic coastal zones via efficient adaptations of early developmental phases of seaweeds are central to envision the future seaweed diversity in Antarctica (see Chap. 2 by Oliveira et al. and Chap. 5 by Pellizzari et al.).

Acknowledgments This work was supported by CONICYT-Chile: Fondecyt Iniciación N°11160520, Anillo ART1101, and Centro Fondap-IDEAL 15150003.

References

Amsler C, Reed D, Neushul M (1992) The microclimate inhabited by macroalgal propagules. Eur J Phycol 27:253–270

Bargagli R (2008) Environmental contamination in Antarctic ecosystems. Sci Total Environ 400:2012–2226

Becker S, Graeve M, Bischof K (2010) Photosynthesis and lipid composition in the Antarctic rhodophyte *Palmaria decipiens*: effects of changing light and temperature levels. Polar Biol 33:945–955

Beer S, Björk M, Gademan R, Ralph P (2001) Measurements of photosynthesis rate in seagrass. In: Short FT, Cole R (eds) Global seagrass research methods. Elsevier Publishing, Amsterdam, pp 183–198

Bischof K, Gómez J, Molis M et al (2006) UV radiation shapes seaweed communities. Rev Environ Sci Biotechnol 5:141–166

Bouchard JN, Roy S, Campbell DA (2006) UVB effects on the photosystem II-D1 protein of phytoplankton and natural phytoplankton communities. Photochem Photobiol 82:936–951

Britt AB (1995) DNA damage and repair in plants. Annu Rev Plant Physiol Plant Mol Biol 47:75–100

Campana GL, Zacher K, Fricke A et al (2009) Drivers of radiation colonization and succession in polar benthic macro- and microalgal communities. Bot Mar 52:655–667

Clayton MN (1988) Evolution and life histories of brown algae. Bot Mar 31:379–387

Clayton MN (1994) The evolution of the Antarctic marine benthic algal flora. J Phycol 30:897–904

Clayton MN, Wiencke C (1990) The anatomy, life history and development of the Antarctic brown alga *Phaeurus antarcticus* (Desmarestiales, Phaeophyceae). Phycologia 29:303–315

Coelho SM, Rijstenbil JW, Brown MT (2000) Impacts of anthropogenic stresses on the early development stages of seaweeds. J Aquat Ecosyst Stress Recover 7:317–333

Contreras L, Medina MH, Andrade S et al (2007) Effects of copper on early developmental stages of *Lessonia nigrescens* Bory (Phaeophyceae). Environ Pollut 145:75–83

Cormaci M, Furnari G, Scammacca B (1992) The benthic alga flora of Terra Nova Bay (Ross Sea, Antarctica). Bot Mar 35:541–552

Crame JA (1993) Latitudinal range fluctuations in the marine realm through geological time. Trends Ecol Evol 8:162–266

Davison IR (1991) Environmental effects on algal photosynthesis: temperature. J Phycol 27:2–8

de Reviers B, Délépine R (1981) Biological data on a southern Phaeophyceae *Adenocystis utricularis* (Bory) Skottsberg, possible material for alginate industry. Proc 10th Int Seaweed Symp, p 345–350

Dummermuth AL, Wiencke C (2003) Experimental investigation of seasonal development in six Antarctic red macroalgae. Antarct Sci 15:449–457

Eggert A (2012) Seaweed responses to temperature. In: Wiencke C, Bischof K (eds) Seaweed biology: novel insights into ecophysiology, ecology and utilization. Ecological studies, vol 219. Springer, Berlin, pp 47–66

Eggert A, Wiencke C (2000) Adaptation and acclimation of growth and photosynthesis of five Antarctic red algae to low temperatures. Polar Biol 23:609–618. https://doi.org/10.1007/s003000000130

Gerstein AC, Cleathero LA, Mandegar MA, Otto SP (2010) Haploids adapt faster than diploids across a range of environments. J Evol Biol 24:531–540

Gómez I, Wiencke C (1996) Photosynthesis, dark respiration and pigment contents of gametophytes and sporophytes of the Antarctic brown alga *Desmarestia menziesii*. Bot Mar 39:149–157

Gómez I, Wiencke C (1997) Seasonal growth and photosynthetic performance of the Antarctic macroalga *Desmarestia menziesii* (Phaeophyceae) cultured under fluctuating Antarctic daylengths. Bot Acta 110:25–31

Gómez I, Thomas DN, Wiencke C (1995) Longitudinal profiles of growth, photosynthesis and light independent carbon fixation in the Antarctic brown alga *Ascoseira mirabilis*. Bot Mar 38:157–164

Gómez I, Wiencke C, Thomas DN (1996) Variations in photosynthetic characteristics of the Antarctic marine brown alga *Ascoseira mirabilis* in relation to thallus age and size. Eur J Phycol 31:167–172

Gómez I, Weykam G, Klöser H, Wiencke C (1997) Photosynthetic light requirements, metabolic carbon balance and zonation of sublittoral macroalgae from King George Island (Antarctica). Mar Ecol Prog Ser 148:281–229

Gómez I, Wulff A, Roleda M et al (2009) Light and temperature demands of marine benthic microalgae and seaweeds in polar regions. Bot Mar 52:593–608

Gómez I, Navarro NP, Huovinen P (2019) Bio-optical and physiological patterns in Antarctic seaweeds: a functional trait based approach to characterize vertical zonation. Prog Oceanogr 174:17–27

Goutte A, Chevreuil M, Alliot F et al (2013) Persistent organic pollutants in benthic and pelagic organisms off Adélie land, Antarctica. Mar Pollut Bull 77:82–89

Grzymski J, Johnsen G, Sakshaug E (1997) The significance of intracellular self-shading on the bio-optical properties of brown, red and green macroalgae. J Phycol 33:408–414

Huovinen P, Gómez I (2012) Cold-temperate seaweed communities of the Southern Hemisphere. In: Wiencke C, Bischof K (eds) Seaweed biology: Novel insights into ecophysiology, ecology and utilization. Ecological studies, vol 219. Springer, Berlin, pp 293–313

Huovinen P, Gómez I (2013) Photosynthetic characteristics and UV stress tolerance of Antarctic seaweeds along the depth gradient. Polar Biol 36:1319–1332

Huovinen P, Gómez I (2015) UV sensitivity of vegetative and reproductive tissues of three Antarctic macroalgae is related to differential allocation of phenolic substances. Photochem Photobiol 91:1382–1388

Huovinen PS, Oikari AOJ, Soimasuo MR, Cherr GN (2000) Impact of UV radiation on the early development of the giant kelp (*Macrocystis pyrifera*) gametophytes. Photochem Photobiol 72:308–313

Huovinen P, Ramírez J, Gómez I (2016) Underwater optics in sub-Antarctic and Antarctic coastal ecosystems. PLoS One 11(5):e0154887

Hurd CL, Hepburn CD, Currie KI et al (2009) Testing the effects of ocean acidification on algal metabolism: considerations for experimental designs. J Phycol 45:1236–1251

Jacob A, Kirst GO, Wiencke C, Lehmann H (1991) Physiological responses of the Antarctic green alga *Prasiola crispa* ssp. *antarctica* to salinity stress. J Plant Physiol 139:57–62

Jacob A, Lehmann H, Kirst GO, Wiencke C (1992a) Changes in the ultrastructure of *Prasiola crispa* ssp. *antarctica* under salinity stress. Plant Biol 105:41–46

Jacob A, Wiencke C, Lehmann H, Kirst GO (1992b) Physiology and ultrastructure of desiccation in the green alga *Prasiola crispa* from Antarctica. Bot Mar 35:297–303

Kain JM (1989) The seasons in the subtidal. Eur J Phycol 24:203–215

Karsten U (2012) Seaweed acclimation to salinity and desiccation stress. In: Wiencke C, Bischof K (eds) Seaweed biology: Novel insights into ecophysiology, ecology and utilization. Ecological studies, vol 219. Springer, Berlin, Heidelberg, pp 87–107

Karsten U, Wiencke C, Kirst GO (1991a) The effect of salinity changes upon physiology of eulittoral green macroalgae from Antarctica and southern Chile. I. Cell viability, growth, photosynthesis and dark respiration. J Plant Physiol 138:667–673

Karsten U, Wiencke C, Kirst GO (1991b) The effect of salinity changes upon physiology of eulittoral green macroalgae from Antarctica and southern Chile. II. Inorganic ions and organic compounds. J Exp Bot 42:1533–1539

Karsten U, Sawall T, West J, Wiencke C (2000) Ultraviolet sunscreen compounds in epiphytic red algae from mangroves. Hydrobiologia 432:159–171

Karsten U, Wulff A, Roleda MY, Müller R, Steinhoff F, Fredersdorf J, Wiencke C (2009) Physiological responses of polar benthic micro- and macroalgae to ultraviolet radiation. Bot Mar 52:639–654

Kirst GO (1990) Salinity tolerance of eukaryotic marine algae. Annu Rev Plant Physiol Plant Mol Biol 41:21–53

Kornmann P, Sahling PH (1977) Meeresalgen von Helgoland. Benthische Grün-, Braun-und Rotalgen. Helgolander Meeresun 29:1–289

Kylin H, Skottsberg C (1919) Zur Kenntnis der Subantarktischen und Antarktischen Meeresalgen, II. Rhodophyceen. In: Nordenskjöld O (ed) Wissenschaftliche Ergebnisse der schwedischen Südpolarexpedition 1901–1903, Band IV, Lieferung 15. Stockholm, Lithographisches Institut des Generalstabs, 88 pp

Lamb IM, Zimmerman MH (1977) Benthic marine algae of the Antarctic Peninsula. Antarct Res Ser 23:129–229

Lignell A, Roomans GM, Pedersen M (1982) Localization of absorbed cadmium in *Fucus vesiculosus* by X-ray microanalysis. J Plant Physiol 105:103–109

Lorenz M, Schubert H, Forster RM (1997) *In vitro*- and *in vivo*-effects of ultraviolet-B radiation on the energy transfer in phycobilisomes. Photosynthetica 33:517–527

Lüning K, Kadel P (1993) Daylength range for circannual rhythmicity in *Pterygophora californica* (Alariaceae, Phaeophyta) and synchronization of seasonal growth by daylength cycles in several other brown algae. Phycologia 32:379–387

Lüning K, tom Dieck I (1989) Environmental triggers in algal seasonality. Bot Mar 32:389–397

Maier I (1993) Gamete orientation and induction of gametogenesis by pheromones in algae and plants. Plant Cell Environ 16:891–907

Marcías ML, Deregibus D, Saravia LA et al (2017) Life between tides: spatial and temporal variations of an intertidal macroalgal community at Potter Peninsula, South Shetland Islands, Antarctica. Estuar Coast Shelf Sci 187:193–203

Müller DG, Westermeier R, Peters A, Boland W (1990) Sexual reproduction of the Antarctic brown alga *Ascoseira mirabilis* (Ascoseirales, Phaeophyceae). Bot Mar 33:251–255

Navarro NP, Huovinen P, Gómez I (2016) Stress tolerance of Antarctic macroalgae in the early life stage. Rev Chil Hist Nat 89:5

Navarro NP, Huovinen P, Gómez I (2019) Photosynthetic characteristics of geographically disjunct seaweeds: a case study on the early life stages of Antarctic and Subantarctic species. Prog Oceanogr 174:28–36

Otto SP, Gerstein AC (2008) The evolution of haploidy and diploidy. Curr Biol 18:R1121–R1124

Quartino ML, Deregibus D, Campana GL et al (2013) Evidence of macroalgal colonization on newly ice-free areas following glacial retreat in Potter Cove (South Shetland Islands), Antarctica. PLoS One 8(3):e58223

Raper JR, Flexer AS (1970) The road to diploidy with emphasis on a detour. Symposium of the Society of General Microbiology 20:401–432

Rautenberger R, Bischof K (2006) Impact of temperature on UV-susceptibility of two *Ulva* (Chlorophyta) species from Antarctic and Subantarctic regions. Polar Biol 28:988–996

Renger G, Voss M, Graber P, Schulze A (1986) Effect of UV irradiation on differential partial reactions of the primary processes of photosynthesis. In: Worrest RC, Caldwell MM (eds) Stratospheric ozone reduction, solar ultraviolet radiation and plant life. NATO ASI series, vol G8. Springer, Heidelberg, pp 171–184

Richter M, Rühle W, Wild A (1990) Studies on the mechanism of photosystem II photoinhibition I. a two-step degradation of D1-protein. Photosynth Res 24:229–235

Roleda M, Zacher K, Wulff A et al (2007) Photosynthetic performance, DNA damage and repair in gametes of the endemic Antarctic brown alga *Ascoseira mirabilis* exposed to ultraviolet radiation. Austral Ecol 32:917–926

Roleda MY, Zacher K, Wulff A et al (2008) Susceptibility of spores of different ploidy levels from Antarctic *Gigartina skottsbergii* (Gigartinales, Rhodophyta) to ultraviolet radiation. Phycologia 47:361–370

Roleda MY, Campana G, Wiencke C et al (2009) Sensitivity of Antarctic *Urospora penicilliformis* (Ulotrichales, Chlorophyta) to ultraviolet radiation is life stage dependent. J Phycol 45:600–609

Roleda MY, Morris JN, McGraw CM, Hurd CL (2011) Ocean acidification and seaweed reproduction: increased CO_2 ameliorates the negative effect of lowered pH on meiospore germination in the giant kelp *Macrocystis pyrifera* (Laminariales, Phaeophyceae). Glob Chang Biol 18:854–864

Salles S, Aguilera J, Figueroa FL (1996) Light field in algal canopies: changes in spectral light ratios and growth of *Porphyra leucosticta*. Thur. In Le Jol. Sci Mar 60:29–38

Skottsberg C (1953) On two collections of Antarctic marine algae. Arkiv Botanik Serie 2 2:531–566

Swanson AK, Druehl LD (2000) Differential meiospore size and tolerance of ultraviolet light stress within and among kelp species along a depth gradient. Mar Biol 136:657–664

Tom Dieck I (1989) Vergleichende Untersuchungen zur Ökophysiologie und Kreuzbarkeit innerhalb der digitaten Sektion der Gattung *Laminaria*. PhD thesis, University of Hamburg, Hamburg, Germany, 168 pp

Vass I (1997) Adverse effects of UV-B light on the structure and function of the photosynthetic apparatus. In: Pessarakli M (ed) Handbook of photosynthesis. Marcel Dekker Inc., New York, pp 931–949

Weykam G, Wiencke C (1996) Seasonal photosynthetic performance of the endemic Antarctic alga *Palmaria decipiens* (Reinsch) Ricker. Polar Biol 16:357–361

Weykam G, Thomas DN, Wiencke C (1997) Growth and photosynthesis of the Antarctic red algae *Palmaria decipiens* (Palmariales) and *Iridaea cordata* (Gigartinales) during and following extended periods of darkness. Phycologia 36:395–405

Wiencke C (1990a) Seasonality of brown macroalgae from Antarctica – a long-term culture study under fluctuating Antarctic daylengths. Polar Biol 10:589–600

Wiencke C (1990b) Seasonality of red and green macroalgae from Antarctica – a long-term culture study under fluctuating Antarctic daylengths. Polar Biol 10:601–607

Wiencke C, Clayton MN (1990) Sexual reproduction, life history, and early development in culture of the Antarctic brown alga *Himantothallus grandifolius* (Desmarestiales, Phaeophyceae). Phycologia 29:9–18

Wiencke C, Clayton MN (2002) In: Wägele JW, Sieg J (eds) Antarctic seaweeds. Synopses of the Antarctic benthos, vol 9. Gantner, Ruggell, Lichtenstein. 239 pp

Wiencke C, Stolpe U, Lehmann H (1991) Morphogenesis of the brown alga *Desmarestia antarctica* cultivated under seasonally fluctuating Antarctic daylengths. Ser Cient INACH 41:65–78

Wiencke C, Clayton MN, Schulz D (1995) Life history, reproductive morphology and development of the Antarctic brown alga *Desmarestia menziesii* J. Agardh. Plant Biol 108:201–208

Wiencke C, Clayton MN, Langreder C (1996) Life history and seasonal morphogenesis of the endemic Antarctic brown alga *Desmarestia anceps* Montagne. Bot Mar 39:435–444

Wiencke C, Gómez I, Pakker H, Flores-Moya A, Altamirano M, Hanelt D, Bischof K, Figueroa F-L (2000) Impact of UV radiation on viability, photosynthetic characteristics and DNA of brown algal zoospores: implications for depth zonation. Mar Ecol Prog Ser 197:217–229

Wiencke C, Roleda MY, Gruber A et al (2006) Susceptibility of zoospores to UV radiation determines upper depth distribution limit of Arctic kelps: evidence through field experiments. J Ecol 94:455–463

Wiencke C, Clayton MN, Gómez I et al (2007) Life strategy, ecophysiology and ecology of seaweeds in polar waters. Rev Environ Sci Biotechnol 6:95–126

Wiencke C, Gómez I, Dunton K (2009) Phenology and seasonal physiological performance of polar seaweeds. Bot Mar 52:585–559

Wünschmann G, Brand JJ (1992) Rapid turnover of a component required for photosynthesis explains temperature dependence and kinetics of photoinhibition in a cyanobacterium, *Synechococcus* 6301. Planta 186:426–433

Wynne MJ (1982) Observations on four species of Delesseriaceae (Rhodophyta) from the South Sandwich Islands, the Antarctic. Contr Univ Mich Herb 15:325–337

Zacher K (2014) The susceptibility of spores and propagules of Antarctic seaweeds to UV and photosynthetically active radiation — field versus laboratory experiments. J Exp Mar Biol Ecol 458:57–63

Zacher K, Roleda MY, Hanelt D, Wiencke C (2007) UV effects on photosynthesis and DNA in propagules of three Antarctic seaweeds (*Adenocystis utricularis*, *Monostroma hariotii* and *Porphyra endiviifolium*). Planta 225:1505–1516

Zacher K, Roleda MY, Wulff A et al (2009) Responses of Antarctic *Iridaea cordata* (Rhodophyta) tetraspores exposed to ultraviolet radiation. Phycol Res 57:186–193

Chapter 11
Form and Function in Antarctic Seaweeds: Photobiological Adaptations, Zonation Patterns, and Ecosystem Feedbacks

Iván Gómez and Pirjo Huovinen

Abstract Morpho-functional traits of Antarctic seaweeds are modeled by different physical and biological factors. Due to the extreme seasonality, which imposes light limitation for extended periods, Antarctic seaweeds are shade-adapted organisms that are physiologically able to thrive at considerable depths down to 40 m. This vertical distribution is defined by a suite of bio-optical and morphological features that allow algae occupying habitats with different environmental conditions in the water column. However, various species can also colonize the highly perturbed intertidal zone where environmental setting, e.g. ice scouring, high solar radiation, extremely variable temperature, limit growth, and reproduction. In the maritime Antarctic region, large endemic brown algae attaining a massive (leathery) morphology and perennial life history dominate at depths below 10 m or less. Here, they coexist with perennial highly shade-adapted coarsely branched rhodophytes, which show understory characteristics. At shallower locations, various annual species with very rapid growth can be found. The intertidal zone, characterized by a depauperate diversity, is populated mostly by ephemeral and delicate green algae. In the present chapter, form and function of seaweeds is revisited in the context of a changing Antarctic environment. Here, the functional groups display different acclimation mechanisms, which can operate at different temporal scales and consequently with variable impact on the biogeochemical coastal processes. The role of canopy-forming algae, whose "bioengineering" processes alleviate the impact of environmental variability, is fundamental in determining the fate of the benthic communities in the coastal system.

Keywords Canopy-forming algae · Life strategy · Light absorption · Morpho-functional traits · Vertical zonation

---

I. Gómez (✉) · P. Huovinen
Instituto de Ciencias Marinas y Limnológicas, Facultad de Ciencias, Universidad Austral de Chile, Valdivia, Chile

Research Center Dynamics of High Latitude Marine Ecosystems (IDEAL), Valdivia, Chile
e-mail: igomezo@uach.cl; pirjo.huovinen@uach.cl

I. Gómez, P. Huovinen (eds.), *Antarctic Seaweeds*,
https://doi.org/10.1007/978-3-030-39448-6_11

11.1 Brief Overview of Form and Function in Seaweeds

In contrast to unicellular algae, organisms bearing multiple cells (from simple pluricellular colonies to more advanced forms attaining, e.g. parenchyma) have to synchronize a more complex structural organization, characterized by different function-specific elements that follow a morpho-genetic-based program. Thus, multicellular algae adjust form and function through an intricate molecular network that allows them to interact with their physical, chemical, and biotic environment (Grosberg and Strathmann 2007). A suite of morphological models found in benthic organisms (e.g. seaweeds, sponges, corals, bryozoa) coexist in a physical environment, which raises a question of how similar convergent forms have evolved even in species phylogenetically very distant or, alternatively, how related organisms display completely different growth patterns and shape (Kaandorp and Kübler 2001).

A striking characteristic of almost all seaweeds is their morphological plasticity, i.e. although the basic thallus plan is based on a determined morphogenetic design, body shape can change through the life span of an organism or within the life history sequence. These variations in the morphological traits within a genotype can be subtle or drastic depending on the intensity of the endogenous and environmentally driven shifts (Innes 1984; Taylor and Hay 1984). Especially in sites where the physical perturbations are extreme, seaweeds display complex mechanisms to adjust form and function to the prevailing environmental condition (Hay 1986). Phenotypic plasticity, one of the most well-known types of intrinsic morphological variability, is normally prompted by environmental conditions and thereby complicates efforts to identify the routes of morpho-functional responses within a multi-specific assemblage: intrinsic properties at an organismal level mask the morpho-functional differences at community scale (Steneck and Dethier 1994). Changes in morphology due to ontogenetic development and heteromorphic phase expression within of life cycle are also important to characterize form and function in seaweeds. For example, in the brown alga *Himantothallus grandifolius*, the largest Antarctic seaweed, the thallus undergoes considerable changes with development: while juvenile individuals are characterized by partial cortication and coarsely branched morphology, adult plants are characterized by a thick leathery strap-like anatomical structure, where lateral branches are absent (Moe and Silva 1981; Wiencke and Clayton 1990).

Traditionally, functional groups of seaweeds (which could also be applied to other groups of benthic marine organisms) are defined by their thallus architecture (also called life form). This concept implies intuitively a series of intrinsic properties of an organism, which can or not be shared by other unrelated organisms. Clearly this gives a high value to the anatomical features (form) and less emphasis on the function. Such conceptual framework represented the basis on, which Littler and Littler (1980) and Steneck and Watling (1982) developed their general functional form models, where functionality of algae, e.g. resistance to biotic disturbance, was inferred from gross morphology. Although the general applicability of these models has been questioned since similar morphologies often show different functional responses to, e.g. disturbance gradients (Phillips et al. 1997; Ingólfsson

2005; Padilla and Allen 2000), functionality, which depends on extrinsic factors (e.g. resource utilization, disturbance, biotic interactions, etc.), has an unavoidable expression in the morphology. The rationale to understand in what ways form and function is modeled by the environment, known as "the holy grail framework", considers necessarily different areas of knowledge, e.g. genomics, physiology, ecology, demography, etc., and has been used with different emphasis to explain the structure of different types of vegetation, both in terrestrial and aquatic realms (Littler and Littler 1980; Grime 1981; Steneck and Dethier 1994; Lavorel and Garnier 2002). Thus, if one assumes that different habitats have a different environmental setting, it is possible to argue that they host assemblages of organisms with similar morphology but different functional attributes. For example, in littoral stress, tolerance of macroalgae depends strongly on fine photochemical adjustments, which are related to their position on the shore and less with functional form groups (Aguilera et al. 1999; Gómez and Huovinen 2011; Balata et al. 2011). This has been commonly found in different types of terrestrial vegetation where environmental tolerance, and not gross morphology, defines functional groups (Grime 1981; Ackerly and Reich 1999; Poorter and Bongers 2006).

11.2 Functional Groups of Seaweeds in the Antarctic

The coastal systems around Antarctica may be regarded as highly inhospitable for life, where physical disturbance associated with ice scouring, extreme light limitation, and low temperatures imposes severe restrictions for marine organisms. However, benthic algae thrive in these habitats displaying different functional strategies and morphologies (Fig. 11.1). Based on various surveys, 131 species of seaweeds (Fig. 11.2) are distributed among different types of functional groups: filamentous and finely branched (45); foliose (9); coarsely branched, including corticated species (48); thick leathery, including terete forms (11); and postrate species (18) (Fig. 11.2). However, when the different functional forms are grouped according to the major phylogenetic categories, it is possible to observe that 64% of green algae are filamentous, while practically the totality of thick leathery forms belong to brown algae. In red algae, 58% of the known species can be recognized as finely and coarsely branched morphs. In the case of Chlorophyta, most of the species attain delicate filamentous or sheet-like morphs, and with the exception of the endemic *Lambia antarctica* and *Monostoma hariotii* (Wiencke and Clayton 2002; see also Chap. 2 by Oliveira et al. and Chap. 5 by Pellizzari et al.), all are restricted to intertidal zones. However, it should be emphasized that many species cannot be easily assigned to these major functional categories. For example, the brown algae *Adenocystis utricularis* and *Utriculidium durvillaei* are the only species with a saccate morphology (and thus were not included in this analysis). Moreover, the number of postrate species, which may include crustose, calcareous, and endophytic life forms, is largely underrepresented. In fact, these algae have been very little studied,

Fig. 11.1 Diversity of gross morphologies in Antarctic seaweeds. (**a**) Crustose Rhodophyta; (**b**) Saccate morphology (*Adenocystis utricularis*); (**c**) Filamentous tubular (*Ulva intestinalis*); (**d**) finely branched *Plocamium cartilagineum*; (**e**) coarsely branched (*Trematocarpus antarcticus*); (**f**) Thick leathery (*Ascoseira mirabilis*). (Photos **a**, **d**, **e** and **f** by Ignacio Garrido; **b** and **c** by Iván Gómez)

due to mostly that they are not ubiquitous (e.g. encrusting morphs) or can inhabit deeper locations (e.g. calcified coralline algae) (Alongi et al. 2002).

The absence of a marked dominance of a given seaweed gross morphology in the Antarctic can only be explained in terms of the distribution of these different functional forms in the mosaic of benthic habitats. In fact, the arrangement of species across different environmental gradients implies also an ordination of organismal traits that can be classified in different functional entities (e.g. gross morphology, size, life forms, physiological responses). Thus, it is possible to understand why similar morpho-functional "solutions" are exploited by different taxa, many of them phylogenetically unrelated. For example, of the 25 species of green algae recorded in the Antarctic, 56% correspond to widely distributed taxa, which normally display

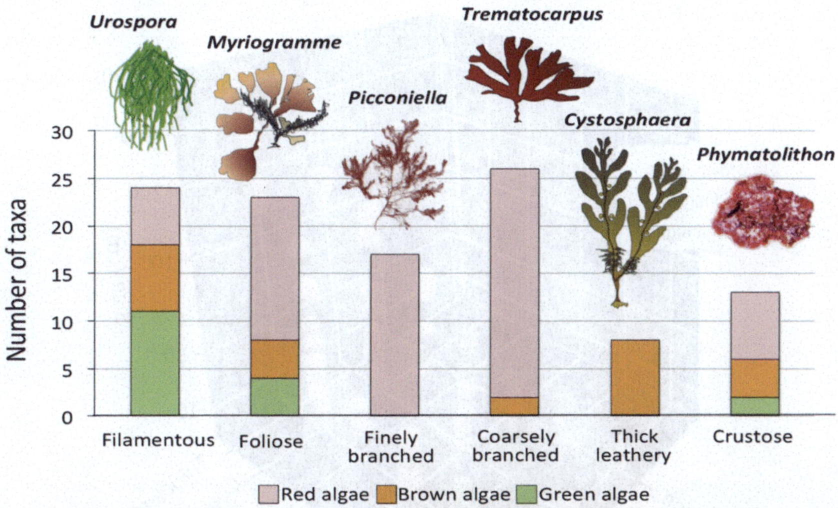

RED ALGAE		BROWN ALGAE	GREEN ALGAE
Filamentous & Finely branched	**Coarsely branched**	**Flamentous & finely branched**	**Filamentous & Finely branched**
Antarcticothamnion polysporum	*Ahnfeltia plicata*	*Elachista antarctica*	*Acrosiphonia arcta*
Audouinella purpurea	*Austropugetia crassa*	*Geminocarpus austrogeorgiae*	*Blidingia minima*
Audouinella concrescens	*Callophyllis atrosanguinea*	*Geminocarpus geminatus*	*Chaetomorpha irregularis*
Audouinella membranacea	*Callophyllis tenera*	*Leptonematella falklandica*	*Chaetomorpha mawsonii*
Ballia callitricha	*Callophyllis variegata*	*Scytosiphon lomentaria*	*Cladophora coelothrix*
Ballia sertularioides	*Curdiaea racovitzae*	*Pylaiella littoralis*	*Cladophora repens*
Bangia atropurpurea	*Delisea pulchra*	**Coarsely branched**	*Lambia antarctica*
Erythrotrichia carnea	*Georgiella confluens*	*Chordaria linearis*	*Rhizoclonium tortuosum*
Ceramium involutum	*Gracilaria sp.*	*Halopteris corymbosa*	*Spongomorpha pacifica*
Cystoclonium obtusangulum	*Gymnogongrus antarcticus*	*Halopteris obovata*	*Ulothrix australis*
Dasyptilon harveyi	*Gymnogongrus turquetii*	*Microzonia australe*	*Ulothrix flacca*
Griffithsia antarctica	*Hymenena sp.*	*Phaeurus antarcticus*	*Ulothrix implexa*
Meiodiscus concrescens	*Hymenocladiopsis prolifera*	*Scytothamnus fasciculatus*	*Ulothrix subflaccida*
Pantoneura plocamioides	*Hymenocladiopsis sp.*	**Crustose/postrate/endophytic**	*Urospora penicilliformis*
Picconiella plumosa	*Iridaea cordata*	*Australofilum incommodum*	*Ulva hookeriana*
Rhodochorton purpureum	*Iridaea mawsonii*	*Lithoderma antarcticum*	*Ulva intestinalis*
Plocamium cartilagineum	*Microrhinus carnosus*	*Petroderma maculiformis*	**Foliose**
Plocamium hookeri	*Myriogramme livida*	*Ralfsia australis*	*Monostroma hariotii*
Plocamium secundatum	*Myriogramme manginii*	**Saccate**	*Prasiola crispa*
Plumariopsis pensinsularis	*Nereoginkgo adiantifolia*	*Adenocystis utricularis*	*Ulva compressa*
Polysiphonia abscissa	*Nereoginkgo populifolia*	*Utriculidium durvillaei*	*Ulva lactuca*
Pterothamnion antarcticum	*Neuroglossum delesseriae*	*Antarctosaccion applanatum*	*Ulva rigida*
Pterothamnion simile	*Notophycus fimbriatus*	**Thick leathery**	**Postrate/endophytic**
Foliose	*Pachymenia orbicularis*	*Ascoseira mirabilis*	*Endophyton atroviride*
Leniea lubrica	*Palmaria decipiens*	*Cystosphaera jacquinotii*	*Entocladia maculans*
Pyropia endiviifolia	*Palmaria georgica*	*Desmarestia anceps*	*Entocladia viridis*
Pyropia plocamiestris	*Paraglossum amsleri*	*Desmarestia antarctica*	*Ulvella viridis*
Crustose/postrate/endophytic	*Phycodrys antarctica*	*Desmarestia confervoides*	
Antarctocolax lambii	*Phycodrys austrogeorgica*	*Desmarestia menziesii*	
Clathromorphum lemoineanum	*Phycodrys quercifolia*	*Himantothallus grandifolius*	
Clathromorphum obtectulum	*Phyllophora ahnfeltioides*	**Foliose**	
Gainia mollis	*Phyllophora antarctica*	*Petalonia fascia*	
Hildenbrandia lecannellieri	*Plumariopsis eatonii*		
Hydrolithon subantarcticum	*Ptilonia magellanica*		
Synarthrophyton patena	*Rhodokrambe laingioides*		
Synarthrophyton schmitzii	*Rhodymenia coccocarpa*		
Phymatolithon foecundum	*Rubrointrusa membranacea*		
Pseudolithophyllum fueguianum	*Sarcodia montagneana*		
Thick leathery	*Sarcothalia papillosa*		
Gigartina skottsbergii	*Sarcothalia circumcincta*		
	Trematocarpus antarcticus		
	Varimenia macropustulosa		

Fig. 11.2 Major functional-form groups in different divisions of Antarctic seaweeds based on different surveys (Clayton et al. 1997; Wiencke and Clayton 2002; Hommersand et al. 2009; Ramírez 2010; Charles Amsler, personal communication)

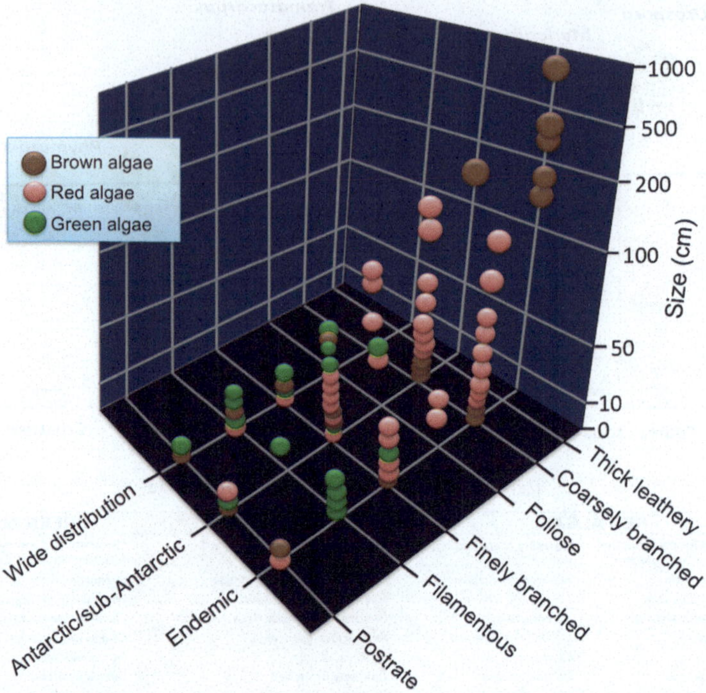

Fig. 11.3 Antarctic seaweeds and their organization in relation to form and function, size, geographic affinity, and taxonomy. Number of taxa extracted from Fig. 11.2

filamentous or sheet-like forms (Fig. 11.3). According to Gómez et al. (2019), form and function and biogeographic affinity are highly correlated in Antarctic green algae. In contrast, only 12% of Rhodophyta can be regarded as widely distributed, contrasting with the high prevalence of endemic (42%) and Antarctic/sub-Antarctic taxa (46%). Similar pattern can be observed in brown algae where endemic and Antarctic/sub-Antarctic species (characterized by thick leathery and saccate morphs) account by 88% of the total numbers of recorded taxa (Fig. 11.2).

11.3 The Vertical Zonation of Antarctic Seaweeds: A Paradigm of Spatial Distribution of Different Morpho-functional Traits

Knowledge on zonation and in general the structure of the submarine landscape in the shallow sublittoral in the Antarctic began to increase in the 1960s and 1970s, along with scuba diving–based surveys (Neushul 1965; Delépine et al. 1966;

Zaneveld 1966; DeLaca and Lipps 1976; Lamb and Zimmerman 1977). These studies across different geographical zones demonstrated that the vertical distribution of macroalgae could be defined in terms of functional groups, which apparently do not follow uniform patterns, principally due to differences in latitude, substrate, influence of ice, associated fauna, etc. The vertical distribution of Antarctic seaweeds has been much more studied in the Western Antarctic Peninsula and adjacent islands, an eco-region known as the Maritime Antarctic. Due to the relatively milder climatic conditions, seaweed assemblages reach their maximal development in terms of abundance and diversity in the north-western part of the Antarctic Peninsula, decreasing the macroalgal biodiversity towards the southern regions (Wiencke et al. 2014; Mystikou et al. 2014). The zonation in the Maritime Antarctic can be characterized by a dominance of large canopy-forming endemic species of the order Desmarestiales (*Desmarestia menziesii*, *D. anceps*, and *Himantothallus grandifolius*) between 10 and 40 m or greater depth (Fig. 11.4a). These three species have thick leathery and terete gross morphology and can alternate their dominance depending on the substrate characteristics, whose consolidation can vary considerably depending on closeness to glaciers, slope of the vertical profile, terrestrial runoff, etc. (Klöser et al. 1994). Coexisting at this level, it is possible to found delicate understory red algae, e.g. *Myriogramme*, *Gymnogongrus*, and *Georgiella* (Amsler et al. 1995). Between 0 and 5 m depth, a zone marked by ice abrasion and waves, the substrate is colonized by fast-growing species, algae with an ability for re-sprout from basal shoots and crustose forms. In contrast, the intertidal rocky shores are dominated by ephemeral, turf species, mainly filamentous Chlorophyceans (e.g. *Urospora*, *Ulva*, *Ulothrix*) and the saccate brown alga *Adenocystis utricularis* (Huovinen and Gómez 2013; Marcías et al. 2017).

In areas outside the Western Antarctic Peninsula, e.g. around the Ross Sea and some sites along the East Antarctica, the diversity and abundance of seaweeds decreases and their vertical distribution is much more constrained by available substrata and the longer permanence of sea ice cover (Zaneveld 1966; Miller and Pearse 1991; Gambi et al. 1994; Johnston et al. 2007; Clark et al. 2011). In these sites, although some large Desmarestiales (e.g. *H. grandifolius* and *D. menziesii*) can be found at deeper locations, in general the coarsely branched red algae *Iridaea cordata* and *Phyllophora antarctica* are the dominant assemblages, especially at intermediate depths (between 2 and 20 m) (Cormaci et al. 2000) (Fig. 11.4b). Another particular feature of these ecosystems is the massive presence of crustose coralline red algae at deeper locations, especially of *Phymatolithon foecundum* (Hommersand et al. 2009), which can cover >70% of the available substrate under the canopy of red and brown algae (Irving et al. 2005). Remarkably, algae have to adapt to very low light conditions for primary productivity, irrespective of their functional form architecture. In fact, due to their extreme shade adaptation, these species can reach considerable depths and live with <2% of surface irradiances (Schwarz et al. 2003, 2005; see also Chap. 7 by Huovinen and Gómez).

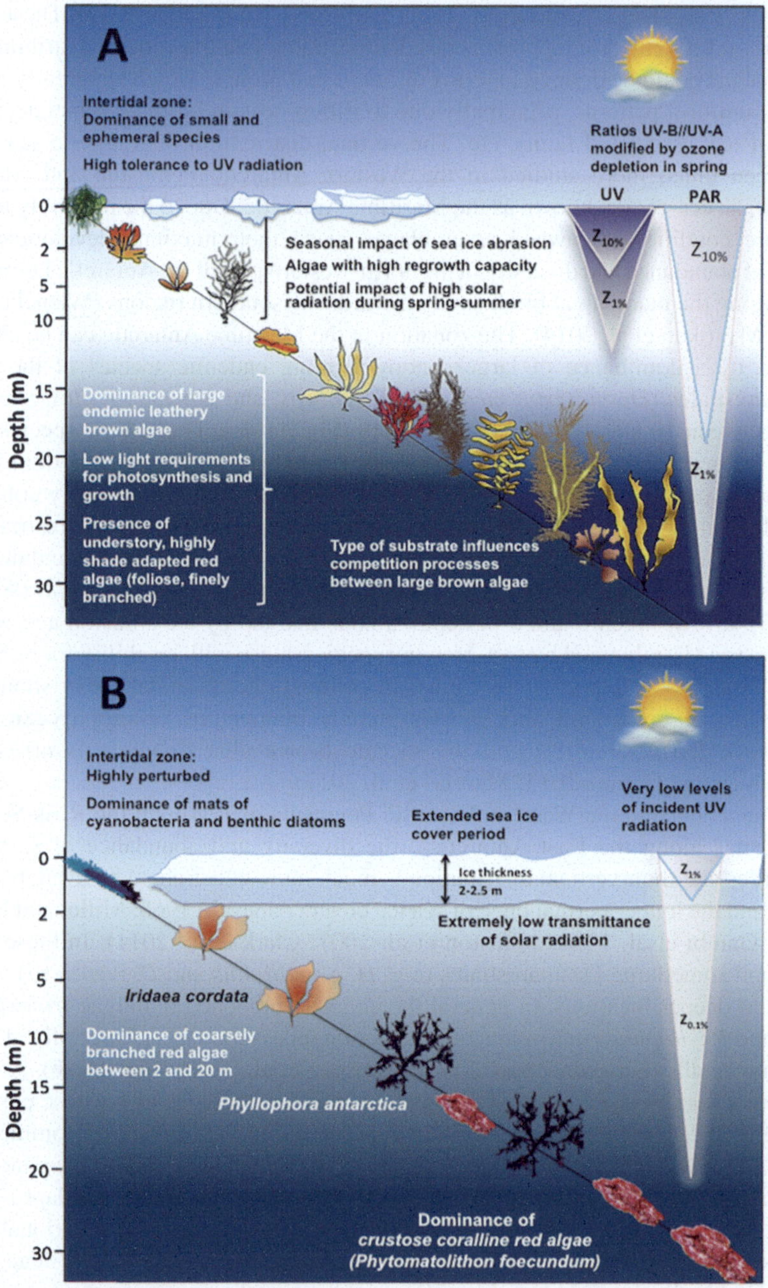

Fig. 11.4 Patterns of vertical distribution of seaweeds in Western Antarctic Peninsula (**a**) and Eastern Antarctica (**b**) indicating major functional groups and photobiological processes. Synthetic schemes from observations reported in Gambi et al. (1994), Johnston et al. (2007), and Huovinen and Gómez (2013)

11.4 Light Use Characteristics as a Major Factor Delineating Physiological Thallus Anatomy of Seaweeds

The arrangement of different functional forms along the depth gradient is strongly determined by different physico-chemical and biological factors. However, spatial and temporal availability of light is probably the most relevant factor by which zonation of Antarctic seaweeds can be explained. Because light governs the primary processes of photosynthesis, and, hence, primary productivity and biomass formation, Antarctic seaweeds, irrespective of their morphological organization, display efficient mechanisms of light harvesting. In fact, in the Antarctic, the marked seasonality in light availability defines strongly an intrinsic shade adaptation of macroalgae. Here the whole phenology of algae is closely tuned with the Antarctic light regime, which exposes organisms to darkness in winter (Wiencke et al. 2009). However, although in summer algae can be exposed to very high doses of solar radiation due to extended daylengths of up to 24 h at the highest latitudes, they do not acclimate and retain the capacity for very low light requirements for metabolism (Gómez and Huovinen 2015). This ability to use very low irradiances for photosynthesis and an intrinsic positive metabolic carbon balance (an indicator of compensation of carbon losses due to respiration) (Gómez et al. 1997; Deregibus et al. 2016) has important implications for the spatial dimension of the algal zonation: it allows Antarctic algae to colonize shaded locations, especially deeper sites. As a consequence, many Antarctic species can occupy extended ranges of depth and hence different light fields (Gómez et al. 1997). This situation contrasts with zonation patterns of various cold and temperate coasts, where the different algal groups are arranged in well-defined "belts" (Lüning 1990). Light trapping, especially under very limited conditions of irradiance, requires not only a specific pigment configuration but also morphological features such as thickness and thallus translucency, which are important in terms of absorptance of the different wavelengths (Gómez and Huovinen 2011). Algae increase light trapping through their thallus architecture, which can result in different in vivo spectral absorptance (Lüning and Dring 1985; Gómez et al. 2019). In Fig. 11.5, the spectral characteristics of several Antarctic macroalgae with different functional form and thickness are exemplified. In the case of thick leathery and coarsely branched morphs (e.g. *Himantothallus*, *Desmarestia*, *Iridaea*), attaining thallus thickness >500 μm, show high absorptance practically along the whole spectrum. In contrast, delicate morphs with thickness <100 μm, mostly foliose and finely branched such as *Monostroma*, *Pantoneura*, *Myriogramme*, and *Pyropia*, show decreased absorptance between 550 and 650 nm. Interestingly, some thin filamentous algae, e.g. *Ulva intestinalis*, *Acrosiphonia*, and *Urospora*, can exhibit high absorptance at these wavelengths, which is related to their turf arrangement, i.e. the overlapping of different filaments equals the several cell layers of thicker algae. Overall, these patterns are related to algal taxonomy and distribution in the zonation profile. In fact, thick leathery forms commonly belong to the brown algae, and their efficient light absorption over an extended range of wavelengths allows using the impoverished light field at higher depths (Gómez and

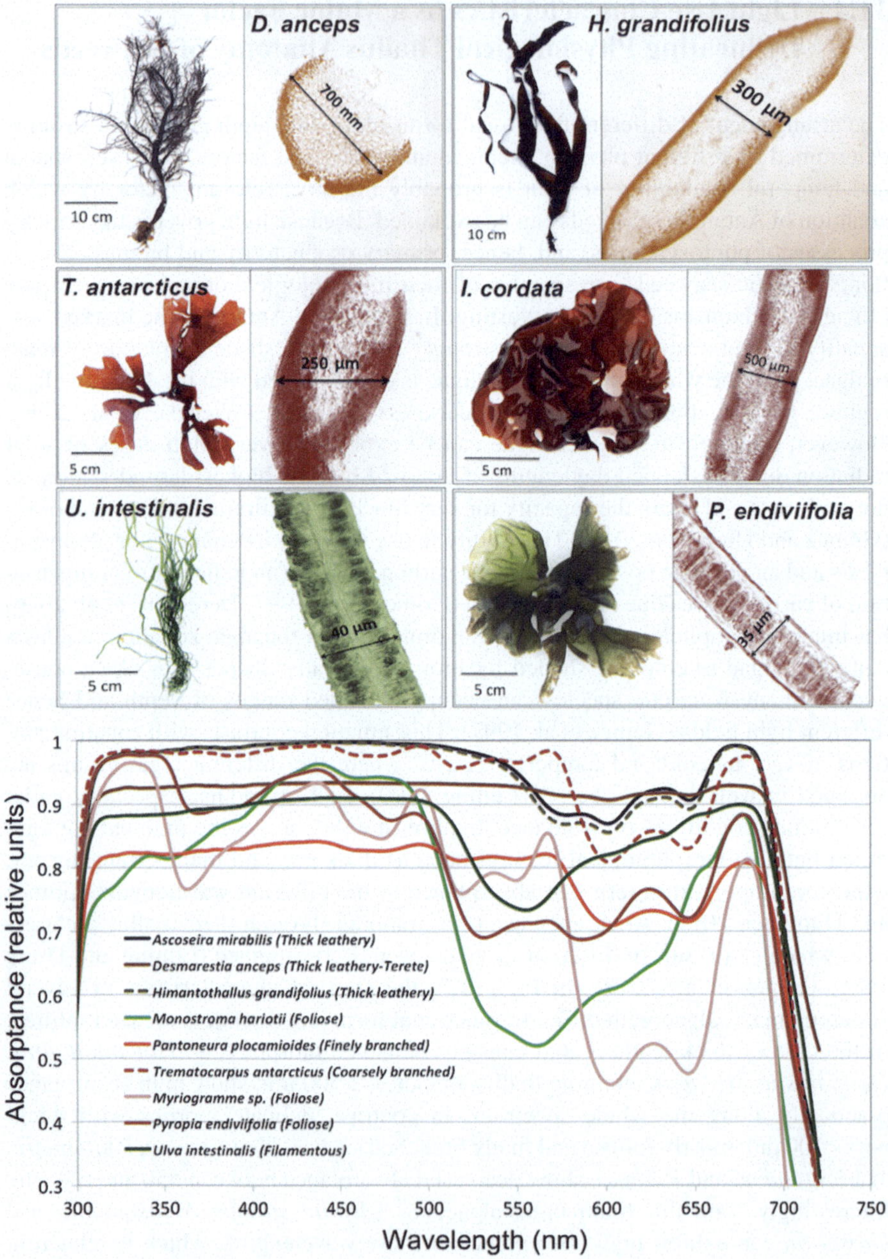

Fig. 11.5 Patterns of spectral in vivo absorptance of several Antarctic seaweeds with different morpho-functional organization. Examples of thallus gross morphology, cross section, and thickness are indicated for selected species. Cross-sectional microphotographs courtesy of Nelso Navarro

Huovinen 2015). Similarly, understory coarsely branched and finely branched red algae, inhabiting deep sites with high absorption between 400 and 500 nm, are well suited to live at these depths (Gómez et al. 2019). Considering that, irrespective of their thallus shape and taxonomical affinity, Antarctic seaweeds are shade-adapted organisms, whose morphological and optical traits (e.g. thickness), pigment composition, and intrinsic photochemical capacities are superimposed in their vertical distribution (Huovinen and Gómez 2013; Gómez et al. 2019).

11.5 Form and Function in the Context of Life Strategies and Stress Tolerance

Since the publication of MacArthur and Wilson in 1967 on r and k selection, which put into a context the evolutionary divergences of organisms in relation to the production and care of offsprings, many studies have tried to expand these concepts to other adaptive traits (concept discussed in Pianka 1970). Because r and k strategies involve normally differences in energy allocation and, hence, body size, the form and function concept could be easily integrated in the theory (Grime 1981). For the case of marine macroalgae, in the 1980s, Joanna Kain used the terms "season responder" and "season anticipator" to describe the different phenological responses of seaweeds to seasonal changes in the environment (Kain 1987). Accordingly, "season responders" correspond to organisms that grow and reproduce under favourable environmental conditions, which could be analogous to r strategists, while "season anticipators" are organisms whose development is triggered by environmental factors at which they anticipate. The latter classification resembles k-selection strategy. Although not a strict rule, most of the season responders identified in the literature seem to correspond to delicate, small-sized forms, which normally exhibit an opportunistic strategy. In contrast a number of season anticipators can be associated with long-lived (perennial) seaweeds attaining normally large thalli (Kain 1989). However, although intuitively one may argue that differences in thallus complexity (including size) are correlated with differential responses to season, in general they are complex and depend on different environmental factors, type of biological indicator (growth, photosynthesis, reproduction, etc.), age and life history phase, and endogenous rhythms, all which can show complementary or divergent patterns (Kain 1986; Lüning and Kadel 1993; see also Chap. 10 by Navarro et al.).

In the Antarctic benthos, seaweeds are exposed to a marked seasonality, and thus, the concepts of "responders" and "anticipators" could explain well the different phenological patterns found in Antarctic seaweeds. In fact, various Antarctic species, e.g. *Iridaea cordata*, *Ulva intestinalis*, *Acrosiphonia arcta*, and *Adenocystis utricularis*, have been regarded as "season anticipators", while Antarctic Desmarestiales, *Ascoseira mirabilis* and *Palmaria decipiens*, can be considered as "season anticipators" (Wiencke 1990a, b). Although these classifications were based mostly on growth responses to the Antarctic light regime, it has been shown that photosynthetic light use characteristics can respond in the same seasonal manner. For example, the brown alga *Adenocystis utricularis* and the red alga *Iridaea cordata*, two species regarded as responders, maintain high photosynthetic functionality still in winter, when light is very limited (Gutkowski and Maleszewski 1989; Weykam et al. 1997). This strategy is completely different in large Desmarestiales and *Ascoseira mirabilis*, as well as the red alga *Palmaria decipiens*, which activate their photosynthetic apparatus during early spring to optimize the available irradiance after the ice break-up (Gómez et al. 1995a, b; Wiencke et al. 2009). For large brown algae whose thalli can have length of various meters, these responses have important morpho-functional implications: Firstly, photosynthesis and growth during early spring are strongly synchronized to potentiate the use of newly fixed and stored carbon. Secondly, there is spatial separation between carbon production and sink zones with different metabolic activity, which can also be exposed to very contrasting light fields. Because these massive thick leathery species require compensation for the enhanced carbon burning due to high rates of dark respiration during the rapid biomass formation, the lamina elongation is powered by carbon stored in the previous season (Gómez and Wiencke 1998), similarly as in high-latitude kelps (e.g. *Laminaria*, *Saccharina*) (Dunton and Schell 1986). A well-studied case is *Ascoseira mirabilis*, which grows through the action of an intercalary meristem and presents "conducting channels" in medullary cell regions (Clayton and Ashburner 1990; Gómez et al. 1995b). In this species, during the growth phase, carbon stored as laminarin in distal parts is remobilized through the conducting cells (normally as mannitol and some amino acids) towards the meristem to replenish carbon substrates in the so-called light-independent carbon fixation (LICF) reactions (Kremer 1981; Gómez and Huovinen 2012). Such morpho-functional strategies have not only been demonstrated in large brown algae: in the perennial, coarsely branched red alga *Palmaria decipiens*, LICF reactions accounting 9% of the total fixed carbon have been reported (Weykam et al. 1997), suggesting that this type of mechanisms are operating in algae with complex thallus anatomy and season anticipation phenology, which allow them thriving at high depths and under extreme seasonality in the Antarctic.

Form and function in the context of stress tolerance have been revisited in the last years. Interestingly, several anatomical traits related to resistance to physical disturbance, e.g. multilayered architecture, thickness, and large size, are also functional to increase light trapping, e.g. efficient absorptance (Gómez and Huovinen 2011). Recently it was claimed that populations of three species of Desmarestiales (*D. anceps*, *D. menziesii*, and *H. grandifolius*) and *Ascoseira mirabilis* extending between 5 and 30 m depth show similar photosynthetic characteristics along the depth profile (Gómez and Huovinen 2015). However, not only the efficient and highly conserved light use across different irradiances but also an intrinsic capacity for UV stress tolerance was shown in these algae (see Chap. 7 by Huovinen and Gómez). Although all these traits conferring UV shielding show a strong overlapping with other factors, e.g. competence for space and overgrowth, scape from herbivores, there appears to exist a trade-off between photoprotection against enhanced solar UV radiation, mostly due to an increased thallus cross section (low area/weight ratio) and ultrastructural compounds, and highly efficient shade adaptation (Gómez and Huovinen 2015). A key element explaining this feature in algae rarely exposed to UV radiation is their constitutively high levels of phenolics (phlorotannins) (Flores-Molina et al. 2016). These secondary metabolites in Antarctic brown algae represent multifunctional compounds with putative roles in, e.g. resistance to grazing, temperature, and UV radiation (Amsler et al. 2005; Iken et al. 2009; Huovinen and Gómez 2013; Rautenberger et al. 2015) (for a description of functional roles of phlorotannins in Antarctic seaweeds, see Chap. 17 by Amsler et al. and Chap. 18 by Gómez and Huovinen).

In the case of delicate morphs, mostly filamentous and finely branched green algae, the opportunistic life strategy of these organisms allows them to respond rapidly to environmental stressors and, in virtue of their high metabolic rates per weight, to restore the homeostasis at short term (Holzinger and Karsten 2013). Albeit stress tolerance of seaweeds living in the intertidal zone would rely on highly efficient metabolic adjustments (Holzinger and Lutz 2006; Karsten et al. 2009; Gómez and Huovinen 2011), some structural adaptations have been described. For example, in *Urospora penicilliformis* a dense cell wall, presence of mucilage and external mineral deposition provide efficient shielding from high solar radiation and desiccation (Roleda et al. 2010). In many cases, filamentous green algae can form mats or turf-like structures that are effective to minimize the harmful effects of changing environment (Bischof et al. 2006). In all, in terms of photosynthetic characteristics and physiological responses to stress, form and function of some Antarctic seaweed assemblages have been related to biogeographic affinity and depth. Based on 31 species from King George Island, three major groups of species were defined: (a) coarsely branched Rhodophyta are mostly found at shallow subtidal sites and have an Antarctic-sub-Antarctic origin; (b) endemic Antarctic brown algae are dominant at depths between 10 and 30 m and practically all attain thick leathery morphology; and (c) filamentous and sheet-like green algae, mostly intertidal species, normally can be categorized as algae with wide geographic distribution (Gómez et al. 2019).

11.6 Functional Traits of Seaweeds and Properties of Benthic Communities

Seaweeds in practically all cold-temperate and polar coastal ecosystems represent foundational organisms, whose processes and fate determine key community indicators, such as structure and functional and taxonomic richness (Chapman 1987; Lüning 1990). In King George Island, the distribution and composition of different functional groups in both intertidal and subtidal sites are regulated by different factors, which are defined by some species that account by 90% of the dissimilarities between depth strata (Valdivia et al. 2014). When representative taxa are analysed, effects are scale dependent: variance components increase at the finer scale of variation (from centimeters to meters) compared to shore level (hundreds of meters) (Valdivia et al. 2014). In the intertidal system dominated by filamentous and finely branched morphs, the grazing by the limpet *Nacella concinna* is probably one of the most important biological interactions (Kim 2001; Segovia-Rivera and Valdivia 2016). Apart from green algae, *N. concinna* exerts control on periphyton, thus determining far-reaching ecological processes, e.g. the fate of re-colonization and succession in these systems (Campana et al. 2009; Valdivia et al. 2019; see also Chap. 12 by Campana et al. and Chap. 13 by Valdivia).

At the subtidal zone, facilitative interactions held by large brown algae through bioengineering seem relevant for the structure and maintenance of the benthic communities (Valdivia et al. 2015). These canopy-forming seaweeds are important as they shelter other species of algae and invertebrates from harmful environmental conditions and thus have an important effect on the community biomass of the whole ecosystem (Valdivia et al. 2015; Ortiz et al. 2016; see also Chap. 15 by Momo et al. and Chap. 16 by Ortiz et al.). However, in locations exposed to severe impact of physical disturbance, small organisms can be favoured while canopy-forming algae would be more sensitive (Smale 2007). For example, in eastern Antarctica where ice cover can be considerably extended through spring, canopy-forming macroalgae were only abundant at sites where sea-ice cover break-up occurs during spring, but absent at sites that retained ice cover until summer (Johnston et al. 2007). Thus, these organisms appear to respond slowly to the changing environment due, for example, to enhanced warming. For example, in new ice-free areas originated from glacier retreat where enhanced sediment input limits light penetration, establishment of large brown algae is highly constrained (Quartino et al. 2013). In these highly perturbed sites, ice scouring and unconsolidated substrate affect considerably the presence of canopy-forming algae and hence the taxonomic richness (Klöser et al. 1994; Smale 2007; Smale et al. 2008; Valdivia et al. 2015). On the other hand, environmental shifts driven by climate change can affect the morpho-functional responses of Antarctic species. For example, physiology of canopy-forming algae (e.g. *Desmarestia* spp.) may have consequences for the whole benthic community (Schoenrock et al. 2015). In the case of crustose species, fleshy encrusting forms (*Hildenbrandia*) could be favoured in scenarios of changing pH and temperature compared to calcified Coralline species (*Clathromorphum*) (Schoenrock et al. 2016). In general, morpho-functional and anti-stress mechanisms of macroalgae to cope with sharp physical gradients percolate towards upper hierarchies through insurance of functional richness in the community, which set high degree of resilience to physical perturbation (Ortiz et al. 2016, 2017) or to minimize the impact of alien species (Arenas et al. 2006; see Chap. 16 by Ortiz et al.).

Considering some functional form models for marine seaweeds, similitudes and analogies with terrestrial vegetation strategies can be identified. For example, according to the functional groups described by Grime (1981) for terrestrial vegetation, opportunistic green algae growing at the intertidal zone could correspond to the "ruderal" species, permanently subjected to strongly physical perturbation. In contrast, large endemic brown algae, which thrive in sites with lower physical

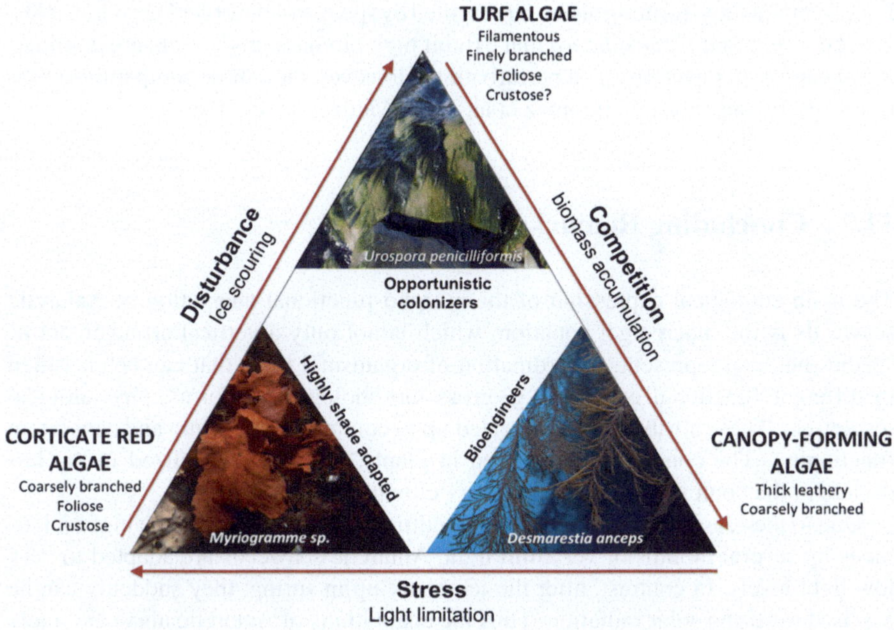

Fig. 11.6 CSR Grime's triangle describing the main strategies and environmental trade-off of the major functional groups of Antarctic seaweeds. (Photos *Myriogramme* and *Desmarestia* by Ignacio Garrido; *Urospora* by Iván Gómez)

perturbation, can be analogue to "competitive species" in virtue of their exuberant canopy and perennial characteristics. For many temperate ecosystems, the life history traits conferring advantages under high levels of disturbance are convergent in different types of algal assemblages, suggesting that some patterns could be generalized (Steneck and Dethier 1994). However, in Antarctic communities some factors associated with disturbance and stress require adjustment to the extreme Antarctic conditions. In the conceptual framework in Fig. 11.6, three major functional groups of Antarctic seaweeds (turf algae, dominated by intertidal green algae; canopy-forming algae, especially large brown algae; and corticated red algae, grouping many understory species) can be oriented through the three axes following a Grime's CSR triangle schema. Here, the extreme action of ice (perturbation), light limitation (stress), and biomass (competition) dimensions determine the separation among algal groups. Corticate red algae in virtue of their extreme shade adaptation represent the stress tolerant group. Here, many crustose species growing at very low light conditions in the eastern Antarctic can also be added to this group. In the perturbation axis, filamentous and finely branched green algae and some little saccate brown algae (*Adenocystis utricularis*) exemplify the colonizers, well adapted to occupy sites highly perturbed by ice, terrestrial run-off, and high solar radiation. Under these conditions abundance and species richness are less influenced by biological interactions (Valdivia et al. 2014; Segovia-Rivera and Valdivia 2016).

Finally, the canopy-forming algae, represented by species of the order Desmarestiales, and *Ascoseira* and *Cystosphaera* that exhibit high biomass production, are dominating at sites with lower physical perturbation. However, they show competitive abilities for light and substrate (Gómez et al. 1997; Valdivia et al. 2015).

11.7 Concluding Remarks

The main ecological expression of the morpho-functional adaptation of Antarctic seaweeds is the macroalgal zonation, which is not only a vertical arrangement of species but also represents an ordination of organismal traits that can be classified in different functional entities (e.g. gross morphology, life forms, physiological responses). These attributes can be scaled up to community structure and ecosystem functioning. The concept, well studied in plants, has been revitalized in the last decade in the context of the contemporary climate change.

Due to the seasonally changing light conditions, characterized in the highest latitudes by several months of very dim light, Antarctic seaweeds are adapted to very low light levels. In contrast, after the ice break-up in spring, they suddenly can be exposed to strong solar radiation. Thus the adaptations of Antarctic algae are finely tuned with the daylength, changes in water turbidity, and ice perturbations. This environmental variability is fully exploited by seaweeds in virtue of their efficient morpho-functional adaptations. However, due to climate change, the environmental settings in which Antarctic seaweeds have evolved for millions of years are changing. In these new scenarios, the adaptive capacities of these organisms as well as the ecosystem functions they provide will be challenged (Constable et al. 2014; Gutt et al. 2015). Although one can recognize that polar seaweeds are particularly susceptible to these changes with unpredictable consequences for the whole coastal ecosystem, we have still a limited understanding on how physiological and morphological traits respond and how they will be integrated in, for example, molecular mechanisms of environmental tolerance and stress resilience.

Acknowledgements The authors acknowledge the financial support from Conicyt Chile (FONDAP 15150003, Anillo-PIA ART1101, Fondecyt 1161129) and INACH T-20-09 from the Instituto Antártico Chileno. The helpful assistance and collaboration of the members of our laboratories at Universidad Austral de Chile as well as the staff of the Instituto Antártico Chileno for during various Antarctic expeditions are acknowledged.

References

Ackerly DD, Reich PB (1999) Convergence and correlations among leaf size and function in seed plants: a comparative test using independent contrasts. Am J Bot 86:1272–1128

Aguilera J, Karsten U, Lippert H, Vögele B, Philipp E, Hanelt D, Wiencke C (1999) Effects of solar radiation on growth, photosynthesis and respiration of marine macroalgae from the Arctic. Mar Ecol Prog Ser 191:109–119

Alongi G, Cormaci M, Furnari G (2002) The Corallinaceae (Rhodophyta) from the Ross Sea (Antarctica): a taxonomic revision rejects all records except *Phymatolithon foecundum*. Phycologia 41:140–146

Amsler CD, Rowley RJ, Laur DR, Quetin LB, Ross RM (1995) Vertical distribution of Antarctic peninsular macroalgae: cover, biomass and species composition. Phycologia 34:424–430

Amsler CD, Iken K, McClintock JB, Amsler MO, Peters KJ, Hubbard JM, Furrow FB, Baker JB (2005) Comprehensive evaluation of the palatability and chemical defenses of subtidal macroalgae from the Antarctic Peninsula. Mar Ecol Prog Ser 294:141–159

Arenas F, Sánchez I, Hawkins SJ, Jenkins SRJ (2006) The invasibility of marine algal assemblages: role of functional diversity and identity. Ecology 87(11):2851–2861

Balata D, Piazzi L, Rindi F (2011) Testing a new classification of morphological functional groups of marine macroalgae for the detection of responses to stress. Mar Biol 158:2459–2469

Bischof K, Rautenberger R, Brey L, Pérez-Lloréns JL (2006) Physiological acclimation to gradients of solar irradiance within mats of the filamentous green macroalga *Chaetomorpha linum* from southern Spain. Mar Ecol Prog Ser 306:165–175

Campana GL, Zacher K, Fricke A, Molis M, Wulff A, Quartino ML, Wiencke C (2009) Drivers of colonization and succession in polar benthic macro- and microalgal communities. Bot Mar 52:655–667

Chapman ARO (1987) Population and community ecology of seaweeds. Adv Mar Biol 23:1–161

Clark GF, Stark JS, Perrett LA, Hill NA, Johnston EL (2011) Algal canopy as a proxy for the disturbance history of understorey communities in East Antarctica. Polar Biol 34:781–790

Clayton MN, Ashburner CM (1990) The anatomy and ultrastructure of "conducting channels" in *Ascoseira mirabilis* (Ascoseirales, Phaeophyceae). Bot Mar 33:63–70

Clayton M, Wiencke C, Klöser H (1997) New records of temperate and sub Antarctic marine benthic macroalgae from Antarctica. Polar Biol 17:141–179

Constable AJ, Melbourne-Thomas J, Corney SP, Arrigo KR, Barbraud C, Barnes DK, Bindoff NL et al (2014) Climate change and Southern Ocean ecosystems. I: how changes in physical habitats directly affect marine biota. Glob Chang Biol 20:3004–3025. https://doi.org/10.1111/gcb.12623

Cormaci M, Furnari G, Scammacca B (2000) The macrophytobenthos of Terra Nova Bay. In: Faranda FM et al (eds) Ross Sea ecology. Italian Antarctic expeditions (1987–1995). Springer, Berlin, pp 493–502

DeLaca TE, Lipps JH (1976) Shallow-water marine associations, Antarctic Peninsula. Antarct J US 11:12–20

Delépine R, Lamb JM, Zimmermann MH (1966) Preliminary report on the vegetation of the Antarctic Peninsula. Proc Int Seaweed Symp 5:107–116

Deregibus D, Quartino ML, Campana GL, Momo FR, Wiencke C, Zacher K (2016) Photosynthetic light requirements and vertical distribution of macroalgae in newly ice-free areas in Potter Cove, South Shetland Islands, Antarctica. Polar Biol 39:153–166

Dunton KH, Schell DM (1986) Seasonal carbon budget and growth of *Laminaria solidungula* in the Alaskan high Arctic. Mar Ecol Prog Ser 31:57–66

Flores-Molina MR, Muñoz P, Rautenberger R, Huovinen P, Gómez I (2016) Stress tolerance to UV radiation and temperature of the endemic Antarctic brown alga *Desmarestia anceps* is mediated by high concentrations of phlorotannins. Photochem Photobiol 92:455–466

Gambi M, Lorent IM, Russo G, Scipione M (1994) Benthic associations of the shallow hard bottoms off Terra Nova Bay, Ross Sea: zonation, biomass and population structure. Antarct Sci 6:449–462

Gómez I, Huovinen P (2011) Morpho-functional patterns and zonation of south Chilean seaweeds: the importance of photosynthetic and bio-optical traits. Mar Ecol Prog Ser 422:77–91

Gómez I, Huovinen P (2012) Morpho-functionality of carbon metabolism in seaweeds. In: Wiencke C, Bischof K (eds) Seaweed biology: novel insights into ecophysiology, ecology and utilization. Springer, Berlin, pp 25–46

Gómez I, Huovinen P (2015) Lack of physiological depth patterns in conspecifics of endemic Antarctic brown algae: a trade-off between UV stress tolerance and shade adaptation? PLoS One 10(8):e0134440

Gómez I, Wiencke C (1998) Seasonal changes in C, N and major organic compounds, and their significance to morpho-functional processes in the endemic Antarctic brown alga *Ascoseira mirabilis*. Polar Biol 19:115–124

Gómez I, Thomas DN, Wiencke C (1995a) Longitudinal profiles of growth, photosynthesis and light independent carbon fixation in the Antarctic brown alga *Ascoseira mirabilis*. Bot Mar 38:157–164

Gómez I, Wiencke C, Weykam G (1995b) Seasonal photosynthetic characteristics of *Ascoseira mirabilis* (Ascoseirales, Phaeophyceae) from King George Island, Antarctica. Mar Biol 123:167–172

Gómez I, Weykam G, Klöser H, Wiencke C (1997) Photosynthetic light requirements, daily carbon balance and zonation of sublittoral macroalgae from King George Island (Antarctica). Mar Ecol Prog Ser 148:281–293

Gómez I, Navarro NP, Huovinen P (2019) Bio-optical and physiological patterns in Antarctic seaweeds: a functional trait based approach to characterize vertical zonation. Prog Oceanogr 174:17–27

Grime JP (ed) (1981) Plant strategies and vegetation processes. Wiley, Chichester, p 222

Grosberg RK, Strathmann RR (2007) The evolution of multicellularity: a minor major transition? AREES 38:621–654. https://doi.org/10.1146/annurev.ecolsys.36.102403.114735

Gutkowski R, Maleszewski S (1989) Seasonal changes of the photosynthetic capacity of the Antarctic macroalga *Adenocystis utricularis* (Bory) Skottsberg. Polar Biol 10:145–148

Gutt J, Bertler N, Bracegirdle TJ, Buschmann A, Comiso J, Hosie G, Isla E, Schloss IR et al (2015) The Southern Ocean ecosystem under multiple climate change stresses – an integrated circumpolar assessment. Glob Chang Biol 21:1434–1453. https://doi.org/10.1111/gcb.12794

Hay ME (1986) Functional geometry of seaweeds: ecological consequences of thallus layering and shape in contrasting light environments. In: Givnish TJ (ed) On the economy of plant form and function. Proceedings of the Sixth Maria Moors Cabot Symposium, "Evolutionary Constraints on Primary Productivity: Adaptive Patterns of Energy Capture in Plants", Harvard Forest, August 1983. Cambridge University Press, Cambridge

Holzinger A, Karsten U (2013) Desiccation stress and tolerance in green algae: consequences for ultrastructure, physiological, and molecular mechanisms. Plant Sci 22:327. https://doi.org/10.3389/fpls.2013.00327

Holzinger A, Lütz C (2006) Algae and UV irradiation: effects on ultrastructure and related metabolic functions. Micron 37:190–207. https://doi.org/10.1016/j.micron.2005.10.015

Hommersand MH, Moe RL, Amsler CD, Fredericq S (2009) Notes on the systematics and biogeographical relationships of Antarctic and sub-Antarctic Rhodophyta with descriptions of four new genera and five new species. Bot Mar 52:509–534

Huovinen P, Gómez I (2013) Photosynthetic characteristics and UV stress tolerance of Antarctic seaweeds along the depth gradient. Polar Biol 36:1319–1332

Iken K, Amsler CD, Amsler MO, McClintock JB, Baker BJ (2009) Field studies on deterrent roles of phlorotannins in Antarctic brown algae. Bot Mar 52:547–557

Ingólfsson A (2005) Community structure and zonation patterns of rocky shores at high latitudes: an interocean comparison. J Biogeogr 32:169–182

Innes DJ (1984) Genetic differentiation among populations of marine algae. Helgoländer Meeresunters 38:401–417

Irving A, Connell S, Johnston E, Pile A, Gillanders B (2005) The response of encrusting coralline algae to canopy loss: an independent test of predictions on an Antarctic coast. Mar Biol 147:1075–1083

Johnston EL, Connell SD, Irving AD, Pile AJ, Gillanders BM (2007) Antarctic patterns of shallow subtidal habitat and inhabitants in Wilke's land. Polar Biol 30:781–788

Kaandorp JA, Kübler JE (eds) (2001) The algorithmic beauty of seaweeds, sponges and corals. Springer, Heidelberg, p 194
Kain JM (1986) Plant size and reproductive phenology of six species of Rhodophyta in subtidal Isle of Man. Br Phycol J 21:129–138
Kain JM (1987) Seasonal growth and photoinhibition in *Plocamium cartilagineum* (Rhodophyta) off the Isle of Man. Phycologia 26:88–99
Kain JM (1989) The seasons in the subtidal. Br Phycol J 24:203–215
Karsten U, Wulff A, Roleda MY, Müller R, Steinhoff FS, Fredersdorf J, Wiencke C (2009) Physiological responses of polar benthic algae to ultraviolet radiation. Bot Mar 52:639–654
Kim D (2001) Seasonality of marine algae and grazers of an Antarctic rocky intertidal, with emphasis on the role of the limpet *Nacella concinna* Strebel (Gastropoda: Patellidae). Fachbereich Biologie/Chemie, Ph.D. thesis, University of Bremen, Germany
Klöser H, Mercuri G, Laturnus F, Quartino ML, Wiencke C (1994) On the competitive balance of macroalgae at Potter Cove (King George Island, South Shetlands). Polar Biol 14:11–16
Kremer BP (1981) Aspects of carbon metabolism in marine macroalgae. Oceanogr Mar Biol Annu Rev 19:41–94
Lamb IM, Zimmermann MH (1977) Benthic marine algae of the Antarctic Peninsula. Antarct Res Ser 23:130–229
Lavorel S, Garnier E (2002) Predicting changes in community composition and ecosystem functioning from plant traits: revisiting the holy grail. Funct Ecol 16:545–556
Littler MM, Littler DS (1980) The evolution of thallus form and survival strategies in benthic marine macroalgae: field and laboratory tests of a functional form model. Am Nat 116:25–44
Lüning K (ed) (1990) Seaweeds: their environment, biogeography and ecophysiology. Wiley, New York, p 527
Luning K, Dring MJ (1985) Action spectra and spectral quantum yield of photosynthesis in marine macroalgal with thin and thick thalli. Mar Biol 87:119–129
Lüning K, Kadel P (1993) Daylength range for circannual rhythmicity in *Pterygophora californica* (Alariaceae, Phaeophyta) and synchronization of seasonal growth by daylength cycles in several other brown algae. Phycologia 32:379–387
Marcías ML, Deregibus D, Saravia LA, Campana GL, Quartino ML (2017) Life between tides: spatial and temporal variations of an intertidal macroalgal community at Potter Peninsula, South Shetland Islands, Antarctica. Estuar Coast Shelf Sci 187:193–203
Miller KA, Pearse JS (1991) Ecological studies of seaweeds in McMurdo Sound, Antarctica. Am Zool 31:35–48
Moe RL, Silva PC (1981) Morphology and taxonomy of *Himantothallus* (including *Phaeoglossum* and *Phyllogigas*), an Antarctic member of the Desmarestiales (Phaeophyceae). J Phycol 17:15–29. https://doi.org/10.1111/j.1529-8817.1981.tb00814.x
Mystikou A, Peters AF, Asensi AO, Fletcher KF, Brickle P, van West P, Convey P, Küpper FC (2014) Seaweed biodiversity in the south-western Antarctic Peninsula: surveying macroalgal community composition in the Adelaide Island/Marguerite Bay region over a 35-year time span. Polar Biol 37(11):1607–1619
Neushul M (1965) Diving observations of subtidal Antarctic marine vegetation. Bot Mar 8:234–243
Ortiz M, Berríos F, González J, Rodríguez-Zaragoza FF, Gómez I (2016) Macroscopic network properties and short-term dynamic simulations in coastal ecological systems at Fildes Bay (King George Island, Antarctica). Ecol Complex 28:145–157. https://doi.org/10.1016/j.ecocom.2016.06.003
Ortiz M, Hermosillo-Nuñez B, González J, Rodríguez-Zaragoza F, Gómez I, Jordán F (2017) Quantifying keystone species complexes: ecosystem-based conservation management in the King George Island (Antarctic Peninsula). Ecol Indic 81:453–460. https://doi.org/10.1016/j.ecolind.2017.06.016
Padilla DK, Allen BJ (2000) Paradigm lost: reconsidering functional form and group hypotheses in marine ecology. J Exp Mar Biol Ecol 250:207–221

Phillips JC, Kendrick GA, Lavery PS (1997) A test of a functional group approach to detecting shifts in macroalgal communities along a disturbance gradient. Mar Ecol Prog Ser 153:125–138
Pianka ER (1970) On r and K selection. Am Nat 104(940):592–597
Poorter L, Bongers F (2006) Leaf traits are good predictors of plant performance across 53 rain forest species. Ecology 87:1733–1743
Quartino ML, Deregibus D, Campana GL, Edgar G, Latorre J, Momo FR et al (2013) Evidence of macroalgal colonization on newly ice-free areas following glacial retreat in Potter Cove (South Shetland Islands), Antarctica. PLoS One 8:e58223
Ramírez ME (2010) Flora marina bentónica de la región austral de Sudamérica y la Antártica. An Inst Patagon 38(1):57–71
Rautenberger R, Huovinen P, Gómez I (2015) Effects of increased seawater temperature on UV-tolerance of Antarctic marine macroalgae. Mar Biol 162:1087–1109
Roleda MY, Lütz-Meindl U, Wiencke C, Lütz C (2010) Physiological, biochemical, and ultrastructural responses of the green macroalga *Urospora penicilliformis* from Arctic Spitsbergen to UV radiation. Protoplasma 243:105–116
Schoenrock KM, Schram JB, Amsler CD, McClintock JB, Angus RA (2015) Climate change impacts on overstory *Desmarestia* spp. from the western Antarctic Peninsula. Mar Biol 162:377–389. https://doi.org/10.1007/s00227-014-2582-8
Schoenrock KM, Schram JB, Amsler CD, McClintock JB, Angus RA, Vohra YK (2016) Climate change confers a potential advantage to fleshy Antarctic crustose macroalgae over calcified species. J Exp Mar Biol Ecol 474:58–66
Schwarz A, Hawes I, Andrew N, Norkko A, Cummings V, Thrush S (2003) Macroalgal photosynthesis near the southern global limit for growth, Cape Evans, Ross Sea, Antarctica. Polar Biol 26:789–799
Schwarz A-M, Hawes I, Andrew N, Mercer S, Cummings V, Thrush S (2005) Primary production potential of non-geniculate coralline algae at Cape Evans, Ross Sea, Antarctica. Mar Ecol Prog Ser 294:131–140
Segovia-Rivera V, Valdivia N (2016) Independent effects of grazing and tide pool habitats on the early colonisation of an intertidal community on western Antarctic Peninsula. Rev Chil Hist Nat 89:3
Smale DA (2007) Ice disturbance intensity structures benthic communities in nearshore Antarctic waters. Mar Ecol Prog Ser 349:89–102
Smale DA, Brown KM, Barnes DKA, Fraser KPP, Clarke A (2008) Ice scour disturbance in Antarctic waters. Science 321:371
Steneck RS, Dethier MN (1994) A functional group approach to the structure of algal-dominated communities. Oikos 69:476–498
Steneck RS, Watling L (1982) Feeding capabilities and limitations of herbivorous molluscs: a functional group approach. Mar Biol 68:299–319
Taylor PR, Hay ME (1984) Functional morphology of intertidal seaweeds: adaptive significance of aggregate vs. solitary forms. Mar Ecol Prog Ser 18:295–302
Valdivia N, Díaz MJ, Holtheuer J, Garrido I, Huovinen P, Gómez I (2014) Up, down, and all around: scale-dependent spatial variation in rocky-shore communities of Fildes Peninsula, King George Island, Antarctica. PLoS One 9(6):e100714. https://doi.org/10.1371/journal.pone.0100714
Valdivia N, Díaz MJ, Garrido I, Gómez I (2015) Consistent richness-biomass relationship across environmental stress gradients in a marine macroalgal-dominated subtidal community on the Western Antarctic Peninsula. PLoS One 10(9):e0138582
Valdivia N, Huovinen P, Gómez I, Macaya E, Pardo LM (2019) Different ecological mechanisms lead to similar grazer controls on the functioning of periphyton Antarctic and sub-Antarctic communities. Prog Oceanogr 174:7–16. https://doi.org/10.1016/j.pocean.2018.01.008
Weykam G, Thomas DN, Wiencke C (1997) Growth and photosynthesis of the Antarctic red algae *Palmaria decipiens* (Palmariales) and *Iridaea cordata* (Gigartinales) during and following extended periods of darkness. Phycologia 36:395–405

Wiencke C (1990a) Seasonality of brown macroalgae from Antarctica – a long-term culture study under fluctuating Antarctic daylengths. Polar Biol 10:589–600

Wiencke C (1990b) Seasonality of red and green macroalgae from Antarctica – a long-term culture study under fluctuating Antarctic daylengths. Polar Biol 10:601–607

Wiencke C, Clayton MN (1990) Sexual reproduction, life history, and early development in culture of the Antarctic brown alga *Himantothallus grandifolius* (Desmarestiales, Phaeophyceae). Phycologia 29:9–18

Wiencke C, Clayton MN (2002) Antarctic seaweeds. In: Wägele JW (ed) Synopses of the Antarctic benthos, vol 9. A.R.G Gantner Verlag KG, Ruggell, p 239

Wiencke C, Gómez I, Dunton K (2009) Phenology and seasonal physiological performance of polar seaweeds. Bot Mar 52:585–592

Wiencke C, Amsler CD, Clayton MN (2014) Chapter 5.1. Macroalgae. In: De Broyer C, Koubbi P, Griffiths H, Raymond B, Udekem d'Acoz C et al (eds) Biogeographic atlas of the Southern Ocean, vol 8. Scientific Committee on Antarctic Research, Cambridge, pp 66–73

Zaneveld JS (1966) Vertical zonation of Antarctic and subantarctic benthic marine algae. Antarct J US 1(5):211–213

Part IV
Biological Interactions and Ecosystem Processes

Chapter 12
Successional Processes in Antarctic Benthic Algae

Gabriela L. Campana, Katharina Zacher, Fernando R. Momo, Dolores Deregibus, Juan Ignacio Debandi, Gustavo A. Ferreyra, Martha E. Ferrario, Christian Wiencke, and María L. Quartino

Abstract Despite the importance of benthic algal communities to Antarctic coastal ecosystems, much information about their dynamics is still needed. Primary succession processes in the Antarctic benthos are frequently initiated by ice-mediated disturbance and by the creation of denuded substrate following glacier retreat, both expected to increase in the future. Primary succession of benthic algae starts with rapid colonization by diatoms, ephemeral green algal filaments and propagules of annual and pseudoperennial macroalgae. Early stages of macroalgae can be particularly vulnerable to environmental stress factors, being critical for the structure of mature communities. The Antarctic environment is severely affected by global change, and successional patterns can change due to species-specific susceptibilities to abiotic and biotic drivers, introducing changes in the matter and energy flow in the coastal food webs.

This chapter summarizes new advances in our knowledge on the successional dynamics of benthic primary producers in the Antarctic hard-bottom benthos. Manipulative experiments on the effects of grazing and ultraviolet (UV) radiation as drivers of the succession at early stages and long-term experiments carried out at sites with different environmental conditions are compiled. The gathered information can contribute to achieve a deeper knowledge of these key communities and their structure and functioning in a changing environment.

G. L. Campana (✉)
Departamento de Biología Costera, Instituto Antártico Argentino (IAA) - Dirección Nacional del Antártico (DNA), Buenos Aires, Argentina

Departamento de Ciencias Básicas, Universidad Nacional de Luján (UNLu), Luján, Buenos Aires, Argentina
e-mail: gcampana@dna.gov.ar

K. Zacher · C. Wiencke
Alfred-Wegener-Institute, Helmholtz Centre for Polar and Marine Research (AWI), Bremerhaven, Germany
e-mail: katharina.zacher@awi.de; christian.wiencke@awi.de

I. Gómez, P. Huovinen (eds.), *Antarctic Seaweeds*,
https://doi.org/10.1007/978-3-030-39448-6_12

Keywords Early colonizers · Ecological succession · Grazing · Microphytobenthos · Rocky substrate · UV radiation

12.1 Introduction

The Antarctic benthos is frequently exposed to ice-mediated disturbance leading to successional processes and to the coexistence of patches of different developmental stages (Gutt 2001; Quartino et al. 2005; Barnes and Conlan 2007; Teixidó et al. 2007). Furthermore, climate change-related phenomena have accelerated the retreat of glaciers opening newly available substrate formerly covered by ice and, hence, initiating primary succession processes (Quartino et al. 2013; Sahade et al. 2015; Lagger et al. 2017, 2018). Even though the successional process in marine rocky coasts has been subject to a great number of studies worldwide, they were conceived mainly on temperate habitats (Noël et al. 2009; Benincà et al. 2015), whereas successional processes in polar regions are still less known.

Studies on the successional patterns on the Antarctic benthos have mainly been focussed on sessile faunal assemblages (Barnes and Conlan 2007; Dayton et al. 2016; Lagger et al. 2017, 2018, and references therein). These studies showed a

F. R. Momo
Instituto de Ciencias, Universidad Nacional de General Sarmiento (UNGS), Los Polvorines, Argentina

Departamento de Ciencias Básicas, Universidad Nacional de Luján (UNLu), Luján, Buenos Aires, Argentina
e-mail: fmomo@campus.ungs.edu.ar

D. Deregibus
Departamento de Biología Costera, Instituto Antártico Argentino (IAA) - Dirección Nacional del Antártico (DNA), Buenos Aires, Argentina

Consejo Nacional de Investigaciones Científicas y Técnicas (CONICET), Buenos Aires, Argentina
e-mail: dderegibus@dna.gov.ar

J. I. Debandi
Departamento de Ciencias Básicas, Universidad Nacional de Luján (UNLu), Luján, Buenos Aires, Argentina

G. A. Ferreyra
Centro Austral de Investigaciones Científicas (CADIC-CONICET), Ushuaia, Argentina
e-mail: gferreyra@cadic-conicet.gob.ar

M. E. Ferrario
Facultad de Ciencias Naturales y Museo (FCNyM), Universidad Nacional de La Plata (UNLP), La Plata, Argentina
e-mail: meferra@fcnym.unlp.edu.ar

M. L. Quartino
Departamento de Biología Costera, Instituto Antártico Argentino (IAA) - Dirección Nacional del Antártico (DNA), Buenos Aires, Argentina
e-mail: lquartino@dna.gov.ar

general low cover of the substrate even after several years, postulating predictable successional processes as a result of slow growth and high seasonality (Bowden 2005; Bowden et al. 2006). However, the occurrence of high interannual changes as a result of episodic events of massive colonization was also reported (Dayton 1989; Dayton et al. 2016). More recent studies revealed a rapid colonization by a few species after ice shelves collapse (Raes et al. 2010; Gutt et al. 2011; Fillinger et al. 2013) and the discovery of complex communities in newly ice-free areas (Lagger et al. 2017, 2018).

For Antarctic hard-bottom coasts dominated by macroalgae, the successional stages of the assemblages were frequently indirectly assessed, by relating it to the identity and relative abundance of the dominant taxa and to the disturbance regime of the site (Klöser et al. 1996; Quartino et al. 2005; Valdivia et al. 2014). For instance, the presence of the green seaweed *Monostroma hariotii* Gain and colonial diatoms in habitats severely affected by ice scouring was signalled as indicators of early stages of succession (Klöser et al. 1996). In the study carried out by Quartino et al. (2005), the cause of higher macroalgal diversity recorded at intermediate depth ranges (5–10 m) was attributed to the coexistence of patches in different successional stages. More recent studies performed in newly ice-free hard-bottom areas demonstrated the capability of macroalgae to colonize them rapidly (Quartino et al. 2013). The complexity of these communities, in terms of cover and richness, was positively correlated to the elapsed time since the generation of the space and the lower level of stress and disturbance imposed by a retreating glacier (Quartino et al. 2013; see also Chap. 8 by Quartino et al.).

In Potter Cove, a small fjord at the western Antarctic Peninsula (South Shetland Islands), quite a number of studies were focussed to unravel the successional patterns of benthic algal communities at the inter- and subtidal and over different time spans (months to years) (Fig. 12.1). This chapter summarizes these works and the advances in our knowledge on the successional process of primary producers in the hard-bottom Antarctic benthos. We review in situ primary succession studies that have been performed at this site and that have allowed for the description of the structure of benthic primary producer communities during succession. Early stages of succession are characterized through studies performed over the first 2–3 months of colonization in the rocky substrate in the intertidal (Zacher et al. 2007a, b) and the upper subtidal, at approximately 2 m depth (Campana et al. 2008a, b; Zacher and Campana 2008; Campana 2018) (Fig. 12.1). These studies included a multifactorial design to test the effects of ultraviolet radiation and grazing on the developing communities. Furthermore, experiments performed over a 4-year period with monthly surveys at the subtidal (3–5 m depth) allowed for the characterization of communities over a longer time span, at a site with no glacial influence, located at the outer part of the cove and at an inner cove site, in close proximity to a retreating glacier (Campana et al. 2018) (Fig. 12.1). We also discuss the importance of abiotic and biotic drivers on community succession, the importance of their interactions and the experimental approaches applied so far, with an emphasis on the effects of global change-related phenomena.

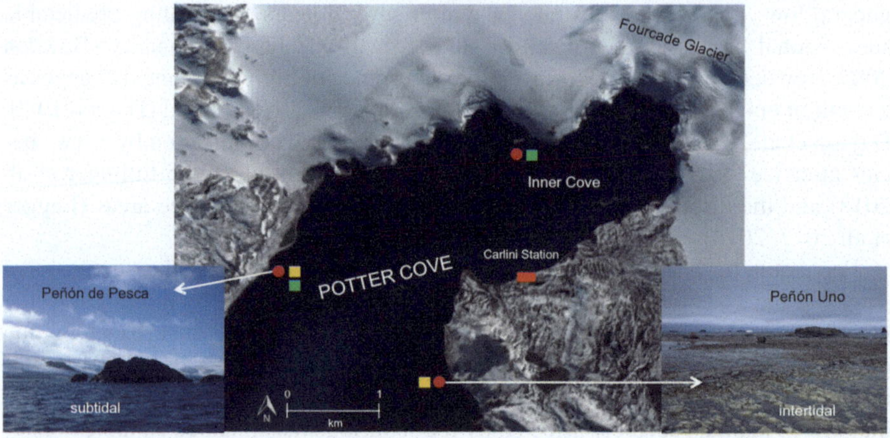

Fig. 12.1 Map of Potter Cove (South Shetland Islands, western Antarctic Peninsula), showing the location of successional experiments performed at the upper subtidal (at Peñón de Pesca, situated in the outer part of the cove and at a site located in the inner cove, in close proximity to a retreating glacier) and at an intertidal platform (Peñón Uno, located on Potter Peninsula). The effects of ultraviolet radiation and grazing on developing communities were studied at the subtidal of Peñón de Pesca and the intertidal of Peñón Uno (yellow square); long-term successional studies were performed at Peñón de Pesca and at the inner cove over 4 years (green square). (Satellite image of Potter Cove: Google Earth 2016). (Photos: left by Gabriela L. Campana; right by Katharina Zacher)

12.2 Structural Patterns and Changes in Algal Community Composition during Succession

Along the Antarctic hard-bottom coastlines studied in Potter Cove, the primary succession of benthic algae begins with a rapid colonization by diatoms, filamentous green algae and propagules of annual and pseudoperennial macroalgae (Zacher et al. 2007a, b; Campana et al. 2011; Campana 2018) (Fig. 12.2). This general taxonomic composition is similar to studies carried out at the Arctic for early stages of succession (Fricke et al. 2008, 2011).

As early colonizers in the marine benthos, diatoms may enhance or inhibit macroalgal recruitment and growth (Huang and Boney 1985; Noël et al. 2009). As an example, it was shown for the Arctic that diatoms apparently facilitate the establishment of macroalgal propagules (Fricke et al. 2008). In the Antarctic benthos, few investigations have focussed on the simultaneous study of microalgae and macroalgae in the successional process, particularly including detailed taxonomic analyses (Wahl et al. 2004; Zacher et al. 2007a, b; Zacher and Campana 2008; Campana et al. 2008a, b; Campana 2018). Furthermore, the ecology and taxonomy of Antarctic benthic diatoms are generally scarce when compared to macroalgae studies (Wulff et al. 2011).

Over the first months of colonization, diatom assemblages were mainly dominated by pennate, typically benthic diatom species (Zacher et al. 2007b; Campana

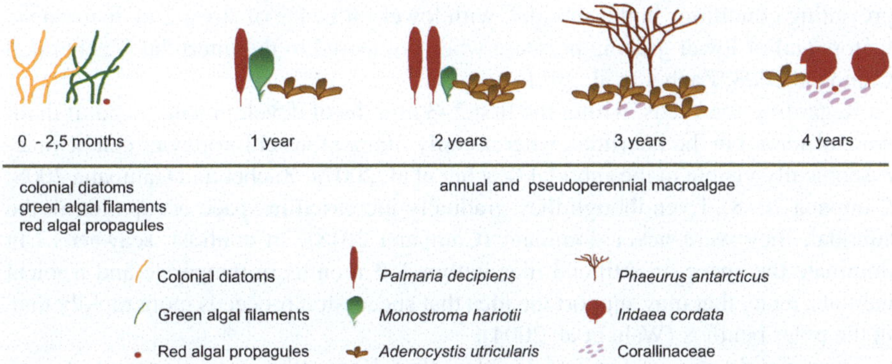

Fig. 12.2 The composition of the community of benthic primary producers over succession in an Antarctic upper subtidal rocky coast. The scheme compiles the information obtained by in situ studies focussed on the effects of ultraviolet radiation and grazing at early succession (0–2.5 months) and long-term successional studies carried out from 1 to 4 years (Campana et al. 2008a, b, 2018; Campana 2018). Both studies were performed at the same subtidal site (Peñón de Pesca, Fig. 12.1), at low depths (less than 3 m), and were started during summer (December and February, respectively). The succession of the bacterial community during biofilm development is not described (see Lee et al. 2016). The new substrate is initially colonized by pioneer groups that occupy rapidly the space; the colonization starts with dominance of diatoms, and a gradual higher space occupation by green algal filaments occurs. Propagules of red algae are established since the beginning of the colonization, *P. decipiens* being the more abundant. These young sporelings probably develop the sporophyte during the following spring, which coincides with the appearance of individuals in communities of 1 year of colonization, during summer. At longer time scales, annual and pseudoperennial macroalgae dominate the substrate. *Monostroma hariotii* and *P. decipiens* are identified in these stages, after 1 and 2 years of colonization, whereas *P. antarcticus* and *I. cordata* were more abundant after 3 and 4 years, respectively. The dominance of *A. utricularis* was evident since the first year, and a higher space occupation by red crustose algae occurs over time. The 4-year-old communities were similar to the surrounding communities at low depths (less than 5 m), with absence of large thalli of Desmarestiales that usually form dense forests at deeper depths

et al. 2008b; Campana 2018). The small naviculoid *Navicula perminuta* Grunow was conspicuous in the intertidal and the subtidal assemblage (Zacher et al. 2007b; Campana 2018). The assemblages further included colonial forms, some of which reached macroscopic dimensions in the subtidal, such as *Fragilaria striatula* Lyngbye and *Fragilaria islandica* Grunow var. *adeliae* Manguin; pedunculated forms, such as *Pseudogomphonema kamtschaticum* (Grunow) Medlin and the species of the genus *Licmophora*; and adnate forms, which have a firm adhesion to the substrate through most of the valvar surface, such as the species of the genus *Cocconeis* (Zacher et al. 2007b; Campana et al. 2008b; Campana 2018).

These benthic diatoms are probably more relevant to the early stages of succession and were shown to contribute less to the differences between years over longer time scales, when macroalgae dominate the substrate (Campana et al. 2018) (Fig. 12.2). During these early stages, diatom abundance can differ drastically when comparing intertidal and subtidal assemblages, being much higher in the subtidal (Zacher and Campana 2008). This pattern could be related to the more stable prevailing conditions in the subtidal, with lower influence of stress and disturbance factors and/or lower grazing pressure when compared to the intertidal (Zacher and Campana 2008; Zacher et al. 2011).

Regarding seaweeds, within the first 2–3 months of development, no adult thalli were observed in the intertidal, whereas only filamentous chlorophytes can be macroscopically visible in the subtidal (Zacher et al. 2007a; Zacher and Campana 2008; Campana 2018). Even though they gradually increased in space occupation in the subtidal, they were never dominant (Campana 2018). In contrast, seaweeds can dominate the space on artificial tiles within 2–3 months in temperate and tropical habitats, a fact that may support the idea that succession proceeds more rapidly than in the polar benthos (Wahl et al. 2004).

During early stages of succession (less than 10 months of development), young algal recruits of *M. hariotii* and *Palmaria decipiens* (Reinsch) Ricker were registered in intertidal (Zacher et al. 2007a) and subtidal assemblages (Campana et al. 2011; Debandi et al. 2015) (Fig. 12.2). Both species were the only macroalgae inhabiting a newly ice-free area severely affected by a retreating glacier, showing their ability not only to colonize newly open space but also to grow in areas subjected to high sedimentation, low light penetration and high ice disturbance (Quartino et al. 2013). *Monostroma hariotii* is an annual species that can be very abundant in intertidal habitats, which are subject to high stress and disturbance intensity (Wiencke and Clayton 2002; Kim 2001; Marcías et al. 2017).

It has also been proposed as a pioneer species in the subtidal, as it has been recorded in sites repeatedly affected by ice scouring (Klöser et al. 1996). Similarly, *P. decipiens* can cope with marked changes in salinity and light availability and can grow rapidly in intertidal sites subjected to ice scouring (Becker et al. 2011; Deregibus et al. 2016; Marcías et al. 2017). For these reasons, it was signalled as having a high competitive capacity and considered to be an opportunistic species in the colonization process (Becker et al. 2011). The recruits of this species probably develop the sporophyte during the next spring, which coincides with their appearance in the communities after 1 year of colonization (Campana et al. 2018) (Fig. 12.2).

The brown seaweed *Adenocystis utricularis* (Bory) Skottsberg is another species that can also be present at early stages of succession, as small recruits were recorded after only 1 month of colonization in the intertidal (Zacher et al. 2007a) and 10 months in the subtidal (Debandi et al. 2015). However, it was dominant over the whole experimental exposure at a four-year successional study performed at the upper subtidal, at 3 m depth (Campana et al. 2018) (Fig. 12.2). *Adenocystis utricularis* is an aseasonal annual species in Antarctic intertidal habitats that can be very abundant (de Reviers and Délépine 1981; Wiencke and Clayton 2002). It has a crustose phase that can survive winter, and it is biannual in deeper subtidal habitats (Wiencke 1990). The studies performed so far revealed that this species tends to monopolize the substrate and, thus, may impede the recruitment—or resist the invasion—by other species (Sousa 1979).

The brown seaweed *Phaeurus antarcticus* Skottsberg and the red *Iridaea cordata* (Turner) Bory de Saint-Vincent were found to reach their maximum cover after 3 and 4 years of colonization, respectively (Fig. 12.2). A dense canopy of *P. antarcticus* was found after 3 years of colonization, when the highest macroalgal cover and space occupation were attained (Campana et al. 2018). The branched thalli of this species generated several layers and a general more complex three-dimensional structure of the communities (Campana et al. 2018). Between the third and the fourth year, competitive interactions may result in the liberation of space, and the red alga *I. cordata* reached its maximum cover, during the following summer. These four-year-old communities were also composed of the crustose phase of *A. utricularis* and red coralline algae, which showed a higher space occupation over time (Campana et al. 2018). The same study has also shown that coralline algae have the potential to establish and grow on the crusts of *A. utricularis* (Campana 2018; Campana et al. 2018).

A great space occupation by crustose red algae can be a characteristic trait of mature communities at the same study site, at these low depths (Klöser et al. 1994). Besides, they can be indicators of low disturbance in the Antarctic subtidal (Barnes et al. 1996), where they are frequently found under the canopies of Desmarestiales (Irving et al. 2005; Clark et al. 2011). Long-term studies could reveal the role of these algae in successional processes (Barner et al. 2016) as they have been proposed to favour the recruitment of late colonizers by providing structurally simple habitats, with higher rugosity than the natural substrate (Maggi et al. 2011).

It has been suggested that communities in the upper subtidal of polar regions are permanently in early successional stages, with low species diversity (Witman and Dayton 2001). In the long-term experiment carried out at Potter Cove, the experimental tiles were colonized by fast-growing macroalgae, with absence of large thalli of perennial species, and after 4 years of colonization, these communities resembled the surrounding community at low depths (Quartino et al. 2005). Even though recruits of perennial species were found in the colonization tiles, they did not grow until maturity and reached only a few centimetres (Campana et al. 2018). Repeated ice abrasion, interspecific competition with fast-growing species and/or high irradiance conditions in spring if no canopies are present were signalled as possible drivers of these communities that deflect the successional trajectory and maintain the communities in early stages of succession at these low depths (less than 5 m) (Campana et al. 2018).

In macroalgal dominated Antarctic subtidal sites, climax communities can be identified by the abundance of large Desmarestiales that form extensive and persistent patches that seem to be stable over time (Campana et al. 2011). Among these species, some representatives can be the season anticipators *Desmarestia menziesii* J. Agardh, *D. anceps* Montagne and *Himantothallus grandifolius* (A. Gepp and E.S. Gepp) Zinova, as they are terete to leathery, perennial macroalgae with high biomass and a complex three-dimensional structure and are habitat providers and chemically defended (Wiencke and Clayton 2002; Amsler et al. 2005; Quartino et al. 2005, 2008; Huang et al. 2006, 2007; see also Chap. 11 by Gomez and Huovinen for a description of form and function in Antarctic seaweeds). These species gather many of the attributes of species recorded in mature communities in other environments (sensu Littler and Littler 1980; Noël et al. 2009; Wiencke and

Amsler 2012; Gómez et al. 2019). They can also be considered as foundation species (sensu Dayton 1972) that provide structurally complex habitats favouring the presence of smaller organisms that live associated to them, such as amphipods and gastropods (Huang et al. 2007; Amsler et al. 2015). Whereas certain macroalgal species are found as epiphytes on these species, it is noteworthy the low abundance of epiphytic algae in this Antarctic subtidal forests (Wiencke and Clayton 2002; Peters 2003). In contrast, there is a high occurrence of filamentous endophytes (Peters 2003) that were shown to be highly palatable to sympatric amphipods (Amsler et al. 2009; see also Chap. 17 by Amsler et al.).

12.3 Ecological Factors Influencing Antarctic Algal Succession

12.3.1 Ultraviolet Radiation

Ultraviolet-B (UV-B) is an environmental stress factor that can limit the development of benthic primary producers causing inhibition of photosynthesis and damage to biomolecules such as DNA, proteins and lipids, among other damaging effects (Villafañe et al. 2003; Bischof et al. 2006; Häder et al. 2011; Karsten et al. 2011). In fact, it has been well established that Antarctic seaweed assemblages show species-specific sensitivity towards UV, which determine major aspects of the ecology of these organisms (Bischof et al. 1998, 2006; Wiencke et al. 2007). Thus, colonization, establishment and further development of the benthic algae are affected by UV radiation. As an example, green algae that inhabit the intertidal have a rapid and high acclimation potential to UV, whereas some red algae that are found under the canopy provided by other species are very UV-sensitive (Bischof et al. 2006). Besides, certain species can show high phenotypic plasticity revealing a correspondence between UV-B tolerance to their vertical distribution or growth sites (Rautenberger et al. 2013; see Chap. 7 by Huovinen and Gómez).

The experiments carried out in the intertidal and the subtidal of Potter Cove revealed that UV is an important structuring driver of the benthic primary producers community in Antarctica at early stages of succession (Zacher et al. 2007a, b; Campana et al. 2008a, b; Zacher and Campana 2008; Campana 2018). Ultraviolet radiation was shown to affect algal groups differently: whereas diatoms were mostly unaffected, the establishment and/or growth of green and red algal germlings was limited by ambient UV radiation (Zacher et al. 2007a, b; Campana et al. 2008b; Campana 2018). Antarctic benthic diatoms from soft-bottom habitats have shown a high resistance to UV, with low levels of photoinhibition and efficient repair mechanisms (Wulff et al. 2008). In these habitats, vertical migration can be a mechanism of UV avoidance (Karsten et al. 2011). However, as revealed by experiments performed in rocky coasts, a high resistance of these algae was also shown where this mechanism of avoidance is not possible (Zacher et al. 2007b; Campana et al. 2008b).

Particularly for the subtidal, this group dominates the substrate at early stages of succession, and when grazers were excluded (Fig. 12.3a), they provided a UV-resistant canopy that could positively influence macroalgal recruitment and/or growth (Molis and Wahl 2004).

Early life stages of macroalgae in polar environments can be more vulnerable to UV radiation than their adult thalli (Bischof et al. 2006; Wiencke et al. 2007; Roleda et al. 2009; Karsten et al. 2011). For instance, filaments of Antarctic specimens of *Urospora penicilliformis* (Roth) Areschoug have high light requirements, but their spores are more UV-B sensitive than adult stages (Roleda et al. 2009), being their settlement possibly limited by UV-B radiation (Campana 2018). Besides, the negative UV-B effects on the photosynthesis and DNA of propagules can be correlated to the depth collection of the adult thalli (Wiencke et al. 2000). In Antarctica, propagules of macroalgae dwelling in the intertidal have a high resistance to UV, with high recovery capacity and scarce or even null DNA damage (Zacher et al. 2007c). In contrast, the germination of spores of species that inhabit the subtidal can be severely affected by ambient UV, an effect that can be potentiated by UV-B in laboratory conditions (Zacher 2014). Besides, UV was also shown to cause higher photosynthesis inhibition in spores of subtidal species compared to intertidal ones (Navarro et al. 2016; see also Chap. 10 by Navarro et al.).

The available information suggest that at the beginning of the successional process, during the colonization of the substrate, the differential sensitivity to UV radiation among different groups of benthic algae (viz. diatoms and macroalgae), among different seaweed species and/or different developmental stages of a species, can shape the structure and functioning of the communities (Lotze et al. 2002; Villafañe et al. 2003; Zacher et al. 2007a, b; Campana et al. 2008a; Zacher 2014), which seem to be more UV-resistant over time (Lotze et al. 2002; Wahl et al. 2004; Molis and Wahl 2004, 2009).

12.3.2 Grazing

According to the available information, grazing effects in the Antarctic benthos are expected to be intense in early stages of algal succession, when communities are composed of more vulnerable life forms such as diatoms and certain early life stages of macroalgae (Brêthes et al. 1994; Kim 2001; Zacher et al. 2007a, b), a pattern that has been recorded in other environments as well (Sousa 1979; Farrell 1991; Lubchenco 1983; Sousa and Connell 1992). For instance, the conspicuous limpet *Nacella concinna* Strebel exerts a high influence in structuring the assemblages of primary producers in the intertidal, particularly at low shore levels (Kim 2001; Segovia-Rivera and Valdivia 2016; Valdivia et al. 2019; see also Chap. 13 by Valdivia). On their upwards migration from the subtidal to the intertidal, *N. concinna* feeds on microphytobenthos and green filamentous algae at low shore levels (Brêthes et al. 1994; Kim 2001), more severely affecting the early life stages of macroalgae compared to adults (Kim 2001). In fact, grazing was identified as a

Fig. 12.3 (**a**) Experimental unit used to determine the effects of UV and grazing on developing algal communities; the same experiment was performed at the subtidal of Peñón de Pesca and the intertidal site Peñón Uno. Note tiles (100 cm^2) obtained from a (i) caged (grazer-excluded treatment) and (ii) an uncaged treatment, at the subtidal. (**b**) Underwater view of a set-up used to allow colonization over 10 months; tiles (25 cm^2) were subsequently transported to the laboratory to perform further experiments. Note a detail of the obtained communities, composed of young germlings of the red seaweed *P. decipiens* of different age (i and ii), brown *A. utricularis* (iii) and green *M. hariotii* (iv). (**c**) Underwater view of a long-term successional study performed in Peñón de Pesca and at an inner cove site exposed to the influence of a retreating glacier (Fig. 12.1); note bigger-sized colonization tiles (500 cm^2) applied to perform the study over 4 years. (Photos: (**a**) left panel by Katharina Zacher, (**b**) left panel by Argentine Army diving crew, (**c**) left panel by Marcelo Mammana; (a) and (c) right panel by Gabriela L. Campana, (b) right panel by Gabriela L. Campana and Juan I. Debandi)

Fig. 12.3 (continued)

strong driver of the structure of developing algal communities in the studies performed in Potter Cove. Biomass reduction was mainly caused by the gastropods *N. concinna* and *Laevilacunaria antarctica* Martens (Zacher et al. 2007a, b; Campana et al. 2008a, b; Campana 2018). At both depth ranges (intertidal and subtidal), grazers significantly consumed diatoms, and particularly for the subtidal site, feeding on colonial diatoms exerted a drastic change in the physiognomy of the assemblages (Campana 2018) (Fig. 12.3a). This is a trait that was also observed for freshwater (Steinman 1996) and other marine habitats (Nicotri 1977; Sommer 1999a, b; Hillebrand et al. 2000), where the consumption of canopy algae results in an increase in the relative abundance of smaller taxa and prostrate forms (Nicotri 1977; Hillebrand et al. 2000).

A differential susceptibility towards grazing was detected among different macroalgal species, which resulted in changes in species composition (Zacher et al. 2007a; Campana 2018). Among green algae, *Urospora penicilliformis* (Roth) Areschoug was not affected by grazing, but early stages of *M. hariotii* could be severely affected (Zacher et al. 2007a, Campana 2018). Besides, the reduction of the abundance of the dominant taxa (diatoms) was shown to be beneficial for the establishment of certain red algal species whose early stages are firmly attached to the substrate, such as *P. decipiens* (Zacher et al. 2007a; Campana 2018). The diminishing space pre-emption and/or the reduction of shading may result in a higher abundance of red algae, causing increased evenness and, thus, diversity of the community (Campana et al. 2011; Campana 2018). A similar effect was observed at the intertidal site at Potter Cove where a higher macroalgal diversity was attributed to the increased spatial heterogeneity (Sommer 2000) caused by the simultaneous presence of untouched biofilms and areas visibly consumed by gastropods (Zacher et al. 2007a).

In contrast to vulnerable early successional stages, chemically defended macroalgae can be characteristic at later successional stages. Dominant brown macroalgae (except for *Desmarestia antarctica* R.L. Moe and P.C. Silva) are chemically defended against the three most common sympatric consumers: the sea star *Odontaster validus* Koehler, the fish *Notothenia coriiceps* J. Richardson and the amphipod *Gondogeneia antarctica* Chevreux (Amsler et al. 2005). Macroalgal palatability on the sea urchin *Sterechinus neumayeri* Meissner was not tested due to its unsuitability for feeding bioassays (Amsler et al. 2005), but studies in McMurdo revealed that both dominant macroalgae in that area are chemically protected against this consumer (Amsler et al. 1998, 1999). These authors propose that macroalgae are commonly unpalatable to sympatric consumers mainly as a result of chemical defences, as nor physical properties such as the toughness of the thallus or the nutritional content appeared to be related to the algal palatability (Amsler et al. 2005; Peters et al. 2005; see also Chap. 17 by Amsler et al.). In this sense, several works have stressed that there is a high macroalgal contribution to the food webs through the detrital pathway (Fischer and Wiencke 1992; Amsler et al. 2005; Seefeldt et al. 2017; Braeckman et al. 2019). However, amphipod grazing can probably be considered a biological force that also shapes more mature seaweed communities (see Huang et al. 2006; Amsler et al. 2009; Aumack et al. 2011; Bucolo et al. 2011). Indeed, amphipod grazing is presumably responsible for the exclusion of subtidal filamentous algae in the western Antarctic Peninsula (Peters 2003), and it is hypothesized that they live in a mutualistic relationship with macroalgae, cleaning potentially harmful epiphytes in a chemically defended habitat (Amsler et al. 2014; see Chap. 13 by Valdivia and Chap. 17 by Amsler et al.).

It is important to point out that grazing effects are not unidirectional and interactions with other biotic and abiotic stressors are expected to occur (e.g. Bothwell et al. 1994). For instance, grazing by a limpet had a positive effect on Arctic macroalgae germlings under intermediate levels of sedimentation (Zacher et al. 2016a). In the subtidal experiment carried out at Potter Cove, some of the grazing effects were more intense in UV or UV-B shielded communities: a direct effect on algae causing lower palatability or a negative effect on grazers that reduce their activity or density when this radiation is present was postulated as explanations for the observed tendencies (Campana et al. 2008a).

12.3.3 Glacier Retreat

Climate change has already shown to have a strong influence on Antarctic benthic communities (Smale and Barnes 2008; Pasotti et al. 2015; Sahade et al. 2015; Moon et al. 2015). Antarctic macroalgae are cold-water-adapted organisms, and temperature stress may limit their development (Wiencke et al. 2007). Furthermore, the rapid glacier retreat observed over the western Antarctic Peninsula has opened newly ice-free areas where reduced light penetration caused by increased sedimentation are the prevailing conditions (Rückamp et al. 2011; Quartino et al. 2013; Deregibus et al. 2016). Primary succession patterns may be affected by high sedimentation rates, which may reduce macroalgal propagules survival or even prevent the spore settlement on the rocky substrate (Airoldi 2003; Zacher et al. 2016a). In fact, in Potter Cove, the cover and diversity of the assemblages of macroalgae colonizing newly ice-free areas were inversely correlated to the level of stress and disturbance imposed by the retreating glacier, being the lowest in sites close to the glacier with high sedimentation rates, lowest light penetration and high ice disturbance (Quartino et al. 2013; see Chap. 8 by Quartino et al. and Chap. 9 by Deregibus et al.).

Long-term successional patterns were also evaluated by performing a colonization experiment in close proximity to this retreating glacier in Potter Cove (Campana et al. 2018). On one hand, this study showed convergent patterns to the observed for the site not affected by glacier influence (Campana et al. 2018). The assemblages were also dominated by algae, in particular a few opportunistic species, with *A. utricularis* reaching a similar cover—approximately of 70%—after 2 years. Besides, there were significant interannual changes in the assemblages and an increase in cover over time until the third year, followed by a significant decline between the third and fourth year. As both experiments were carried out with a year of difference in their starting points, these convergent patterns can point to a predictable successional process for subtidal macroalgal communities in the Antarctic benthos (Campana et al. 2018).

On the other hand, the communities located close to the retreating glacier showed lower macroalgal richness and a decreased diversity trend over time, which was attributed to (i) a lower spore availability in more simple communities established in newly ice-free areas, (ii) a higher sedimentation causing direct abrasion or burial of propagules and reduced light penetration and/or (iii) a higher ice disturbance caused by ice block landslides from the glacier (Quartino et al. 2013; Deregibus et al. 2016; Campana et al. 2018). More recent colonization studies performed in newly ice-free areas with different glacial influence showed similar patterns, with an inverse relationship between the algal cover and diversity, and the degree of sedimentation (Deregibus 2017; see Chap. 9 by Deregibus et al.).

The combined effects of increased temperature, sedimentation and grazing on the early succession of benthic algae were also studied for subtidal communities developed over 10 months in Potter Cove (Debandi et al. 2015). These studies revealed that increased sedimentation might favour the growth of *P. decipiens* and *M. hariottii* and have neutral effects on brown algal early colonizers (Debandi 2019).

12.4 Experimental Approaches to Study In Situ Succession of Antarctic Benthic Algae

Different experimental approaches have been applied so far in order to study primary succession at the polar benthos (Barnes and Conlan 2007; Campana et al. 2011; Dayton et al. 2016). In most of the cases, artificial substrates were used to allow the colonization by benthic algae (Zacher et al. 2007a, b; Zacher and Campana

2008; Campana et al. 2008a, b; Campana 2018) (Fig. 12.3). The use of settlement tiles was shown to be a useful tool to tackle these studies, particularly for polar regions, where relatively easy installation and retrieval are very much needed (Stanwell-Smith and Barnes 1997; Campana et al. 2011). This experimental approach allows uniform settlement conditions and standardized replicates so that several-factor designs and the possibility of deploying the same experiment simultaneously at different sites can be achieved (e.g. Wahl et al. 2004; Zacher and Campana 2008).

In order to assess the effects of abiotic and biotic drivers of communities at the very early stages of succession in Antarctica, artificial substrates of relatively small dimensions (25 to 100 cm^2) were shown to be adequate (e.g. Zacher et al. 2007a, b; Campana et al. 2008a, b; Debandi et al. 2015) (Fig 12.3a), as they were in other polar regions (Fricke et al. 2008, 2011). Besides the mentioned advantages of their use, these tiles can be transferred to the laboratory, where detailed analyses of diatoms and early stages of seaweeds, as well as quantifications of biomass, cell densities and percentage of cover of algae on known areas, can be assessed (Foster and Sousa 1985; Stanwell-Smith and Barnes 1997; Campana et al. 2011). Moreover, these assemblages can be applied to physiological studies and further experimentation (Fig. 12.3b).

When studying successional patterns over a longer time scale (> a year), bigger tiles, directly fixed to rocky substrate, were chosen (500 cm^2) (Campana et al. 2018) (Fig. 12.3c). These bigger dimensions still permitted an easy manipulation and replication of a somehow simpler design, and importantly, they allowed an adequate assessment of community structure as organisms grew. For instance, the dense canopy formed by *P. antarcticus* included organisms reaching a maximum length of 1.2 m (Campana et al. 2018). In this experiment, tiles were monitored by two methods, as photographic samplings were performed on a monthly basis and detailed laboratory analyses were done in spring and summer. Both methods revealed the same patterns of succession and gave complementary information (Campana 2018). Laboratory analyses applied the "point quadrat" method, which resulted to be the most adequate to determine ecological indexes and community cover particularly at later stages of succession, when several layers occur (Foster et al. 1991). This sampling method gave information about the canopy and the understory algae and, thus, about the three-dimensional characteristics of the analysed communities. On the other hand, photographic samplings allowed for the detection of faster seasonal changes that occurred over winter and autumn, achieving a higher time resolution of the sampling.

12.5 Concluding Remarks and Perspectives

Even though preforming in situ experimental studies in the marine Antarctic benthos can involve logistic constraints and difficulties, they are fundamental to elucidate mechanisms and to evaluate the influence of specific factors on biological processes such as succession (Benedetti-Cecchi 2000; Meiners et al. 2015; Barner et al. 2016). The studies performed so far reveal that continuous monitoring of environmental conditions is essential to better explain the successional patterns of algae in the Antarctic benthos. Light availability—including UV radiation—can be an important driver of successional changes and is modified by season, ice cover, phytoplankton blooms and sediment input during the warmer seasons (Campana et al. 2011; Gómez et al. 2011). Furthermore, biological drivers such as grazers should also be monitored; for instance, the increasing grazing pressure by *N. concinna* at the end of winter can be signalled as an important factor in the control of the dynamics of these communities in the rocky bottoms (Brêthes et al. 1994; Kim 2001). Non-selective grazing by this gastropod could be seasonal, exerting a higher pressure during its migration from the deep subtidal to the intertidal at the end of winter and early spring (Brêthes et al. 1994; Kim 2001; Zacher et al. 2007b; Campana 2018).

Early stages of succession were shown to be particularly vulnerable to UV radiation and grazing and are probably controlled by complex interactions among abiotic factors and biological interactions. Considering that UV radiation is an environmental stress factor for Antarctic ecosystems, it is important to explore the effects of these wavelengths on algae-grazers interactions, such as the effects of UV on algal palatability (e.g. Pavia et al. 1997; Macaya et al. 2005; Fairhead et al. 2006) and on the physiology and behaviour of consumers (Sommaruga 2003; Obermüller et al. 2007). Furthermore, the interactions with climate change processes such as increased temperature and acidification should also be considered (Rautenberger et al. 2015; Flores-Molina et al. 2016; Häder 2018).

Antarctica, and particularly the Antarctic Peninsula, is one of the regions most seriously affected by climate change (Turner et al. 2009), where a number of associated phenomena can exert a strong influence on the structure and functioning of benthic communities (Quartino et al. 2013; Sahade et al. 2015; Moon et al. 2015; Torre et al. 2017; Häder 2018; see Chap. 8 by Quartino et al.). Some of these phenomena are bound to initiate colonization processes and affect the following stages of succession. On one hand, glacier retreat on the western Antarctic Peninsula has originated newly ice-free areas that were colonized by macroalgae (Quartino et al. 2013) and invertebrates (Sahade et al. 2015; Lagger et al. 2017, 2018). Furthermore, recent blooms of benthic diatoms have been registered on the soft substrate at sites exposed to glacier melting (Ahn et al. 2016), pointing to a substantial change in benthic communities. On the other hand, ice abrasion is postulated to increase due to a lower formation of fast ice that allows higher movement of existing icebergs and by the existence of new ice blocks and icebergs originated from retreating glaciers (Barnes 2017; Deregibus et al. 2017).

Besides, the temporal and spatial reduction in fast ice formation may result in higher light availability in the water column during winter and early spring, leading to a higher primary production (Johnston et al. 2007; McClintock et al. 2008; Clark et al. 2013; Deregibus et al. 2016) and favouring the colonization of deeper areas (Miller and Pearse 1991; Gómez and Huovinen 2015). However, this will imply a higher penetration and time exposure to damaging UV radiation (Gómez and Huovinen 2015; see also Chap. 7 by Huovinen and Gómez). Besides, increased sedimentation during warm months may counteract negative UV effects but simultaneously reduce photosynthetically active radiation (PAR, 400–700 nm) availability, leading to changes in community structure and affecting the carbon balance of macroalgae (Quartino et al. 2013; Deregibus et al. 2016; see Chap. 9 by Deregibus et al.). As already mentioned, higher sedimentation can also affect the survival or interfere with the establishment of macroalgal propagules (Zacher et al. 2016a), benefitting the settlement of early space occupiers such as *P. decipiens* and *M. hariotii* (Quartino et al. 2013)

Overall, in this Antarctic environment seriously affected by global change phenomena, it is important to assess the patterns of community development during succession and how they can be modified due to a differential sensitivity of algae to abiotic changes (i.e. a higher space availability but, simultaneously, higher temperatures, changes in PAR availability, higher ice disturbance, higher sedimentation, acidification) that may lead to changes in biological interactions, such as grazing and competition (Schoenrock et al. 2015; Navarro et al. 2016; Zacher et al. 2016a, b; Schram et al. 2017). Besides, studies using natural substrata, at greater depths, and larger temporal and spatial scales are necessary to reveal the underlying mechanisms and interactions (Campana et al. 2011). The gathered information can contribute to achieve a deeper knowledge of these communities, whose structure and functioning are key to the changing Antarctic coastal systems.

Acknowledgements A significant part of the work reviewed here was performed within the framework of the scientific collaboration between Instituto Antártico Argentino (IAA)/Dirección Nacional del Antártico (DNA) and the Alfred Wegener Institute Helmholtz Centre for Polar and Marine Research (AWI). We thank the support of Carlini Station and Dallmann Laboratory crews over the several expeditions involved. Field assistance of Oscar González and Alejandro Ulrich was always invaluable, as was the work of summer and overwinter expedition scientists and of the Argentine Army and German diving crews. We acknowledge the financial support by grants from CONICET, DNA-IAA, ANPCyT-DNA (PICTO 0116/2012–2015, PICT 2017-2691), AWI, Deutsche Forschungsgemeinschaft (DFG, grant Za735/1-1) and MINCyT-BMBF Program (AL/17/06-01DN18024). This chapter also presents an outcome of the International Research Network IMCONet funded by the Marie Curie Action IRSES IMCONet (FP7 IRSES, Action No. 318718) and CoastCarb (Funding ID 872690, H2020-MSCA-RISE-2019-Research and Innovation Staff Exchange).

References

Ahn I-Y, Moon H-W, Jeon M, Kang S-H (2016) First record of massive blooming of benthic diatoms and their association with megabenthic filter feeders on the shallow seafloor of an Antarctic fjord: does glacier melting fuel the bloom? Ocean Sci J 51:273–279. https://doi.org/10.1007/s12601-016-0023-y

Airoldi L (2003) The effects of sedimentation on rocky coast assemblages. In: Gibson RN, Atkinson RJA (eds) Oceanography and marine biology, an annual review, vol 41. Taylor & Francis, Boca Raton, pp 161–236

Amsler CD, McClintock JB, Baker BJ (1998) Chemical defense against herbivory in the Antarctic marine macroalgae *Iridaea cordata* and *Phyllophora antarctica* (Rhodophyceae). J Phycol 34:53–59. https://doi.org/10.1046/j.1529-8817.1998.340053.x

Amsler CD, McClintock JB, Baker BJ (1999) An Antarctic feeding triangle: defensive interactions between macroalgae, sea urchins, and sea anemones. Mar Ecol Prog Ser 183:105–114. https://doi.org/10.3354/meps183105

Amsler CD, Iken K, McClintock JB, Amsler MO, Peters KJ, Hubbard JM, Furrow FB, Baker BJ (2005) Comprehensive evaluation of the palatability and chemical defenses of subtidal macroalgae from the Antarctic Peninsula. Mar Ecol Prog Ser 294:141–159. https://doi.org/10.3354/meps294141

Amsler CD, Amsler MO, McClintock JB, Baker BJ (2009) Filamentous algal endophytes in macrophytic Antarctic algae: prevalence in hosts and palatability to mesoherbivores. Phycologia 48:324–334. https://doi.org/10.2216/08-79.1

Amsler CD, McClintock JB, Baker BJ (2014) Chemical mediation of mutualistic interactions between macroalgae and mesograzers structure unique coastal communities along the Western Antarctic Peninsula. J Phycol 50:1–10. https://doi.org/10.1111/jpy.12137

Amsler MO, Huang YM, Engl W, McClintock MJ, Amsler CD (2015) Abundance and diversity of gastropods associated with dominant subtidal macroalgae from the western Antarctic Peninsula. Polar Biol 38:1171–1181. https://doi.org/10.1007/s00300-015-1681-4

Aumack CF, Amsler CD, McClintock JB, Baker BJ (2011) Impacts of mesograzers on epiphyte and endophyte growth associated with chemically defended macroalgae from the Western Antarctic Peninsula: a mesocosm experiment. J Phycol 47:36–41. https://doi.org/10.1111/j.1529-8817.2010.00927.x

Barner AK, Hacker SD, Menge BA, Nielse KJ (2016) The complex net effect of reciprocal interactions and recruitment facilitation maintains an intertidal kelp community. J Ecol 104:33–43. https://doi.org/10.1111/1365-2745.12495

Barnes DKA (2017) Iceberg killing fields limit huge potential for benthic blue carbon in Antarctic shallows. Glob Chang Biol 23:2649–2659. https://doi.org/10.1111/gcb.13523

Barnes DKA, Conlan KE (2007) Disturbance, colonization and development of Antarctic benthic communities. Philos Trans R Soc Lond B Biol Sci 362:11–38. https://doi.org/10.1098/rstb.2006.1951

Barnes DKA, Rothery P, Clarke A (1996) Colonisation and development in encrusting communities from the Antarctic intertidal and sublittoral. J Exp Mar Biol Ecol 196:251–265. https://doi.org/10.1016/0022-0981(95)00132-8

Becker SB, Quartino ML, Campana GL, Bucolo P, Wiencke C, Bischof K (2011) The biology of an Antarctic rhodophyte, *Palmaria decipiens*: recent advances. Antarct Sci 23:419–430. https://doi.org/10.1017/S0954102011000575

Benedetti-Cecchi L (2000) Predicting direct and indirect interactions during succession in a mid-littoral rocky shore assemblage. Ecol Monogr 70:45–72. https://doi.org/10.2307/2657167

Benincà E, Ballantine B, Ellner SP, Huisman J (2015) Species fluctuations sustained by a cyclic succession at the edge of chaos. PNAS 112:6389–6394. https://doi.org/10.1073/pnas.1421968112

Bischof K, Hanelt D, Wiencke C (1998) UV-radiation can affect depth zonation of Antarctic macroalgae. Mar Biol 131:597–605. https://doi.org/10.1007/s002270050351

Bischof K, Gómez I, Molis M, Hanelt D, Karsten U, Lüder UH, Roleda MY, Zacher K, Wiencke C (2006) Ultraviolet radiation shapes seaweeds communities. Rev Environ Sci Biotechnol 5:141–166. https://doi.org/10.1007/s11157-006-0002-3

Bothwell ML, Sherbot DMJ, Pollock CM (1994) Ecosystem response to solar ultraviolet-B radiation: influence of trophic-level interactions. Science 265:97–100. https://doi.org/10.1126/science.265.5168.97

Bowden DA (2005) Seasonality of recruitment in Antarctic sessile marine benthos. Mar Ecol Prog Ser 297:101–118. https://doi.org/10.3354/meps297101

Bowden DA, Clarke A, Peck LS, Barnes DKA (2006) Antarctic sessile marine benthos: colonisation and growth on artificial substrata over three years. Mar Ecol Prog Ser 316:1–16. https://doi.org/10.3354/meps316001

Braeckman U, Pasotti F, Vázquez S, Zacher K, Hoffmann R, Elvert M, Marchant H, Buckner C, Quartino ML, Mac Cormack W, Soetaert K, Wenzhöfer F, Vanreusel A (2019) Degradation of macroalgal detritus in shallow coastal Antarctic sediments. Limnol Oceanogr 64:1423–1441. https://doi.org/10.1002/lno.11125

Brêthes JC, Ferreyra GA, de la Vega S (1994) Distribution, growth and reproduction of the limpet *Nacella* (Patinigera) *concinna* (Strebel 1908) in relation to potential food availability, in Esperanza bay (Antarctic Peninsula). Polar Biol 14:161–170. https://doi.org/10.1007/BF00240521

Bucolo P, Amsler CD, McClintock JB, Baker BJ (2011) Palatability of the Antarctic rhodophyte *Palmaria decipiens* (Reinsch) RW Ricker and its endo/epiphyte *Elachista antarctica* Skottsberg to sympatric amphipods. J Exp Mar Biol Ecol 396:202–206. https://doi.org/10.1016/j.jembe.2010.10.023

Campana GL (2018) Microalgas y macroalgas bentónicas en un sistema costero Antártico (Caleta Potter, Islas Shetland del Sur): Efectos de la radiación ultravioleta y el pastoreo sobre la colonización y sucesión. PhD Thesis, Universidad Nacional de La Plata, Argentina

Campana GL, Momo FR, Quartino ML, Ferreyra GA (2008a) Effects of UVR and grazing on biomass and primary production of subtidal benthic algae in Antarctica. Ber Polarforsch Meeresforsch 571:278–286

Campana GL, Quartino ML, Yousif A, Wulff A (2008b) Effects of UV radiation and grazing on the structure of a subtidal benthic diatom assemblage in Antarctica. Ber Polarforsch Meeresforsch 571:302–310

Campana GL, Zacher K, Fricke A, Molis M, Wulff A, Quartino ML, Wiencke C (2011) Drivers of colonization and succession in polar benthic macro- and microalgal communities. In: Wiencke C (ed) Biology of polar benthic algae. Walter de Gruyter GmbH & Co, Berlin, pp 399–320

Campana GL, Zacher K, Deregibus D, Momo FR, Wiencke C, Quartino ML (2018) Succession of Antarctic benthic algae (Potter Cove, South Shetland Islands): structural patterns and glacial impact over a four-year period. Polar Biol 41:377–396. https://doi.org/10.1007/s00300-017-2197-x

Clark GF, Stark JS, Perrett LA, Hill NA, Johnston EL (2011) Algal canopy as a proxy for the disturbance history of understorey communities in East Antarctica. Polar Biol 34:781–790. https://doi.org/10.1007/s00300-010-0931-8

Clark GF, Stark JS, Johnston EL, Runcie JW, Goldsworthy PM, Raymond B, Riddle MJ (2013) Light-driven tipping points in polar ecosystems. Glob Chang Biol 19:3749–3761. https://doi.org/10.1111/gcb.12337

Dayton PK (1972) Toward an understanding of community resilience and the potential effects of enrichments to the benthos at McMurdo Sound, Antarctica. In: Parker BC (ed) Proceedings of the colloquium on conservation problems. Allen Press, Lawrence, pp 81–96

Dayton PK (1989) Interdecadal variation in an Antarctic sponge and its predators from oceanographic climate shifts. Science 245:1484–1486. https://doi.org/10.1126/science.245.4925.1484

Dayton PK, Jarrell S, Kim S, Thrush S, Hammerstrom K, Slattery M, Parnell E (2016) Surprising episodic recruitment and growth of Antarctic sponges: implications for ecological resilience. J Exp Mar Biol Ecol 482:38–55. https://doi.org/10.1016/j.jembe.2016.05.001

de Reviers B, Délépine R (1981) Biological data on a southern Phaeophyceae *Adenocystis utricularis* (Bory) Skottsberg, possible material for alginate industry. In: Levring T (ed) Proceedings on the 10th International Seaweed Symposium. Walter de Gruyter, Berlin, pp 345–350

Debandi JI (2019) Efectos del calentamiento climático y el pastoreo sobre una comunidad de algas marinas bentónicas antárticas, en etapas tempranas de la sucesión. Thesis, Universidad Nacional de Luján, Argentina

Debandi JI, Zacher K, Campana GL, Deregibus D, Quartino ML (2015) Estudio experimental del efecto de la temperatura, la sedimentación y el pastoreo sobre comunidades de algas marinas bentónicas antárticas. In: Riccialdelli L (ed) Abstracts of the IX Jornadas Nacionales de Ciencias del Mar, Ushuaia, Tierra del Fuego. CADIC-CONICET-University of Tierra del Fuego, Tierra del Fuego, p 187. https://jornadasdelmar2015.files.wordpress.com/2015/11/libro-de-resc3bamenes-final.pdf

Deregibus D (2017) Efecto del retroceso glaciario inducido por el cambio climático sobre la comunidad de algas marinas bentónicas en nuevas áreas libres de hielo en un ecosistema costero antártico (Caleta Potter, I. 25 de Mayo, I. Shetland del Sur). PhD Thesis, Universidad de Buenos Aires, Argentina

Deregibus D, Quartino ML, Campana GL, Momo FR, Wiencke C, Zacher K (2016) Photosynthetic light requirements and vertical distribution of macroalgae in newly ice-free areas in potter cove, South Shetland Islands, Antarctica. Polar Biol 39:153–166. https://doi.org/10.1007/s00300-015-1679-y

Deregibus D, Quartino ML, Zacher K, Campana GL, Barnes DKA (2017) Understanding the link between sea ice, ice scour and Antarctic benthic biodiversity; the need for cross station and nation collaboration. Polar Rec 53:143–152. 10.1017/S0032247416000875

Fairhead V, Amsler CD, McClintock JB, Baker BJ (2006) Lack of defense or phlorotannin induction by UV radiation or mesograzers in *Desmarestia anceps* and *D. menziesii* (Phaeophyceae). J Phycol 42:1174–1183. https://doi.org/10.1111/j.1529-8817.2006.00283.x

Farrell TM (1991) Models and mechanisms of succession: an example from a rocky intertidal community. Ecol Monogr 61:95–113. https://doi.org/10.2307/1943001

Fillinger L, Janussen D, Lundälv T, Richter C (2013) Rapid glass sponge expansion after climate-induced Antarctic ice shelf collapse. Curr Biol 23:1330–1334. https://doi.org/10.1016/j.cub.2013.05.051

Fischer G, Wiencke C (1992) Stable carbon isotope composition, depth distribution and fate of macroalgae from the Antarctic peninsula region. Polar Biol 12:341–348. https://doi.org/10.1007/BF00243105

Flores-Molina MR, Rautenberger R, Muñoz P, Huovinen P, Gómez I (2016) Stress tolerance of the endemic Antarctic brown alga *Desmarestia anceps* to UV radiation and temperature is mediated by high concentrations of phlorotannins. Photochem Photobiol 92:455–466. https://doi.org/10.1111/php.12580

Foster MS, Sousa WP (1985) Succession. In: Littler MM, Littler DS (eds) Handbook of phycological methods – ecological field methods: macroalgae. Cambridge University Press, Cambridge, pp 269–290

Foster MS, Harrold C, Hardin DD (1991) Point vs. photo quadrat estimates of the cover of sessile marine organisms. J Exp Mar Biol Ecol 146:193–203. https://doi.org/10.1016/0022-0981(91)90025-R

Fricke A, Molis M, Wiencke C, Valdivia N, Chapman AS (2008) Natural succession of macroalgal-dominated epibenthic assemblages at different water depths and after transplantation from deep to shallow water on Spitsbergen. Polar Biol 31:1191–1203. https://doi.org/10.1007/s00300-008-0458-4

Fricke A, Molis M, Wiencke C, Valdivia N, Chapman AS (2011) Effects of UV radiation on the structure of Arctic macrobenthic communities. Polar Biol 34:995–1009. https://doi.org/10.1007/s00300-011-0959-4

Gómez I, Huovinen P (2015) Lack of physiological depth patterns in conspecifics of endemic Antarctic brown algae: a trade-off between UV stress tolerance and shade adaptation? PLoS One 10(8):e0134440. https://doi.org/10.1371/journal.pone.0134440

Gómez I, Wulff A, Dunton K, Karsten U, Roleda M, Wiencke C (2011) Light and temperature demands of benthic microalgae and seaweeds in polar regions. In: Wiencke C (ed) Biology of polar benthic algae. Walter de Gruyter GmbH & Co, Berlin, pp 195–220

Gómez I, Navarro NP, Huovinen P (2019) Bio-optical and physiological patterns in Antarctic seaweeds: a functional trait based approach to characterize vertical zonation. Prog Oceanogr 174:17–27. https://doi.org/10.1016/j.pocean.2018.03.013

Gutt J (2001) On the direct impact of ice on marine benthic communities, a review. Polar Biol 24:553–564. https://doi.org/10.1007/s003000100262

Gutt J, Barratt I, Domack E, d'Udekem d'Acoz C, Dimmler W et al (2011) Biodiversity change after climate-induced ice-shelf collapse in the Antarctic. Deep Sea Res Part II 58:74–83. https://doi.org/10.1016/j.dsr2.2010.05.024

Häder D-P (2018) Polar Macroalgae. In: Häder D-P, Gao K (eds) Aquatic ecosystems in a changing climate. CRC Press, Princeton, pp 253–267

Häder D-P, Helbling EW, Williamson CE, Worrest RC (2011) Effects of UV radiation on aquatic ecosystems and interactions with climate change. Photochem Photobiol Sci 10:242–260. https://doi.org/10.1039/c0pp90036b

Hillebrand H, Worm B, Lotze HK (2000) Marine microbenthic community structure regulated by nitrogen loading and grazing pressure. Mar Ecol Prog Ser 204:27–38. https://doi.org/10.3354/meps204027

Huang R, Boney AD (1985) Individual and combined interactions between littoral diatoms and sporelings of red algae. J Exp Mar Biol Ecol 85:101–111. https://doi.org/10.1016/0022-0981(85)90136-4

Huang Y, McClintock JB, Amsler CD, Peters KJ, Baker BJ (2006) Feeding rates of common Antarctic gammarid amphipods on ecologically important sympatric macroalgae. J Exp Mar Biol Ecol 329:55–65. https://doi.org/10.1016/j.jembe.2005.08.013

Huang Y, Amsler MO, McClintock JB, Amsler CD, Baker BJ (2007) Patterns of gammaridean amphipod abundance and species composition associated with dominant subtidal macroalgae from the western Antarctic Peninsula. Polar Biol 30:1417–1430. https://doi.org/10.1007/s00300-007-0303-1

Irving AD, Connell SD, Johnston EL, Pile AJ, Gillanders BM (2005) The response of encrusting coralline algae to canopy loss: an independent test of predictions on an Antarctic coast. Mar Biol 147:1075–1083. https://doi.org/10.1007/s00227-005-0007-4

Johnston EL, Connell SD, Irving AD, Pile AJ, Gillanders BM (2007) Antarctic patterns of shallow subtidal habitat and inhabitants in Wilke's land. Polar Biol 30:781–788. https://doi.org/10.1007/s00300-006-0237-z

Karsten U, Wulff A, Roleda MY, Müller R, Steinhoff FS, Fredersdorf J, Wiencke C (2011) Physiological responses of polar benthic algae to ultraviolet radiation. In: Wiencke C (ed) Biology of polar benthic algae. Walter de Gruyter GmbH & Co, Berlin, pp 195–220

Kim D (2001) Seasonality of marine algae and grazers of an Antarctic rocky intertidal, with emphasis on the role of the limpet *Nacella concinna* Strebel (Gastropoda: Patellidae). Ber Polarforsch Meeresforsch 397:1–120

Klöser H, Mercuri G, Laturnus F, Quartino ML, Wiencke C (1994) On the competitive balance of macroalgae at Potter Cove (King George Island, South Shetlands). Polar Biol 14:11–16. https://doi.org/10.1007/BF00240266

Klöser H, Quartino ML, Wiencke C (1996) Distribution of macroalgae and macroalgal communities in gradients of physical conditions in potter cove, King George Island, Antarctica. Hydrobiologia 333:1–17. https://doi.org/10.1007/BF00020959

Lagger C, Servetto N, Torre L, Sahade R (2017) Benthic colonization in newly ice-free soft-bottom areas in an Antarctic fjord. PLoS One 12(11):e0186756. https://doi.org/10.1371/journal.pone.0186756

Lagger C, Nime M, Torre L, Servetto N, Tatián M, Sahade R (2018) Climate change, glacier retreat and a new ice-free island offer new insights on Antarctic benthic responses. Ecography 40:1–12. https://doi.org/10.1111/ecog.03018

Lee YM, Cho KH, Hwang K, Kim EH, Kim M, Hong SG, Lee HK (2016) Succession of bacterial community structure during the early stage of biofilm development in the Antarctic marine environment. Korean J Microbiol 52:49–58. https://doi.org/10.7845/kjm.2016.6005

Littler MM, Littler DS (1980) The evolution of thallus form and survival strategies in benthic marine macroalgae: field and laboratory tests of a functional form model. Am Nat 116:25–44. https://doi.org/10.1086/283610

Lotze HK, Worm B, Molis M, Wahl M (2002) Effects of UV radiation and consumers on recruitment and succession of a marine macrobenthic community. Mar Ecol Prog Ser 243:57–66. https://doi.org/10.3354/meps243057

Lubchenco J (1983) *Littorina* and *Fucus*: effects of herbivores, substratum heterogeneity, and plant escapes during succession. Ecology 64:1116–1123. https://doi.org/10.2307/1937822

Macaya EC, Rothäusler E, Thiel M, Molis M, Wahl M (2005) Induction of defenses and within-alga variation of palatability in two brown algae from the northern-central coast of Chile: effects of mesograzers and UV radiation. J Exp Mar Biol Ecol 325:214–227

Maggi E, Bertocci I, Vaselli S, Benedetti-Cecchi L (2011) Connell and Slayter's models of succession in the biodiversity era. Ecology 92:1399–1406. https://doi.org/10.1890/10-1323.1

Marcías ML, Deregibus D, Saravia LA, Campana GL, Quartino ML (2017) Life between tides: spatial and temporal variations of an intertidal macroalgal community at Potter Peninsula, South Shetland Islands, Antarctica. Estuar Coast Shelf Sci 187:193–203. https://doi.org/10.1016/j.ecss.2016.12.023

McClintock JB, Ducklow H, Fraser W (2008) Ecological responses to climate change on the Antarctic Peninsula. Am Sci 96:302–310. https://doi.org/10.1511/2008.73.302

Meiners SJ, Cadotte MW, Fridley JD, Pickett STA, Walker LR (2015) Is successional research nearing its climax? New approaches for understanding dynamic communities. Funct Ecol 29:154–164. https://doi.org/10.1111/1365-2435.12391

Miller KA, Pearse JS (1991) Ecological studies of seaweeds in McMurdo Sound, Antarctica. Am Zool 31:35–48. https://doi.org/10.1093/icb/31.1.35

Molis M, Wahl M (2004) Transient effects of solar ultraviolet radiation on the diversity and structure of a field-grown epibenthic community at Lüderitz, Namibia. J Exp Mar Biol Ecol 302:51–62. https://doi.org/10.1016/j.jembe.2003.10.003

Molis M, Wahl M (2009) Comparison of the impacts of consumers, ambient UV, and future UVB irradiance on mid-latitudinal macroepibenthic assemblages. Glob Chang Biol 15:1833–1845. https://doi.org/10.1111/j.1365-2486.2008.01812.x

Moon H-W, Hussin WMRW, Kim H-C, Ahn I-Y (2015) The impacts of climate change on Antarctic nearshore mega-epifaunal benthic assemblages in a glacial fjord on King George Island: responses and implications. Ecol Indic 57:280–292. https://doi.org/10.1016/j.ecolind.2015.04.031

Navarro NP, Huovinen P, Gómez I (2016) Stress tolerance of Antarctic macroalgae in the early life stages. Rev Chil Hist Nat 89:5. https://doi.org/10.1186/s40693-016-0051-0

Nicotri ME (1977) Grazing effects of four marine intertidal herbivores on the microflora. Ecology 58:1020–1032. https://doi.org/10.2307/1936922

Noël LM-LJ, Griffin JN, Moschella PS, Jenkins SR, Thompson RC, Hawkins SJ (2009) Changes in diversity and ecosystem functioning during succession. In: Wahl M (ed) Marine hard bottom communities: patterns, dynamics, diversity, and change. Ecological studies, vol 206. Springer, Heidelberg, pp 213–223

Obermüller B, Puntarulo S, Abele D (2007) UV-tolerance and instantaneous physiological stress responses of two Antarctic amphipod species *Gondogeneia antarctica* and *Djerboa furcipes* during exposure to UV radiation. Mar Environ Res 64:267–285. https://doi.org/10.1016/j.marenvres.2007.02.001

Pasotti F, Manini E, Giovannelli D, Wölfl A-C, Monien D, Verleyen E, Braeckman U, Abele D, Vanreusel A (2015) Antarctic shallow water benthos in an area of recent rapid glacier retreat. Mar Ecol 36:716–733. https://doi.org/10.1111/maec.12179

Pavia H, Cervin G, Lindgren A, Åberg P (1997) Effects of UV-B radiation and simulated herbivory on phlorotannins in the brown alga *Ascophyllum nodosum*. Mar Ecol Prog Ser 157:139–146. https://doi.org/10.3354/meps157139

Peters AF (2003) Molecular identification, distribution and taxonomy of brown algal endophytes, with emphasis on species from Antarctica. In: Chapman ARO, Anderson RJ, Vreeland V, Davidson IR (eds) Proceedings on the 17th international seaweed symposium. Oxford University Press, New York, pp 293–302

Peters KJ, Amsler CD, Amsler MO, McClintock JB, Dunbar RB, Baker BJ (2005) A comparative analysis of the nutritional and elemental composition of macroalgae from the western Antarctic Peninsula. Phycologia 44:453–463. https://doi.org/10.2216/0031-8884(2005)44[453:ACAOTN]2.0.CO;2

Quartino ML, Zaixso H, Boraso de Zaixso A (2005) Biological and environmental characterization of marine macroalgal assemblages in Potter Cove, South Shetland Islands, Antarctica. Bot Mar 48:187–197. https://doi.org/10.1515/BOT.2005.029

Quartino ML, Boraso de Zaixso A, Momo FR (2008) Macroalgal production and the energy cycle of Potter Cove. Ber Polar Meeresforsch 571:68–74

Quartino ML, Deregibus D, Campana GL, Latorre GEJ, Momo FR (2013) Evidence of macroalgal colonization on newly ice-free areas following glacial retreat in Potter Cove (South Shetland Islands), Antarctica. PLoS One 8(3):e58223. https://doi.org/10.1371/journal.pone.0058223

Raes M, Rose A, Vanreusel A (2010) Response of nematode communities after large-scale ice-shelf collapse events in the Antarctic Larsen area. Glob Chang Biol 16:1618–1631. https://doi.org/10.1111/j.1365-2486.2009.02137.x

Rautenberger R, Wiencke C, Bischof K (2013) Acclimation to UV radiation and antioxidative defence in the endemic Antarctic brown macroalga *Desmarestia anceps* along a depth gradient. Polar Biol 36:1779–1789. https://doi.org/10.1007/s00300-013-1397-2

Rautenberger R, Huovinen P, Gómez I (2015) Effects of increased seawater temperature on UV tolerance of Antarctic marine macroalgae. Mar Biol 162:1087–1097. https://doi.org/10.1007/s00227-015-2651-7

Roleda MY, Campana GL, Wiencke C, Hanelt D, Quartino ML, Wulff A (2009) Sensitivity of Antarctic *Urospora penicilliformis* (Ulotrichales, Chlorophyta) to ultraviolet radiation is life-stage dependent. J Phycol 45:600–609. https://doi.org/10.1111/j.1529-8817.2009.00691.x

Rückamp M, Braun M, Suckro S, Blindow N (2011) Observed glacial changes on the King George Island ice cap, Antarctica, in the last decade. Glob Planet Change 79:99–109. https://doi.org/10.1016/j.gloplacha.2011.06.009

Sahade R, Lagger C, Torre L, Momo FR, Monien P, Schloss IR, Barnes DKA, Servetto N, Tarantelli S, Tatián M, Zamboni N, Abele D (2015) Climate change and glacier retreat drive shifts in an Antarctic benthic ecosystem. Sci Adv 1(10):e1500050. https://doi.org/10.1126/sciadv.1500050

Schoenrock KM, Schram JB, Amsler CD, McClintock JB, Angus RA (2015) Climate change impacts on overstory *Desmarestia spp.* from the western Antarctic Peninsula. Mar Biol 162:377–389. https://doi.org/10.1007/s00227-014-2582-8

Schram JB, Schoenrock KM, McClintock JB, Amsler CD, Angus RA (2017) Ocean warming and acidification alter Antarctic macroalgal biochemical composition but not amphipod grazer feeding preferences. Mar Ecol Prog Ser 581:45–56. https://doi.org/10.3354/meps1230

Seefeldt MA, Campana GL, Deregibus D, Quartino ML, Abele D, Tollrian R, Held C (2017) Different feeding strategies in Antarctic scavenging amphipods and their implications for colonisation success in times of retreating glaciers. Front Zool 14:59–74. https://doi.org/10.1186/s12983-017-0248-3

Segovia-Rivera V, Valdivia N (2016) Independent effects of grazing and tide pool habitats on the early colonisation of an intertidal community on western Antarctic peninsula. Rev Chil Hist Nat 89:3. https://doi.org/10.1186/s40693-016-0053-y

Smale DA, Barnes DKA (2008) Likely responses of the Antarctic benthos to climate-related changes in physical disturbance during the 21st century, based primarily on evidence from the West Antarctic Peninsula region. Ecography 31:289–305. https://doi.org/10.1111/j.0906-7590.2008.05456.x

Sommaruga R (2003) UVR and its effects on species interactions. In: Helbling EW, Zagarese H (eds) UV effects in aquatic organisms and ecosystems. The Royal Society of Chemistry, Cambridge, pp 485–508. https://doi.org/10.1039/9781847552266-00485

Sommer U (1999a) The impact of herbivore type and grazing pressure on benthic microalgal diversity. Ecol Lett 2:65–69. https://doi.org/10.1046/j.1461-0248.1999.22052.x

Sommer U (1999b) Periphyton architecture and susceptibility to grazing by periwinkles (*Littorina littorea*, Gastropoda). Int Rev Hydrobiol 84:197–204. https://doi.org/10.1002/iroh.199900020

Sommer U (2000) Benthic microalgal diversity enhanced by spatial heterogeneity of grazing. Oecologia 122:284–287. https://doi.org/10.1007/PL00008857

Sousa WP (1979) Experimental investigations of disturbance and ecological succession in a rocky intertidal algal community. Ecol Monogr 49:227–254. https://doi.org/10.2307/1942484

Sousa WP, Connell JH (1992) Grazing and succession in marine algae. In: John DM, Hawkins SJ, Price JH (eds) Plant-animal interactions in the marine benthos. Clarendon Press, Oxford, pp 425–441

Stanwell-Smith D, Barnes DKA (1997) Benthic community development in Antarctica: recruitment and growth on settlement panels at Signy Island. J Exp Mar Biol Ecol 212:61–79. https://doi.org/10.1016/S0022-0981(96)02754-2

Steinman AD (1996) Effects of grazers on freshwater benthic algae. In: Stevenson RJ, Bothwell ML, Lowe RL (eds) Algal ecology – freshwater benthic ecosystems. Academic Press, San Diego, pp 341–373

Teixidó N, Garrabou J, Gutt J, Arntz WE (2007) Iceberg disturbance and successional spatial patterns: the case of the shelf Antarctic benthic communities. Ecosystems 10:143–158. https://doi.org/10.1007/s10021-006-9012-9

Torre L, Carmona Tabares PC, Momo FR, Meyer JFCA, Sahade R (2017) Climate change effects on Antarctic benthos: a spatially explicit model approach. Clim Chang 141:733–746. https://doi.org/10.1007/s10584-017-1915-2

Turner J, Bindschadler R, Convey P, di Prisco G, Fahrbach E, Gutt J, Hodgson D, Mayewski P, Summerhayes C (2009) Antarctic climate change and the environment. Scientific Committee on Antarctic Research, Cambridge

Valdivia N, Díaz MJ, Holtheuer J, Garrido I, Huovinen P, Gómez I (2014) Up, down, and all around: scale-dependent spatial variation in rocky-shore communities of Fildes Peninsula, King George Island, Antarctica. PLoS One 9(6):e100714. https://doi.org/10.1371/journal.pone.0100714

Valdivia N, Pardo LM, Macaya EC, Huovinen P, Gómez I (2019) Different ecological mechanisms lead to similar grazer controls on the functioning of periphyton Antarctic and sub-Antarctic communities. Prog Oceanogr 174:7–16. https://doi.org/10.1016/j.pocean.2018.01.008

Villafañe VE, Sundbäck K, L Figueroa F, Helbling EW (2003) Photosynthesis in the aquatic environment as affected by UVR. In: Helbling EW, Zagarese H (eds) UV effects in aquatic organisms and ecosystems. The Royal Society of Chemistry, Cambridge, pp 357–397. https://doi.org/10.1039/9781847552266-00357

Wahl M, Molis M, Davis A, Dobretsov S, Dürr S, Johansson J, Kinley J, Kirugara D, Langer M, Lotze HK, Thiel M, Thomason JC, Worm B, Zeevi Ben-Yosef D (2004) UV effects that come and go: a global comparison of marine benthic community level impacts. Glob Chang Biol 10:1962–1972. https://doi.org/10.1111/j.1365-2486.2004.00872.x

Wiencke C (1990) Seasonality of brown macroalgae from Antarctica – a long-term culture study under fluctuating Antarctic daylengths. Polar Biol 10:589–600. https://doi.org/10.1007/BF00239370

Wiencke C, Amsler CD (2012) Seaweeds and their communities in polar regions. In: Wiencke C, Bischof K (eds) Seaweed biology: novel insights into ecophysiology, ecology and utilization. Ecological studies, vol 219, Springer, Heidelberg, pp 265–291

Wiencke C, Clayton MN (eds) (2002) Antarctic seaweeds. A.R.G. Gantner Verlag KG, Ruggell

Wiencke C, Gómez I, Pakker H, Flores-Moya A, Altamirano M, Hanelt D, Bischof K, L Figueroa F (2000) Impact of UV-radiation on viability photosynthetic characteristics and DNA of brown algal zoospores: implications for depth zonation. Mar Ecol Prog Ser 197:217–229. https://doi.org/10.3354/meps197217

Wiencke C, Clayton MN, Gómez I, Iken K, Lüder UH, Amsler CD, Karsten U, Hanelt D, Bischof K, Dunton K (2007) Life strategy, ecophysiology and ecology of seaweeds in polar waters. Rev Environ Sci Biotechnol 6:95–126. https://doi.org/10.1007/s11157-006-9106-z

Witman JD, Dayton PK (2001) Rocky subtidal communities. In: Bertness MD, Gaines SD, Hay ME (eds) Marine community ecology. Sinauer Associates Inc, Sunderland, pp 339–366

Wulff A, Roleda MY, Zacher K, Wiencke C (2008) UV radiation effects on pigments, photosynthetic efficiency and DNA of an Antarctic marine benthic diatom community. Aquat Biol 3:167–177. https://doi.org/10.3354/ab00076

Wulff A, Iken K, Quartino ML, Al-Handal AY, Wiencke C, Clayton MN (2011) Biodiversity, biogeography and zonation of benthic micro- and macroalgae in the Arctic and Antarctic. In: Wiencke C (ed) Biology of polar benthic algae. Walter de Gruyter GmbH & Co, Berlin, pp 23–52

Zacher K (2014) The susceptibility of spores and propagules of Antarctic seaweeds to UV and photosynthetically active radiation – field versus laboratory experiments. J Exp Mar Biol Ecol 458:57–63. https://doi.org/10.1016/j.jembe.2014.05.007

Zacher K, Campana GL (2008) UV and grazing effects on an intertidal and subtidal algal assemblage: a comparative study. Ber Polar Meeresforsch 571:287–294

Zacher K, Wulff A, Molis M, Hanelt D, Wiencke C (2007a) Ultraviolet radiation and consumer effects on a field-grown intertidal macroalgal assemblage in Antarctica. Glob Change Biol 13:1201–1215. https://doi.org/10.1111/j.1365-2486.2007.01349.x

Zacher K, Hanelt D, Wiencke C, Wulff A (2007b) Grazing and UV radiation effects on an Antarctic intertidal microalgal assemblage: a long-term field study. Polar Biol 30:1203–1212. https://doi.org/10.1007/s00300-007-0278-y

Zacher K, Roleda MY, Hanelt D, Wiencke C (2007c) UV effects on photosynthesis and DNA in propagules of three Antarctic seaweeds (*Adenocystis utricularis*, *Monostroma hariotii* and *Porphyra endiviifolium*). Planta 225:1505–1516. https://doi.org/10.1007/s00425-006-0436-4

Zacher K, Rautenberger R, Hanelt D, Wulff A, Wiencke C (2011) The abiotic environment of polar marine benthic algae. In: Wiencke C (ed) Biology of polar benthic algae. Walter de Gruyter GmbH & Co, Berlin, pp 9–21

Zacher K, Bernard M, Bartsch I, Wiencke C (2016a) Survival of early life history stages of Arctic kelps (Kongsfjorden, Svalbard) under multifactorial global change scenarios. Polar Biol 39:2009–2020. https://doi.org/10.1007/s00300-016-1906-1

Zacher K, Savaglia V, Bartsch I (2016b) Effects of temperature and interspecific competition on growth and photosynthesis of two endemic Antarctic *Desmarestia* species. Algol Stud 151:103–122. https://doi.org/10.1127/algol_stud/2016/0269

Chapter 13
Seaweed-Herbivore Interactions: Grazing as Biotic Filtering in Intertidal Antarctic Ecosystems

Nelson Valdivia

Abstract Consumers constitute a key component of the environmental filters that restrict the establishment of colonists into local assemblages. Thus, the trophic activity of consumers, particularly grazers, can be pivotal to control the development of potential algal invaders in Antarctic coasts. Here, the consumptive effects of coastal macrobenthic grazers on algal communities are reviewed to assess the degree to which these consumers can mediate the introduction of seaweeds in intertidal Antarctic communities. Gastropods and amphipods have strong consumptive effects on algal communities. Yet, amphipods are sensitive to climate change factors such as warming and acidification, which could hamper their ability to control native and alien macroalgae. Alien macroalgae that modify the abiotic environment, such as the gutweed *Ulva intestinalis* in tidepools, represent potential superior competitors in Antarctic ecosystems. In this line, simulations based on a simple probabilistic model showed that intermediate to high levels of frequency-dependent consumption seem to be fundamental to allow for stable coexistence when the alien species is competitively superior. With this work, I hope to stimulate further manipulative research to assess the role of benthic consumers in mediating the coexistence (or lack thereof) between alien and native seaweeds under multiple climate change scenarios.

Keywords Competition · Environmental filters · Grazing · Alien macroalgae

N. Valdivia (✉)
Instituto de Ciencias Marinas y Limnológicas, Facultad de Ciencias, Universidad Austral de Chile, Valdivia, Chile

Research Center Dynamics of High Latitude Marine Ecosystems (IDEAL), Valdivia, Chile
e-mail: nelson.valdivia@uach.cl

I. Gómez, P. Huovinen (eds.), *Antarctic Seaweeds*,
https://doi.org/10.1007/978-3-030-39448-6_13

13.1 Biological Invasions and Their Impact on the Ecology of Antarctic Coastal Systems

Biological invasions are one of the major anthropogenic threats to biodiversity across multiple spatiotemporal scales (Simberloff et al. 2013). The negative consequences of invasions range from local extinction of native species to the impairment of ecosystem functioning and from alterations of interaction webs to the loss of ecosystem services (Vitousek et al. 1997; Bulleri et al. 2010; White and Shurin 2011; Silva et al. 2019). In marine communities, for example, the introduction of consumers into a bottom-up-controlled ecosystem can lead to severe compositional shifts and significant declines in diversity (Kotta et al. 2018). Invasive macroalgae, in addition, have been shown to reduce local diversity through interference competition that involves drastic environmental changes (Björk et al. 2005). The effects of invasive species can be particularly severe on ecologically isolated ecosystems, because of low functional redundancy and restricted niche coverage (Hughes and Convey 2014). Antarctica is an iconic example for ecologically isolated ecosystems—this isolation has influenced the evolution of a high proportion of endemic species, which makes the Antarctic ecosystems a highly ranked scientific priority (Kennicutt et al. 2014). Accordingly, the study of biological invasions, especially in ecologically isolated ecosystems like Antarctica, is of widespread relevance for conservation ecology.

Due to its well-described biological isolation and adverse abiotic conditions, Antarctica provides a unique opportunity to improve our mechanistic understanding of biological invasions (Chown et al. 2015; McCarthy et al. 2019). This is particularly accurate for Antarctic coastal marine rocky communities, because of the multiple biotic and abiotic environmental changes that are currently increasing the risk of non-native marine species in these ecosystems (McCarthy et al. 2019). For instance, increasing ship activity and high transportability have been suggested as critical factors during the first stages of the invasion process of marine macroalgae and invertebrates (Blackburn et al. 2011; McCarthy et al. 2019). Associated to the transport of marine invaders, recent evidence indicates that rafting macroalgae are able to cross the Antarctic Circumpolar Current and arrive to Antarctica (King George Island) from Subantarctic and cold-temperate source populations (Fraser et al. 2017; Fraser et al. 2018). In addition, the establishment of alien species in Antarctic marine rocky communities could well be facilitated by the current scenario of increasing water temperature and decreasing ice cover (Blunden et al. 2013; Clark et al. 2013; Quartino et al. 2013; McCarthy et al. 2019). These alterations represent the weakening of dispersal limitations and also of abiotic filtering that would benefit aliens over Antarctic natives (Duffy et al. 2017; Griffiths et al. 2017; see also Chap. 3 by Fraser et al. and Chap. 4 by Macaya et al.).

In addition to dispersal and abiotic environmental filtering, local community assembly also depends upon chance (particularly when population numbers are low) and biotic filters that restrict the establishment of some taxa (HilleRisLambers et al. 2012; Briski et al. 2018). As such, both factors are relevant during the establishment of aliens. Once settlers have passed through the abiotic filtering (e.g. environmental extremes in Antarctica), negative biotic interactions, such as competition, predation and herbivory, encompass local biotic filters that finally determine the success of settled organisms (e.g. Guisan and Thuiller 2005; HilleRisLambers et al. 2012; King and Howeth 2019). Although invasion theory predicts that generalist enemies (e.g. herbivores) have stronger impacts on native than alien competitors, the analysis of local adaptation of consumers to native resources provides limited evidence that generalist enemies are better adapted to attacking natives than exotics (Keane and Crawley 2002). In addition, there are several examples of marine Antarctic macroalgae that have developed chemical defences to local herbivores (e.g. Amsler et al. 2005, 2009b; Aumack et al. 2010), which may allow them to have fitness advantages over alien macroalgae (see Chap. 17 by Amsler et al.). Thus, negative biotic interactions in Antarctic coastal marine communities can have an important role as local filters mediating the process of invasion in these ecosystems.

13.2 Recent Introductions of Exotic Macroalgae in Antarctica

Examples of alien macroalgae in Antarctica include green algae such as the gutweed *Ulva intestinalis*, which was described as *Enteromorpha intestinalis* together with the brown alga *Petalonia fascia* and the red alga *Rhodymenia subantarctica* in South Shetland Islands (Clayton et al. 1997; see also Frenot et al. 2005; Campana et al. 2009). In addition, other green macroalgae have been suggested to be recent introductions to South Shetland Island shores, namely, the filamentous algae *Urospora penicilliformis* and *Ulothrix* sp. (Gómez 2015). Considering the growing human activity and ship traffic to South Shetland Islands (Frenot et al. 2005; Bender et al. 2016), the occurrence of exotic and potentially invasive species in these locations is not surprising. Recent multivariate analyses suggest, moreover, a trend of biotic homogenisation among seaweed assemblages of South Patagonia and both West and East Antarctic Peninsula (Sanches et al. 2016). In this line, but not necessarily related to recent biological invasions, a small subset of widely distributed macroalgae dominate the biogeographic structure of intertidal rocky shores in the Southern Ocean, likely as a result of rafting across in the Antarctic Circumpolar Current (Griffiths and Waller 2016; see also Pellizzari et al. 2017). Although the recent increases of Antarctic seaweed diversity could well be the result of improved and more efficient techniques of sampling and molecular taxonomical methods (e.g. Dubrasquet et al. 2018), the role of human-mediated transport and climate change in modifying the biogeography of macroalgae in the Southern Ocean should not be ruled out (Pellizzari et al. 2017; McCarthy et al. 2019; see Chap. 5 by Pellizzari et al.).

13.3 Can Grazers Control Alien Macroalgae in Antarctica?

Worldwide, marine herbivores have profound and chiefly negative effects on the abundance of primary producers—in particular, these effects are strongest in rocky intertidal habitats (Lubchenco 1978; Hawkins and Hartnoll 1983; Poore et al. 2012). In this line, manipulative and observational studies suggest that grazers can exert a significant effect on the structure of algal communities in Antarctica. For example, field-based manipulative experiments show a strong control of intertidal grazers, namely, the limpet *Nacella concinna* (Fig. 13.1), on intertidal periphyton assemblages in Fildes Bay, King George Island (Segovia-Rivera and Valdivia 2016). An important outcome of this work is that the overall negative effects of *N. concinna* on the abundance of periphyton taxa were consistent across intertidal microhabitats (i.e. emergent rocks and tidepools). Similarly, Zacher et al. (2007b) demonstrate that *N. concinna*'s effects on early-succession algal communities are consistent across multiple levels of ultraviolet radiation (UVR) exposure. Moreover, the magnitude and sign of the effects of this grazer on benthic periphyton in King George Island can be similar to those of congeneric limpets in Chilean South Patagonia (CSP), albeit channelled through different ecological mechanisms—while the effects of Antarctic limpets appear to be frequency-dependent, the effects of Magellan limpets seem to be related to niche complementarity in a rich community of grazers (Aldea and Rosenfeld 2011; Valdivia et al. 2019). Generally speaking, these results may indicate that the effects of *N. concinna* on algal abundance and diversity may be consistent across abiotic environmental conditions related to emersion time, desiccation, and photobiotic and osmotic stress. This conclusion is well in line with a major meta-analysis that shows only a little influence of environmental conditions (i.e. latitude or mean annual water temperature) on the effects of grazers on the abundance of primary producers (Poore et al. 2012).

Mesograzers, on the other hand, have also been suggested as strong top-down controllers of Antarctic seaweed communities. Field experiments demonstrate that

Fig. 13.1 Aggregations of individuals of *Nacella concinna* in Antarctic rocky tidepools. (Photo by Nelson Valdivia, Proyecto Anillo ART1101)

the abundant amphipod fauna rapidly consumes intertidal filamentous algae transplanted to the subtidal in West Antarctic Peninsula (WAP; Amsler et al. 2012). The remarkable abundance of amphipods (e.g. up to 20 ind/g algal wet weight; Huang et al. 2007) suggests that amphipod assemblages do have a significant population-level control on benthic algal communities. Indeed, amphipods and subtidal macroalgae have developed mutualist interactions in which the former benefit from predator-sheltered habitats and the latter from reduced fouling (Amsler et al. 2014). In this line, amphipods and other mesograzers are proposed to control the occurrence of filamentous macroalgae, which can explain the absence of these species in Antarctic subtidal habitats (Peters 2003; Amsler et al. 2009a). Albeit less abundant than amphipods, small-sized Antarctic gastropods can also play an important role as top-down controls of epiphytic microalgae on large pseudo-kelps, as shown in mesocosm experiments (Amsler et al. 2015, 2019). Thus, it is highly likely that mesograzers, through their consumptive activities, can be relevant for the assembly of local macrobenthic communities in WAP (see Chap. 17 by Amsler et al.).

The results of the experiments described above are supported by early and recent observational evidence. Kim (2001), for example, found a strong association between the seasonal variation in the abundance of *N. concinna* and that of intertidal filamentous algae in King George Island—although the role of ice scouring in mediating this association cannot be ruled out. In addition, the analysis of stomach contents and stable isotopes strongly supports the idea that amphipods are central in WAP coastal food webs (Aumack et al. 2017; Zenteno et al. 2019). On the other hand, observational evidence shows that notothenioid fish, like *Notothenia rossi* and *N. coriiceps*, actively select for macroalgae as food (Casaux et al. 1990; Barrera-Oro et al. 2019). Despite correlation does not imply causality, the results of these observational studies agree with field- and lab-based manipulative evidence of the central role of herbivores in coastal Antarctic food webs.

Could these grazers prevent the establishment, or at least control the abundance, of exotic seaweeds? As introduced above, theory predicts that parasites and generalist predators and herbivores (i.e. 'enemies') can have stronger impacts on native than alien competitors, allowing the latter to expand their spatial distribution and adopt an invasive behaviour (i.e. the Enemy Release Hypothesis, reviewed in Keane and Crawley 2002). In addition, the current trend of seawater warming would also reduce the physiological constrains imposed by Antarctic environmental extremes to temperate seaweeds. This may picture a scenario of improving biotic and abiotic environmental conditions for the establishment and spread of invasive seaweeds in Antarctica. Yet, sophisticated anti-herbivory defences have evolved in several Antarctic seaweeds (Amsler et al. 1998, 2009b, 2019; Aumack et al. 2010), and today it is proposed that small-sized herbivores have actually positive effects on macroalgae owing their antifouling consumptive activity (Amsler et al. 2014, 2019). This could provide native seaweeds with consumer-mediated competitive advantages over exotic seaweeds, provided that the latter are not equipped with anti-herbivory defences. Indeed, grazing has been shown to mediate the competitive interaction between native and alien seaweeds elsewhere (Noè et al. 2018). For

Antarctic communities, however, further empirical research is needed to assess the role of consumers on competitive seaweed interactions.

How can climate change influence the potential effects of grazers on an invasion process in Antarctica? The answer to this question seems to depend on the species or functional type analysed. For instance, Antarctic grazing gastropods, including *N. concinna* and *Margarella antarctica*, have been shown to resist the combined effects of decreased pH and seawater warming (Schram et al. 2014), hinting for certain level of resistance of the grazing function to this stressor. However, the authors warn that the slow growth rates and longevity of the analysed gastropods could mask long-term sublethal effects of warming and acidification. On the other hand, ocean acidification can significantly increase the mortality rates of amphipods, while warming can have sublethal effects in terms of increased whole-body protein content of those organisms (Schram et al. 2016). Acute warming, moreover, is shown to modify the feeding preferences of amphipods (Schram et al. 2015). As changes in consumer abundances and prey shift have been proposed as major causes of food web variations (e.g. Lopez et al. 2017), ocean acidification and warming can have profound indirect effects on the structure of coastal Antarctic communities. In addition, recent evidence hints for indirect effects of climate change on grazer populations, as seawater warming and freshening can have significant and independent effects on consumption rates of predatory fish (Navarro et al. 2019). Sedimentation, which relates to warming-associated glacier melting, has been shown to mediate the effects of grazers on the germination and development of young sporophytes of Arctic kelps (Zacher et al. 2016). Finally, ocean warming and increased pCO_2 could mediate macroalgal competitive interactions, as it is demonstrated that both factors in combination favour fleshy over crustose forms (Schoenrock et al. 2016). In this way, the role of grazing and competition as biotic filtering in the assembly of local Antarctic communities should be assessed in combination to climate change stressors.

13.4 *Ulva intestinalis* as a Case Study in a Simple, Two-Species Assembly Model

The gutweed *Ulva intestinalis* was described inhabiting Antarctic shores in 1997 (Clayton et al. 1997). Elsewhere, this species is able to generate adverse abiotic conditions for potential competitors in tidepools, which involve conditions of high pH and low inorganic carbon concentrations—at the same time, *U. intestinalis* is able to capture HCO_3^- under these conditions (Larsson et al. 1997). This ability is suggested to provide *U. intestinalis* with competitive advantages over other macroalgae in high-intertidal tidepools, explaining the dominance of this species observed in Swedish Atlantic tidepools (Björk et al. 2004, 2005). In Antarctica, however, *U. intestinalis* is not a dominant species like in Sweden, and its spatial distribution is usually confined to high-intertidal tidepools (Clayton et al. 1997;

Gómez 2015). A first explanation for the restricted distribution of *U. intestinalis* would be the climatic environmental conditions that prevail in WAP, particularly average seawater temperatures that are below its optimum for somatic growth (Bischoff and Wiencke 1993). According to the high consumption rates demonstrated for Antarctic intertidal benthic grazers, including abundant populations of amphipods (Huang et al. 2006) and the limpet *N. concinna* (e.g. Zacher et al. 2007a; Segovia-Rivera and Valdivia 2016), a possible, non-exclusive explanation could be that grazer activity defines the lower intertidal limit of this species. Intertidal herbivores have been shown to control the abundance of fast-growing macroalgae in tidepools of WAP (Segovia-Rivera and Valdivia 2016) and elsewhere (e.g. Noël et al. 2009).

To improve our understanding of an invasion process in Antarctica, I used *U. intestinalis* introduction into high-intertidal tidepools as a model system. The scenario, therefore, includes two sites connected by dispersal (WAP and Chilean South Patagonia, CSP), tidepools as local habitats, *U. intestinalis* as a competitively superior alien (see previous paragraph) and the corticated red seaweed *Iridaea cordata* as a native competitor that is frequent in tidepools (Valdivia et al. 2014); for simplicity, both species will be referred to as *Ulva* and *Iridaea*, respectively. To this aim, I used a simple probabilistic model known as Moran model, which was originally generated to understand the temporal changes in allele frequencies in populations (Moran 1958). This model has been lately used and extended to simulate the temporal dynamics of two-species communities (Hubbell 2001; Vellend 2016). The basic structure of the model assumes a neutral, closed assemblage without speciation, in which there are J individuals that belong to one of two species, either species A or species B (either *Ulva* or *Iridaea* in this case). Since J is assumed to be fixed (i.e. a zero-sum dynamic), there are j individuals of species A and $J - j$ individuals of B. At each time, one individual of the community is selected at random to die. At the next time, an individual is selected at random to produce one offspring that replaces the dead individual. Each individual is chosen to be A or B with probabilities

$$p_j = jJ^{-1}, \text{ and}$$

$$q_j = (J - j)J^{-1},$$

for species A and B, respectively (Moran 1958). The community is then described as a Markov chain in which the state (community structure at a given time) is defined by the abundance of species (j) and a transition probabilities p_{jk} from state j to k:

$$p_{jk} = \binom{J}{k} p_j^k q_j^{J-k}.$$

In this work, and as previously done by Hubbell (2001) and Vellend (2016), I took advantage of the simplicity of the Moran model to describe how differences in competitive abilities, reflected in differences between probabilities p_j and q_j, can lead to differing community patterns. These differences were expressed as *average fitness ratios*, in which larger ratios will indicate that species *A* (*Ulva*) generates a larger offspring than species *B* (*Iridaea*); i.e. the former is competitively superior than the latter (Vellend 2016). In addition, the simulations of this work were done in a metacommunity context, in which the reproducing individual of the Moran model was chosen at random from the entire set of habitat patches.

The rationale of these simulations was to resemble an invasion process in which *Ulva* (species *A*) is an invasive and competitively superior species that already colonised high-intertidal tidepools in the WAP from the nearest continental shore, i.e. CSP. *Iridaea*, on the other hand, represents a native competitor in tidepools. The simulations, thus, considered a simple metacommunity of two local sites that are linked by dispersal (*m* in Fig. 13.2). Previous unpublished data was used to estimate the initial abundances (in terms of frequency) of *Ulva*. Thus, the simulation included empirical data only to set the initial abundances of *Ulva*. In addition, I assumed that *Ulva* would be competitively dominant in high-intertidal tidepools in WAP but not in CSP, because of the lower seaweed diversity observed in Antarctica that would provide more niche opportunities to alien species. These differences were expressed as fitness ratios >1 (1.3) in WAP and <1 (1/1.1) in CSP.

A central aspect of these simulations is the incorporation of negative density (or frequency) dependence of competitive advantages as the result of grazing on the abundance and fitness of competitors. If fitness ratio always is >1, that is, if species *A* has consistently greater fitness than species *B*, then the former species will tend to competitively exclude the latter, and no stable coexistence will take place (Chesson 2000; Letten et al. 2017). However, if the fitness advantages of species *A* over species *B* are greater when species *A* is rare, and vice versa, then each species will have relative advantage when rare, and there should be a stable coexistence (equilibrium) point (Chesson 2000). This negative frequency dependence of growth rates can be generated by, for instance, negative effects of consumers (predators and prey) that become harsher with increasing prey abundances (i.e. enemy-mediated coexistence; Holt et al. 1994). Therefore, the simulations in this work were done across a range of strengths of negative frequency dependence (*b* in Fig. 13.2) to resemble potential effects of grazers on the abundance of macroalgal competitors in Antarctic high-intertidal tidepools. These values were set to 0, −0.1, and −0.5, representing absence and intermediate and high negative density dependence. All simulations were conducted in R programming environment (R Core Team 2019).

The simulations showed that the complete absence of dispersal in the metacommunity resulted in extinction in both regions, even at intermediate levels of negative frequency dependence (Fig. 13.2a,b). Stronger negative density dependence would allow for coexistence at low abundances of *Ulva* only in CPS (segmented line in Fig. 13.2c). Interestingly, dominance, but not competitive exclusion, in Antarctica would be apparent even in the absence of significant negative frequency dependence (i.e. grazing) and when dispersal is intermediate (Fig. 13.2d). As expected, high levels of dispersal would lead to biotic homogenisation of the region (Fig. 13.2g–i). Intermediate dispersal and increasing negative density dependence led to reduced

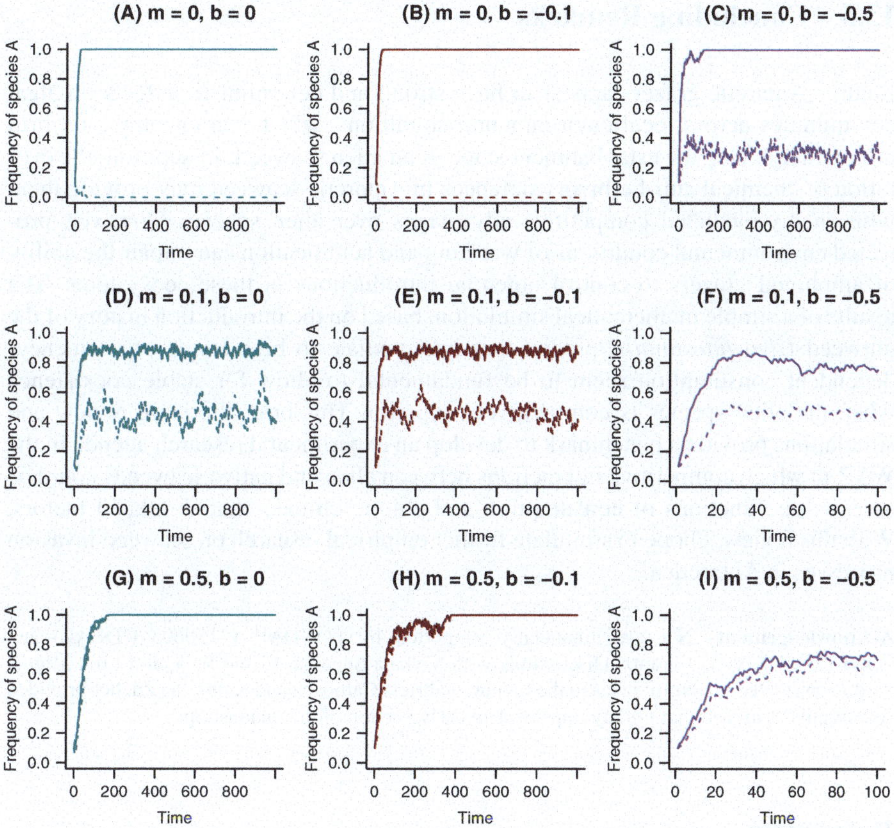

Fig. 13.2 Simulations of two-species communities in two sites connected by dispersal. The lines represent the frequency of one of the species (species *A*) in each site. The solid and segmented lines represent WAP and CSP, respectively. Since the number of individuals in each site is fixed (*J*), then the frequency of species B is equal to 1 – freq. species *A*. The parameters *m* and *b* represent the dispersal parameter and negative frequency dependency, respectively. Competitive exclusion is evident when species *A* reaches a frequency of 1 or 0. It was assumed that *Ulva* is competitively dominant in high-intertidal tidepools in WAP but not in CSP, which was expressed as fitness ratios >1 (1.3) and <1 (1/1.1), respectively. The *b* parameter resembles the effects of grazers on the frequency of competitors and was set to 0, −0.1, and −0.5, representing absence and intermediate and high negative density dependence

dominance (Fig. 13.2d–f). Competitive exclusion by *Ulva* was expected under the scenario of high dispersal and no or intermediate negative density dependence (Fig. 13.2g, h). An important outcome of the simulation is that stable coexistence can be reached when both dispersal between CPS and WAP *and* negative frequency dependence are high (Fig. 13.2i). This could be caused by, on the one hand, 'rescue effect' of immigrants in both sites (e.g. Altermatt et al. 2011) and, on the other hand, by the controlling effect of consumers on the numerically dominant competitor.

13.5 Concluding Remarks

Benthic Antarctic grazers appear to have strong and deterministic effects on algal communities across local environmental conditions, which can encompass a firm biotic filter during the establishment stage of an alien seaweed. In addition, the evolution of chemical anti-herbivory defences in Antarctic seaweed may provide them with enemy-mediated competitive advantages over alien species. However, projected environmental conditions of warming and acidification can impair the ability of amphipod grazers to control potential introductions in these ecosystems. The results of a simple mathematical simulation, based on the introduction history of the gutweed *Ulva intestinalis*, predict that intermediate to high levels of frequency-dependent consumption seem to be fundamental to allow for stable coexistence when the alien species is competitively superior. This brief literature review and simulations provide a benchmark to develop an experimental research agenda in the WAP, in which competitive interactions between alien and native seaweeds could be assessed as functions of consumption and abiotic climate change-related factors. With this review, I hope to stimulate further empirical research on seaweed invasion processes in Antarctica.

Acknowledgements NV was financially supported by FONDAP 15150003 (IDEAL) and FONDECYT 1190529 grants. Discussions with Viviana Segovia, Eliseo Fica, and Luis Miguel Pardo contributed insightful ideas to the review. Gabriela Campana and Katharina Zacher provided constructive criticism that greatly improved an early version of this manuscript.

References

Aldea C, Rosenfeld S (2011) Macromoluscos intermareales de sustratos rocosos de la playa Buque Quemado, Estrecho de Magallanes, sur de Chile. Rev Biol Mar Oceanogr 46(2):115–124

Altermatt F, Bieger A, Carrara F, Rinaldo A, Holyoak M (2011) Effects of connectivity and recurrent local disturbances on community structure and population density in experimental metacommunities. PLoS One 6(4):e19525

Amsler CD, McClintock JB, Baker BJ (1998) Chemical defense against herbivory in the Antarctic marine macroalgae *Iridaea cordata* and *Phyllophora antarctica* (Rhodophyceae). J Phycol 34(1):53–59

Amsler CD, Katrin I, James BM, Margaret OA, Kevin JP, Joanna MH, Furrow FB, Bill JB (2005) Comprehensive evaluation of the palatability and chemical defenses of subtidal macroalgae from the Antarctic Peninsula. Mar Ecol Prog Ser 294:141–159

Amsler CD, Amsler MO, McClintock JB, Baker BJ (2009a) Filamentous algal endophytes in macrophytic Antarctic algae: prevalence in hosts and palatability to mesoherbivores. Phycologia 48(5):324–334

Amsler CD, Iken K, McClintock JB, Baker BJ (2009b) Defenses of polar macroalgae against herbivores and biofoulers. Bot Mar 52(6):535–545

Amsler CD, McClintock JB, Baker BJ (2012) Amphipods exclude filamentous algae from the Western Antarctic Peninsula benthos: experimental evidence. Polar Biol 35(2):171–177

Amsler CD, McClintock JB, Baker BJ (2014) Chemical mediation of mutualistic interactions between macroalgae and mesograzers structure unique coastal communities along the western Antarctic Peninsula. J Phycol 50(1):1–10

Amsler MO, Huang YSM, Engl W, McClintock JB, Amsler CD (2015) Abundance and diversity of gastropods associated with dominant subtidal macroalgae from the western Antarctic Peninsula. Polar Biol 38(8):1171–1181

Amsler CD, Amsler MO, Curtis MD, McClintock JB, Baker BJ (2019) Impacts of gastropods on epiphytic microalgae on the brown macroalga *Himantothallus grandifolius*. Antarct Sci 31(2):89–97

Aumack CF, Amsler CD, McClintock JB, Baker BJ (2010) Chemically mediated resistance to mesoherbivory in finely branched macroalgae along the western Antarctic Peninsula. Eur J Phycol 45(1):19–26

Aumack CF, Lowe AT, Amsler CD, Amsler MO, McClintock JB, Baker BJ (2017) Gut content, fatty acid, and stable isotope analyses reveal dietary sources of macroalgal-associated amphipods along the western Antarctic Peninsula. Polar Biol 40(7):1371–1384

Barrera-Oro E, Moreira E, Seefeldt MA, Valli Francione M, Quartino ML (2019) The importance of macroalgae and associated amphipods in the selective benthic feeding of sister rockcod species *Notothenia rossii* and *N. coriiceps* (Nototheniidae) in West Antarctica. Polar Biol 42(2):317–334

Bender NA, Crosbie K, Lynch HJ (2016) Patterns of tourism in the Antarctic Peninsula region: a 20-year analysis. Antarct Sci 28(3):194–203

Bischoff B, Wiencke C (1993) Temperature requirements for growth and survival of macroalgae from Disko Island (Greenland). Helgoländer Meeresun 47(2):167–191

Björk M, Axelsson L, Beer S (2004) Why is *Ulva intestinalis* the only macroalga inhabiting isolated rockpools along the Swedish Atlantic coast? Mar Ecol Prog Ser 284:109–116

Björk M, Axelsson L, Beer S (2005) Photosynthetic traits of one alga can limit competition with others: the case of *Ulva Intestinalis* growing alone in rockpools. Phycologia 44(4):7–7

Blackburn TM, Pyšek P, Bacher S, Carlton JT, Duncan RP, Jarošík V, Wilson JRU, Richardson DM (2011) A proposed unified framework for biological invasions. Trends Ecol Evol 26(7):333–339

Blunden J, Arndt DS, Achberger C, Ackerman SA, Albanil A, Alexander P et al (2013) State of the climate in 2012. Bull Am Meteorol Soc 94(8):S1–S258

Briski E, Chan FT, Darling JA, Lauringson V, MacIsaac HJ, Zhan A, Bailey SA (2018) Beyond propagule pressure: importance of selection during the transport stage of biological invasions. Front Ecol Environ 16(6):345–353

Bulleri F, Balata D, Bertocci I, Tamburello L, Benedetti-Cecchi L (2010) The seaweed *Caulerpa racemosa* on Mediterranean rocky reefs: from passenger to driver of ecological change. Ecology 91(8):2205–2212

Campana GL, Zacher K, Fricke A, Molis M, Wulff A, Quartino ML, Wiencke C (2009) Drivers of colonization and succession in polar benthic macro- and microalgal communities. Bot Mar 52(6):655–667

Casaux RJ, Mazzotta AS, Barrera-Oro ER (1990) Seasonal aspects of the biology and diet of nearshore Nototheniid fish at Potter Cove, South Shetland Islands, Antarctica. Polar Biol 11(1):63–72

Chesson P (2000) Mechanisms of maintenance of species diversity. Annu Rev Ecol Syst 31:343–366

Chown SL, Clarke A, Fraser CI, Cary SC, Moon KL, McGeoch MA (2015) The changing form of Antarctic biodiversity. Nature 522(7557):431–438

Clark GF, Stark JS, Johnston EL, Runcie JW, Goldsworthy PM, Raymond B, Riddle MJ (2013) Light-driven tipping points in polar ecosystems. Glob Chang Biol 19(12):3749–3761

Clayton MN, Wiencke C, Klöser H (1997) New records of temperate and sub Antarctic marine benthic macroalgae from Antarctica. Polar Biol 17(2):141–149

Dubrasquet H, Reyes J, Sanchez RP, Valdivia N, Guillemin ML (2018) Molecular-assisted revision of red macroalgal diversity and distribution along the Western Antarctic Peninsula and South Shetland Islands. Cryptogam Algol 39(4):409–429

Duffy GA, Coetzee BWT, Latombe G, Akerman AH, McGeoch MA, Chown SL (2017) Barriers to globally invasive species are weakening across the Antarctic. Divers Distrib 23(9):982–996

Fraser CI, Kay GM, Plessis M, Ryan PG (2017) Breaking down the barrier: dispersal across the Antarctic polar front. Ecography 40(1):235–237

Fraser CI, Morrison AK, Hogg AM, Macaya EC, van Sebille E, Ryan PG et al (2018) Antarctica's ecological isolation will be broken by storm-driven dispersal and warming. Nat Clim Chang 8(8):704–708

Frenot Y, Chown SL, Whinam J, Selkirk PM, Convey P, Skotnicki M, Bergstrom DM (2005) Biological invasions in the Antarctic: extent, impacts and implications. Biol Rev 80(1):45–72

Gómez I (ed) (2015) Flora marina antártica: patrimonio de biodiversidad. Ediciones Kultrún, Valdivia

Griffiths HJ, Waller CL (2016) The first comprehensive description of the biodiversity and biogeography of Antarctic and sub-Antarctic intertidal communities. J Biogeogr 43(6):1143–1155

Griffiths HJ, Meijers AJS, Bracegirdle TJ (2017) More losers than winners in a century of future Southern Ocean seafloor warming. Nat Clim Chang 7(10):749–754

Guisan A, Thuiller W (2005) Predicting species distribution: offering more than simple habitat models. Ecol Lett 8(9):993–1009

Hawkins SJ, Hartnoll RG (1983) Grazing of intertidal algae by marine invertebrates. Oceanogr Mar Biol 21:195–282

HilleRisLambers J, Adler PB, Harpole WS, Levine JM, Mayfield MM (2012) Rethinking community assembly through the lens of coexistence theory. Annu Rev Ecol Evol Syst 43:227–248

Holt RD, Grover J, Tilman D (1994) Simple rules for interspecific dominance in systems with exploitative and apparent competition. Am Nat 144(5):741–771

Huang YM, McClintock JB, Amsler CD, Peters KJ, Baker BJ (2006) Feeding rates of common Antarctic gammarid amphipods on ecologically important sympatric macroalgae. J Exp Mar Biol Ecol 329(1):55–65

Huang YM, Amsler MO, McClintock JB, Amsler CD, Baker BJ (2007) Patterns of gammaridean amphipod abundance and species composition associated with dominant subtidal macroalgae from the western Antarctic Peninsula. Polar Biol 30(11):1417–1430

Hubbell SP (2001) The unified neutral theory of biodiversity and biogeography. Princeton University Press, Princeton

Hughes KA, Convey P (2014) Alien invasions in Antarctica-is anyone liable? Polar Res 33:1–13

Keane RM, Crawley MJ (2002) Exotic plant invasions and the enemy release hypothesis. Trends Ecol Evol 17(4):164–170

Kennicutt MC, Chown SL, Cassano JJ, Liggett D, Massom R, Peck LS et al (2014) Six priorities for Antarctic science. Nature 512(7512):23–25

Kim D (2001) Seasonality of marine algae and grazers of an Antarctic rocky intertidal, with emphasis on the role of the limpet *Nacella concinna* Strebel (Gastropoda: Patellidae). PhD Thesis, University of Bremen, Bremen

King GE, Howeth JG (2019) Propagule pressure and native community connectivity interact to influence invasion success in metacommunities. Oikos 128:1549–1564

Kotta J, Wernberg T, Jänes H, Kotta I, Nurkse K, Pärnoja M, Orav-Kotta H (2018) Novel crab predator causes marine ecosystem regime shift. Sci Rep 8(1):4956–4963

Larsson C, Axelsson L, Ryberg H, Beer S (1997) Photosynthetic carbon utilization by *Enteromorpha intestinalis* (Chlorophyta) from a Swedish rockpool. Eur J Phycol 32(1):49–54

Letten AD, Ke PJ, Fukami T (2017) Linking modern coexistence theory and contemporary niche theory. Ecol Monogr 87(2):161–177

Lopez DN, Camus PA, Valdivia N, Estay SA (2017) High temporal variability in the occurrence of consumer-resource interactions in ecological networks. Oikos 126(12):1699–1707

Lubchenco J (1978) Plant species diversity in a marine intertidal community: importance of herbivore food preference and algal competitive abilities. Am Nat 112(983):23–39

McCarthy AH, Peck LS, Hughes KA, Aldridge DC (2019) Antarctica: the final frontier for marine biological invasions. Glob Chang Biol 25(7):2221–2241

Moran PAP (1958) Random processes in genetics. Math Proc Camb Philos Soc 54(1):60–71

Navarro JM, Paschke K, Ortiz A, Vargas-Chacoff L, Pardo LM, Valdivia N (2019) The Antarctic fish *Harpagifer antarcticus* under current temperatures and salinities and future scenarios of climate change. Prog Oceanogr 174:37–43

Noè S, Badalamenti F, Bonaviri C, Musco L, Fernández TV, Vizzini S, Gianguzza P (2018) Food selection of a generalist herbivore exposed to native and alien seaweeds. Mar Pollut Bull 129(2):469–473

Noël L, Hawkins SJ, Jenkins SR, Thompson RC (2009) Grazing dynamics in intertidal rockpools: connectivity of microhabitats. J Exp Mar Biol Ecol 370(1–2):9–17

Pellizzari F, Silva MC, Silva EM, Medeiros A, Oliveira MC, Yokoya NS, Pupo D, Rosa LH, Colepicolo P (2017) Diversity and spatial distribution of seaweeds in the South Shetland Islands, Antarctica: an updated database for environmental monitoring under climate change scenarios. Polar Biol 40(8):1671–1685

Peters AF (2003) Molecular identification, taxonomy and distribution of brown algal endophytes, with emphasis on species Antarctica. In: Chapman ARO, Anderson RJ, Vreeland V, Davison IR (eds) Proceedings of the 17th International Seaweed Symposium. Oxford University Press, New York, pp 293–302

Poore AGB, Campbell AH, Coleman RA, Edgar GJ, Jormalainen V, Reynolds PL et al (2012) Global patterns in the impact of marine herbivores on benthic primary producers. Ecol Lett 15(8):912–922

Quartino ML, Deregibus D, Campana GL, Latorre GEJ, Momo FR (2013) Evidence of macroalgal colonization on newly ice-free areas following glacial retreat in Potter Cove (South Shetland Islands), Antarctica. PLoS One 8(3):e58223

R Core Team (2019) R: a language and environment for statistical computing, 3.0.3 edn. R Foundation for Statistical Computing, Vienna

Sanches PF, Pellizzari F, Horta PA (2016) Multivariate analyses of Antarctic and sub-Antarctic seaweed distribution patterns: an evaluation of the role of the Antarctic Circumpolar Current. J Sea Res 110:29–38

Schoenrock KM, Schram JB, Amsler CD, McClintock JB, Angus RA, Vohra YK (2016) Climate change confers a potential advantage to fleshy Antarctic crustose macroalgae over calcified species. J Exp Mar Biol Ecol 474:58–66

Schram JB, Schoenrock KM, McClintock JB, Amsler CD, Angus RA (2014) Multiple stressor effects of near-future elevated seawater temperature and decreased pH on righting and escape behaviors of two common Antarctic gastropods. J Exp Mar Biol Ecol 457:90–96

Schram JB, McClintock JB, Amsler CD, Baker BJ (2015) Impacts of acute elevated seawater temperature on the feeding preferences of an Antarctic amphipod toward chemically deterrent macroalgae. Mar Biol 162(2):425–433

Schram JB, Schoenrock KM, McClintock JB, Amsler CD, Angus RA (2016) Seawater acidification more than warming presents a challenge for two Antarctic macroalgal-associated amphipods. Mar Ecol Prog Ser 554:81–97

Segovia-Rivera V, Valdivia N (2016) Independent effects of grazing and tide pool habitats on the early colonisation of an intertidal community on western Antarctic Peninsula. Rev Chil Hist Nat 89:1–9

Silva R, Vinagre C, Kitahara MV, Acorsi IV, Mizrahi D, Flores AAV (2019) Sun coral invasion of shallow rocky reefs: effects on mobile invertebrate assemblages in southeastern Brazil. Biol Invasions 21(4):1339–1350

Simberloff D, Martin JL, Genovesi P, Maris V, Wardle DA, Aronson J et al (2013) Impacts of biological invasions: what's what and the way forward. Trends Ecol Evol 28(1):58–66

Valdivia N, Díaz MJ, Holtheuer J, Garrido I, Huovinen P, Gómez I (2014) Up, down, and all around: scale-dependent spatial variation in rocky-shore communities of Fildes Peninsula, King George Island, Antarctica. PLoS One 9(6):e100714

Valdivia N, Pardo LM, Macaya EC, Huovinen P, Gómez I (2019) Different ecological mechanisms lead to similar grazer controls on the functioning of periphyton Antarctic and sub-Antarctic communities. Prog Oceanogr 174:7–16

Vellend M (2016) The theory of ecological communities (MPB-57). Princeton University Press, Princeton

Vitousek PM, DAntonio CM, Loope LL, Rejmanek M, Westbrooks R (1997) Introduced species: a significant component of human-caused global change. N Z J Ecol 21(1):1–16

White LF, Shurin JB (2011) Density dependent effects of an exotic marine macroalga on native community diversity. J Exp Mar Biol Ecol 405(1–2):111–119

Zacher K, Hanelt D, Wiencke C, Wulff A (2007a) Grazing and UV radiation effects on an Antarctic intertidal microalgal assemblage: a long-term field study. Polar Biol 30(9):1203–1212

Zacher K, Wulff A, Molis M, Hanelt D, Wiencke C (2007b) Ultraviolet radiation and consumer effects on a field-grown intertidal macroalgal assemblage in Antarctica. Glob Chang Biol 13(6):1201–1215

Zacher K, Bernard M, Bartsch I, Wiencke C (2016) Survival of early life history stages of Arctic kelps (Kongsfjorden, Svalbard) under multifactorial global change scenarios. Polar Biol 39(11):2009–2020

Zenteno L, Cárdenas L, Valdivia N, Gómez I, Höfer J, Garrido I, Pardo LM (2019) Unraveling the multiple bottom-up supplies of an Antarctic nearshore benthic community. Prog Oceanogr 174:55–63

Chapter 14
Diversity and Functioning of Antarctic Seaweed Microbiomes

Juan Diego Gaitan-Espitia and Matthias Schmid

Abstract Antarctic macroalgae are important primary producers and habitat-forming species that play fundamental roles in Antarctic coastal habitats and sustain important communities of benthic organisms, including a not well-known microbiota. Anthropogenic pressures, e.g., increasing ocean temperatures and extreme events, have threatened the ecological integrity of several seaweed species and also can modify the range shifts (e.g., introduction of *Durvillaea antarctica* in King George Island), cause local extinctions, and alter the structure of these associations in their natural habitats. However, understanding and prediction of the responses of seaweeds to changing environment and rapid anthropogenic-driven change cannot be done without considering the associated microbiome. These complex microbial communities are intricately involved in the host health, defense, growth, and development of seaweeds, thus with far-reaching implications for the ecology of the whole coastal ecosystem. For most Antarctic seaweeds, the microbiome comprises a stable core as well as microbes whose presence depends on local conditions and transient microbial associates that are responsive to biotic and abiotic processes across spatial and temporal scales. In this chapter, we will explore the ecological and genetic diversity of microbiomes in Antarctic seaweeds and their functional connections.

Keywords Bacteria · Microbial communities · Genetic diversity · Holobionts · Microbiota

J. D. Gaitan-Espitia (✉)
The Swire Institute of Marine Science and The School of Biological Sciences, The University of Hong Kong, Hong Kong, SAR, China
e-mail: jdgaitan@hku.hk

M. Schmid
Institute for Marine and Antarctic Studies, University of Tasmania, Hobart, TAS, Australia
e-mail: matthias.schmid@utas.edu.au

I. Gómez, P. Huovinen (eds.), *Antarctic Seaweeds*,
https://doi.org/10.1007/978-3-030-39448-6_14

14.1 Introduction: Environment and Antarctic Seaweed Host-Microbiome

From very abundant and complex microbial communities to larger eukaryotes such as seaweeds and marine mammals, Antarctic organisms have evolve a variety of physiological, life history, and molecular adaptations that allow them to cope with this challenging environment (Hoyer et al. 2002; Clark et al. 2004; Marx et al. 2007; Rogers 2007). In Antarctic waters, the marine benthic flora is expected to be highly susceptible to global changes (i.e., invasive species, ocean warming, ocean acidification) due to its high degree of niche conservatism, reduced phenotypic plasticity, and the low species richness (Clayton 1994; see also Chap. 11 by Gómez and Huovinen). In order to survive in this harsh environment, Antarctic seaweeds have evolved structural and functional adaptations such as high synthesis of photoprotective substances and antioxidant activity for mitigation of photodamage, molecular adaptations of enzymes in order to maintain sufficient rates of enzyme-catalyzed reactions of key metabolic processes, the evolution of cold shock and antifreeze proteins, and different effective strategies for inorganic C acquisition and assimilation (Morgan-Kiss et al. 2006; Karsten et al. 2009; Gómez et al. 2009, 2019; Huovinen and Gómez 2013; Hurd et al. 2014). While there has been a lot of interest in documenting and understanding these functional adaptations, only scattered information is available about other potential adaptive mechanisms that influence functional regulation of Antarctic seaweeds. In fact, we know that seaweeds are not independent biological units. They rely on tight relationships with their associated microbiota for basic functions such as morphological development, growth, health, defense, nutrient supply, and adaptation/acclimation to environmental stress (Egan et al. 2013; Wichard 2015; Dittami et al. 2016; Singh and Reddy 2016). This suggests that seaweeds and their microbiome interact as a unified functional entity or holobiont (Egan et al. 2013). Therefore, it is essential to gain better understanding of the adaptive role of seaweed host-microbiome interactions in changing/extreme oceans and the mechanisms underlying their eco-evolutionary dynamics in Antarctica.

The complex microbial communities (mutualistic symbionts and hazardous pathogens) are intricately involved in health, defense, growth, and development of seaweeds (Friedrich 2012; Egan et al. 2013; Martin et al. 2014; Egan and Gardiner 2016; Singh and Reddy 2016). For most eukaryotic hosts, the microbiome comprises a stable core as well as microbes whose presence depends on local conditions and transient microbial associates that are responsive to biotic and abiotic processes across spatial and temporal scales (Vandenkoornhuyse et al. 2015; Hernandez-Agreda et al. 2016). Nevertheless, for seaweeds, the notion of host-specific microbial taxa in the core microbiome is not a rule of thumb: some species have specific microbial functional genes rather than taxonomic affiliations of microbial populations (Burke et al. 2011a, b). On the contrary, in other seaweeds, microbial taxonomic diversity can be unique to each type of seaweed host but with high degree of functional redundancy (Roth-Schulze et al. 2016). Whether or not these patterns (i.e., host-specific vs variable microbial diversity, functional redundancy) apply more broadly to species from other seaweed phyla in extreme polar regions, and considering temporal and spatial scales, is unknown.

14.2 Functional Interactions of Antarctic Seaweeds and Their Associated Microbiota

The seaweed microbiome represents a remarkably diverse array of microorganisms that includes bacteria, archaea, fungi, and other eukaryotic unicellular organisms and even viruses (Singh and Reddy 2016). Evidence suggests that the composition of the seaweed microbiome is likely to be modulated, at least in part, by the host because it is significantly different to the microbial community that is found in the surrounding environment (Brodie et al. 2016). These microorganisms exhibit strong seasonal and spatial changes in diversity and abundance (Tujula et al. 2010; Lachnit et al. 2011; Campbell et al. 2015) that are also influenced by day length, nutrient availability, and temperature (Gilbert et al. 2012; Moran 2015). Seaweeds can control and maintain communication with their associated microbes by producing secondary metabolites and exudates such as sugars and amino acids that serve as an energy source, as well as for a variety of functions including antimicrobials, allopathic molecules, and pathogen defenses (Friedrich 2012; Egan et al. 2013; Hollants et al. 2013; Rout 2014). The seaweed microbiome, in turn, can exert influence on trait expression by controlling growth and morphogenesis (Wichard 2015), acclimation and physiological responses of the host to environmental gradients (Dittami et al. 2016), survival and settlement of propagules (Morris et al. 2016), competition among seaweeds by inhibiting the germination of algal spores (Egan et al. 2001), and rapid defense adaptation (chemical controls) to the new bacterial epibionts and pathogens during range shifts (Saha et al. 2016; Arnaud-Haond et al. 2017). All of these effects influenced by associated microbial communities have important implications in the ecosystem services provided by the seaweed host (Rout 2014). Although bacteria are by far the most abundant members of the seaweed microbiome, our knowledge about epiphytic and endophytic bacterial communities living in Antarctic seaweeds is quite limited. Very few studies have explored this component of the seaweed microbiome. To date, a broad diversity of pigmented, Gram-positive epiphytic bacteria has been reported (mainly affiliated to Actinobacteria) (Leiva et al. 2015), some of them with antibiotic activity that may influence microbial dynamics (i.e., competition and colonization) of bacterial epibionts (Alvarado et al. 2018). On the contrary, other components of the seaweed microbiome such as fungi are very well documented. This group is characterized by the presence of symbiont, saprobe, and parasitic species that form fungal assemblages on the seaweed host. The structure and dynamics of these assemblages are controlled by environmental factors (e.g., availability of dissolved oxygen and organic matter), the seaweed host, and the capacity of the fungi to tolerate or detoxify the array of antifungal metabolites produced by the seaweed (Ogaki et al. 2019). The typical fungal community structure in Antarctic seaweeds is based on filamentous fungi and yeasts belonging to the genera *Geomyces*, *Antarctomyces*, *Oidiodendron*, *Penicillium*, *Phaeosphaeria*, *Aureobasidium*, *Cryptococcus*, *Leucosporidium*, *Metschnikowia*, and *Rhodotorula* (Loque et al. 2010). These assemblages are dominated by very few species (e.g., the filamentous fungus *Pseudogymnoascus pannorum* and the yeast *Metschnikowia australis*) (Loque et al. 2010; Godinho et al. 2013; Furbino et al. 2014; Ogaki et al. 2019), some of which have the potential to degrade algal biomass through agarolytic and carrageenolytic activities (Furbino et al. 2017).

14.3 Deciphering the Structure and Diversity of Seaweed Microbiomes

Pioneering studies of microbial communities in Antarctic marine environments employed cultivation-based methods, which are based on techniques aiming to grow microorganisms under controlled laboratory conditions (Pham and Kim 2012). This approach has several limitations as it only provides a reduced representation of the microbial community diversity because most of the species in environmental samples do not grow on standard media under laboratory conditions (Stewart 2012). However, in the last few years, a major progress in microbial research has been achieved, thanks to the development of several culture-independent molecular techniques (e.g., polymerase chain reaction (PCR), hybridization, fingerprinting, Sanger sequencing). These approaches have been routinely applied to study microbial communities in Antarctic soils, water, ice, and biota (Loque et al. 2010; Godinho et al. 2013; Furbino et al. 2014; Leiva et al. 2015; Moreno-Pino et al. 2016; Alvarado et al. 2018; Castro-Sowinski 2019; Chua et al. 2018; Ogaki et al. 2019). From these, the most powerful molecular approach to assess the structure and diversity of microbial communities relies in the use of high-throughput sequencing (HTS) of housekeeping genetic markers (amplicon sequencing) such as the internal transcribed spacer region (ITS; fungi), the 18S gene (fungi and other eukaryotes), viral RNA (viruses), and the 16 ribosomal RNA gene (16S rRNA; bacteria and archaea; Fig. 14.1). The main drawback of microbial community profiling using these universal markers is that they cannot directly identify metabolic or other functional capabilities of the microorganisms under study (Janda and Abbott 2007; Rausch et al. 2019). This can be done using shotgun metagenomic sequencing (Fig. 14.1). However, amplicon sequencing still offers many other advantages such as cost-efficiency, high precision, and fast speed characterization of taxonomic composition and phylogenetic diversity of microbial communities (Langille et al. 2013; Rausch et al. 2019). Although amplicon/targeted sequencing and shotgun sequencing strategies differ in the type of information produced, phylogeny and biomolecular function are strongly correlated. For instance, phylogenetic trees based on 16S rRNA closely resemble clusters obtained based on shared gene content, and researchers

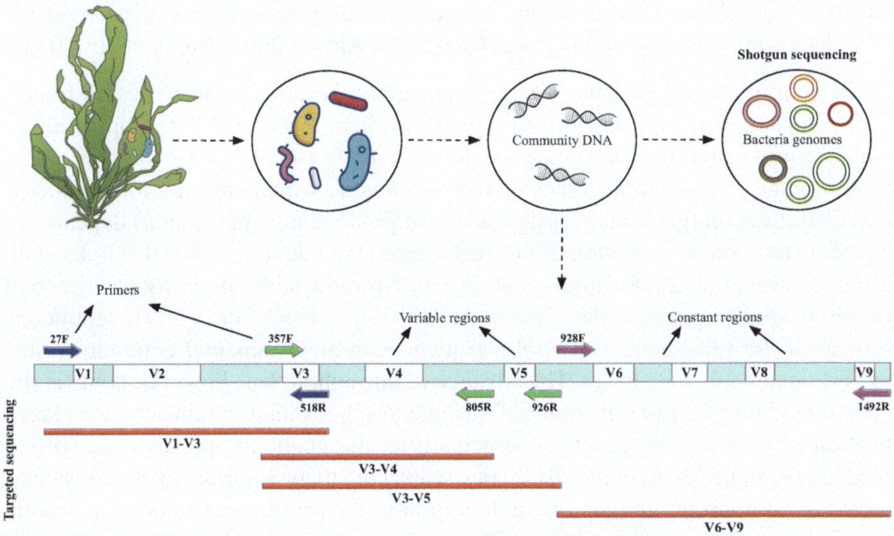

Fig. 14.1 Schematic representation of the workflow for bacterial community assessment. First, epiphytes are collected from the seaweed host for posterior total DNA extraction. The DNA from the microbiome is then used for targeted, amplicon sequencing (16S; down arrow) and/or shotgun metagenome sequencing (right arrow). For the targeted approach, different combinations (red lines) of variable (white boxes) and conserved (cyan boxes) 16S regions are sequenced via PCR with traditional primers (blue, green, and pink arrows). For the shotgun sequencing approach, DNA is fragmented and then sequenced using Illumina platforms, and the reads are finally sorted and assembled into contigs and into circular genomes

often infer properties of uncultured organisms from cultured relatives (Langille et al. 2013). Here, we will focus on targeted sequencing of bacterial 16S ribosomal RNA gene (Fig. 14.1), as this approach has been frequently used to characterize dynamics of microbial communities in Antarctic marine areas. Molecular approaches used for characterizing other groups, such as fungi, are well described in Ogaki et al. (2019).

The 16S rRNA gene is the most popular target gene in many molecular methods (e.g., fluorescence in situ hybridization, FISH; quantitative PCR; terminal restriction fragment length polymorphism, T-RFLP; denaturing-gradient gel electrophoresis, DGGE) of microbial studies (Bukin et al. 2019; Fuks et al. 2018; Janda and Abbott 2007; Sambo et al. 2018). This gene offers several advantages for the study of bacterial phylogeny and taxonomy. Such advantages are as follows:

(i) 16S rRNA is present in all prokaryotes.
(ii) Its function over time has not changed, suggesting that random sequence changes are a more accurate measure of time (evolution).
(iii) It is approximately 1600 base pairs long and includes nine hypervariable, fast-evolving regions (V1–V9; Fig. 14.1), which can be used to classify organisms at different taxonomic levels (genus or species).

(iv) The conserved, slow-evolving regions of this gene, which can be used for determining higher-ranking taxa (Janda and Abbott 2007; Bukin et al. 2019).

These conserved regions have structural characteristics that allow to design broad-spectrum, degenerated primers for polymerase chain reaction (PCR) amplification, which in turn can be used to isolate species-specific fast-evolving regions (Fig. 14.1) (Sambo et al. 2018). The accuracy of 16S rRNA gene sequencing as a tool in microbial identification (taxonomic assignment and phylogenetic placement) depends, in a great extent, on the selection of the 16S region (Pootakham et al. 2017; Fuks et al. 2018; Sambo et al. 2018; Bukin et al. 2019). Nowadays, the majority of microbial profiling studies utilizes the short-read V3–V4, V4–V5, or V5–V6 amplicons instead of the full-length 16S rRNA sequences in environmental community surveys (Pootakham et al. 2017). The advance in throughput has, however, come at the cost of read length, and this trade-off has inevitably resulted in less accurate classification of partial 16S sequences, especially at the genus or species level (Bukin et al. 2019; Pootakham et al. 2017). For seaweeds, there is currently no consensus on the most appropriate hypervariable region(s) for profiling associated microbial communities. Some of the studies using amplicon sequencing have targeted the 16S hypervariable regions V1–V3 (Campbell et al. 2015), whereas others have used the V3 (Lachnit et al. 2009), V4 (Lemay et al. 2018; Marzinelli et al. 2015), V3–V4 (Martin et al. 2015; Parrot et al. 2019), V6 (Brodie et al. 2016), or the full-length 16S (Leiva et al. 2015; Kumar et al. 2016; Alvarado et al. 2018; Serebryakova et al. 2018; Morrissey et al. 2019). Overall, and considering the percentage of sequences that retrieve hits from public databases (e.g., Greengenes, NCBI, RDP, and SILVA), it seems that hypervariable regions targeting V3–V4 (90%), V4 (91%), V4–V5 (88%), and V1–V9 (full-length; >99%) produce more reproducible results than V1–V2 (30%) or V1–V3 (40%). These findings have been also documented in microbial communities associated to other environmental and biological systems (Pootakham et al. 2017; Almeida et al. 2018; Pollock et al. 2018).

14.4 Variation of Bacterial Community Diversity in Antarctic Seaweeds

In Antarctica, the diversity of marine prokaryotic (bacteria) communities in seawater correlates to both physical (Wilkins et al. 2013) and chemical oceanographic conditions (Giudice and Azzaro 2019). These factors can be highly variable in space and time, shaping life in this polar region. As a result, prokaryotes have evolved a wide range of pro-survival mechanisms such as habitat selection, life cycle strategies, changes in cellular composition and enzyme activity, and the production of extracellular polymeric substances (De Maayer et al. 2014). These adaptations might provide some functional benefits for the seaweed hosts. However, microbial communities in seawater and sediments can differ in diversity and structure (as well as in their functional properties) from those associated to seaweeds (Egan et al.

2013). In coastal and marine sediments along the west part of the Antarctic Peninsula and Antarctic islands, the predominant bacteria include representatives of the phyla Proteobacteria, Actinobacteria, Chloroflexi, Bacteroidetes, Verrucomicrobia, and Firmicutes, followed by Acidobacteria and Cyanobacteria (Chua et al. 2018; Flocco et al. 2019). In seawater, on the other hand, the dominant phyla are Proteobacteria (mainly Alphaproteobacteria and Gammaproteobacteria) and Bacteroidetes (Gentile et al. 2006; Moreno-Pino et al. 2016). Although there are some similarities at the higher taxonomic level, major differences in bacterial community diversity can be found between sediments and seawater at the at genus level. In sediments, for example, the most predominant genera is Sphingomonas, while in seawater the more abundant bacteria belong to the family Rhodobacteraceae and the genera *Psychromonas*, *Pseudoalteromonas*, and *Balneatrix* (Gentile et al. 2006; Moreno-Pino et al. 2016; Chua et al. 2018; Flocco et al. 2019; Giudice and Azzaro 2019; Lo Giudice et al. 2019). This is also evidenced in bacterial communities associated to Antarctic seaweeds (Fig. 14.2) (Leiva et al. 2015; Alvarado et al. 2018; Gaitan-Espitia et al. unpublished). For instance, in a recent work, Gaitan-Espitia et al. (unpublished data) used the full-length 16S rRNA gene and PacBio SMRT sequencing in order to assess the diversity of microbiomes associated to some of the most abundant seaweeds in the north of the Antarctic Peninsula. This study included representative of the phylum Ochrophyta (*Adenocystis utricularis*, *Geminocarpus geminatus*, *Ascoseira mirabilis*, *Desmarestia antarctica*, *D. menziesii*,

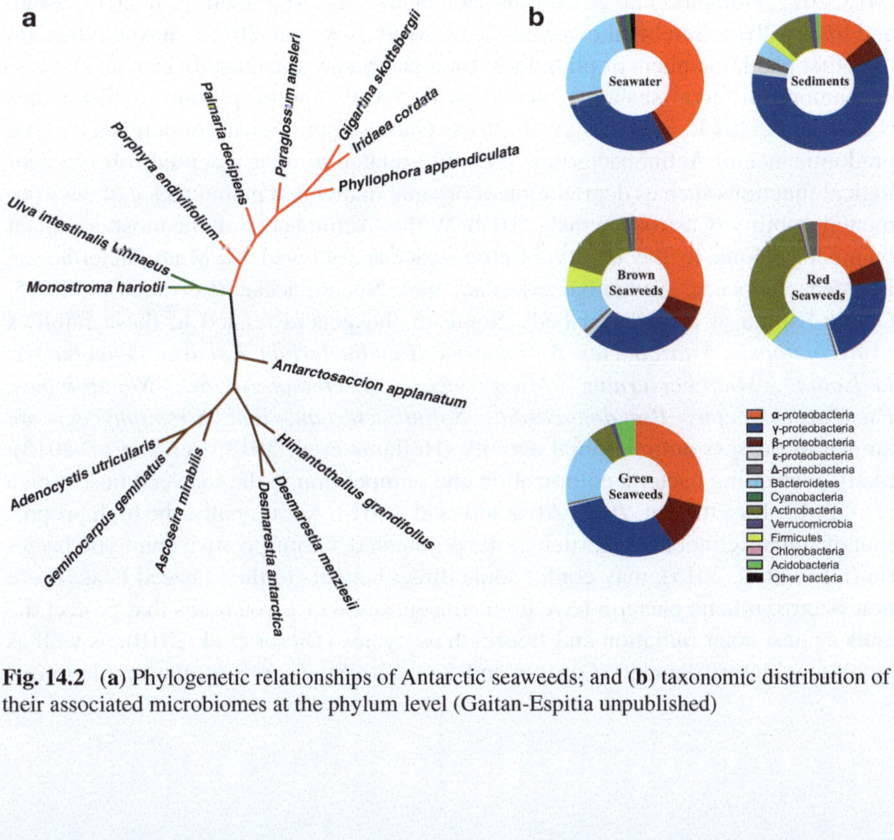

Fig. 14.2 (**a**) Phylogenetic relationships of Antarctic seaweeds; and (**b**) taxonomic distribution of their associated microbiomes at the phylum level (Gaitan-Espitia unpublished)

Himantothallus grandifolius, *Antarctosaccion applanatum*), Chlorophyta (*Monostroma hariotii*, *Ulva intestinalis* Linnaeus), and Rhodophyta (*Porphyra endiviifolium*, *Paraglossum amsleri*, *Phyllophora appendiculata*, *Iridaea cordata*, *Gigartina skottsbergii*, *Palmaria decipiens*) (Fig. 14.2). These species are characterized by different evolutionary histories, physiological and genomic architectures, and distribution along the vertical zonation of Antarctic shores (Klöser et al. 1996; Wiencke et al. 2007; Valdivia et al. 2014; Pellizzari et al. 2017). Overall, the findings indicate that microbial community composition (at the family and genera level) differs among species of seaweeds despite shared distribution in the vertical zonation (Fig. 14.2). This suggests that for Antarctic seaweeds, the composition of epiphytic bacteria is, at least in part, likely regulated by the host (Gaitan-Espitia et al. unpublished).

Studies analyzing the epiphytic and endophytic bacterial communities of temperate and tropical seaweeds have shown different phylogenetic patterns across ecological (e.g., temporal, spatial) and evolutionary (e.g., host) scales (Burke et al. 2011b; Lachnit et al. 2011; Egan et al. 2013, 2017; Hollants et al. 2013; Campbell et al. 2015; Marzinelli et al. 2015; Singh and Reddy 2016; Alvarado et al. 2018; Serebryakova et al. 2018; Morrissey et al. 2019). This is consistent with findings reported for Antarctic seaweed microbiomes (Alvarado et al. 2018; Leiva et al. 2015; Gaitan-Espitia et al. unpublished). However, in tropical and temperate seaweeds, Proteobacteria and Firmicutes are generally the most abundant bacteria associated with seaweed hosts (Burke et al. 2011b; Lachnit et al. 2011; Egan et al. 2013, 2017; Hollants et al. 2013; Campbell et al. 2015; Marzinelli et al. 2015; Singh and Reddy 2016; Serebryakova et al. 2018; Morrissey et al. 2019). In Antarctica, on the other hand, members of phylum Actinobacteria are the most diverse and persistent among different seaweed species, with a very low proportion of Firmicutes (Leiva et al. 2015; Alvarado et al. 2018; Gaitan-Espitia et al. unpublished). The predominance of Actinobacteria is probably related to their essential roles in ecological functions such as degradation of organic matter and maintenance of environmental stability (Castro-Sowinski 2019). Within Actinobacteria, the most dominant members belong to the family Micrococcaceae followed by Microbacteriaceae, Dermabacteriaceae, Sanguibacteriaceae, and Nocardiaceae (Leiva et al. 2015; Gaitan-Espitia et al. unpublished). Some of the genera related to these families (*Amycolatopsis*, *Arthrobacter*, *Agrococcus*, *Brevibacterium*, *Kocuria*, *Leucobacter*, *Leifsonia*, *Microbacterium*, *Micrococcus*, *Micromonospora*, *Nocardiopsis*, *Pseudarthrobacter*, *Pseudonocardia*, *Salinibacterium*, *and Streptomyces*) are known to possess antimicrobial activity (Hollants et al. 2013; Leiva et al. 2015), likely controlling bacteria colonization and competition in the seaweed host (Leiva et al. 2015; Busetti et al. 2017; Alvarado et al. 2018). Additionally, the high proportion of some Actinobacteria, such as the pigmented, Gram-positive epiphytic bacteria (Leiva et al. 2015), may confer some direct benefits to the seaweed host. These non-photosynthetic bacteria have high concentration of carotenoids that protect the cells against solar radiation and freeze–thaw cycles (Dieser et al. 2010), as well as against oxidative damage (Glaeser and Klug 2005). Therefore, the regulation of biofilm formation of these epiphytic bacteria may represent an adaptive mechanism to enhance tolerances and survival of Antarctic seaweeds.

14.5 Conclusions and Future Perspectives

Species diversity is widely recognized as an essential adaptive ecological property that increases the robustness and evolvability of biological systems (Kahilainen et al. 2014). This applies to the framework of the seaweed holobiont, because the diversity of microorganisms that live and interact with the seaweed host is known to play a fundamental role in its health and resilience, particularly when faced with environmental stress (Friedrich 2012; Egan et al. 2013). Most studies exploring seaweed microbiomes have determined the diversity (species and genes) of microbes living in association with a seaweed host (mainly on surfaces = epiphytic communities) (Egan et al. 2001, 2013; Burke et al. 2011a, b; Friedrich 2012; Hollants et al. 2013; Rout 2014; Wichard 2015; Dittami et al. 2016; Morris et al. 2016; Saha et al. 2016; Arnaud-Haond et al. 2017). These studies reveal a seaweed microbiota that is complex and very diverse and consists of a number of partners of different origins and evolutionary trajectories (Friedrich 2012; Egan et al. 2013). However, diversity without context provides limited insights into the mechanisms underpinning the assembly of the microbiome, community patterns, and the benefits for the seaweed host. High diversity is not necessarily "better" or "healthier," whereas lower diversity is not necessarily indicative of less stable or less "healthy" community (Shade 2016). In fact, microbial diversity can change across latitudinal–temporal scales linked to biotic and abiotic factors rather than to the health or resilience of the host (Gilbert et al. 2012; Koskella et al. 2017). Therefore, moving away from taxonomic diversity toward functional diversity linked to the seaweed host phenotype and performance under characterized environmental conditions could help us to understand ecological drivers that shape the Antarctic seaweed microbiome and its functional stability/plasticity in changing environments.

Acknowledgments The work outlined in this review was partially supported by the Chilean Antarctic Institute (INACH), the Commonwealth Scientific and Industrial Research Organisation (CSIRO), the University of Hong Kong, and the Institute for Marine and Antarctic Studies (IMAS) from the University of Tasmania. JDGE was supported by the Research Grants Council of Hong Kong via the Early Career Scheme (Project ECS-27124318). MS was funded through the Deutsche Forschungsgemeinschaft (DFG, grant ID: SCHM 3335/1-1).

References

Almeida A, Mitchell AL, Tarkowska A, Finn RD (2018) Benchmarking taxonomic assignments based on 16S rRNA gene profiling of the microbiota from commonly sampled environments. Gigascience 7:1–10

Alvarado P, Huang Y, Wang J, Garrido I, Leiva S (2018) Phylogeny and bioactivity of epiphytic gram-positive bacteria isolated from three co-occurring Antarctic macroalgae. Antonie van Leeuwenhoek Int J Gen Mol Microbiol 111:1543–1555

Arnaud-Haond S, Aires T, Candeias R, Teixeira S, Duarte C, Valero M, Serrão E (2017) Entangled fates of holobiont genomes during invasion: nested bacterial and host diversities in *Caulerpa taxifolia*. Mol Ecol 26:2379–2391

Brodie J, Williamson C, Barker GL, Walker RH, Briscoe A, Yallop M (2016) Characterising the microbiome of *Corallina officinalis*, a dominant calcified intertidal red alga. FEMS Microbiol Ecol 92:1–12

Bukin YS, Galachyants YP, Morozov IV, Bukin SV, Zakharenko AS, Zemskaya TI (2019) The effect of 16s rRNA region choice on bacterial community metabarcoding results. Sci Data 6:1–14

Burke C, Steinberg P, Rusch DB, Kjelleberg S, Thomas T (2011a) Bacterial community assembly based on functional genes rather than species. Proc Natl Acad Sci U S A 108:14288–14293

Burke C, Thomas T, Lewis M, Steinberg P, Kjelleberg (2011b) Composition, uniqueness and variability of the epiphytic bacterial community of the green alga *Ulva australis*. ISME J 5:590–600

Busetti A, Maggs CA, Gilmore BF (2017) Marine macroalgae and their associated microbiomes as a source of antimicrobial chemical diversity. Eur J Phycol 52:452–465

Campbell AH, Marzinelli EM, Gelber J, Steinberg PD (2015) Spatial variability of microbial assemblages associated with a dominant habitat-forming seaweed. Front Microbiol 6:1–10

Castro-Sowinski S (ed) (2019) The ecological role of micro-organisms in the Antarctic environment. Springer Polar Sciences, Springer, Cham

Chua CY, Yong ST, González MA, Lavin P, Cheah YK, Tan GYA, Wong CMVL (2018) Analysis of bacterial communities of King George and Deception Islands, Antarctica using high-throughput sequencing. Curr Sci 115:1701–1705

Clark MS, Clarke A, Cockell CS, Convey P, Detrich HW, Fraser KPP, Johnston IA, Methe BA et al (2004) Antarctic genomics. Comp Funct Genomics 16:230–238

Clayton MN (1994) Evolution of the Antarctic marine benthic algal flora. J Phycol 30:897–904

De Maayer P, Anderson D, Cary C, Cowan DA (2014) Some like it cold: understanding the survival strategies of psychrophiles. EMBO Rep 15:508–517

Dieser M, Greenwood M, Foreman CM (2010) Carotenoid pigmentation in Antarctic heterotrophic bacteria as a strategy to withstand environmental stresses. Arct Antarct Alp Res 42:396–405

Dittami SM, Duboscq-Bidot L, Perennou M, Gobet A, Corre E, Boyen C, Tonon T (2016) Host-microbe interactions as a driver of acclimation to salinity gradients in brown algal cultures. ISME J 10:51–63

Egan S, Gardiner M (2016) Microbial dysbiosis: rethinking disease in marine ecosystems. Front Microbiol 7:1–8

Egan S, James S, Holmström C, Kjelleberg S (2001) Inhibition of algal spore germination by the marine bacterium *Pseudoalteromonas tunicata*. FEMS Microbiol Ecol 35:67–73

Egan S, Harder T, Burke C, Steinberg P, Kjelleberg S, Thomas T (2013) The seaweed holobiont: understanding seaweed-bacteria interactions. FEMS Microbiol Rev 37:462–476

Egan S, Kumar V, Nappi J, Gardiner M (2017) Microbial diversity and symbiotic interactions with macroalgae. In: Grube M, Seckbach J, Muggia L (eds) Algal and cyanobacteria symbioses. World Scientific Publishing Europe Ltd, London/Singapore, pp 493–546

Flocco C, Cormack W, Smalla K (2019) Antarctic soil microbial communities in a changing environment: their contributions to the sustainability of Antarctic ecosystems and the bioremediation of anthropogenic pollution. In: Castro-Sowinski S (ed) The ecological role of micro-organisms in the Antarctic environment. Springer Polar Sciences, Springer, Cham, pp 133–161

Friedrich MW (2012) Bacterial communities on macroalgae. In: Wiencke C, Bischof K (eds) Seaweed biology: novel insights into ecophysiology, ecology and utilization. Springer, Berlin, pp 189–201

Fuks G, Elgart M, Amir A, Zeisel A, Turnbaugh PJ, Soen Y, Shental N (2018) Combining 16S rRNA gene variable regions enables high-resolution microbial community profiling. Microbiome 6:1–13

Furbino LE, Godinho VM, Santiago IF, Pellizari FM, Alves TMA, Zani CL, Junior PAS et al (2014) Diversity patterns, ecology and biological activities of fungal communities associated with the endemic macroalgae across the Antarctic Peninsula. Microb Ecol 67:775–787

Furbino LE, Pellizzari FM, Neto PC, Rosa CA, Rosa LH (2017) Isolation of fungi associated with macroalgae from maritime Antarctica and their production of agarolytic and carrageenolytic activities. Polar Biol 41:527–535

Gentile G, Giuliano L, D'Auria G, Smedile F, Azzaro M, De Domenico M, Yakimov MM (2006) Study of bacterial communities in Antarctic coastal waters by a combination of 16S rRNA and 16S rDNA sequencing. Environ Microbiol 8:2150–2161

Gilbert JA, Steele JA, Caporaso JG, Steinbrück L, Reeder J, Temperton B, Huse S, McHardy AC, Knight R et al (2012) Defining seasonal marine microbial community dynamics. ISME J 6:298–308

Giudice A, Azzaro M (2019) Diversity and ecological roles of prokaryotes in the changing Antarctic marine environment. In: Castro-Sowinski (ed) the ecological role of micro-organisms in the Antarctic environment. Springer Polar Sciences, Springer, Cham, pp 109–131

Glaeser J, Klug G (2005) Photo-oxidative stress in *Rhodobacter sphaeroides*: protective role of carotenoids and expression of selected genes. Microbiology 151:1927–1938

Godinho VM, Furbino LE, Santiago IF, Pellizzari FM, Yokoya NS, Pupo D, Alves TMA, Junior PAS et al (2013) Diversity and bioprospecting of fungal communities associated with endemic and cold-adapted macroalgae in Antarctica. ISME J 7:1434–1451

Gómez I, Wulff A, Roleda MY, Huovinen P, Karsten U, Quartino ML, Dunton K, Wiencke C (2009) Light and temperature demands of marine benthic microalgae and seaweeds in polar regions. Bot Mar 52:593–608

Gómez I, Navarro NP, Huovinen P (2019) Bio-optical and physiological patterns in Antarctic seaweeds: a functional trait based approach to characterize vertical zonation. Prog Oceanogr 174:17–27

Hernandez-Agreda A, Leggat W, Bongaerts P, Ainsworth T (2016) The microbial signature provides insight into the mechanistic basis of coral success across reef habitats. MBio 7:1–10

Hollants J, Leliaert F, De Clerck O, Willems A (2013) What we can learn from sushi: a review on seaweed-bacterial associations. FEMS Microbiol Ecol 83:1–16

Hoyer K, Karsten U, Wiencke C (2002) Induction of sunscreen compounds in Antarctic macroalgae by different radiation conditions. Mar Biol 141:619–627

Huovinen P, Gómez I (2013) Photosynthetic characteristics and UV stress tolerance of Antarctic seaweeds along the depth gradient. Polar Biol 36:1319–1332

Hurd C, Harrison P, Bischof K, Lobban C (eds) (2014) Seaweed ecology and physiology. Cambridge University Press, Cambridge

Janda JM, Abbott SL (2007) 16S rRNA gene sequencing for bacterial identification in the diagnostic laboratory: pluses, perils, and pitfalls. J Clin Microbiol 45:2761–2764

Kahilainen A, Puurtinen M, Kotiaho JS (2014) Conservation implications of species-genetic diversity correlations. Glob Ecol Conserv 2:315–323

Karsten U, Wulff A, Roleda MY, Müller R, Steinhoff FS, Fredersdorf J, Wiencke C (2009) Physiological responses of polar benthic algae to ultraviolet radiation. Bot Mar 52:639–654

Klöser H, Quartino ML, Wiencke C (1996). Distribution of macroalgae and macroalgal communities in gradients of physical conditions in Potter Cove, King George Island, Antarctica. Hydrobiologia 333: 1–17

Koskella B, Hall LJ, Metcalf CJE (2017) The microbiome beyond the horizon of ecological and evolutionary theory. Nat Ecol Evol 1:1606–1615

Kumar V, Zozaya-Valdes E, Kjelleberg S, Thomas T, Egan S (2016) Multiple opportunistic pathogens can cause a bleaching disease in the red seaweed *Delisea pulchra*. Environ Microbiol 18:3962–3975

Lachnit T, Blümel M, Imhoff JF, Wahl M (2009) Specific epibacterial communities on macroalgae: phylogeny matters more than habitat. Aquat Biol 5:181–186

Lachnit T, Meske D, Wahl M, Harder T, Schmitz R (2011) Epibacterial community patterns on marine macroalgae are host-specific but temporally variable. Environ Microbiol 13:655–665

Langille M, Zaneveld J, Caporaso J, McDonald D, Knights D, Reyes J, Clemente J et al (2013) Predictive functional profiling of microbial communities using 16S rRNA marker gene sequences. Nat Biotechnol 31:814

Leiva S, Alvarado P, Huang Y, Wang J, Garrido I (2015) Diversity of pigmented gram-positive bacteria associated with marine macroalgae from Antarctica. FEMS Microbiol Lett 362:1–6

Lemay MA, Martone PT, Hind KR, Lindstrom SC, Wegener-Parfrey L (2018) Alternate life history phases of a common seaweed have distinct microbial surface communities. Mol Ecol 27:3555–3568

Lo Giudice A, Caruso G, Rizzo C, Papale M, Azzaro M (2019) Bacterial communities versus anthropogenic disturbances in the Antarctic coastal marine environment. Environ Sustain 2:297–310

Loque CP, Medeiros AO, Pellizzari FM, Oliveira EC, Rosa CA, Rosa LH (2010) Fungal community associated with marine macroalgae from Antarctica. Polar Biol 33:641–648

Martin M, Portetelle D, Michel G, Vandenbol M (2014) Microorganisms living on macroalgae: diversity, interactions, and biotechnological applications. Appl Microbiol Biotechnol 98:2917–2935

Martin M, Barbeyron T, Martin R, Portetelle D, Michel G, Vandenbol M (2015) The cultivable surface microbiota of the brown alga *Ascophyllum nodosum* is enriched in macroalgal-polysaccharide-degrading bacteria. Front Microbiol 6:1–14

Marx JC, Collins T, D'Amico S, Feller G, Gerday C (2007) Cold-adapted enzymes from marine Antarctic microorganisms. Mar Biotechnol 9:293–304

Marzinelli E, Campbell A, Zozaya E, Vergés A, Nielsen S, Wernberg T, de Bettignies T et al (2015) Continental-scale variation in seaweed host-associated bacterial communities is a function of host condition, not geography. Environ Microbiol 17:4078–4088

Moran M (2015) The global ocean microbiome. Science 80:350, aac84551–aac84556

Moreno-Pino M, De la Iglesia R, Valdivia N, Henríquez-Castilo C, Galán A, Díez B, Trefault N (2016) Variation in coastal Antarctic microbial community composition at sub-mesoscale: spatial distance or environmental filtering? FEMS Microbiol Ecol 92:1–13

Morgan-Kiss RM, Priscu JC, Pocock T, Gudynaite-Savitch L, Huner NPA (2006) Adaptation and acclimation of photosynthetic microorganisms to permanently cold environments microbiol. Mol Biol Rev 70:222–252

Morris M, Haggerty J, Papudeshi B, Vega A, Edwards M, Dinsdale E (2016) Nearshore pelagic microbial community abundance affects recruitment success of giant kelp, *Macrocystis pyrifera*. Front Microbiol 7:1–12

Morrissey KL, Çavas L, Willems A, De Clerck O (2019) Disentangling the influence of environment, host specificity and thallus differentiation on bacterial communities in siphonous green seaweeds. Front Microbiol 10:1–12

Ogaki MB, de Paula MT, Ruas D, Pellizzari FM, García-Laviña CX, Rosa LH (2019) Marine fungi associated with Antarctic macroalgae. In: Castro-Sowinski S (ed) The ecological role of microorganisms in the Antarctic environment. Springer Polar Sciences, Springer, Cham, pp 239–255

Parrot D, Blümel M, Utermann C, Chianese G, Krause S, Kovalev A, Gorb SN, Tasdemir D (2019) Mapping the surface microbiome and metabolome of brown seaweed *Fucus vesiculosus* by amplicon sequencing, integrated metabolomics and imaging techniques. Sci Rep 9:1–17

Pellizzari F, Silva MC, Silva EM, Medeiros A, Oliveira MC, Yokoya NS, Pupo D, Rosa LH, Colepicolo P (2017) Diversity and spatial distribution of seaweeds in the South Shetland Islands, Antarctica: an updated database for environmental monitoring under climate change scenarios. Polar Biol 40:1671–1685

Pham VHT, Kim J (2012) Cultivation of unculturable soil bacteria. Trends Biotechnol 30:475–484

Pollock J, Glendinning L, Wisedchanwet T, Watson M (2018) The madness of microbiome: attempting to find consensus. Appl Environ Microbiol 84:1–12

Pootakham W, Mhuantong W, Yoocha T, Putchim L, Sonthirod C, Naktang C, Thongtham N, Tangphatsornruang S (2017) High resolution profiling of coral-associated bacterial communities using full-length 16S rRNA sequence data from PacBio SMRT sequencing system. Sci Rep 7:1–14

Rausch P, Hermes B, Doms S, Dagan T, Domin H, Fraune S, Humeida UH, Heinsen F, Jahn M, Jaspers C et al (2019) Comparative analysis of amplicon and metagenomic sequencing methods reveals key features in the evolution of animal metaorganisms. Microbiome 7:1–19

Rogers AD (2007) Evolution and biodiversity of Antarctic organisms: a molecular perspective. Philos Trans R Soc B Biol Sci 362:2191–2214

Roth-Schulze AJ, Zozaya-Valdés E, Steinberg PD, Thomas T (2016) Partitioning of functional and taxonomic diversity in surface-associated microbial communities. Environ Microbiol 18:4391–4402

Rout ME (2014) The plant microbiome. In: Paterson AH (ed) Genomes of herbaceous land plants. Advances in Botanical Research. Elsevier Ltd., London, pp 279–309

Saha M, Wiese J, Weinberger F, Wahl M (2016) Rapid adaptation to controlling new microbial epibionts in the invaded range promotes invasiveness of an exotic seaweed. J Ecol 104:969–978

Sambo F, Finotello F, Lavezzo E, Baruzzo G, Masi G, Peta E, Falda M, Toppo S, Barzon L, Di Camillo B (2018) Optimizing PCR primers targeting the bacterial 16S ribosomal RNA gene. BMC Bioinformatics 19:1–10

Serebryakova A, Aires T, Viard F, Serrão EA, Engelen AH (2018) Summer shifts of bacterial communities associated with the invasive brown seaweed *Sargassum muticum* are location and tissue dependent. PLoS One 13:1–18

Shade A (2016) Diversity is the question, not the answer. ISME J 4:e2287v1

Singh R, Reddy C (2016) Unraveling the functions of the macroalgal microbiome. Front Microbiol 6:1–8

Stewart EJ (2012) Growing unculturable bacteria. J Bacteriol 194:4151–4160

Tujula NA, Crocetti GR, Burke C, Thomas T, Holmström C, Kjelleberg S (2010) Variability and abundance of the epiphytic bacterial community associated with a green marine ulvacean alga. ISME J 4:301–311

Valdivia N, Díaz MJ, Holtheuer J, Garrido I, Huovinen P, Gómez I (2014) Up, down, and all around: scale-dependent spatial variation in rocky-shore communities of Fildes Peninsula, King George Island, Antarctica. PLoS One 9:1–12

Vandenkoornhuyse P, Quaiser A, Duhamel M, Le Van A, Dufresne A (2015) The importance of the microbiome of the plant holobiont. New Phytol 206:1196–1206

Wichard T (2015) Exploring bacteria-induced growth and morphogenesis in the green macroalga order Ulvales (Chlorophyta). Front Plant Sci 6:1–19

Wiencke C, Clayton MN, Gómez I, Iken K, Lüder UH, Amsler CD, Karsten U, Hanelt D, Bischof K, Dunton K (2007) Life strategy, ecophysiology and ecology of seaweeds in polar waters. Rev Environ Sci Biotechnol 6:95–126

Wilkins D, Yau S, Williams TJ, Allen MA, Brown MV, Demaere MZ, Lauro FM, Cavicchioli R (2013) Key microbial drivers in Antarctic aquatic environments. FEMS Microbiol Rev 37:303–335

Chapter 15
Seaweeds in the Antarctic Marine Coastal Food Web

Fernando R. Momo, Georgina Cordone, Tomás I. Marina, Vanesa Salinas, Gabriela L. Campana, Mariano A. Valli, Santiago R. Doyle, and Leonardo A. Saravia

Abstract Antarctic macroalgae are the basis of marine food webs in most coastal environments, especially the more confined ones such as bays and fjords. Whether through direct consumption or via detritus, their role in maintaining biodiversity is

F. R. Momo (✉)
Instituto de Ciencias, Universidad Nacional de General Sarmiento, Los Polvorines, Buenos Aires, Argentina

Departamento de Ciencias Básicas, Universidad Nacional de Luján (UNLu), Luján, Buenos Aires, Argentina
e-mail: fmomo@campus.ungs.edu.ar

G. Cordone
Centro Nacional Patagónico (CCT CONICET-CENPAT), Centro para el Estudio de Sistemas Marinos (CESIMAR), Puerto Madryn, Chubut, Argentina
e-mail: gcordone@cenpat-conicet.gob.ar

T. I. Marina
Consejo Nacional de Investigaciones Científicas y Técnicas (CONICET), Centro Austral de Investigaciones Científicas (CADIC), Ushuaia, Tierra del Fuego, Argentina
e-mail: tomasimarina@cadic-conicet.gob.ar

V. Salinas · L. A. Saravia
Instituto de Ciencias, Universidad Nacional de General Sarmiento, Los Polvorines, Buenos Aires, Argentina
e-mail: vsalinas@campus.ungs.edu.ar; lsaravia@campus.ungs.edu.ar

G. L. Campana
Departamento de Biología Costera, Instituto Antártico Argentino, Buenos Aires, Argentina

Departamento de Ciencias Básicas, Universidad Nacional de Luján (UNLu), Luján, Buenos Aires, Argentina
e-mail: gcampana@dna.gov.ar

M. A. Valli
Departamento de Ciencias Básicas, Universidad Nacional de Luján (UNLu), Luján, Buenos Aires, Argentina

S. R. Doyle
Instituto de Ecología y Desarrollo Sustentable (INEDES), Universidad Nacional de Luján (UNLu), Luján, Buenos Aires, Argentina
e-mail: sdoyle@campus.ungs.edu.ar

I. Gómez, P. Huovinen (eds.), *Antarctic Seaweeds*,
https://doi.org/10.1007/978-3-030-39448-6_15

essential. However, their relevance is due not only by direct trophic interactions but also by indirect feedbacks, since macroalgae act as a habitat and refuge for multiple benthic organisms and as a substrate for epiphytic microalgae. Macroalgae also establish relations of exploitative competition, apparent competition, and mutualism. The control over the biomass and diversity of the macroalgae itself does not seem to be due to trophic interactions (top-down control) but rather to competition and diverse abiotic factors such as substrata and light availability or physical disturbances (ice scouring). The extreme connectivity of trophic networks linked to algae and their detritus determines that food webs are robust to local extinctions; however, non-trophic interactions indicate that changes that affect the growth, biomass, and distribution of macroalgae can have dramatic effects on the diversity of their associated fauna and, indirectly, on the networks of consumers of that fauna. In this chapter, we present a detailed description of macroalgae relationship networks and analyze the stability of the Antarctic community using food web theory.

Keywords Biological interactions · Community structure · Ecological networks · Trophic level

15.1 Introduction

Macroalgae and their detritus or fragments are the basal energy source of Antarctic coastal food webs. Approximately 150 species of macroalgae have been reported at Antarctica (Wiencke et al. 2014; Pellizzari et al. 2017; see also Chap. 2 by Oliveira et al.). Several local studies proved that macroalgae together with microphytobenthos constitute a direct pathway of energy and matter into organisms that feed on them (Wiencke et al. 2014; Ahn et al. 2016) and indirectly through detritus pathway (Iken et al. 1997; Tatián et al. 2004; Amsler et al. 2005, 2012; Quartino et al. 2008; Campana et al. 2018). Besides this, benthic algae in some locations can have high primary productivity similar to or higher than phytoplankton production (Fogg 1977). Furthermore, Antarctic macroalgae constitute structural and even chemical refuges from predation (Amsler et al. 1999, 2009), habitat for a variety of associated fauna (Huang et al. 2007) that provide a large fraction of the secondary production to the benthos (Gómez et al. 2009), and substrate for epiphytic communities (Majewska et al. 2016). Thus, macroalgae play a fundamental role in the food web being able to influence ecosystem dynamics and stability through propagation of direct and indirect effects.

The network of interspecific relationships involving seaweeds is very complex and includes not only consumption of algae but also several types of relations. Direct interactions are, for instance, competition among different macroalgae species for space (Smale and Barnes 2008; Quartino et al. 2013; Campana et al. 2009,

2018), when hard substrate is a limiting resource, or for light (Klöser et al. 1994, 1996; Deregibus et al. 2016). On the other hand, seaweeds can establish relationships with potentially harmful light-blocking epiphytes, whereas macroalgae provide a substrate for these organisms; however, seaweeds are chemically defended against mesograzers (and other herbivores; see Chap. 18 by Amsler et al.) that, in turn, use them as a refuge from predation: the macroalgae benefit in return because the mesograzers remove epiphytic algae (Amsler et al. 2014), and in this way, commensalism can become mutualism.

However, not only direct effects play an important role: several seaweed species share predators and establish a complex dynamics that may be observed as much as apparent competition or "apparent" mutualism depending on the density of predators and the closeness of macroalgae species involved. When a group of species establish different types of interactions (i.e., trophic, commensalism, mutualism) in an intricate manner, it is challenging to decide if the regulation of numbers or biomass is mainly controlled by top-down, bottom-up, or wasp-waist effects or by different and nontrivial combinations of them. In consequence, it is necessary to take into account the complexity of the ecosystem and all its ecological interactions (trophic and non-trophic) to understand its stability in response to environmental changes. In particular, macroalgal species are exposed to different environmental disturbances such as sediment input or ice scouring (Sahade et al. 2015; Quartino et al. 2013; Deregibus et al. 2016), resulting in a complex mosaic of effects. In this sense, the understanding of the effects of climate change in the western Antarctic Peninsula and a possible losses and gains in biodiversity urgently needs a deep knowledge of the relations between species' functional roles and the ecosystem structure (Woodward et al. 2010).

15.2 Food Webs and Seaweeds

The complex network of interactions varies in space and time, and not all interactions take place simultaneously (Poisot et al. 2015), but the interactions should not be studied separately but be understood as a whole. For this purpose, we have a powerful tool: the food web theory (Delmas et al. 2018).

After comparing 19 food web properties, Dunne et al. (2004) concluded that the excessively low percentage of basal taxa in marine food webs compared to other systems is clearly an artifact due to the poor resolution of primary producers and consumer links to them. One of the methodological strengths of the food web studied here is the high taxonomic resolution of basal nodes. A good taxonomic resolution of the lower trophic levels, such as the macroalgal community, is essential to understand ecosystem functioning, since there seems to be a species-specific selective consumption (Iken et al. 1997). Implications for ecosystem functioning and stability are only possible to elucidate in food webs where the species involved in energy and matter transfer processes are well represented.

A first step to identify the seaweeds embedded in the functional context of their community is to visualize the food web in which they are integrated. For example, Fig. 15.1 shows the food web of the Potter Cove ecosystem (Marina et al. 2018). We can see that the base of the pyramid is filled with several macroalgal species accompanied by fresh and old detritus and microalgae. Arrows indicate matter and energy fluxes through predation; the node size is proportional to its number of connections; the placement of the node on the vertical coordinate is related to its weighted trophic level. Also, macroalgae contribute to the detritivore pathway by decomposition and accumulation in the cove, generating a link between macroalgae species and detritus that is not represented in Fig. 15.1. However, we consider it in our studies and conclusions.

An understanding of the relations between species functional roles and ecosystem structure is an indispensable step toward the comprehension of change in the western Antarctic Peninsula and subsequent biodiversity loss and gain (Woodward et al. 2010). Several network properties are commonly used to describe food webs (Dunne et al. 2002): (1) number of species, *S*; (2) total number of interactions or trophic links, *L*; (3) number of interactions per species or linkage density, *L/S*; (4) connectance or trophic links divided by total number of possible interactions, $C = L/S^2$; (5) percentage of top species (species with preys, but without predators); (6) intermediate species (species having preys and predators); (7) basal species (species with predators/consumers, but without preys); and (8) percentage of omnivores (species eating prey from more than one trophic level). Additionally, the topology of the food web can be studied by measuring three more properties: (9) characteristic path length (ChPath), which is the average shortest path length between all pairs of species; (10) clustering coefficient (CC), which is the average fraction of pairs of species one link away from a species that are also linked to each other; and (11) distribution degree, which is the fraction of trophic species P(*k*) that have *k* or more trophic links (both predator and prey links) (Albert and Barabási 2002). These last three metrics give us information related to the degree of self-organization shown by the web.

The trophic levels (*TL*) of species were calculated using the short-weighted *TL* (Williams and Martinez 2004). Short-weighted trophic level is defined as the average of the shortest *TL* and prey-averaged *TL*. Shortest *TL* of a consumer in a food web is equal to 1+ the shortest chain length from this consumer to any basal species. Prey-averaged *TL* is equal to 1+ the mean *TL* of all consumer's trophic resources, calculated as follows:

$$TLj = 1 + \sum_{i=1}^{S} lij \frac{TLi}{nj.}$$

where *TLj* is the trophic level of species *j*; *S* is the total number of species in the food web; *lij* are the elements of the connection matrix with *S* rows and *S* columns; for column *j* and row *i*, *lij* is 1 if species *j* consumes species *i* and 0 if not; and *nj* is the number of prey species in the diet of species *j*. Therefore, short-weighted *TL* yields a minimum estimate of *TL* and assumes a value of 1.0 for basal species (Williams and Martinez 2004). We considered the mean *TL* of the web as the average of all species' *TL*.

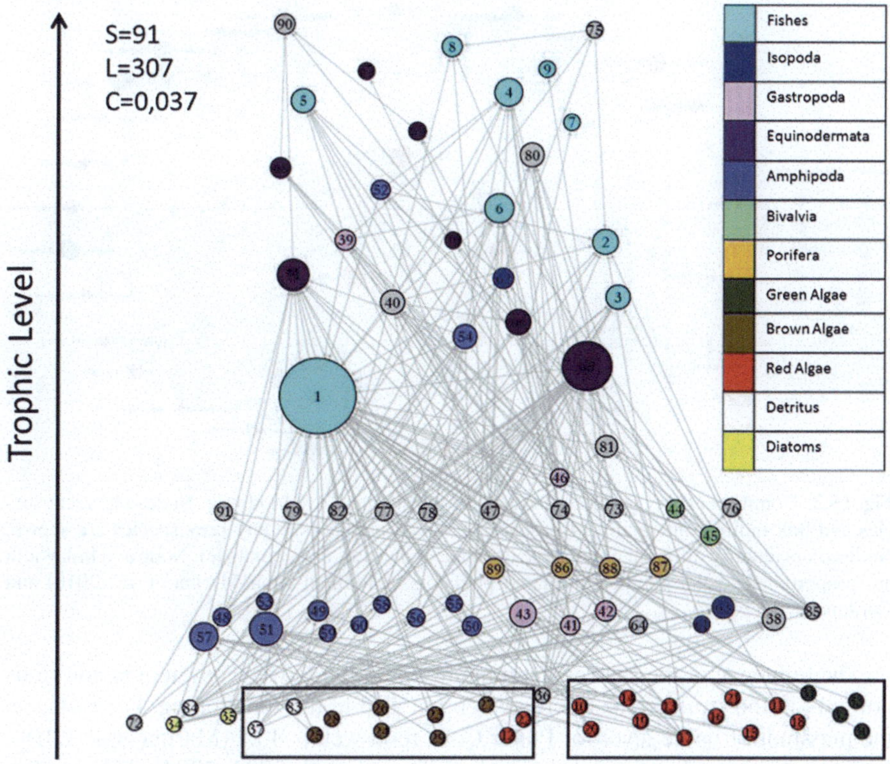

Fig. 15.1 Food web of Potter Cove. Vertical position is related to trophic level. Node size is proportional to the total degree (in and out). Node colors are by functional group. Nodes inside the boxes are the basal species. Network was plotted with Visone (version 2.9.2). [1]*Notothenia coriiceps*, [2]*Notothenia rossii*, [3]*Lepidonotothen nudifrons*, [4]*Trematomus newnesi*, [5]*Trematomus bernacchii*, [6]*Harpagifer antarcticus*, [7]*Parachaenichthys charcoti*, [8]*Chaenocephalus aceratus*, [9]*Protomyctophum* sp., [10]*Callophyllis atrosanguinea*, [11]*Curdiea racovitzae*, [12]*Georgiella confluens*, [13]*Gigartina skottsbergii*, [14]*Iridaea cordata*, [15]*Myriogramme manginii*, [16]*Neuroglossum delesseriae*, [17]*Palmaria decipiens*, [18]*Pantoneura plocamioides*, [19]*Picconiella plumosa*, [20]*Plocamium cartilagineum*, [21]*Porphyra plocamiestris*, [22]*Trematocarpus antarcticus*, [23]*Adenocystis utricularis*, [24]*Ascoseira mirabilis*, [25]*Desmarestia anceps*, [26]*Desmarestia antarctica*, [27]*Desmarestia menziesii*, [28]*Geminocarpus geminatus*, [29]*Phaeurus antarcticus*, [30]*Lambia antarctica*, [31]*Monostroma hariotii*, [32]*Urospora penicilliformis*, [33]*Ulothrix* sp., [34]Epiphytic diatoms, [35]Benthic diatoms, [36]Phytoplankton, [37]Aged detritus, [38]Nereidae, [39]*Margarella antarctica*, [40]*Austrodoris kerguelensis*, [41]*Eatoniella* sp., [42]*Nacella concinna*, [43]*Laevilacunaria antarctica*, [44]*Dacrydyum* sp., [45]*Laternula elliptica*, [46]*Neobuccinum eatoni*, [47]*Euphausia superba*, [48]*Paradexamine* sp., [49]*Eurymera monticulosa*, [50]*Pontogeneiella* sp., [51]*Gondogeneia antarctica*, [52]Hyperiids, [53]*Pariphimedia integricauda*, [54]*Bovallia gigantea*, [55]*Cheirimedon femoratus*, [56]*Gitanopsis antarctica*, [57]*Prostebbingia gracilis*, [58]*Waldeckia obesa*, [59]Hippo-Orcho (*Hippomedon kergueleni* and *Orchomene plebs* collapsed), [60]*Oradarea bidentata*, [61]*Serolis* sp., [62]*Glyptonotus antarcticus*, [63]*Plakarthrium puncattissimum*, [64]*Hemiarthrum setulosum*, [65]*Ophionotus victoriae*, [66]*Odontaster validus*, [67]*Diplasterias brucei*, [68]*Odontaster meridionalis*, [69]*Perknaster fuscus antarticus*, [70]*Perknaster aurorae*, [71]*Sterechinus neumayeri*, [72]squids, [73]Copepods, [74]Ascidians, [75]*Octopus* sp., [76]Oligochaetes, [77]Hydrozoa, [78]Bryozoa, [79]Priapulids, [80]*Parborlasia corrugatus*, [81]Salpidae, [82]Mysida, [83]fresh detritus, [84]necromass, [85]zooplankton, [86]*Haliclonidae* sp., [87]Stylo-Myca (*Stylocordila borealis* and *Mycale acerata* collapsed), [88]*Rosella* sp., [89]*Dendrilla antarctica*, [90]*Urticinopsis antarctica*, [91]*Malacobelemnon daytoni*. (Modified from Marina et al. 2018)

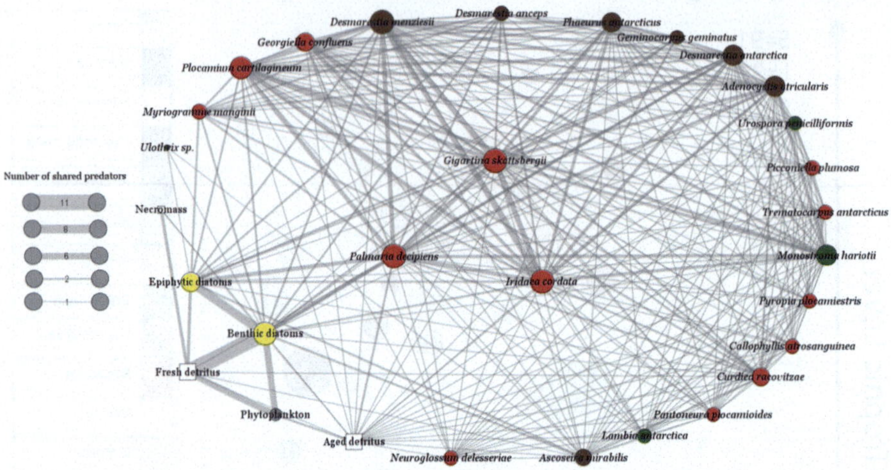

Fig. 15.2 Common-enemy graph for basal species of Potter Cove food web. Nodes represent species and link indirect interactions (presence of shared predators). Only prey species are shown. Node colors identify different preys (type of algae, or detritus, or microalgae). Node and link width are proportional to the number of shared predators. (Modified from Marina et al. (2018) and Cordone et al. (2018))

The influence of the seaweed community in Potter Cove ecosystem is notorious not only in the structure of the food web but also in the functioning. For instance, the maximum trophic level for Potter Cove food web is 4.27 (Marina et al. 2018), lower than most other food webs studied (Dunne et al. 2002, 2004). This implies that top and basal species (dominated by macroalgae) are separated by few intermediate taxa. Therefore, the transfers of energy or nutrients from the base to the top of Potter Cove food web is small, so that the number of times chemical energy is transformed from a consumer's diet into a consumer's biomass along the food web is also small. The mean trophic level for the mentioned food web is also low (2.10), which is a consequence of the fact that most predators at intermediate levels (e.g., amphipods, isopods, bivalves, *Notothenia coriiceps*) feed predominantly on algae species and/or detritus, being mainly the product of dead and decomposed macroalgae in Potter Cove (Iken et al. 1998; Quartino et al. 2008; see also Chap. 8 by Quartino et al.). The macroalgal detritus decomposes and is eaten by detritivores and suspensivores (e.g., sponges, ascidians, bryozoans, cnidarians), supporting an important amount of the secondary production (Tatián et al. 2004). The obtained low mean trophic level for Potter Cove food web clearly shows what species-specific and/or community studies have suggested. These characteristics of ecological communities have a high impact on ecosystem functioning, such as nutrient and carbon cycling, and trophic cascades (Post 2002).

However, our interest here is focused on macroalgae. To explore mainly their interactions, we can use another useful information due to the secondary graph of common-enemy graph (Fig. 15.2). This graph represents prey species linked with each other according the existence of shared predators. In the representation of

Fig. 15.2, the thickness of each link joining a couple of nodes is proportional to the number of enemies shared by these nodes. The common-enemy graph is hyperconnected having a high-density value of interactions per node (Cordone et al. 2018); this is an indicator of the existence of multiple alternative energy paths. This hyperconnectivity makes the prediction of indirect relationships between species very difficult, as each interaction involves positive and negative effects between species abundances (Holt and Lawton 1994).

15.3 Network Dynamics and Robustness

An immediate question that we can try to answer is the following: can we predict the behavior of the community against the (local) extinction of some algae? This important question cannot be easily answered by field observations or experiments; however, we can carry out in silico experiments using our food web (Cordone et al. 2018). We made a virtual experiment trying to determinate if successive extinctions of base species in Potter Cove could produce phase transitions in the emergent properties of the food web. With this objective, we simulated extinctions by deleting nodes (seaweeds, detritus, microalgae) according to different sequences established by the degree of each macroalgae (total number of trophic interactions per node) and their biomass in Potter Cove. Four sequences of extinctions were used: random order, degree in ascending and descending order, and biomass in ascending order.

The method quantifies secondary extinctions taking into account the existence of different extinction thresholds (Schleuning et al. 2016). That means that a species suffers a secondary extinction when it loses a given percentage of its preys (Solé and Montoya 2001). We define the threshold (v) as the minimum level of energy necessary for species' survival. After each node removal, the fraction of original incoming energy $e(i)$ is calculated for each species i, and when this fraction is equal or less than the threshold, the species is secondarily lost. The classical topological approach assumes that v is equal to 0, so only when the energy inflow is null a species goes extinct (Bellingeri and Bodini 2013).

The results of the in silico experiment considering different thresholds for extinction and the four sequences of primary elimination are indicated in Fig. 15.3. We found that the Potter Cove food web is relatively stable to macroalgae loss, but a significant number of secondary extinctions were obtained beyond a 50% threshold (Cordone et al. 2018). The elimination of macroalgae species from the Potter Cove food web does not seems to generate a catastrophic cascade of secondary extinctions, suggesting that the Potter Cove food web could be more robust than other similar networks (Allesina et al. 2006). Most connected macroalgae shared more predators, which could indicates that these macroalgae species are functionally redundant (i.e., species with equivalent trophic interactions). Functional redundancy has important consequences in potential cascade extinctions, since it increases food web resistance by means of availability of alternative preys (Borrvall et al. 2000; Petchey et al. 2008). However, these results are tied to the limitations

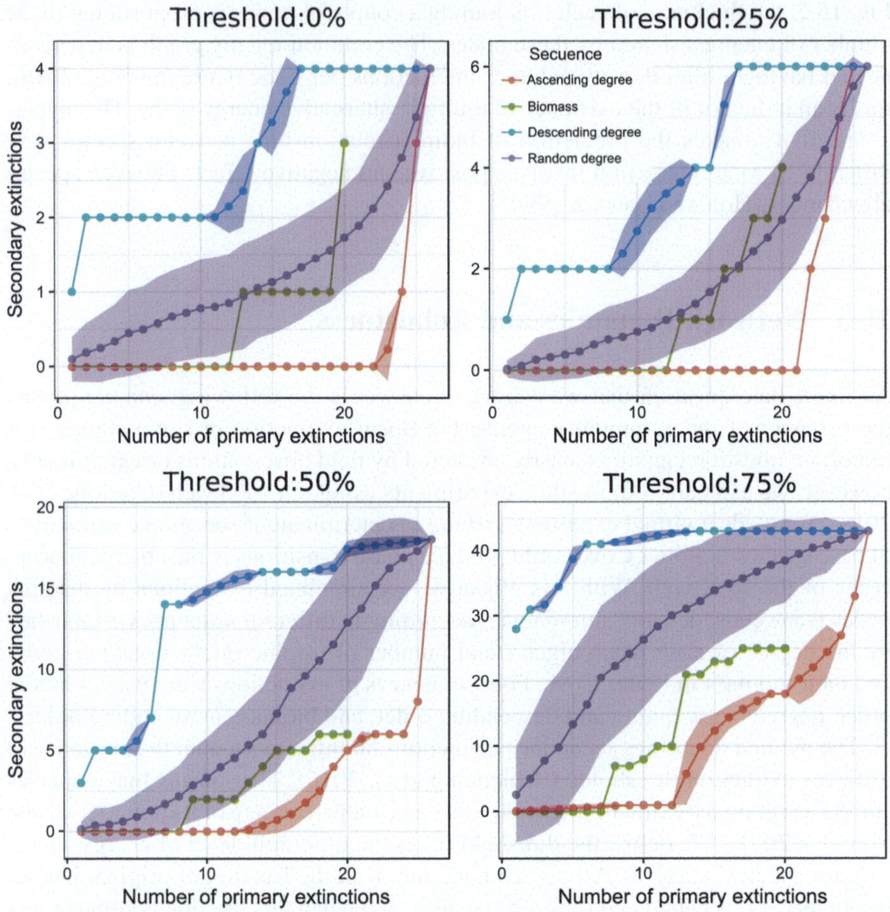

Fig. 15.3 Secondary extinctions vs extinctions (total, primary plus secondary) when macroalgae species and detritus are removed. The secondary extinctions after primary extinctions take place or not depending on the numbers of preys that the predator needs to survive. This number is determined by the threshold of extinctions as the number of original preys multiplied by threshold. Each box corresponds to a simulation in which the threshold of extinctions was constant and different (0%, 25%, 50%, and 75%). The macroalgae loss was made following the four different sequences: red, ascending degree; blue, descending degree; violet, random degree – mean and interval confidence; and green, biomass – in ascending order

of our method. Typically, the topological approach underestimates the actual number of secondary extinctions and in consequence overestimates food web robustness. The introduction of the threshold effect gives us a more accurate prediction (Fig. 15.4).

Despite this, the topological approach enables the analysis of complex food webs, since it only requires knowledge of network structure and could be used as a proxy of food web stability (Eklöf et al. 2013). Furthermore, we have to

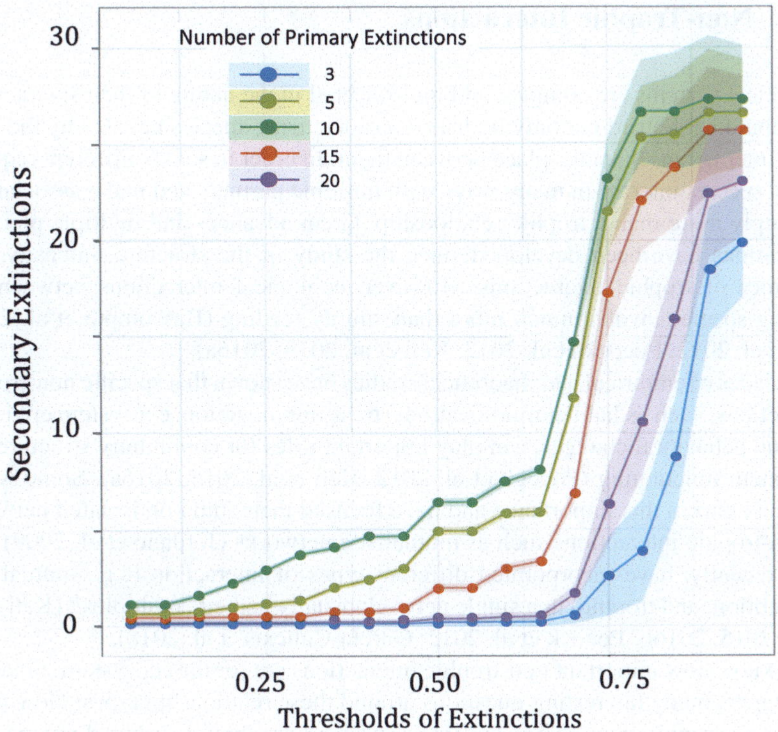

Fig. 15.4 This figure summarizes the effect of varying thresholds in the number of secondary extinctions recorded. We tested thresholds from 5% to 95% by 5% when macroalgae are primary lost (blue, 3; brown, 5; green, 10; red, 15; pink, 20). Points are means of secondary extinctions when x number of primary extinctions took place at a particular threshold (by random simulations). Shaded area represents confidence interval of the series. Secondary extinctions recorded at a fixed threshold (0.25, 0.50, 0.75) from Fig. 15.3. (Modified from Cordone et al. 2018)

consider that this food web includes all recorded trophic interactions, though these interactions do not always occur simultaneously in time and space. We cannot ignore the potentially confounding effects of seasonality and spatial sampling (Ings et al. 2009). In fact, the network is an average representation across seasons and different habitats, and dynamic stability depends more on how interactions are materialized in time and space. Some nodes are pulsatile, such as the massive influx of Krill (Fuentes et al. 2016), which is difficult to reflect in a static description of the network topology. The consideration of all these factors might change our estimation of the fragility of Potter Cove food web. Our current efforts are focused on incorporating spatial and temporal variability into the modeling of Potter Cove food web to achieve a more realistic picture of this ecosystem and its macroalgae species.

15.4 Non-Trophic Interactions

Ecological systems are complex and multifaceted in the nature of their interactions and should be defined not only by lists of co-occurring species but also by the variety of interactions that take place between them. In order to satisfy different requirements, species interact in many ways with multiple partners and make associations that imply more than a trophic relationship. Great advances and multiple practical applications have been developed since the study of the structure, intensity, and dynamics of trophic interactions. However, ecological interactions between co-existing species involve much more than simply feeding (Bascompte et al. 2003; Kéfi et al. 2012; Pocock et al. 2012; Kéfi et al. 2015, 2016a).

Decades of empirical and theoretical studies have shown that specific non-trophic interactions, such as habitat modification, stress minimization, ecosystem engineering, and behavioral changes, can play important roles for community structure and ecosystem functioning (Pocock et al. 2012; Kéfi et al. 2015, 2016a). Some works have recognized this importance and have focused their study on isolated networks of non-trophic interactions such as mutualistic networks (Jordano et al. 2009) and, more recently, have incorporated different types of interaction (e.g., mutualistic, competition, and trophic) in a single network usually named "multiplex" (Kéfi et al. 2012, 2015, 2016b; Pocock et al. 2012; García-Callejas et al. 2018).

To know how important non-trophic interactions are for the ecosystem, functioning triggers many interesting questions around these relationship types: How many direct non-trophic interactions are there and how are they distributed among species? Do the topological properties of trophic and non-trophic networks differ? Can simple species attributes help predict the type of interaction between two species? Can trophic characteristics help predict the non-trophic webs (Kéfi et al. 2015)? Here, we have focused on the first question to describe non-trophic interactions that involve macroalgae species from the Potter Cove food web.

The number of non-trophic interactions identified in Potter Cove marine ecosystem duplicates the number of trophic relationships: 1091 versus 454, respectively. Here, non-trophic interactions include competition or negative interactions (−/−) considered when a pair of species share at least one prey and represent the 74% of the total number of non-trophic interactions; mutualism or positive interactions that involve a benefit of both interacting species (+/+) that represent the 13.6% of the total number of non-trophic interactions; and commensalism (+/0) and amensalism (−/0) or neutral interactions in which one species benefit or not-benefit, respectively, and the other species is not affected by the link, and these interactions represent the 12.4% of the total number of non-trophic interactions. The distribution of these interactions (Fig. 15.5) depends on the trophic classification of the species in basal, intermediate, or top species, e.g., competition, as we defined here, occurs between intermediate and/or top species, which means that most non-trophic interactions in Potter Cove ecosystem occur between species of high trophic levels. So, we should wonder what the role of macroalgae in the non-trophic relationships is.

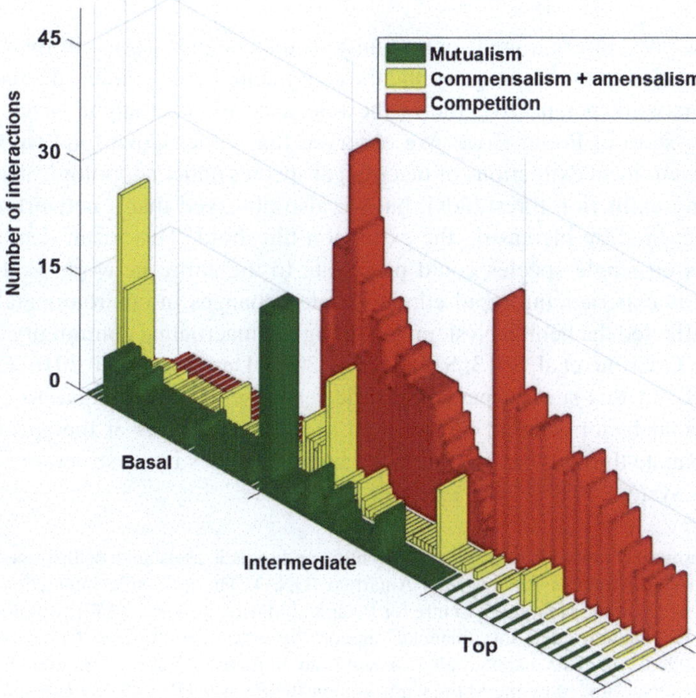

Fig. 15.5 Number of non-trophic interactions for basal, intermediate, and top species. Each color represents a type of interaction

Many works highlight the importance of studying the structure of non-trophic networks versus the trophic network by suggesting that, in general, trophic networks tend to be compartmentalized, while positive non-trophic networks are more nested (Bascompte et al. 2003; Thébault and Fontaine 2010), which minimizes competition, increases the number of co-existing species, and makes the community more robust against random extinctions and diversity loss (Memmott et al. 2004; Fortuna and Bascompte 2006).

In food webs, it is argued that modular patterns tend to increase the stability of the network by retaining the impacts of disturbances within each module and minimizing that impact on other modules in the network. In the particular case of Potter Cove, it was observed that the food web has not modular characteristics, reason to suppose that the network should be not very robust against random disturbances. However, it has been shown that the trophic network is highly robust against different types of disturbances, which makes it possible to assume that positive non-trophic interactions could have a stabilizing role in the Potter Cove ecosystem. Then, we should not fail to recognize that macroalgae are a key component of the system because they represent, directly or indirectly, a stabilizing species.

15.5 Final Remarks

Network science offers a great opportunity to understand species in their context and their relation to the community and the ecosystem. In our studies, the incorporation of a network perspective shows the relevance of macroalgae species to the entire ecosystem of Potter Cove. We observed that Potter Cove food web is relatively robust to local extinctions of macroalgal species under a topological approach (considering extinction thresholds), but we also observed that a network collapse could be reached by increasing the extinction threshold. This result demonstrates that effects on single species could propagate to the entire network. Potter Cove ecosystem is experiencing rapid climate-related changes in environmental factors that have affected the benthic system, including the macroalgal community (Schloss et al. 2012; Quartino et al. 2013; Sahade et al. 2015; Deregibus et al. 2016; Campana et al. 2018). In this sense, species-specific responses of macroalgae to changing environmental factors should be explored as negative effects at the species level could propagate through the food web, leading to changes in the structure and function of this system.

Acknowledgments We wish to thanks the scientific and logistic assistance of Instituto Antártico Argentino (IAA)/Dirección Nacional del Antártico (DNA). The academic interaction with the Alfred Wegener Institute Helmholtz Centre for Polar and Marine Research (AWI) was essential for this work. We acknowledge the financial support by grants from CONICET-UNGS (PIO 14420140100035CO). This chapter also presents an outcome of the international Research Network IMCONet funded by the Marie Curie Action IRSES IMCONet (FP7 IRSES, Action No. 318718). Many thanks to C. Gardel and I. Corsini for their assistance.

References

Ahn I-Y, Moon H-W, Jeon M, Kang S-H (2016) First record of massive blooming of benthic diatoms and their association with megabenthic filter feeders on the shallow seafloor of an Antarctic fjord: does glacier melting fuel the bloom? Ocean Sci J 51:273–279. https://doi.org/10.1007/s12601-016-0023-y

Albert R, Barabási A (2002) Statistical mechanics of complex networks. Rev Mod Phys 74:47–97. https://doi.org/10.1103/RevModPhys.74.47

Allesina S, Bodini A, Bondavalli C (2006) Secondary extinctions in ecological networks: bottlenecks unveiled. Ecol Model 194:150–161. https://doi.org/10.1016/j.ecolmodel.2005.10.016

Amsler CD, McClintock JB, Baker BJ (1999) An Antarctic feeding triangle: defensive interactions between macroalgae, sea urchins, and sea anemones. Mar Ecol Prog Ser 183:105–114. https://doi.org/10.3354/meps183105

Amsler CD, Iken K, McClintock JB, Amsler MO, Peters KJ, Hubbard JM, Furrow FB, Baker BJ (2005) Comprehensive evaluation of the palatability and chemical defenses of subtidal macroalgae from the Antarctic peninsula. Mar Ecol Prog Ser 294:141–159. https://doi.org/10.3354/meps294141

Amsler CD, Iken K, McClintock JB, Baker BJ (2009) Defenses of polar macroalgae against herbivores and biofoulers. Bot Mar 52:535–545. https://doi.org/10.1515/BOT.2009.070

Amsler CD, McClintock JB, Baker BJ (2012) Palatability of living and dead detached Antarctic macroalgae to consumers. Antarct Sci 24:589–590. https://doi.org/10.1017/S0954102012000624

Amsler CD, McClintock JB, Baker BJ (2014) Chemical mediation of mutualistic interactions between macroalgae and mesograzers structure unique coastal communities along the Western Antarctic Peninsula. J Phycol 50:1–10. https://doi.org/10.1111/jpy.12137

Bascompte J, Jordano P, Melián CJ, Olesen JM (2003) The nested assembly of plant–animal mutualistic networks. PNAS 100:9383–9387. https://doi.org/10.1073/pnas.1633576100

Bellingeri M, Bodini A (2013) Threshold extinction in food webs. Theor Ecol 6:143–152. https://doi.org/10.1007/s12080-012-0166-0

Borrvall C, Ebenman B, Jonsson T (2000) Biodiversity lessens the risk of cascading extinction in model food webs. Ecol Lett 3:131–136. https://doi.org/10.1046/j.1461-0248.2000.00130.x

Campana GL, Zacher K, Fricke A, Molis M, Wulff A, Quartino ML (2009) Drivers of colonization and succession in polar benthic macro- and microalgal communities. Bot Mar 52:655–667. https://doi.org/10.1515/BOT.2009.076

Campana GL, Zacher K, Deregibus D, Momo FR, Wiencke C, Quartino ML (2018) Succession of Antarctic benthic algae (Potter Cove, South Shetland Islands): structural patterns and glacial impact over a four-year period. Polar Biol 41:377–396. https://doi.org/10.1007/s00300-017-2197-x

Cordone G, Marina TI, Salinas V, Doyle SR, Saravia LA, Momo FR (2018) Effects of macroalgae loss in an Antarctic marine food web: applying extinction thresholds to food web studies. Peer J 6:e5531. https://doi.org/10.7717/peerj.5531

Delmas E, Besson M, Brice MH, Burkle LA, Dalla Riva GV, Fortin MJ, Gravel D, Guimarães PR Jr, Hembry DH, Newman EA, Olesen JM, Pires MM, Yeakel JD, Poisot T (2018) Analysing ecological networks of species interactions. Biol Rev 94:16–36. https://doi.org/10.1111/brv.12433

Deregibus D, Quartino ML, Campana GL, Momo FR, Wiencke C, Zacher K (2016) Photosynthetic light requirements and vertical distribution of macroalgae in newly ice-free areas in Potter Cove, South Shetland Islands, Antarctica. Polar Biol 39:153–166. https://doi.org/10.1007/s00300-015-1679-y

Dunne JA, Williams RJ, Martinez ND (2002) Food-web structure and network theory: the role of connectance and size. PNAS 9920:12917–12922. https://doi.org/10.1073/pnas.192407699

Dunne JA, Williams RJ, Martinez ND (2004) Network structure and robustness of marine food webs. Mar Ecol Prog Ser 273:291–302. https://doi.org/10.3354/meps273291

Eklöf A, Tang S, Allesina S (2013) Secondary extinctions in food webs: a Bayesian network approach. Methods Ecol Evol 4:760–770. https://doi.org/10.1111/2041-210X.12062

Fogg GE (1977) Aquatic primary production in the Antarctic. Philos Trans R Soc Lond B Biol Sci 279:27–38. https://doi.org/10.1098/rstb.1977.0069

Fortuna MA, Bascompte J (2006) Habitat loss and the structure of plant–animal mutualistic networks. Ecol Lett 9:281–286. https://doi.org/10.1111/j.1461-0248.2005.00868.x

Fuentes V, Alurralde G, Meyer B, Aguirre GE, Canepa A, Wölfl A-C, Hass HC, Williams GN, Schloss IR (2016) Glacial melting: an overlooked threat to Antarctic krill. Sci Rep 6:27234. https://doi.org/10.1038/srep27234

García-Callejas D, Molowny-Horas R, Araújo MB (2018) Multiple interactions networks: towards more realistic descriptions of the web of life. Oikos 127:5–22. https://doi.org/10.1111/oik.04428

Gómez I, Wulff A, Roleda MY, Huovinen P, Karsten U, Quartino ML, Dunton K, Wiencke C (2009) Light and temperature demands of marine benthic microalgae and seaweeds in polar regions. Bot Mar 52:593–608. https://doi.org/10.1515/BOT.2009.073

Holt RD, Lawton JH (1994) The ecological consequences of shared natural enemies. Annu Rev Ecol Evol Syst 25:495–520. https://doi.org/10.1146/annurev.es.25.110194.002431

Huang Y, Amsler MO, McClintock JB, Amsler CD, Baker BJ (2007) Patterns of gammaridean amphipod abundance and species composition associated with dominant subtidal macroalgae from the western Antarctic peninsula. Polar Biol 30:1417–1430. https://doi.org/10.1007/s00300-007-0303-1

Iken K, Barrera-Oro ER, Quartino ML, Casaux RJ, Brey T (1997) Grazing by the Antarctic fish *Notothenia coriiceps*: evidence for selective feeding on macroalgae. Antarct Sci 9:386–391. https://doi.org/10.1017/S0954102097000497

Iken K, Quartino ML, Barrera-Oro E, Palermo J, Wiencke C, Brey T (1998) Trophic relations between macroalgae and herbivores. In: Wiencke C, Ferreyra G, Arntz W, Rinaldi C (eds) The Potter Cove coastal ecosystem, Antarctica. Rep Polar Res, vol 299, pp 258–262

Ings TC, Montoya JM, Bascompte J, Blüthgen N, Brown L, Dormann CF, Edwards F, Figueroa D, Jacob U, Jones JI, Lauridsen RB, Ledger ME, Lewis HM, Olesen JM, van Veen FJ, Warren PH, Woodward G (2009) Ecological networks – beyond food webs. J Anim Ecol 78:253–269. https://doi.org/10.1111/j.1365-2656.2008.01460.x

Jordano P, Vázquez D, Bascompte J (2009) Redes complejas de interacciones planta-animal. In: Medel R, Aizen MA, Zamora R (eds) Ecología y evolución de interacciones planta-animal. Editorial Universitaria, Santiago de Chile, pp 17–41

Kéfi S, Berlow EL, Wieters EA, Navarrete SA, Petchey OL, Wood SA, Boit A, Joppa LN, Lafferty KD, Williams RJ, Martinez ND, Menge BA, Blanchette CA, Iles AC, Brose U (2012) More than a meal…integrating non-feeding interactions into food webs. Ecol Lett 15:291–300. https://doi.org/10.1111/j.1461-0248.2011.01732.x

Kéfi S, Berlow EL, Wieters EA, Joppa LN, Wood SA, Brose U, Navarrete SA (2015) Network structure beyond food webs: mapping non-trophic and trophic interactions on Chilean rocky shores. Ecology 96:291–303. https://doi.org/10.1890/13-1424.1

Kéfi S, Holmgren M, Scheffer M (2016a) When can positive interactions cause alternative stable states in ecosystems? Funct Ecol 30:88–97. https://doi.org/10.1111/1365-2435.12601

Kéfi S, Miele V, Wieters EA, Navarrete SA, Berlow EL (2016b) How structured is the entangled bank? The surprisingly simple organization of multiplex ecological networks leads to increased persistence and resilience. PLoS Biol 14(8):e1002527. https://doi.org/10.1371/journal.pbio.1002527

Klöser H, Mercuri G, Laturnus F, Quartino ML, Wiencke C (1994) On the competitive balance of macroalgae at Potter Cover (King George Island, South Shetlands). Polar Biol 14:11–16. https://doi.org/10.1007/BF00240266

Klöser H, Quartino ML, Wiencke C (1996) Distribution of macroalgae and macroalgal communities in gradients of physical conditions in Potter Cove, King George Island, Antarctica. Hydrobiologia 333:1–17. https://doi.org/10.1007/BF00020959

Majewska R, Convey P, De Stefano M (2016) Summer epiphytic diatoms from Terra Nova Bay and Cape Evans (Ross Sea, Antarctica) – a synthesis and final conclusions. PLoS One 11(4):1–30. https://doi.org/10.1371/journal.pone.0153254

Marina TI, Salinas V, Cordone G, Campana GL, Moreira E, Deregibus D, Torre L, Sahade R, Tatián M, Barrera Oro E, De Troch M, Doyle S, Quartino ML, Saravia LA, Momo FR (2018) The food web of Potter Cove (Antarctica): complexity, structure and function. Estuar Coast Shelf Sci 200:141–151. https://doi.org/10.1016/j.ecss.2017.10.015

Memmott J, Waser NM, Price MV (2004) Tolerance of pollination networks to species extinctions. Proc R Soc Lond B 271:2605–2611. https://doi.org/10.1098/rspb.2004.2909

Pellizzari F, Silva MC, Silva EM, Medeiros A, Oliveira MC, Yokoya NS, Pupo D, Rosa LH, Colepicolo P (2017) Diversity and spatial distribution of seaweeds in the South Shetland Islands, Antarctica: an updated database for environmental monitoring under climate change scenarios. Polar Biol 40:1671–1685. https://doi.org/10.1007/s00300-017-2092-5

Petchey OL, Eklöf A, Borrvall C, Ebenman B (2008) Trophically unique species are vulnerable to cascading extinction. Am Nat 171:568–579. https://doi.org/10.1086/587068

Pocock MJ, Evans DM, Memmott J (2012) The robustness and restoration of a network of ecological networks. Science 335:973–977. https://doi.org/10.1126/science.1214915

Poisot T, Stouffer DB, Gravel D (2015) Beyond species: why ecological interactions vary through space and time. Oikos 124:243–251. https://doi.org/10.1111/oik.01719

Post DM (2002) Using stable isotopes to estimate trophic position: models, methods, and assumptions. Ecology 83(3):703–718

Quartino ML, Boraso de Zaixso A, Momo FR (2008) Macroalgal production and the energy cycle of Potter Cove. Rep Polar Mar Res 571:68–74
Quartino ML, Deregibus D, Campana GL, Latorre GEJ, Momo FR (2013) Evidence of macroalgal colonization on newly ice-free areas following glacial retreat in potter cove (South Shetland Islands), Antarctica. PLoS One 8(3):e58223. https://doi.org/10.1371/journal.pone.0058223
Sahade R, Lagger C, Torre L, Momo FR, Monien P, Schloss IR, Barnes DKA, Servetto N, Tarantelli S, Tatián M, Zamboni N, Abele D (2015) Climate change and glacier retreat drive shifts in an Antarctic benthic ecosystem. Sci Adv 1(10):e1500050. https://doi.org/10.1126/sciadv.1500050
Schleuning M, Fründ J, Schweiger O, Welk E, Albrecht J, Albrecht M, Beil M, Blüthgen N, Bruelheide H, Böhning-Gaese K, Dehling DM, Dormann CF, Exeler N, Farwig N et al (2016) Ecological networks are more sensitive to plant than to animal extinction under climate change. Nat Commun 7:13965. https://doi.org/10.1038/ncomms13965
Schloss IR, Abele D, Moreau S, Demers S, Bers AV, González O, Ferreyra G (2012) Response of phytoplankton dynamics to 19-year (1991–2009) climate trends in Potter Cove (Antarctica). J Mar Syst 92:53–66. https://doi.org/10.1016/j.jmarsys.2011.10.006
Smale DA, Barnes DKA (2008) Likely responses of the Antarctic benthos to climate-related changes in physical disturbance during the 21st century, based primarily on evidence from the West Antarctic peninsula region. Ecography 31:289–305. https://doi.org/10.1111/j.0906-7590.2008.05456.x
Solé RV, Montoya M (2001) Complexity and fragility in ecological networks. Proc R Soc Lond B 268:2039–2045. https://doi.org/10.1098/rspb.2001.1767
Tatián M, Sahade R, Esnal GB (2004) Diet components in the food of Antarctic ascidians living at low levels of primary production. Antarct Sci 16:123–128. https://doi.org/10.1017/S0954102004001890
Thébault E, Fontaine C (2010) Stability of ecological communities and the architecture of mutualistic and trophic networks. Science 329:853–856. https://doi.org/10.1126/science.1188321
Wiencke C, Amsler CD, Clayton MN (2014) Macroalgae. In: De Broyer C, Koubbi P, Griffiths HJ, Raymond B, Udekem d'Acoz C et al (eds) Biogeographic atlas of the Southern Ocean. SCAR, Cambridge, pp 66–73
Williams RJ, Martinez ND (2004) Limits to trophic levels and omnivory in complex food webs: theory and data. Am Nat 163:458–468. https://doi.org/10.1086/381964
Woodward G, Benstead JP, Beveridge OS, Blanchard J, Brey T, Brown LE, Cross WF, Friberg N, Ings TC, Jacob U, Jennings S et al (2010) Ecological networks in a changing climate. In: Woodward G (ed) Advances in ecological research, vol 42. Elsevier Ltd, Amsterdam, pp 71–138. https://doi.org/10.1016/B978-0-12-381363-3.00002-2

Chapter 16
Trophic Networks and Ecosystem Functioning

Marco Ortiz, Brenda B. Hermosillo-Núñez, and Ferenc Jordán

Abstract The geographic isolation of the Antarctic continent offers an interesting opportunity to quantify and qualify the actual ecological conditions and the most sensitive components from an ecosystem perspective. Antarctic coastal ecosystems are under severe stress as a consequence of climate change, which could facilitate biological invasions, reduced growth of macroalgal species, and local extinctions. The application of network analysis, representing the interactions among multiple species, allows us to quantify macroscopic (emergent) system properties, to assess overall health, to predict the propagation of direct and indirect effects, and to identify keystone species complexes within these complex ecological systems. Three theoretical frameworks are used here for this analysis: (1) ecological network analysis (ENA) considering thermodynamics and information theory (providing measures such as *Ascendency*), (2) semiquantitative (qualitative) mathematics based on the structure and local stability of community matrices (*Loop Analysis*), and (3) topological studies on interaction networks considering central node sets and defining keystone species complexes (KSCs).

Therefore, the integration of ecosystemic properties and keystone species complexes could help us to facilitate the design and assessment of conservation and monitoring measures, especially when the Antarctic coastal marine ecosystems are being severely stressed. The protection of the Antarctic environment – as a whole – not only should be focused on biological populations and communities but also

---

M. Ortiz (✉)
Instituto de Ciencias Naturales Alexander von Humboldt, Facultad de Ciencias del Mar y Recursos Biológicos, Universidad de Antofagasta, Antofagasta, Chile

Instituto Antofagasta, Universidad de Antofagasta, Antofagasta, Chile
e-mail: marco.ortiz@uantof.cl

B. B. Hermosillo-Núñez
Instituto Antofagasta, Universidad de Antofagasta, Antofagasta, Chile
e-mail: brenda.hermosillo@uantof.cl

F. Jordán
Balaton Limnological Institute, MTA Centre for Ecological Research, Tihany, Hungary

Stazione Zoologica, Napoli, Italy
e-mail: jordan.ferenc@okologia.mta.hu

I. Gómez, P. Huovinen (eds.), *Antarctic Seaweeds*,
https://doi.org/10.1007/978-3-030-39448-6_16

should consider changes in macroscopic properties, the propagation of direct and indirect influences in the networks, and keystone species complexes, which emerge using networks of interacting and coexisting species (or functional groups as system components).

Keywords Ascendency framework · Coastal benthic-pelagic ecosystem · Keystone species · Macroscopic network properties · Multispecies modelling

16.1 Introduction

Different scientific strategies – not necessarily mutually exclusive- can be used to study, assess, and attempt to predict the transformations in natural systems caused by direct and indirect human activities. Within of these strategies, we choose the quantitative and semiquantitative (qualitative) simulations permitting us to determine whole-system properties (of structure and dynamics) of Antarctic coastal benthic-pelagic ecosystems (Pikitch et al. 2004). Antarctic coastal ecosystems face several stressors as the increase in UV radiation, which reduces the growth of macroalgae species (Richter et al. 2008), and the rise in temperature, which could modify the amount of the direct light to macroalgae and other organisms (King 1994; Stark 1994). The impact of these stressors on the species and functional groups could affect the structure of their trophic networks and their functioning of diverse Antarctic ecosystems. Network analyses permit us to evaluate the macroscopic properties at ecosystem scale, describing aspects related with growth, development, complexity, and health (Odum 1969; Ulanowicz 1986; Levins 1998a, b; Costanza and Mageau 1999). Likewise, another related analysis can be performed as the propagation of direct and indirect effects (Hawkins 2004; Levins 1998a) and the determination of species or functional groups that play key roles in these complex ecological systems (Ortiz et al. 2013a).

The ecological concept of *keystone species*, introduced by Paine (1969) and broadened by Power et al. (1996), has become a key issue in numerous research programs in different communities and ecosystems around the planet (Mills et al. 1993; Power et al. 1996), especially given its use in the design and application of conservation management and monitoring measures (Payton et al. 2002; Barua 2011; Ortiz et al. 2013a, b). Power et al. (1996) named a key species as *a species whose effect is large, and disproportionately large relative to its abundance*. Although the concept seems to be sufficiently clear, its determination in communities and ecosystems is complex, requiring observations and experiments that include different spatiotemporal scales, levels of organization, and taxonomic groups (Power et al. 1996; Libralato et al. 2006). Similarly, some purely experimental studies (Pace et al. 1999) have omitted the propagation of direct and indirect effects, despite the ecological importance of these processes it has been recognized (Wootton 1994; Patten 1997; Yodzis 2001).

Various studies have determined the role that different species play in their ecological systems by using different network indices (Jordán et al. 2007; Luczkovich et al. 2003; Jordán and Scheuring 2004; Allesina and Bodini 2004; Libralato et al. 2006; Benedek et al. 2007; Ortiz et al. 2013a, 2017; Valls et al. 2015; Giacaman-Smith et al. 2016). Such analyses offer a complementary way to address some of the limitations in the experimental identification of key groups. Quantitative trophic analysis permits estimations of the strength of interactions between species or functional groups by identifying the presence of *keystone species*, which occupy key positions in the networks (Jordán and Scheuring 2004). At the same time, keystoneness can also be determined using semiquantitative or qualitative loop network analysis. In this case, the key role of a species is defined as a dynamical consequence of its changes, modifying the balance (prevalence) of positive and negative feedbacks and, in turn, the local stability of the network (Ortiz et al. 2013a). Likewise, following field observations (Daily et al. 1993), two independent contributions have proposed methodological extensions toward multispecies approaches to keystones. One was given by Benedek et al. (2007), which is based on the centrality of node sets, and the other proposed by Ortiz et al. (2013a) based on quantitative and semiquantitative multispecies trophic models. In both cases, the *keystone species complexes* (KSCs) consist of a core of species and/or functional groups linked by strong interspecific interactions. These more holistic concepts could facilitate the design of conservation and monitoring programs in ecosystems since it is not guaranteed that always a single species plays the key role.

The aim of this chapter was to determine the macroscopic system properties (e.g., level of development, organization, maturity, and health) derived from *Ascendency* theoretical framework (sensu Ulanowicz 1986, 1997) of a model benthic-pelagic ecosystem at Fildes Bay, King George Island Peninsula (Fig. 16.1a). The subtidal communities are dominated by the brown macroalgae species *Himmantothallus grandifolius* and *Desmarestia anceps*; the red algae species *Gigartina skottsbergii*, *Trematocarpus antarcticus*, and *Plocamium cartilagineum*; the grazer species (gastropod) *Nacella concinna*; the asteroid predator species *Diplasterias brucei*; and a variety of fish, sponge, bryozoan, and ascidian species (Targett 1981; Huovinen and Gómez 2013; Valdivia et al. 2014, 2015). Considering this network, dynamical simulations, local stability, and centrality of node sets (KeyPlayer set) were performed to determine the *keystone species complexes* (KSCs). Finally, the contribution of the components of KSCs on macroscopic system properties will be also evaluated. For this purpose, quantitative and semiquantitative (qualitative) multispecies trophic network models will be used and analyzed. The network model constructed represents exclusively transient states under short-term dynamics.

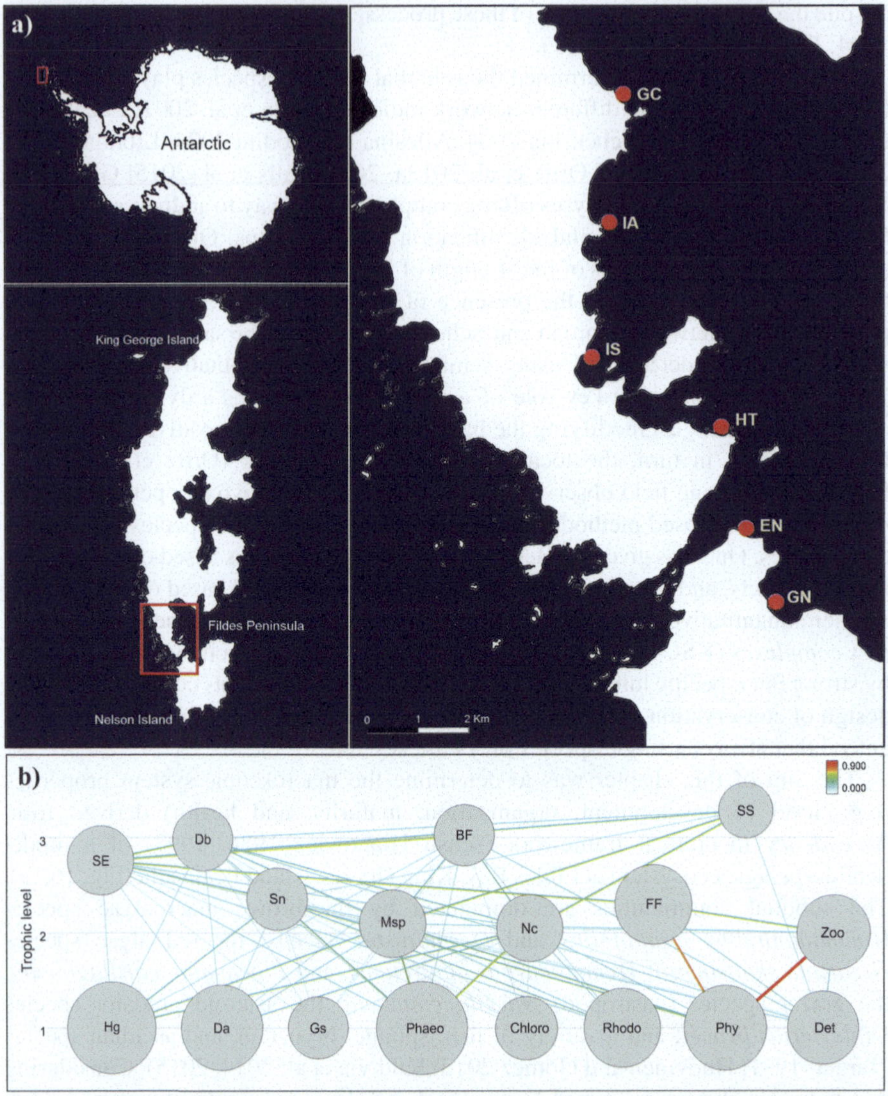

Fig. 16.1 Study area and sampling sites at Fildes Bay (King Georg Island, Antarctica) (*GC* Glacier Collins, *IA* Island Artigas, *IS* Island Shoa, *HT* Half Three, *EN* Estrecho Nelson, *GN* Glacier Nelson) (**a**) and the trophic model for the coastal ecosystem of Fildes Bay (Note: vertical position approximates trophic level, and the circle is proportional to the compartment biomass) (**b**). For details, see Appendix 16.A. (Adapted from Fig. 2 in Ortiz et al. 2017)

16.1.1 Quantitative and Semiquantitative Multispecies Trophic Modelling

Trophic mass balance models were constructed using the *Ecopath* with *Ecosim* (v.5.0) (EwE) software, which was first developed by Polovina (1984) and subsequently extended by Christensen and Pauly (1992) and Walters et al. (1997). *Ecopath* allows the depiction of the flows of matter and/or energy in a stationary state of an ecosystem within a given time, whereas *Ecosim* performs dynamic simulations of the initial conditions (established with *Ecopath*) as a response to perturbations (Christensen and Pauly 1992). The energy mass balance of a species or functional group within a network is represented by linear differential equations for each component in the model (Box 16.1). To employ *Ecosim* (see Walters and Christensen 2007), an extension routine of *Ecopath* is included to define the consumption by compartment which is represented by the control flow equation, allowing to set if the flow control mechanism is top-down, bottom-up, or mixed (Box 16.1). We set the mixed control mechanism following the criteria of Ortiz (2008a, b) and Ortiz et al. (2009, 2010, 2013a, b). We used *Ecopath* with *Ecosim* version 6.4, and details concerning this software package are given in Pauly et al. (2000).

The semiquantitative or qualitative trophic models were built using *Loop Analysis*, which allows for the estimation of the local stability (as a measure of sustainability) of an ecological system and the assessment of the propagation of both direct and indirect effects as a response to external perturbations (Levins 1974, 1998a). This approach has been widely applied in different natural science fields (Puccia and Levins 1991; Levins 1998b; Darmbacher et al. 2009; Ortiz and Levins 2011, 2017). The interactions are shown as signs that indicate the type of influence each variable has upon another (positive, negative, or null). In ecological relationships, the +1/−1 signs denote predator/prey or parasite/host interactions, +1/+1 signs express mutualism, +1/0 signs represent commensalism, and −1/0 signs show amensalism. *Loop Analysis* is based on the relationships of differential equations near equilibrium, *Jacob Levins*' community matrices, and their loop diagrams (Box 16.2). Local stability of the system is quantified using the Routh-Hurwitz criteria, which require the following: (1) all feedbacks (on every level of complexity) must be negative and (2) negative feedbacks on higher levels cannot be too great for comparison with the negative feedbacks on lower levels. Levins (1998a) proposes that the system is more resistant (locally stable), whereas *Fn* (feedback corresponding to higher level of complexity) is more negative (Box 16.2). The semiquantitative or qualitative *Jacob-Levins*' community matrices are based on the diet matrices used for quantitative models (*Ecopath* with *Ecosim*). The polynomial equation required for each community matrix was determined using the software *LoopStability* (Diaz-Ávalos and Ortiz, laboratory uses).

Box 16.1 The Ecopath with Ecosim (EwE) Theoretical Framework

The mass balance of a species or functional group within a trophic network is represented by the following mathematical expression:

$$B_i\left(\frac{P}{B}\right)_i \mathrm{EE}_i - \sum_{j=1}^{n} B_j\left(\frac{Q}{B}\right)_j \mathrm{DC}_{ji} - Y_i - \mathrm{BA}_i - E_i = 0 \quad (16.1)$$

where B_i and B_j are the biomass value of prey i and predator j; P/B_i is the productivity of prey i, which is equivalent to total mortality (Allen 1971); EE_i is the ecotrophic efficiency or the fraction of the total production of a group or species used in the system; Y_i is the fishing production per unit of area and time (Y = fishing mortality × biomass); Q/B_i is the food consumption per biomass unit j; DC_{ji} is the fraction of prey i in the average diet of predator j; BA_i is the biomass accumulation rate for i; and E_i is the net migration of i (emigration minus immigration) (Christensen and Walters 2004).

Based on this equation, the input and output of matter (energy) in each compartment of the system can be balanced. This energy balance is assured for each group with this equation:

$$Q = P + R + \mathrm{UAF} \quad (16.2)$$

where Q is consumption, P is production, R is respiration, and UAF is unassimilated food for each group or species of the system. Given the inclusion of the factors BA_i and E_i in Eq. (16.1), the focus of *Ecopath* is based on a steady state. This situation allows changes in the network compartments when the mathematical expression is expressed in a dynamic form.

To employ *Ecosim* (see Walters and Christensen 2007), an extension routine of *Ecopath* is included to define the consumption by compartment (Q_{ij}), where Q_{ij} is represented by the following equation:

$$Q_{ij} = \frac{a_{ij} v_{ij} B_i B_j}{2v_{ij} + a_{ij} B_j} \quad (16.3)$$

where a_{ij} represents the instantaneous mortality rate on prey i caused by a single unit of predator j biomass. Similarly, a_{ij} can be understood as the rate of effective search by predator j for prey i. Each a_{ij} is estimated directly from the corresponding *Ecopath* models by $a_{ij} = Q_i/(B_iB_j)$, where Q_i is the total consumption of i. The v_{ij} represents the transfer rate between compartments i and j. This parameter determines if the flow control mechanism is top-down, bottom-up, or mixed.

Box 16.2 Loop Analysis Theoretical Framework

Loop Analysis estimates the local stability (as a component of sustainability) of an ecological system and the assessment of the propagation of both direct and indirect effects as a response to external perturbations (Levins 1974, 1998a). The effects are shown as signs that indicate the type of influence each variable has upon another (positive, negative, or null). Interactions are described according to pairwise sign combinations: +/− signs denote predator/prey or parasite/host interactions, +/+ signs express mutualism, +/0 signs represent commensalism, and −/0 signs show amensalism or allelopathy. *Loop Analysis* is based on the relationships of differential equations near equilibrium, community matrices, and their loop diagrams. In a benthic system, the element α_{ij} of the matrix and loop diagrams performs the effect of the variable j in the growing of variable i, and this dynamic performs in the following way:

$$\frac{dX_i}{dt} = f_i\left(X_1, X_2, \ldots, X_n;\ C_1, C_2, \ldots, C_n\right) \tag{16.4}$$

where the change on the time of variable X_i is a function f_i of the other variables – X_n and parameters C_n – which are interconnected. The link of X_j to X_i is similarly to α_{ij} in Levins (1968):

$$\alpha_{ij} = \frac{\partial f_i(X)X^*}{\partial X_j} \tag{16.5}$$

where $x*$ indicates that it has been evaluated in equilibrium. The sign α_{ij} represents the link of j to i where the function of sign X is 1 when $X > 0$, 0 when $X = 0$, and −1 when $X < 0$. Local stability of the system is quantified using the Routh-Hurwitz criteria, which require the following: (1) all feedbacks (on every level of complexity) must be negative and (2) negative feedbacks on higher levels cannot be too great for comparison with the negative feedbacks on lower levels. Levins (1998a) proposes that the system is more resistant (locally stable), whereas F_n (feedback corresponding to higher level of complexity) is more negative:

$$F_0\lambda^n + F_1\lambda^{n-1} + F_2\lambda^{n-2} + \cdots + F_{n-1}\lambda + F_n = 0 \tag{16.6}$$

16.1.2 Selection of Model Components, Sampling Programs, and Data Sources

Three intensive field studies were conducted during the austral summers of 2013, 2014, and 2015 to identify the biological components (species or functional groups) of the system model and to estimate the average biomass (B), average density, and food sources of the selected components (Valdivia et al. 2014). Sampling was performed to directly estimate the average biomass and density of the macrobenthic species (between 5 and 30 m depth) at six stations within Fildes Bay (Fig. 16.1a). The production (P) and turnover rates (productivity) (P/B) were estimated using the following allometric Eq. 16.1:

$$\text{Production} = \left[\left(\frac{\text{Biomass}}{\text{Density}}\right)^{0.73}\right] * \text{Density}$$

where 0.73 is the average exponent regression of annual production on body size for macrobenthic invertebrates (for more details, see Warwick and Clarke 1993).

Food consumption rates were obtained from the literature (Cornejo-Donoso and Antezana 2008; Ortiz 2008a; Ortiz et al. 2015; Pinkerton et al. 2010). To determine the diets of *N. concinna*, *Harpagifer antarcticus*, *Notothenia coriiceps*, *N. rossii*, and the asteroid species, the stomach and guts were revised, and the gut contents were classified to the lowest possible taxonomic level; the frequency of occurrence of each food item was then calculated. Several studies examining the trophic ecology of several benthic and pelagic species were also used to determine the range of food consumed (Cornejo-Donoso and Antezana 2008; Pinkerton et al. 2010; Kaehler et al. 2000; Gili et al. 2001; Corbisier et al. 2004; Jacob et al. 2005; Norkko et al. 2007; Mincks et al. 2008).

A trophic model with 17 components was constructed for Fildes Bay. The components represent the most abundant species or functional groups composed of multiple species. Seven components represented the following individual species: the brown macroalgae *H. grandifolius* and *D. anceps*, the red algae *G. skottsbergii*, the herbivores *N. concinna* and *Margarella* sp., the echinoid *Sterechinus neumayeri*, and the asteroid *D. brucei*. The other components were functional groups that included several species. These groups included multiple species of green (e.g., *Monostroma hariotii*), red (e.g., *T. antarcticus* and *P. cartilagineum*), and brown (e.g., *Ascoseira mirabilis* and *Halopteris obata*) algae, respectively. The filter feeder (FF) group was composed primarily of clams, hydrozoan, bryozoan, and sponge species. The small epifauna (SE) component included gastropod, nematode, and nemertean species. The group of other sea star species (SS) included principally *O. validus* and *Odontaster* sp. The benthic fishes (BF) group was composed primarily of *H. antarcticus*, *N. coriiceps*, and *N. rossii*. The three final groups were the phytoplankton (Phy), zooplankton (Zoo), and detritus (Det) groups (Fig. 16.1b).

All the compartments are trophically linked by detritus – primarily as microbial film and organic matter – because several studies have emphasized the importance of bacteria as food for various species of molluscs (e.g., Epstein 1997; Grossmann and Reichardt 1991; Plante and Mayer 1994; Plante and Shriver 1998), zooplankton (Epstein 1997), and Echinodermata (Findlay and White 1983). The models were constructed to depict the trophic relationships between the most important species or functional groups in the benthic communities of Fildes Bay. Notably, the models excluded the energy flows from epiphytes and the microphytobenthos, in addition to those leading to seals and birds, because insufficient scientific information was available for these groups. Although these exclusions reduced the realism of the model's configuration, the most dominant interdependencies and energy flows are reflected. Moreover, such a system-level error, if consistent, should not impede a comparative analysis of ecological systems placed under similar limitations (e.g., kelp forest of SE pacific coast).

Box 16.3 Macroscopic Network Properties

The macroscopic descriptors are based on Ulanowicz's *Ascendency* analysis that enables quantification of the level of development and organization of ecosystems (Ulanowicz 1986, 1997). This approach stems from information theory:

- *Total System Throughput* (*TST*). Indicates the size of the system, that is, the total number of flows in the system:

$$\mathrm{TST} = \sum_{i}^{n+1}\sum_{j}^{n+2} T_{ij} \quad (16.7)$$

  where TST is the summation of all flows among compartments; *i* and *j* can represent either an arbitrary system component or the environment.
- *Average Mutual Information* (*AMI*). Quantifies the organization of the system in relation to the number and diversity of interactions between components (complexity):

$$\mathrm{AMI} = \sum_{i}^{n+2}\sum_{j}^{n+2} f_{ij} * Q_i * \log_2\left(\frac{f_{ij}}{Q_j}\right) \quad (16.8)$$

  where f_{ij} is the fraction of flow in element *ji* in comparison to the TST; Q_j is the probability that species *j* is the host, calculated as the sum of flows in the *j*th row divided by TST; Q_i is the probability that *i* is the predator, calculated as the sum of flows in the *i*th column divided by TST.
- *Ascendency* (*A*). Measures the growth and development of a system and integrates *TST* and *AMI* of flows:

$$A = \sum_{ij} T_{ij} \log\left(\frac{T_{ij} T_{0,0}}{T_{i,0} T_{0,j}}\right) \quad (16.9)$$

  where *T* is the summation of all flows among compartments (TST); *i* and *j* are the prey and predator, respectively; 0 is the sum of flows of prey or predators; with $T_{i,0}$ the flows from one prey to all their predators, $T_{0,j}$ is the consumption of a predator over all its prey, and $T_{0,0}$ is the total sum of flows over prey and predators.

(continued)

Box 16.3 (continued)

- *Overhead* (*Ov*). Quantifies the degrees of freedom preserved by the network, can be used to estimate the ability of a network to withstand perturbations, and can be estimated from the difference between the *Development Capacity* and the *Ascendency*:

$$O = C - A \tag{16.10}$$

- *Development Capacity* (*C*). Quantifies the upper limit of *Ascendency*:

$$C = A + O \tag{16.11}$$

- *Total Biomass/Total Throughput* (*TB/TST*) ratio. Suggests different states of system maturity (Christensen 1995).
- *A/C* and *Ov/C* ratios. Are used as indicators of ecosystem development and the ability of the system to resist disturbances (Baird and Ulanowicz 1993; Costanza and Mageau 1999; Kaufman and Borrett 2010).
- *Relative Internal Ascendency* (A_i/C_i ratio). Represent well-organized, mature, and efficient systems that are therefore resistant against perturbations (Baird et al. 1991; Baird and Ulanowicz 1993).

The balancing of models was performed on the basis of the following six criteria proposed by Heymans et al. (2016): checking that (1) the ecotrophic efficiency (EE) (for quantifying the proportion of production utilized by the next trophic level through direct predation or fishing) of all compartments was <1.0 (Ricker 1968); (2) the gross efficiency (GE) (gross food conversion efficiency) of all compartments was <0.3 (Christensen and Pauly 1993); and when any inconsistency was detected, the average biomass was modified within the confidence limits (± 1 standard deviation); (3) the net efficiency (value for food conversion after accounting for unassimilated food) of all compartments was >GE; (4) the respiration/assimilation biomass (RA/AS) was <1.0; (5) respiration/biomass (RA/B) values for fish between 1 and 10 years^{-1} and for groups with higher turnover between 50 and 100 years^{-1}; and (6) the production/respiration (P/RA) was <1.0.

16.2 Macroscopic Ecosystem-Network Properties

The macrodescriptors based on Ulanowicz's *Ascendency* (Ulanowicz 1986, 1997) include the *Total Biomass/Total Throughput* (TB/TST) ratio, which suggests different states of system maturity (Christensen 1995); *Total System Throughput* (TST), which indicates the size of the system, that is, the total number of flows in the system; *Average Mutual Information* (AMI), which quantifies the organization of the system in relation to the number and diversity of interactions between components (complexity) (Ulanowicz 1986, 1997); *Ascendency* (A), which measures the growth and development of a system and integrates TST and AMI of flows; and *Overhead* (Ov), which quantifies the degrees of freedom preserved by the network, can be used to estimate the ability of a network to withstand perturbations and can be estimated from the difference between the *Development Capacity* (C) and the *Ascendency* (A) (Ulanowicz 1986): *Development Capacity* (C), which quantifies the upper limit of *Ascendency*, and A/C and Ov/C ratios, which are used as indicators of ecosystem development and the ability of the system to resist disturbances (Ulanowicz 1986, 1997; Baird and Ulanowicz 1993; Costanza and Mageau 1999; Kaufman and Borrett 2010). High values of the *Relative Internal Ascendency* (A_i/C_i ratio) represent efficient systems that are therefore resistant against perturbations (Baird et al. 1991; Baird and Ulanowicz 1993). For more details about macrodescriptors, see Box 16.3.

16.2.1 Macroscopic Properties of Coastal Benthic-Pelagic Ecosystem at Fildes Bay

The coastal benthic/pelagic ecological system of Fildes Bay dominated by brown large macroalgae reached a *Total System Throughput* (TST) equals to 24,234.0 g ww m^{-2} $year^{-1}$; this magnitude was lower than that calculated for a kelp forest off the Antofagasta Peninsula (Ortiz 2008a, 2010). However, compared to the benthic communities of Tongoy Bay (Ortiz and Wolff 2002; Wolff 1994) and different estuaries around the world (Baird and Ulanowicz 1993; Patrício and Marques 2006; Wolff et al. 2000), TST was higher in the Fildes Bay system. The *Development Capacity* (C) accounted for 110,354.4 flow bits, *Ascendency* (A) accounted for 32,953.9 flow bits, and the A/C and Ov/C ratios were 29.8% and 70.1%, respectively (Table 16.1). Notably, the A/C value calculated for the Fildes Bay ecological system was one of the lowest compared to those obtained for other coastal areas along the Chilean coast and around the globe (Table 16.2). The difference between the A/C and the A_i/C_i ratios for the Fildes Bay model may indicate a dependency of this system on external connections (sensu Baird et al. 1991).

Ulanowicz (1997) proposed estimating *Relative Ascendency* by model component to evaluate the contribution of each group to the overall structure and function of the system. In this case, detritus accounted for ~33%, followed by phyto-zooplankton at ~26%, macroalgae at ~19%, filter feeders at ~7%, small epifauna at ~5%, and top predators at ~2%. Moreover, the sea star species groups, Chlorophyta, and the red algae *G. skottsbergii* accounted for the complexity in the system; that is, they exhibited the lowest percentage of *Average Mutual Information* (AMI) within the Fildes model system (Table 16.2).

Macroscopic properties, such as the A/C ratio, Ov/C ratio, and *Redundancy* values, indicate that Fildes Bay would be a less developed system but is more resistant to disturbances than a kelp forest off the Mejillones Peninsula or the

Table 16.1 Macroscopic properties of the trophic network for the coastal benthic-pelagic ecosystem of Fildes Bay (Antarctica)

Property	Fildes Bay
Total System Throughput (TST) (g ww m^{-2} $year^{-1}$)	21,432.00
Total Biomass/Total System Throughput (TB/TST)	0.1312
Ascendency (A) (g ww m^{-2} $year^{-1}$ * bits)	29,758.60
Overhead (Ov) (g ww m^{-2} $year^{-1}$ * bits)	69,355.02
Development Capacity (C) (g ww m^{-2} $year^{-1}$ * bits)	99,114.87
Average Mutual Information (AMI)	1.39
A/C (%)	30.02
A_i/C_i (%)	14.00
Ov/C (%)	69.97

seagrass meadows of Tongoy Bay (Ortiz and Wolff 2002; Ortiz 2008a, 2010) (Tables 16.1 and 16.2). This may be explained by the fact that Fildes Bay is negatively affected by the Antarctic's austral winters, which are characterized by lower temperatures and freezing, leading to a reduction in the herbivore biomass and thereby constraining the flow of energy/matter toward the upper trophic levels. Additionally, as shown in Tables 16.1 and 16.2, the different estuaries and coral reef systems appeared to be more developed (A/C and Ov/C) but less resistant to perturbations compared to Fildes Bay and the benthic ecosystems studied along the Chilean coast. This latter comparison should be taken with a degree of caution because the trophic model constructed for Fildes Bay represents only a narrow temporal window, and unknown system characteristics may emerge during the rest of the year. The difference between the A/C and A_i/C_i ratios in the Fildes Bay model system may primarily be a consequence of the omission of the flows to birds and marine mammals from our analysis. The analysis of relative *Ascendency* by component revealed that those groups that principally contributed to the overall structure and function of the Fildes Bay system (i.e., detritus, the phyto-zooplankton complex, and macroalgae) differed from those that contributed to the kelp forest (Mejillones Peninsula) (Ortiz 2008a, 2010) but were similar to those within the benthic systems of Tongoy Bay and the La Rinconada Marine Reserve (Antofagasta Bay) (Ortiz and Wolff 2002; Ortiz et al. 2010). This outcome indicates that although a significant amount of the Antarctic system's biomass is concentrated in macroalgae, these macroalgae would contribute fewer nutrients to the coastal marine ecosystem than those within kelp forests (Duggins et al. 1989; Ortiz 2008a, 2010).

Table 16.2 Macroscopic properties for system's development and organization derived from *Ascendency* network analysis for Fildes Bay (Antarctic) and other coastal ecosystems

Macroscopic properties of trophic networks						
Ascendency theoretical framework (Ulanowicz 1986, 1997)						
	TST	*C*	*A*	*A*/*C*	Ov	Ov/*C*
Coastal marine ecosystems	(g ww m^{-2} $year^{-1}$)	(flow bits)	(flow bits)	%	(flow bits)	%
Along the Chilean coast (SE Pacific)						
Benthic/pelagic ecological system of Fildes Bay[1]	21,432.00	99,114.87	29,758.60	30.02	69,355.02	69.97
Kelp ecological system dominated by *M. pyrifera*, Antofagasta Peninsula[2]	72,512.0	207,777.4	93,462.6	45.0	112,548.0	55.0
Kelp forest ecological system dominated by *L. trabeculata*, Antofagasta Peninsula[2]	50,105.0	200,609.4	77,613.5	38.7	117,678.9	61.3
Kelp forest ecological system, Antofagasta Peninsula[3]	85,217.0	332,041.6	117,939.7	35.5	211,848.3	64.5
Seagrass habitat ecological system of Tongoy Bay[4]	18,746.6	69,270.4	21,557.8	31.1	46,991.0	68.9
Mud habitat ecological system of Tongoy Bay[4]	17,451.3	59,139.0	19,354.8	32.7	39,433.4	67.3
Benthic/pelagic ecological system of Tongoy Bay[5]	20,834.9	80,689.8	26,312.6	32.6	54,377.2	67.4
La Rinconada Marine Reserve coastal ecological system, Antofagasta Bay[6]	20,124.0	80,321.0	24,375.1	30.3	55,945.9	69.7
Mejillones benthic/pelagic ecological system of Mejillones Bay[7]	29,429.8	142,897.9	34,395.1	24.1	108,353.1	75.9
Antofagasta benthic/pelagic ecological system of Antofagasta Bay[7]	37,539.8	170,237.0	48,574.3	28.5	121,434.8	71.5
Around the world						
Coral reef ecosystem, Chinchorro Bank, México[8]	148,094.1	318,400.0	178,200.0	56.3	139,800.0	43.7
Coral reef ecosystem, Mexican Pacific Coast[9]	194,758.4	890,301.6	308,428.7	34.64	581,872.9	65.35
Mangrove estuary of Caeté, Brazil[10]	10,558.6	44,741.4	12,261.6	27.0	31,129.8	63.0
Zostera meadows of Mondego Estuary, Portugal[11]	10,852.0	39,126.0	16,550.3	42.3	22,575.7	57.7
Ems estuary in the Netherlands[12]	12,980.0	6085.0	2327.0	38.3	3758.0	61.7
Benguela upwelling ecosystem, Namibia[13]	8897.0	36,041.0	17,313.0	48.1	18,728.0	51.9

TST Total System Throughput, *C* System Capacity, *A Ascendency*
[1]Current study, [2]Ortiz (2008a), [3]Ortiz (2010), [4]Ortiz and Wolff (2002), [5]Wolff (1994), [6]Ortiz et al. (2010), [7]Ortiz et al. (2015), [8]Rodriguez-Zaragoza et al. (pers. comm.), [9]Hermosillo-Núñez et al. (2018), [10]Wolff et al. (2000), [11]Patrício and Marques (2006), [12]Baird and Ulanowicz (1993), and [13]Heymans and Baird (2000)

16.3 Keystone Species Complex (KSC)

16.3.1 Functional Keystoneness Indices

Once the trophic model was balanced following the rules given by Heymans et al. (2016), the functional *keystone index* (KS_i) developed by Libralato et al. (2006) was used. This index is an extension of the *mixed trophic impacts* (MTI, Ulanowicz and Puccia 1990). Because every impact can be quantitatively positive or negative, a new measure of the overall effect must be determined for each species or functional group (ε_i) using the following mathematical Eq. 16.2:

$$\varepsilon_i = \sqrt{\sum_{j \neq i}^{n} m_{ij}^2}$$

where m_{ij} corresponds to the elements of the MTI matrix and quantifies the direct and indirect effects that each (affecting) species or group $_i$ has on any (affected) group j of the food web. However, the effect of the change in a group's biomass on the group itself (i.e., m_{ii}) is not included. The contribution of biomass from every species or functional group with respect to the total biomass of the network was estimated using the following Eq. 16.3:

$$p_i = \frac{B_i}{\sum_i^n B_k}$$

where p_i is the proportion of biomass of each species B_i with respect to the sum of the total biomass B_k. Therefore, to balance the overall effect and the biomass, the *keystone index* (KS_i) for each species or functional group was established using the Eq. 16.4, which integrates the Eqs. 16.2 and 16.3 as follows:

$$\mathrm{KS}_i = \log\left[\varepsilon_i * (1 - p_i)\right]$$

This index assigns high values of functional keystoneness to those variables (species) or functional groups that have low biomass and a high overall effect.

The propagation of direct and indirect effects and *system recovery time* (SRT) magnitudes estimated by *Ecosim* were treated in the same way as those obtained with MTI in order to obtain two additional functional keystone indices. The *Ecosim* simulations were used to evaluate the propagation of instantaneous direct and indirect effects and the *system recovery time* (SRT) (as a system resilience measure) in response to a steady increase in the *total mortality* (Z) of all compartments (see Eqs. 16.5 and 16.6) which was set equivalent to 10%, 30%, and 50%. This procedure was done between the first and second year of simulation for all components considered in the model. These three magnitudes (scenarios) were set for prediction purposes as a measure of confidence. As the models studied represent only short-term (transient) dynamics, the propagation of instantaneous effects was determined by evaluating the changes of biomass in the remainder variables in the third year of simulation. All dynamic simulations by *Ecosim* were carried out using the following vulnerabilities (flow control): (1) bottom-up (prey controls the flow), (2) top-down (predator controls the flow), and (3) mixed (both prey and predator control the flow) (Boxes 16.1 and 16.3):

$$Z = M(\text{natural mortality}) + F(\text{fishing mortality})$$

$$\text{Production}(P) = \text{Biomass}(B) * Z$$

After that, Eqs. 16.2, 16.3, and 16.4 were used to obtain one *keystone species index* related to the propagation of direct and indirect effects ($KS_{i\text{Ecosim}1}$), and Eqs. 16.3 and 16.4 were used to obtain another functional *keystone species index* related to SRT magnitudes ($KS_{i\text{Ecosim}2}$). These indices show, similar as the KS_i index (Libralato et al. 2006), high values of keystoneness associated to compartments with low biomass and a high overall effect.

16.3.2 Topological Keystone Index

The *structural keystone index* (K_i) developed by Jordán et al. (1999) and Jordán (2001), following Harary (1961), was also used in this work. This index is applicable for trophic hierarchies (directed acyclic trophic networks) and considers direct and indirect interactions in vertical directions (i.e., bottom-up and top-down). The structural keystone index of the *i*th species or functional group (K_i) is calculated using the following Eq. 16.7:

$$K_i = \sum_{c=1}^{n} \frac{1}{d_c} * (1 + K_{bc}) + \sum_{e=1}^{n} \frac{1}{f_e} * (1 + K_{te})$$

where n is the number of predator species eating species i, d_c is the number of prey of the *c*th predator, K_{bc} is the bottom-up keystone index of the *c*th predator, and symmetrically we have m as the number of prey species eaten by species i, f_e as the number of predators of its *e*th prey, and K_{te} as the top-down keystone index of the *e*th prey. Within this index, the first and second components represent the bottom-up (K_{bc}) and top-down (K_{te}) effects, respectively. Finally, the *keystone index* (K_i) corresponds to the highest value as a product of the addition of bottom-up (K_{bc}) and top-down (K_{te}) components. For more details on this method, see Jordán (2001) and Vasas et al. (2007). The K_i index has been shown to be one of the most robust centrality indices (Fedor and Vasas 2009). It is important to indicate that only bottom-up and top-down components of K_i were used in the current work as a way to compare functional indices obtained using *Ecosim* simulations under comparable flow control mechanisms.

16.3.3 Semiquantitative or Qualitative Keystone Index

A keystoneness index based on qualitative or semiquantitative loop models was also calculated. Once the stabilized trophic matrix with $F_n < 0$ was obtained, the self-dynamics of each variable corresponding to the principal diagonal were modified to estimate two new perturbed magnitudes of local stability F_p. Based on the distance ($\triangle$) between F_n and F_p as shown in Eq. 16.8

$$\triangle = \left| F_n - F_p \right|$$

it was possible to determine the change provoked by each variable on initial stability (F_n), thereby obtaining a first *qualitative keystone species index* (KQ_{iLA1}) (selecting only the largest change by variable). Because *Loop Analysis* does not consider the abundance of the components, the difference (Δ) was treated in similar way to Eq. 16.4 to obtain an additional *keystone index* (KQ_{iLA2}) in which high values of keystoneness corresponded to variables with low biomass and a high overall effect. Due to the qualitative-dialectic character of *Loop Analysis*, the prey-predator interaction is captured as a mixed control mechanism.

16.3.4 Centrality of Node Sets

Field studies suggest that in some situations, a small group of species behave as keystones and they form a keystone species complex (Daily et al. 1993). The importance of this group is typically realized through their interspecific interaction network, so a network approach to better understand multispecies keystone complexes is reasonable. A particular approach was suggested by Borgatti (2003a) in order to find the most central set of *k* nodes in a network. According to this, a topological keystone species complex is defined as a solution of the KeyPlayer Problem (KPP) (*sensu* Borgatti 2003a). The software KeyPlayer 1.44 (Borgatti 2003b) was used to compute the importance of species combinations in maintaining the integrity of a network. The importance of a set of nodes can be calculated by considering either their fragmentation effect (KPP1) or their reachability effect (KPP2). In the first case, we identify which *k* nodes should be deleted from the network of *n* nodes in order to maximally increase its fragmentation. In the second case, we identify from which *k* nodes the largest proportion of the other *n*–*k* nodes are reachable within a certain distance. Based on fragmentation (*F* of KPP1), the best set of the deleted *k* nodes can maximally increase the fragmentation of the network. This means an increase of the number of components and a larger average distance generated within individual components. We used $k = 1, 2$, and 3 with 10,000 simulations for each. We also consider the distance-based reachability approach (R^d of KPP2). We simply count the number of nodes that are reachable within a given distance of $m = 1$ step from a given set of k nodes. We have chosen $m = 2$ steps and increased the size of the KP set from $k = 1$ to $k = 3$. We applied 10,000 runs for each simulation. The outcome was three sets of nodes (for $k = 1, 2, 3$) for each network, containing species codes. For each k, the software presents the percentage of nodes outside the KP set but reachable from it in one step. If this percentage reaches 100%, then the whole network is reachable from the KP set and we cannot create larger KP sets.

16.3.5 Keystone Species Complex in Benthic-Pelagic Ecosystem at Fildes Bay

All individual indices integrated in the holistic *keystone species complex* (KSC_i) detected keystoneness for a variety of different species and functional groups, with some agreement. In general terms, keystoneness properties were detected for species of all different trophic levels, including primary producers, herbivores, and top predators. Notably, the topological-structural (K_i) and the functional (KS_i) indices both identified sea stars (SS) as keystone species even though they are based on different algorithms. This outcome agrees partially with the field observations of Gaymer and Himmelman (2008), who studied dominant sea star species in benthic communities of northern Chile, establishing *Meyenaster gelatinosus* as a keystone species in subtidal systems. Likewise, Ortiz et al. (2013a) determined that most of the keystone species complexes identified by KSC indices in different ecological systems along the Chilean coast include one asteroid species as the top predator. The relevance of the sea star species determined in the present work also coincides with the results described by Ortiz et al. (2009) regarding the longest *system recovery times* (as a measure of "resilience") being obtained in response to perturbations on these species.

Similarly, both the qualitative and semiquantitative keystone indices (KQ_{iLA1} and KQ_{iLA2}) identified small epifauna (SE) to have keystoneness properties. This result is very interesting because loop model predictions respond with a high degree of certainty to external perturbations (Briand and McCauley 1978; Lane and Blouin 1985; Lane 1986; Hulot et al. 2000; Ortiz 2008b). Regarding the functional indices based on *Ecosim* dynamical simulations (under three mortality levels and three types of flow control mechanisms), $KS_{iECOSIM1}$ and $KS_{iECOSIM2}$ both identified the group of Chlorophyta (Chloro), phytoplankton (Phyto), zooplankton (Zoo), the small epifauna (SE), and the species *S. neumayeri* (*Sn*) (grazers) as having keystoneness. It is important to mention that it was not possible to determine species or functional groups with keystoneness properties based on the functional $KS_{iECOSIM2}$ index under a 30% increased mortality and using a top-down flow control mecha-

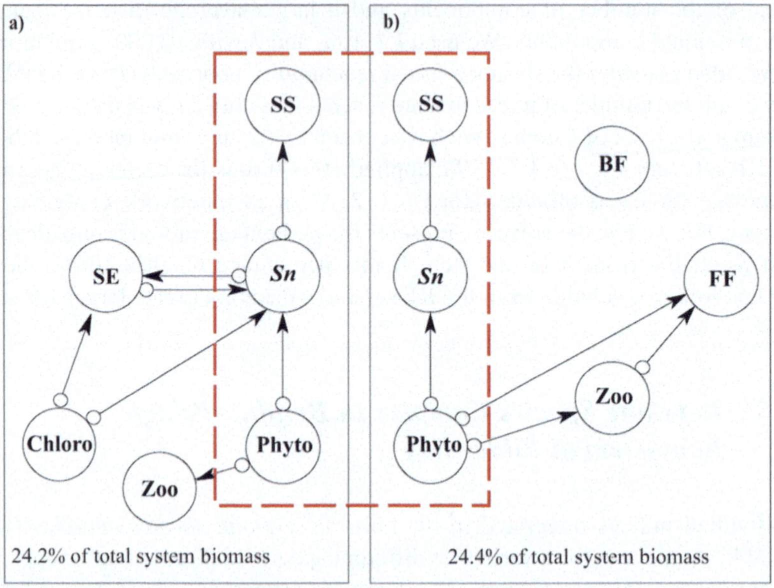

Fig. 16.2 *Keystone species complexes* (KSC) determined for coastal benthic-pelagic ecosystem at Fildes Bay (Antarctica) based on quantitative and semiquantitative networks (**a**) and centrality of node sets (**b**). The share components are highlighted (Note: the circle and arrow mean negative and positive effect, respectively). (Adapted from Fig. 3 in Ortiz et al. 2017)

nism because the model system does not return to initial steady-state conditions, instead oscillating persistently. The core species as indicated by the *keystone species complex* index (KSC_i) for the coastal ecological system of Fildes Bay account by 24.2% of the total system biomass (Fig. 16.2a). Importantly, the KSC includes species and functional groups that make up an ecological path clearly representing three trophic levels.

The composition of the KeyPlayer sets is nested: For $F_{k=1}$, the key group is phytoplankton (Phyto); for $F_{k=2}$, the key groups are Phyto and benthic fishes (BF); and for $F_{k=3}$, the key groups are Phyto, BF, and the sea stars (SS). For $R^d_{k=1}$, the key group is Phyto; for $R^d_{k=2}$, the key groups are Phyto and BF; and for $R^d_{k=3}$, the key groups are Phyto, BF, and *S. neumayeri* (Sn). This means that the Phyto-*Sn*-SS chain as well as BF together composes a core of species in this community (Fig. 16.2b). The keystone species complex obtained by multinode centrality represents 24.4% of the total system biomass, a value quite similar to the previous approach. The multinode approach based on the KP indices thus not only partly reinforces the identity of some key players (e.g., *S. neumayeri*) but also suggests new key organisms (e.g., zooplankton). This latter result is well supported by the literature (e.g., Stibor et al. 2004; Murphy et al. 2007). That phytoplankton that was identified as a component of the KSC could be a consequence of the higher level of primary productivity in Antarctic waters (Smith et al. 2007; Cornejo-Donoso and Antezana 2008).

The *keystone species complex* indices (KSC_i) determined for Fildes Bay integrate fewer components than the one determined for the kelp forest of northern Chile. However, both ecological systems share a sea urchin species (herbivore positioned at intermediate trophic level), a sea star species (top predator), and the Chlorophyta (Chloro) (primary producers). After all, the outcomes obtained show that the components with keystoneness properties in the benthic-pelagic system of Fildes Bay are widely heterogeneous, coinciding with results reported for other ecosystems (Power et al. 1996; Piraino et al. 2002; Libralato et al. 2006). Furthermore, Jordán et al. (2007, 2008) reported similar findings after comparing several structural and functional keystone species indices. Hermosillo-Núñez et al. (2018) showed that in spite the wide trophic heterogeneity of components with keystoneness properties, it is possible to observe that the core set of species and functional groups are trophically linked. Okey (2004) arrived at similar results by defining keystone guilds or clusters of species with keystoneness properties based on a trophic model in Alaska. Thus, we believe that the *keystone species complex* index (KSC) for coastal benthic/pelagic ecological systems of Fildes Bay would facilitate the design and assessment of conservation and monitoring measures, especially when the Antarctic coastal marine ecosystems are being severely stressed by the direct effects of the global warming and UV radiation (Richter et al. 2008; Pessoa 2012). The above notwithstanding, it is necessary also to recognize that the use of the *keystone species complex* indices is still quite difficult because the traditional view of conservation and monitoring efforts is based principally on single species such as keystone and/or niche-builder or bioengineer species. This mindset undoubtedly imposes even greater challenges to understand how global changes, covarying with variability of the natural system, act on networks of interacting species.

16.4 Contribution of Keystone Species Complex to Macroscopic Network Properties

In global terms, the species and functional group belonging to KSCs showed different degrees of contribution to the ecosystem's emergent properties (i.e., growth, organization, development, maturity, and health) (Table 16.3). KSC index and *centrality of node sets* account to lower magnitudes for all macroscopic properties (Table 16.3). The total biomass of KSCs accounted <25% of the total system biomass, coinciding with the classical keystone concept (sensu Power et al. 1996). A particular case is the functional group of phytoplankton, which concentrated a little more biomass, contributing to higher values of TST, AMI, and *Ascendency*. Likewise, the functional groups of Chlorophyta (Chloro) and sea stars (SS) presented the higher percentage magnitudes of Ov/*C* and TB/TST ratios, mainly contributing – in term of flow – to the system's resistance against perturbations and to the whole system's cycling (promoting efficiency). Different indices for the deter-

Table 16.3 Contribution of species and functional groups of KSCs on macroscopic network properties derived from *Ascendency* (Ulanowicz 1986, 1997) of the coastal benthic-pelagic ecosystem of Fildes Bay (Antarctica)

	TB %	TST %	AMI %	A%	Ov/*C* (%)	*A*/*C* (%)	TB/TST (%)
(a) KSC index							
SS	0.658	0.382	0.357	0.357	6.551	3.614	5.953
SE	2.364	4.254	4.743	4.740	5.866	5.943	1.920
Sn	1.226	1.922	2.103	2.104	6.082	5.206	2.204
Chloro	1.369	0.381	0.340	0.340	6.675	3.184	12.402
Phy	10.665	22.283	18.897	18.872	5.398	7.526	1.654
Contribution KSC	16.283	29.223	26.440	26.413	30.572	25.473	24.132
(b) K node sets							
BF	1.161	1.146	1.074	1.077	6.308	4.435	3.502
SS	0.658	0.382	0.357	0.357	6.551	3.614	5.953
Sn	1.226	1.922	2.103	2.104	6.082	5.206	2.204
FF	6.309	6.223	6.904	6.910	5.490	7.217	3.503
Zoo	4.355	7.836	7.154	7.155	5.868	5.931	1.920
Phy	10.665	22.283	18.897	18.872	5.398	7.526	1.654
Contribution KSC	24.375	39.792	36.490	36.474	35.698	33.929	18.735
(c) KSC integrated							
SS	0.658	0.382	0.357	0.357	6.551	3.614	5.953
Sn	1.226	1.922	2.103	2.104	6.082	5.206	2.204
Phy	10.665	22.283	18.897	18.872	5.398	7.526	1.654
Contribution KSC	12.550	24.587	21.357	21.333	18.031	16.346	9.811

SS sea stars, *SE* small epifauna, *Sn S. neumayeri*, *Chloro* Chlorophyta, *Phy* phytoplankton, *BF* benthic fishes, *FF* filter feeders, *Zoo* zooplankton

mination of *keystone species complexes* (KSCs) are suitable when biomass content is considered, and it is consistent when using the total flows of matter and/or energy (TST), the contribution to complexity (AMI), and system development and heath (*Ascendency*).

16.5 Constrains and Perspectives

Although we were well aware that the quantitative trophic model built and analyzed in this study is a partial representation of the overall trophic seascape and interactions underlying the dynamics within Fildes Bay's coastal benthic/pelagic ecological system, such limitations, however, occur in any type of model and are independent of the model's degree of complexity (Levins 1966; Ortiz and Levins 2011). In the present study, the following limitations were identified: (1) The model only represented the austral summer condition, while the annual benthic/pelagic dynamics are unknown; (2) system complexity was reduced in relation to the composition of several functional groups, although the most abundant macroalgae, herbivore, and carnivore species were represented; and (3) regardless of the inherent, well-known limitations and shortcomings of the *Ecopath*, *Ecosim*, *Loop Analysis*, and *KeyPlayer node sets* theoretical frameworks, the constructed model and its dynamic simulations represented underlying system processes based exclusively on short-term or transient dynamics. In spite of these concerns, we claim that the most relevant trophic relationships and energy/matter flows were well reflected in our model. Here, the macroscopic properties and sensitive model compartments of the system were quantified and compared adequately.

Likewise, the core species and functional groups that constituted the keystone species complexes (KSCs) in the coastal ecological systems at Fildes Bay accounted to lower magnitudes of total system biomass and macroscopic properties, coinciding with those groups and species identified as keystones in experimental studies using variations of the original keystone species concept (Menge et al. 1994; Estes et al. 1998; Bond 2001). Several species or functional groups from different trophic levels could have keystone properties. This result should not be considered as ambiguous because populations inhabit heterogeneous and changing environments (Levins 1968). This insight would support the design of putative conservation and monitoring strategies in the Antarctic Peninsula, including a core of species or functional groups linked trophically (Jordán et al. 2019), which could supplement the unique species with keystone properties or those species considered as *niche constructers* (sensu Lewontin and Levins 2007) or *bioengineers* (sensu Jones et al. 1994). We claim that the adequacy of the description of ecosystem properties based on only one species or functional group is limited, especially when the task is to conserve and monitor complex ecosystems (Jordán et al. 2019). In this sense, the determination of the *keystone species complexes* (KSCs) could assist within an ecosystem-based conservation view under an ecological network context (sensu Pikitch et al. 2004). Finally, it is relevant to assess the trajectory of the target species, as they constitute compartments with a relevant role in the structure and trophic functioning of the benthic-pelagic coastal ecosystem at Fildes Bay, Antarctica.

Acknowledgments This research was financed by the grant Proyecto Anillo ART1101 (CONICYT-PIA). We also thank the Instituto Antártico Chileno (INACH) for logistic support. The work of Ferenc Jordán was supported by National Research, Development and Innovation Office – NKFIH, grant OTKA K 116071.

Appendix 16.A

Table 16.4 Parameter values entered (in bold), estimated (standard), and calculated (in italics) by the *Ecopath* with *Ecosim* (EwE) software for the species and functional groups in the coastal benthic-pelagic ecosystem of Fildes Bay (Antarctica)

Model compartments						
Species/functional groups	TL	*B*	*P*/*B*	*Q*/*B*	EE	GE
(1) Benthic fishes (BF)	*3.03*	**32.67**	**2.40**	**8.50**	0.05	0.28
(2) *Diplasterias brucei* (Db)	*2.98*	**37.83**	**1.40**	**5.00**	0.18	0.28
(3) Sea stars (SS)	*3.29*	**18.52**	**1.40**	**5.00**	0.04	0.28
(4) Small epifauna (SE)	*2.76*	**66.50**	**4.60**	**15.50**	0.95	0.30
(5) *Sterechinus neumayeri* (Sn)	*2.35*	**34.50**	**3.70**	**13.50**	0.99	0.27
(6) *Margarella* sp. (Msp)	*2.10*	**59.50**	**3.40**	**11.50**	0.99	0.30
(7) *Nacella concinna* (Nc)	*2.05*	**56.50**	**3.60**	**12.50**	0.99	0.29
(8) Filter feeders (FF)	*2.30*	**177.46**	**1.87**	**8.50**	0.93	0.22
(9) *Himmantothallus grandifolius* (Hg)	*1.00*	**597.67**	2.10	–	0.35	–
(10) *Desmarestia anceps* (Da)	*1.00*	**310.50**	2.30	–	0.38	–
(11) *Gigartina skottsbergii* (Gs)	*1.00*	**35.17**	2.50	–	0.42	–
(12) Phaeophyceae (Phaeo)	*1.00*	**650.50**	2.80	–	0.39	–
(13) Chlorophyta (Chloro)	*1.00*	**38.50**	2.40	–	0.36	–
(14) Rhodophyta (Rhodo)	*1.00*	**174.57**	5.00	–	0.32	–
(15) Zooplankton (Zoo)	*2.00*	**122.50**	**4.60**	**15.50**	0.99	0.30
(16) Phytoplankton (Phy)	*1.00*	**300.00**	**18.00**	–	0.54	–
(17) Detritus (Det)	*1.00*	**100.00**	–	–	0.07	–

TL trophic level, *B* biomass (g wet weight m^{-2}), *P/B* productivity ($year^{-1}$), *Q/B* consumption rate ($year^{-1}$), *EE* ecotrophic efficiency (dimensionless), *GE* gross efficiency (dimensionless)

References

Allen KR (1971) Ralation between production and biomass. J Fish Res Board Can 28:1573–1581
Allesina S, Bodini A (2004) Who dominates whom in the ecosystem? Energy flow bottlenecks and cascading extinctions. J Theor Biol 230:351–358
Baird D, Ulanowicz RR (1993) Comparative study on the trophic structure, cycling and ecosystem properties of four tidal estuaries. Mar Ecol Prog Ser 99:221–237
Baird D, McGlade JJ, Ulanowicz RE (1991) The comparative ecology of six marine ecosystems. Philos Trans R Soc Biol Sci 333:15–29
Barua M (2011) Mobilizing metaphors: the popular use of keystone, flagship and umbrella species concepts. Biodivers Conserv 20:1427–1440
Benedek Z, Jordán F, Báldi A (2007) Topological keystone species complexes in ecological interaction networks. Community Ecol 8:1–7
Bond W (2001) Keystone species – hunting the shark? Science 292:63–64
Borgatti SP (2003a) The key player problem. In: Breiger R, Carley Pattison P (eds) Dynamic social network modeling and analysis: Workshop Summary and Papers. Committee on Human Factors, National Research Council. National Academies Press, Washington, DC, pp 241–252
Borgatti SP (ed) (2003b) KeyPlayer. Analytic technologies. Harvard University Press, Boston

Briand F, McCauley E (1978) Cybernetic mechanisms in lake plankton systems: how to control undesirable algae. Nature 273:228–230
Christensen V (1995) Ecosystem maturity–towards quantification. Ecol Model 77:3–32
Christensen V, Pauly D (1992) Ecopath II: a software for balancing steady–state ecosystem models and calculating network characteristics. Ecol Model 61:169–185
Christensen V, Pauly D (1993) Trophic models in aquatic ecosystem. ICLARM Conf Proc 26:390
Christensen V, Walters CJ (2004) Ecopath with Ecosim: methods, capabilities and limitations. Ecol Model 172:109–139
Corbisier TN, Petti MAV, Skowronski RSP, Brito TAS (2004) Trophic relationships in the near-shore zone of Martel Inlet (King George Island, Antarctica): δ13C stable-isotope analysis. Polar Biol 27:75–82
Cornejo-Donoso J, Antezana T (2008) Preliminary trophic model of the Antarctic Peninsula ecosystem (sub-area CCAMLR 48.1). Ecol Model 218:1–17
Costanza R, Mageau M (1999) What is a healthy ecosystem? Aquat Ecol 33:105–115
Daily GC, Ehrlich PR, Haddad NM (1993) Double keystone bird in a keystone species complex. PNAS 90:592–594
Darmbacher J, Guaghan D, Rochet M, Rossignol P, Trenkel V (2009) Qualitative modelling and indicators of exploited ecosystems. Fish Fish 10:305–322
Duggins DO, Simenstad CA, Estes JA (1989) Magnification of secondary production by kelp detritus in coastal marine ecosystems. Science 245:101–232
Epstein S (1997) Microbial food web in marine sediments: I. Trophic interactions and grazing rates in two tidal flat communities. Microb Ecol 34:188–198
Estes JA, Tinker MT, Williams TM, Doak DF (1998) Killer whale predation on sea otters linking oceanic and near shore ecosystems. Science 282:473–476
Fedor A, Vasas V (2009) The robustness of keystone indices in food webs. J Theor Biol 260:372–378
Findlay R, White D (1983) The effects of feeding by the sand dollar Mellita quinquiesperforata (Leske) on the benthic microbial community. J Exp Mar Biol Ecol 72:25–41
Gaymer C, Himmelman JH (2008) A keystone predatory sea star in the intertidal zone is controlled by a higher-order predatory sea star in subtidal zone. Mar Ecol Prog Ser 370:143–153
Giacaman-Smith J, Neira S, Arancibia H (2016) Community structure and trophic interactions in a coastal management and exploitation area for benthic resources in central Chile. Ocean Coast Manag 119:155–163
Gili JM, Coma R, Orejas C, López-González PJ, Zavala M (2001) Are Antarctic suspension feeding communities different from those elsewhere in the world? Polar Biol 24:473–485
Grossmann S, Reichardt W (1991) Impact of *Arenicola marina* on bacteria in intertidal sediments. Mar Ecol Prog Ser 77:85–93
Harary F (1961) Who eats whom? Gen Syst 6:41–44
Hawkins S (2004) Scaling up: the role of species and habitat patches in functioning of coastal ecosystems. Aquat Conserv Mar Freshwat Ecosyst 14:217–219
Hermosillo-Núñez BB, Ortiz M, Rodríguez-Zaragoza FA (2018) Keystone species complexes in kelp forest ecosystems along the northern Chilean coast (SE Pacific): improving multispecies management strategies. Ecol Indic 93:1101–1111
Heymans J, Baird D (2000) A carbon flow model and network analysis of the northern Benguela upwelling system, Namibia. Ecol Model 126:9–32
Heymans J, Coll M, Link JS, Mackinson S, Steenbeek J, Walters C, Christensen V (2016) Best practice in Ecopath with Ecosim food–web models for ecosystem–based management. Ecol Model 331:173–184
Hulot F, Lacroix G, Lescher-Moutoué F, Loreau M (2000) Functional diversity governs ecosystem response to nutrient enrichment. Nature 405:340–344
Huovinen P, Gómez I (2013) Photosynthetic characteristics and UV stress tolerance of Antarctic seaweeds along the depth gradient. Polar Biol 36:1319–1332

Jacob U, Mintenbeck K, Brey T, Knust R, Beyer K (2005) Stable isotope food web studies: a case for standardized sample treatment. Mar Ecol Prog Ser 287:251–253

Jones C, Lawton J, Shachak M (1994) Organism as ecosystem engineers. Oikos 69:373–386

Jordán F (2001) Trophic fields. Community Ecol 2:181–185

Jordán F, Scheuring I (2004) Network ecology: topological constraints on ecosystem dynamics. Phys Life Rev 1:139–172

Jordán F, Takács-Sánta A, Molnár I (1999) A reliability theoretical quest for keystones. Oikos 86:453–462

Jordán F, Benedek Z, Podani J (2007) Quantifying positional importance in food webs: a comparison of centrality indices. Ecol Model 205:270–275

Jordán F, Okey T, Bauer B, Libralato S (2008) Identifying important species: linking structure and function in ecological networks. Ecol Model 216:75–80

Jordán F, Pereira J, Ortiz M (2019) Mesoscale network properties in ecological system models. Curr Opin Syst Biol 13:122–128

Kaehler S, Pakhomov EA, McQuaid CD (2000) Trophic structure of the marine food web at the Prince Edward Islands (Southern Ocean) determined by $\delta^{13}C$ and $\delta^{15}N$ analysis. Mar Ecol Prog Ser 208:13–20

Kaufman AG, Borrett SR (2010) Ecosystem network analysis indicators are generally robust to parameter uncertainty in a phosphorus model of Lake Sidney Lanier, USA. Ecol Model 221:1230–1238

King JC (1994) Recent climate variability in the vicinity of the Antarctic Peninsula. Int J Climatol 14:357–369

Lane P (1986) Symmetry, change, perturbation, and observing mode in natural communities. Ecology 67:223–239

Lane P, Blouin A (1985) Qualitative analysis of the pelagic food webs of three acid impacted lakes. Int Rev Ges Hydrobiol 70:203–220

Levins R (1966) The strategy of model building in population biology. Am Sci 54:421–431

Levins R (ed) (1968) Evolution in changing environments. Princeton Monographs Series. Princeton University Press, Princeton

Levins R (1974) The qualitative analysis of partially specified systems. Ann NY Acad Sci 231:123–138

Levins R (1998a) Qualitative mathematics for understanding, prediction, and intervention in complex ecosystems. In: Raport D, Costanza R, Epstein P, Gaudet C, Levins R (eds) Ecosystem health. Blackwell Science, MA, pp 178–204

Levins R (1998b) The internal and external in explanatory theories. Sci Cult 7:557–582

Lewontin R, Levins R (2007) Gene, organism, and environment. In: Lewontin R, Levins R (eds) Biology under the influence: dialectical assays on ecology, agriculture, and health. Monthly Review Press, NY, pp 221–234

Libralato S, Christensen V, Pauly D (2006) A method for identifying keystone species in food web models. Ecol Model 195:153–171

Luczkovich J, Borgatti S, Johnson J, Everett M (2003) Defining and measuring trophic role similarity in food webs using regular equivalence. J Theor Biol 220:303–321

Menge BA, Berlow EL, Balchette CA, Navarrete SA, Yamada SB (1994) The keystone species concept: variation in interaction strength in a rocky intertidal habitat. Ecol Monogr 64:249–286

Mills L, Soulè M, Doak F (1993) The key-stone-species concept in ecology and conservation. Bioscience 43:219–224

Mincks S, Smith CR, Jeffreys RM, Sumida PYG (2008) Trophic structure on the West Antarctic Peninsula shelf: detritivory and benthic inertia revealed by $\delta^{13}C$ and $\delta^{15}N$ analysis. Deep-Sea Res II 55:2502–2514

Murphy EJ, Watkins J, Trathan P, Reid K, Meredith M, Thorpe S, Johnston N, Clarke A, Tarling G et al (2007) Spatial and temporal operation of the Scotia Sea ecosystem: a review of large-scale links in a krill centred food web. Philos Trans R Soc Lond B Biol Sci 362:113–148

Norkko A, Thrush SF, Cummings VJ, Gibbs MM, Andrew NL, Norkko J, Schwarz A-M (2007) Trophic structure of coastal Antarctic food webs associated with changes in sea ice and food supply. Ecology 88:2810–2820

Odum EP (1969) The strategy of ecosystem development. Science 164:262–270

Okey T (2004) Shifted community states in four marine ecosystems: some potential mechanisms. PhD thesis, The University of British Columbia, Canada, 173 pp

Ortiz M (2008a) Mass balanced and dynamic simulations of trophic models of kelp ecosystems near the Mejillones Peninsula of Northern Chile (SE Pacific): comparative network structure and assessment of harvest strategies. Ecol Model 216:31–46

Ortiz M (2008b) The effect of a crab predator (*Cancer porteri*) on secondary producers versus ecological model predictions in Tongoy Bay (SE Pacific coast): implications to management and fisheries. Aquat Conserv 18:923–929

Ortiz M (2010) Dynamical and spatial models of kelp forest of *Macrocystis integrifolia* and *Lessonia trabeculata* (SE Pacific) for assessment harvest scenarios: short-term responses. Aquat Conserv 20:494–506

Ortiz M, Levins R (2011) Re–stocking practices and illegal fishing in northern Chile (SE Pacific coast): a study case. Oikos 120:1402–1412

Ortiz M, Levins R (2017) Self–feedbacks determine the sustainability of human interventions in eco–social complex systems: impacts on biodiversity and ecosystem health. PLoS One 12(4):e0176163. https://doi.org/10.1371/journal.pone.0176163

Ortiz M, Wolff M (2002) Trophic models of four benthic communities in Tongoy Bay (Chile): comparative analysis and preliminary assessment of management strategies. J Exp Mar Biol Ecol 268:205–235

Ortiz M, Avendaño M, Campos L, Berrios F (2009) Spatial and mass balanced trophic models of La Rinconada Marine Reserve (SE Pacific coast), a protected benthic ecosystem: management strategy assessment. Ecol Model 220:3413–3423

Ortiz M, Avendaño M, Cantillañez M, Berrios F, Campos L (2010) Trophic mass balanced models and dynamic simulations of benthic communities from La Rinconada Marine Reserve off Northern Chile: network properties and multispecies harvest scenario assessments. Aquat Conserv 20:58–73

Ortiz M, Levins R, Campos L, Berrios F, Campos F, Jordán F, Hermosillo B, González J, Rodríguez-Zaragoza FA (2013a) Identifying keystone trophic groups in benthic ecosystems: implications for fisheries management. Ecol Indic 25:133–140

Ortiz M, Campos L, Berrios F, Rodríguez-Zaragoza FA, Hermosillo-Núñez BB, González J (2013b) Network properties and keystoneness assessment in different intertidal communities dominated by two ecosystem engineer species (SE Pacific coast): a comparative analysis. Ecol Model 250:307–318

Ortiz M, Berrios F, Campos L, Uribe R, Ramírez A, Hermosillo-Nuñez B, González J, Rodríguez-Zaragoza FA (2015) Mass balanced trophic models and short–term dynamical simulations for benthic ecological systems of Mejillones and Antofagasta bays (SE Pacific): comparative network structure and assessment of human impacts. Ecol Model 309–310:153–162

Ortiz M, Hermosillo-Núñez BB, González J, Rodríguez-Zaragoza FA, Gómez I, Jordán F (2017) Quantifying keystone species complexes: ecosystem-based conservation management in the King George Island (Antarctic Peninsula). Ecol Indic 81:453–460

Pace ML, Cole JJ, Carpenter SR, Kitchell JF (1999) Trophic cascades revealed in diverse ecosystems. TREE 14:483–488

Paine RT (1969) A note of tropic complexity and community stability. Am Nat 103:91–93

Patrício J, Marques JC (2006) Mass balanced models of the food web in three areas along a gradient of eutrophication symptoms in the south arm of the Mondego estuary (Portugal). Ecol Model 197:21–34

Patten BC (1997) Synthesis of chaos and sustainability in a nonstationary linear dynamic model of the American black bear (*Ursus americanus* Pallas) in the Adirondack Mountains of New York. Ecol Model 100:11–42

Pauly D, Christensen V, Walters C (2000) Ecopath, Ecosim, and Ecospace as tools for evaluating ecosystem impact of fisheries. ICES J Mar Sci 57(3):697–706

Payton I, Fenner M, Lee W (2002) Keystone species: the concept and its relevance for conservation management in New Zealand. Sci Conserv 203:5–29

Pessoa M (2012) Algae and aquatic macrophytes responses to cope to ultraviolet radiation- a review. Emir J Food Agric 24:527–545

Pikitch EK, Santora C, Babcock EA, Bakun A, Bonfil R, Conover DO, Dayton P, Doukakis P, Fluharty D et al (2004) Ecosystem-based fishery management. Science 305:346–347

Pinkerton MH, Bradford-Grieve JM, Hanchet SM (2010) A balanced model of the food web of the Ross Sea, Antarctica. CCAMLR Sci 17:1–31

Piraino S, Fanelli G, Boero F (2002) Variability of species roles in marine communities: changes of paradigms for conservation priorities. Mar Biol 140:1067–1074

Plante C, Mayer L (1994) Distribution and efficiency of bacteriolysis in the gut of *Arenicola marina* and three additional deposit feeders. Mar Ecol Prog Ser 109:183–194

Plante C, Shriver A (1998) Patterns of differential digestion of bacteria in deposit feeders: a test of resource partitioning. Mar Ecol Prog Ser 163:253–258

Polovina J (1984) Model of a coral reef ecosystem I. ECOPATH model and its application to French Frigate Shoals. Coral Reefs 3:1–11

Power ME, Tilman D, Estes JA, Menge BA, Bond WJ, Mills LS, Daily G, Castilla JC, Lubchengo J, Paine RT (1996) Challenges in the Quest for Keystones. Bioscience 46:609–620

Puccia C, Levins R (1991) Qualitative modelling in ecology: loop analysis, signed digraphs, and time averaging. In: Fishwick PA, Luker PA (eds) Qualitative simulation modeling and analysis. Springer, New York, pp 119–143

Richter A, Wuttke S, Zacher K (2008) Two years of in situ UV measurements at Dallmann Laboratory/Jubany Station. In: Wiencke C, Ferreyra G, Abele D, Marenssi S (eds) The Antarctic ecosystem of Potter Cove, King George Island. Synopsis of research performed 1999–2006 at Dallmann Laboratory and Jubany Station. Reports on Polar and Marine Research, vol 571. AWI, Alfred-Wegener-Institut für Polar- und Meeresforschung, Bremerhaven, pp 12–19

Ricker WE (1968) Food from the sea. In: Committee on Resources and Man (ed) Resource and man. US National Academies of Sciences, EH Freeman, San Francisco, Chap 5, pp 87–108

Smith W, Ainley D, Cattaneo-Vietti R (2007) Trophic interactions within the Ross Sea continental shelf ecosystem. Philos Trans R Soc Lond B Biol Sci 362:95–111

Stark P (1994) Climate warning in the central Antarctic Peninsula area. Weather 49:215–220

Stibor H, Vadstein O, Diehl S, Gelzleichter A, Hansen T, Hantzsche F, Katechakis A, Lippert B, Løseth K, Peters C et al (2004) Copepods act as a switch between alternative trophic cascades in marine pelagic food webs. Ecol Lett 7:321–325

Targett TE (1981) Trophic ecology and structure of coastal Antarctic fish communities. Mar Ecol Prog Ser 4:243–263

Ulanowicz RE (ed) (1986) Growth and development: ecosystems phenomenology. Springer, New York

Ulanowicz R (1997) Ecology, the ascendent perspective. Complexity in ecological systems series. Columbia University Press, New York

Ulanowicz R, Puccia CJ (1990) Mixed trophic impacts in ecosystems. Coenoses 5:7–16

Valdivia N, Díaz M, Holtheuer J, Garrido I, Huovinen P, Gómez I (2014) Up, down, and all around: scale-dependent spatial variation in rocky-shore communities of Fildes Peninsula, King George Island, Antarctica. PLoS One 9(6):e100714. https://doi.org/10.1371/journal.pone.0100714

Valdivia N, Díaz MJ, Garrido I, Gómez I (2015) Consistent richness-biomass relationship across environmental stress gradients in a marine macroalgal-dominated subtidal community on the Western Antarctic Peninsula. PLoS One 10(9):e0138582

Valls A, Coll M, Christensen V, Ellison AM (2015) Keystone species: toward an operational concept for marine biodiversity conservation. Ecol Monogr 85:29–47

Vasas V, Lancelot C, Rousseau V, Jordán F (2007) Eutrophication and overfishing in temperate nearshore pelagic food webs: a network perspective. Mar Ecol Prog Ser 336:1–14

Walters C, Christensen V (2007) Adding realism to foraging arena predictions of trophic flow rates in Ecosim ecosystem models: shared foraging arenas and bout feeding. Ecol Model 209:342–350

Walters C, Christensen V, Pauly D (1997) Structuring dynamic models of exploited ecosystems from trophic mass–balance assessments. Rev Fish Biol Fish 7:139–172

Warwick RM, Clarke KR (1993) Comparing the severity of disturbance: a meta-analysis of marine macrobenthic community data. Mar Ecol Prog Ser 92:221–231

Wolff M (1994) A trophic model for Tongoy Bay – a system exposed to suspended scallop culture (Northern Chile). J Exp Mar Biol Ecol 182:149–168

Wolff M, Koch V, Isaac V (2000) A trophic flow model of the Caeté mangrove estuary (North Brazil) with considerations for the sustainable use of its resources. Estuar Coast Shelf Sci 50:789–803

Wootton JT (1994) Predicting direct and indirect effects: an integrated approach using experiments and path analysis. Ecology 75:151–165

Yodzis P (2001) Must top predators be culled for the sake of fisheries? TREE 16:78–84

Part V
Chemical Ecology

Chapter 17
Chemical Mediation of Antarctic Macroalga-Grazer Interactions

Charles D. Amsler, James B. McClintock, and Bill J. Baker

Abstract Macroalgal forests along the western Antarctic Peninsula (WAP) support dense assemblages of small macroalgal-associated invertebrates, particularly amphipods but also others including gastropods. Most of the macroalgal species, including all the larger, ecologically dominant brown macroalgae, elaborate chemical defenses against herbivory to amphipods as well as fish and sea stars. Consequently, the vast majority of the macroalgal biomass in these forests is unpalatable to potential consumers. A great deal of progress has been made on understanding these relationships during the past decade. Although the macroalgae are seldom consumed by the associated invertebrates and fish, many of the invertebrates, particularly the amphipods, benefit from associating with the chemically defended macroalgae because omnivorous fish avoid feeding on them. The amphipods benefit their macroalgal hosts by greatly reducing biofouling by diatoms and other epiphytic algae. This chapter reviews progress in understanding the chemical defenses of Antarctic macroalgae. It also reviews the community-wide mutualistic interaction between macroalgae and its associated amphipods as well as recent studies examining the extent to which this mutualistic interaction also occurs with macroalgal-associated gastropods.

Keywords Amphipods · Biofouling · Chemical ecology · Herbivory · Mutualism · Trophic dynamics

---

C. D. Amsler (✉) · J. B. McClintock
University of Alabama at Birmingham, Birmingham, AL, USA
e-mail: amsler@uab.edu; mcclinto@uab.edu

B. J. Baker
University of South Florida, Tampa, FL, USA
e-mail: bjbaker@usf.edu

I. Gómez, P. Huovinen (eds.), *Antarctic Seaweeds*,
https://doi.org/10.1007/978-3-030-39448-6_17

17.1 Introduction

Chemistry has many, often central, roles in important biotic interactions of organisms ranging from bacteria to most if not all groups of eukaryotes (e.g., Eisner and Meinwald 1995; Waters and Bassler 2005; Dicke and Takken 2006; Wyatt 2014). Chemical signals mediate a wide variety of forms of communication both within and between species (e.g., Waters and Bassler 2005; Baldwin et al. 2006; Müller-Schwarze 2006; Wyatt 2014), and chemical compounds serve as defenses that help organisms resist predation by herbivores or carnivores (e.g., Müller-Schwarze 2006; Rosenthal and Berenbaum 2012), defenses against competitors (e.g., Inderjit and Mallik 2002), and defenses against pathogens or biofoulers (e.g., Lane and Kubanek 2008). Such relationships are widespread in the marine environment (McClintock and Baker 2001; Breithaupt and Thiel 2011; Brönmark and Hansson 2012; Puglisi and Becerro 2019).

In marine macroalgae, some studies have examined chemical mediation of inter- and intraspecific sensory ecology, but much more attention has been focused on chemical mediation of defensive interactions, particularly defenses against herbivory (Amsler and Fairhead 2006; Amsler 2008, 2012). In the case of Antarctic macroalgae, while chemical roles in sensory ecology have not been ignored (Zamzow et al. 2010; Bucolo et al. 2012), the vast majority of studies to date have focused on the chemical mediation of macroalga-herbivore interactions (Amsler et al. 2008, 2009a, 2014; Avila et al. 2008; Núñez-Pons and Avila 2014; von Salm et al. 2019).

Macroalgae are especially important in shallow water, hard bottom communities along the northern and north-central regions of the western Antarctic Peninsula (WAP) where they form undersea forests covering much of the benthos, often with very high standing biomass (Wiencke and Amsler 2012; Wiencke et al. 2014). However, their dominance decreases markedly toward the southern half of the WAP (DeLaca and Lipps 1976; C.D. Amsler, personal observations), and while they are present at many sites throughout the rest of coastal Antarctica, in these other areas they are typically not present at the high levels of biomass or percent cover observed in the northern WAP (Wiencke and Amsler 2012; Wiencke et al. 2014; Clark et al. 2017). To date, almost all studies of the chemical mediation of macroalga-invertebrate interactions have been performed in macroalgal-dominated communities along the northern portion of the WAP. Consequently, this chapter focuses on the northern WAP but includes what is known from other parts of coastal Antarctica.

17.2 Feeding Bioassay Methodology

One of the most basic questions concerning macroalgal interactions with sympatric invertebrate or vertebrate animals is "do the animals consume the algae?" The standard methodology to determine if a macroalga or any other potential food item is palatable to a potential consumer is, conceptually, quite simple and straightforward.

One offers the potential food to a sympatric and otherwise ecologically relevant potential consumer and asks if it is consumed (Hay et al. 1998). Depending on the size of the consumer relative to the food item, the feeding bioassay can simply determine if food item is eaten vs. not eaten or can measure how much is consumed per unit time, in either case comparing the result to a control food item that is known to be palatable to the animal.

If a macroalga or other food item is consumed by an animal, depending on the overall questions being posed, one might ask a variety of specific questions about it including its nutrient content or digestibility or its relative palatability or nutritive value relative to other potential foods. If the macroalga is not palatable to the animal, a relevant question becomes "why not?" There are at least three reasons a potential food could be unpalatable to a consumer. It might be physically defended against consumption by being mechanically too tough to eat or by possessing spines or other physical impediments to consumption (Littler and Littler 1980; Hay 1997). Although not as common, it is also possible for a potential food item to be of such low nutritive value that it is not energetically worthwhile for many animals to consume (Bullard and Hay 2002). As further discussed throughout this chapter, a very common reason that Antarctic macroalgae are unpalatable to sympatric potential consumers is because of the production of chemical defenses that make them unpalatable.

To establish a chemical role in feeding deterrence of a macroalga (or any other possible food), the potential defensive chemicals, which typically are secondary metabolites (Maschek and Baker 2008), need to be extracted from the macroalga and used in further feeding bioassays (Hay et al. 1998). The first steps involve solvents likely to extract a wide range of compounds, and the resulting mixtures are referred to as *crude extracts*. There are many assumptions underlying extract bioassay methods as well as many variations on the extraction and bioassay protocols, and it is likely that none are "perfect" at reproducing what occurs in nature. Bringing attention to the various bioassay methods and organisms that have been used in examining macroalga-consumer interactions in Antarctica and the inherent strengths and limitations of each is a sub-goal of this chapter. Ideally, the protocol would produce two or more crude extracts likely to capture all possible lipophilic and hydrophilic secondary metabolites or other compounds that have a potential to be responsible for feeding deterrence in intact macroalgal thalli. The assay procedure would then present them to an ecologically relevant potential consumer in an artificial food that is as similar as possible to the chemical and nutritive composition of the original, intact macroalgal thallus.

17.3 Antarctic Macroalgal Resistance to Herbivory

Faced with predation, a macroalga may continue to invest resources in growth and reproduction at similar levels to those without predation, which is often referred to as tolerating herbivory (Núñez-Farfán et al. 2007). Alternately they may invest resources that would otherwise have been used for growth and/or reproduction into resisting herbivores through the production of physical and/or chemical defenses (Herms and Mattson 1992). These strategies are not always mutually exclusive (Arnold and Targett 2003; Núñez-Farfán et al. 2007).

Studies of the palatability of fresh thallus material and/or crude chemical extracts of WAP macroalgae have utilized amphipods, sea stars, and fish as "taste tester" animals (Table 17.1). Of these, the amphipods and fish are probably the most ecologically relevant potential consumers. The rationale for utilizing sea stars in feeding bioassays is discussed further below (Sect. 17.3.3). Overall, almost all of the macroalgae that have been used in fresh thallus bioassays with amphipods and/or fish have been unpalatable to at least one of the animal species (Table 17.1). The exceptions to this are the small, filamentous brown algae *Geminocarpus geminatus* and *Elachista antarctica* and the red algae *Palmaria decipiens* and *Porphyra plocamiestris* (Table 17.1). Clearly the vast majority of the enormous live macroalgal standing biomass on the WAP is unpalatable to these potential and numerous consumers.

17.3.1 Macroalgal Palatability and Resistance to Amphipods

Amphipods are by far the most numerous and the most speciose group of animals that associate with macroalgae on the WAP (Richardson 1971, 1977; Huang et al. 2007; Aumack et al. 2011a) although smaller numbers of gastropods, isopods, copepods, and ostracods are also found in association with WAP macroalgae (e.g., Amsler et al. 2015; Schram et al. 2016). Currently there are at least 564 known Antarctic amphipod species (De Broyer and Jażdżewska 2014), and amphipod abundances are particularly high on finely branched species. Estimated amphipod densities range up to 308,000 and 32,000 individuals m^{-2} benthos in stands of the ecologically dominant, overstory brown macroalgae *Desmarestia menziesii* and *Desmarestia anceps*, respectively (Amsler et al. 2008).

Gondogeneia antarctica (Fig. 17.1a) is a member of family Pontogeneiidae and one of the best-studied shallow water Antarctic amphipods overall (e.g., Obermüller et al. 2007; Doyle et al. 2012; Kang et al. 2015). As one of the most common amphipods which associate with macroalgae (Richardson 1971, 1977; Huang et al. 2007; Aumack et al. 2011a; Barrera-Oro et al. 2019), it has been used extensively in feeding bioassays with WAP macroalgae (Table 17.1) and also with some benthic macroinvertebrates (Ma et al. 2009; Amsler et al. 2009c; authors' unpublished data). In nature, the diet of *G. antarctica* consists largely of diatoms, but it also includes macroalgae, particularly filamentous algae and *Palmaria decipiens*, as an important diet component (Aumack et al. 2017; Zenteno et al. 2019). In feeding bioassays, *G. antarctica* readily consumes filamentous macroalgae; artificial foods using freeze-dried, ground, filamentous macroalgae as a feeding stimulant; and *P. decipiens* (Table 17.1). One disadvantage of the use of this amphipod in feeding bioassays with macroalgal extracts in artificial foods is that in some instances, it is probably

Table 17.1 Summary of fresh WAP macroalgal thalli and crude extract bioassays with amphipods, sea stars, and fish

	Amphipods							Sea star			Fish		
	Gondogeneia antarctica			*Cheirimedon femoratus*	Other species			*Odontaster validus*			*Notothenia coriiceps*		
Macroalgal species	Thal	LpEx	HpEx	LpEx	Thal	LpEx	HpEx	Thal	LpEx	HpEx	Thal	LpEx	HpEx
Class Phaeophyceae													
Adenocystis utricularis		a[3]	r[3]	a[5]				r[3]	r[3] a[8]	a[3]	r[3]	a[3]	a[3]
Ascoseira mirabilis	r[1]	a[3]	a[3a]	r[5]				a[3]	a[3] r[8]	a[3]	r[3]	a[3]	r[6]
Chordaria linearis								r[3]					
Cystosphaera jacquinotii		a[3]	r[3]					r[3]	r[3]	r[3]	r[3]		r[6]
Desmarestia anceps	r[1]	a[3]	r[3]	r[5]	r[7]			r[3]	a[3,8]	r[3]	r[3]	a[3]	r[3,6]
Desmarestia antarctica	r[1]	a[3]	r[3]					r[3]	r[3]	r[3]	r[3]	a[3]	
Desmarestia menziesii		r[3]	a[3]	r[5]	r[7]			r[3]	r[3] a[8]	a[3]	r[3]	r[3]	a[3] r[6]
Elachista antarctica	a[4]				a[4]								
Geminocarpus geminatus	a[1]							a[3]					
Halopteris obovata	r[2]	a[2,3]	a[2,3]		r[2]	a[2]	a[2]	a[3]			r[3]		
Himantothallus grandifolius	r[1]	r[3]	r[3]					r[3]	r[3]	r[3]	r[3]	a[3]	r[3,6]
Phaeurus antarcticus		r[3]	r[3]					a[3]			r[3]		
Class Rhodophyceae													
Austropugetia crassa[b]		a[3]	r[3]					r[3]	a[3]	a[3]	r[3]	r[3]	
Callophyllis atrosanguinea		a[3]	a[3]					r[3]	a[3]	r[3]	r[3]		a[3]
Curdiaea racovitzae	r[1]	a[3]	a[3]					r[3]	r[3]	r[3]	r[3]		
Cystoclonium obtusangulum	r[2]	r[2]	r[2]		r[2,7]	r[2]	r[2]						
Delisea pulchra		r[3]	a[3]		r[7]			r[3]	r[3]	a[3]	r[3]		
Georgiella confluens				r[5]				a[3]	r[8]				
Gigartina skottsbergii	r[1]			r[5]				a[3]	a[8]		a[3]		
Gymnogongrus antarcticus		a[3]	a[3]					a[3]			r[3]		
Gymnogongrus turquetii	r[1]	r[3]	r[3]					r[3]	a[3]	r[3]	r[3]	r[3]	a[3]
Iridaea cordata	r[1]	a[3]	r[3]					r[3]	a[3]	r[3]			

(continued)

Table 17.1 (continued)

	Amphipods							Sea star			Fish		
	Gondogeneia antarctica			*Cheirimedon femoratus*	Other species			*Odontaster validus*			*Notothenia coriiceps*		
Macroalgal species	Thal	LpEx	HpEx	LpEx	Thal	LpEx	HpEx	Thal	LpEx	HpEx	Thal	LpEx	HpEx
Myriogramme mangini	r[1]	a[3]	a[3]					r[3]	r[3]	a[3]	r[3]		
Myriogramme smithii	r[1]	r[3]	a[3]					r[3]	r[3]	r[3]	r[3]	r[3]	a[3]
Nereoginkgo adiantifolia								a[3]					
Palmaria decipiens	a[1,2,3]	a[2]	a[2]	r[5]	r[2,4] a[4]	a[2]	a[2]	a[3]			a[3]		
Pantoneura plocamioides	r[2]	r[2]	r[2]		r[2,7]	r[2]	a[2]	a[3]			r[3]	a[3]	a[3]
Paraglossum amsleri [b]		a[3]	r[3]					r[3]	a[3]	a[3]			
Paraglossum salicifolium [b]		r[3]	r[3]					r[3]	a[3]	r[3]	r[3]	r[3]	a[3]
Phycodrys austrogeorgica								a[3]					
Picconiella plumosa	r[2]	r[2]	r[2]		r[2] a[7]	r[2]	r[2]	a[3]			r[2]	a[3]	
Plocamium cartilagineum	r[2,7]	r[2,3]	r[2,3]		r[2,7] a[7]	r[2]	a[2]	r[3]	r[3]	r[3]	r[3]	r[3]	a[3]
Plumariopsis peninsularis		a[3]	r[3]					r[3]	r[3]	a[3]	r[3]	a[3]	
Porphyra plocamiestris								a[3]			a[3]		
Trematocarpus antarcticus	r[1]	a[3]	a[3]					r[3]	a[3]	a[3]	r[3]	a[3]	a[3]
Varimenia macropustulosa [b]	r[1]	a[3]	a[3]					r[3]	a[3]	a[3]	r[3]		
Class Ulvophyceae													
Lambia antarctica		a[3]	a[3]					r[3]	a[3]	a[3]	a[3]		

Thal fresh thallus bioassays, *LpEx* lipophilic extract bioassays, *HpEx* hydrophilic extract bioassays, *a* thallus or extract in artificial food accepted in bioassay, *r* thallus or extract in artificial food rejected in bioassay. Superscript numbers = references (see table endnote).

[1]Amsler et al. (2009b), [2]Aumack et al. (2010), [3]Amsler et al. (2005), [4]Bucolo et al. (2011), [5]Núñez-Pons et al. (2012), [6]Iken et al. (2009), [7]Amsler et al. (2013), [8]Núñez-Pons and Avila (2014)

Prior names in Amsler et al. (2005): *Austropugetia crassa* as undescribed sp. 1, *Paraglossum amsleri* as *Delesseria lancifolia*, *Paraglossum salicifolium* as *Delesseria salicifolia*, *Varimenia macropustulosa* as *Pachymenia orbicularis* in Amsler et al. (2005) and undescribed species A in Amsler et al. (2009b). *Pachymenia orbicularis* was a misidentification in Amsler et al. (2005). The other changes resulted from taxonomic revisions.

[a]Water-soluble, apparently activated defense inhibits feeding (McDowell et al. 2014b).

[b]All or some bioassay data published under other species name (see table endnote).

Fig. 17.1 Common potential consumers of macroalgae on the western Antarctic Peninsula: (**a**) *Gondogeneia antarctica*; (**b**) *Paradexamine fissicauda*; (**c**) *Odontaster validus*; (**d**) *Notothenia coriiceps*. (Photos (**a**) and (**b**) by M.O. Amsler, (**c**) and (**d**) by B.J. Baker)

conservative in identifying crude extracts with feeding deterrence. In these feeding bioassays, the extracts by themselves – presumably because of primary metabolites therein – are commonly preferred over control foods at statistically significant levels (Amsler et al. 2005; Schram et al. 2015). Indeed, when either lipophilic or hydrophilic crude extracts of the filamentous green alga *Cladophora repens*, which is the feeding stimulant usually used in the artificial foods, are added to the artificial foods, they are significantly preferred to the *C. repens*-based artificial foods without extracts (Amsler et al. 2005). Consequently, if there is no significant difference between feeding rates on control and extract containing artificial foods in a bioassay, could the extracts actually have been deterrent, but not enough to overcome the inherent extra palatability of the extracts in general? *Halopteris obovata* and *Varimenia macropustulosa* are species that were unpalatable to *G. antarctica* as fresh thallus but in which feeding on one or both crude extracts was not significantly different from the control food (Amsler et al. 2005; Aumack et al. 2010). This was also true of *Ascoseira mirabilis* (Amsler et al. 2005), but there is an additional possible explanation for this as discussed below. Fresh thallus feeding bioassays using *G. antarctica* have not been done with *Callophyllis atrosanguinea*, but it is unpalatable as fresh thallus to sea stars and fish (Table 17.1). There was no significant preference of *G. antarctica* for either of its crude extracts (Amsler et al. 2005) suggesting that there might be chemically based deterrence to amphipod herbivory. There is a similar situation with *Lambia antarctica* although it is only unpalatable as fresh thallus to sea stars (not to fish; Table 17.1) and only its hydrophilic crude extract was not significantly different from the control food in feeding bioassay with *G. antarctica* (Amsler et al. 2005).

Cheirimedon femoratus is the second most common amphipod that has been used in feeding bioassays of Antarctic macroalgal extracts (Table 17.1). It is a member of family Lysianassidae, which primarily consists of detritivores and necrovores (Bousfield 1973). Consistent with that, it is commonly found in baited fish traps and probably prefers carrion as a food when available (Bregazzi 1972). However, it also feeds on detrital algal material, particularly as juveniles and ovigerous females (Bregazzi 1972). Based on stable isotope analysis, Zenteno et al. (2019) classified this species as a scavenger and reported a particularly high isotope signature apparently originating from intertidal macroalgal sources. *Cheirimedon femoratus* has been reported to be most common on sandy bottoms but also associated with rocky substrata (Bregazzi 1972). Núñez-Pons et al. (2012) report that it associates with live macroalgae at Deception Island (63°S latitude). Our group has examined literally hundreds of thousands if not more macroalgal-associated amphipods from Anvers Island (64°S latitude) as well as smaller numbers from as far south as 68°S and never found *C. femoratus* – indeed only very few members of its family (Huang et al. 2007) – associated with macroalgae. However, we have sometimes found it from Anvers Is. (but not further south) in airlift samples that included substratum material from macroalgal communities as well as occasionally in baited fish traps (authors, unpublished) suggesting that it is present in the macroalgal communities, probably associated with the substratum, but not on the upright portions of the macroalgae. Richardson (1977) reported very small numbers of *C. femoratus* relative to other species associated with the overstory brown alga *D. anceps* at Signy Island (60°S latitude), but his samples included holdfasts which ours usually did not. At King George Island (62°S latitude), *C. femoratus* has been reported in low numbers from samples including both macroalgae and surrounding substratum but commonly in the guts of fish from these communities (Barrera-Oro et al. 2019).

Even though *Cheirimedon femoratus* does not associate with intact macroalgae in many locations along the WAP and as such may not be a primary factor in selecting for chemical defenses in the algae, the palatability of detached and degrading macroalgal material to this amphipod is likely to be relevant to the flow of carbon into the surrounding communities. Macroalgae, probably as detritus, are an important carbon source to a wide variety of animals in these communities (Dunton 2001; Corbisier et al. 2004; Aumack et al. 2017; Braeckman et al. 2019; Zenteno et al. 2019). Once dead, chemically defended brown macroalgae can begin to become palatable to *G. antarctica* and probably other amphipods after a period of weeks (Reichardt and Dieckmann 1985; Amsler et al. 2012a). However, they can survive a very long time after being detached (Brouwer 1996; Amsler et al. 2012a), and drift thalli probably need to be buried in sediments and begin microbial decomposition to become palatable to many animals (Braeckman et al. 2019).

Cheirimedon femoratus has only been used in feeding bioassays with lipophilic macroalgal extracts, but the results of those bioassays are generally comparable to lipophilic extract feeding bioassays using *Gondogeneia antarctica* (Table 17.1). One difference is that feeding deterrence of *Desmarestia anceps* to *G. antarctica* was only in the hydrophilic extract while a lipophilic extract was unpalatable to *C. femoratus*. It is possible that this is attributable to different extraction procedures in the two studies (Amsler et al. 2005; Núñez-Pons et al. 2012) which could have resulted in the compounds responsible for the response partitioning differently. A second difference is that extracts of *A. mirabilis* were unpalatable to *C. femoratus* but not to *G. antarctica* (Table 17.1), but as noted above, the conservative nature of the *G. antarctica* extract bioassay could have resulted in a false negative. A more perplexing difference is that the extract of *P. decipiens* was unpalatable to *C. femoratus* (Núñez-Pons et al. 2012) since in our experience it is one of the few generally palatable macroalgae in the community. It is readily consumed as fresh thallus by *G. antarctica* (Amsler et al. 2009b, 2013; Aumack et al. 2010; Bucolo et al. 2011), by the amphipod *Oradarea bidentata* (Bucolo et al. 2011), and by sea stars and fish (Table 17.1). It is not palatable as fresh thallus to the amphipods *Prostebbingia gracilis* and *Paraphimedia integricauda* (Aumack et al. 2010; Bucolo et al. 2011), but this is almost certainly because of biophysical limitations since *P. gracilis* preferentially consumes extracts of *P. decipiens* (Aumack et al. 2010) and *P. integricauda*, which consumes primarily diatoms, has a mandible morphology unsuited for grinding macroalgal thalli (Aumack et al. 2017).

Although the vast majority of WAP macroalgae are unpalatable to amphipods, usually because of chemical defenses (Table 17.1), there is one amphipod species known to consume otherwise chemically defended macroalgae. *Plocamium* "*cartilagineum*"[1] is unpalatable as thallus to the amphipods *Gondogeneia antarctica, Prostebbingia gracilis, Paraphimedia integricauda, Oradarea bidentata,* and *Schraderia gracilis* but readily consumed by *Paradexamine fissicauda* (Fig. 17.1b) (Amsler et al. 2013). The unpalatability to most species is clearly due to defensive chemistry as its extracts are highly deterrent in feeding bioassays with *G. antarctica* and *P. gracilis* (Amsler et al. 2005; Aumack et al. 2010) as well as with sea stars and fish (Table 17.1). *Paradexamine fissicauda* is also able to consume *Picconiella plumosa* (Amsler et al. 2013) which is unpalatable both as fresh thallus and as crude extracts to *G. antarctica* and *P. gracilis* (Aumack et al. 2010). However, *P. fissicauda* does not consume the chemically defended red algae *Pantoneura plocamioides, Cystoclonium obtusangulum,* or *Delisea pulchra* or the chemically defended brown algae *Desmarestia menziesii* or *Desmarestia anceps* (Amsler et al. 2013). Predators that are able to consume chemically defended organisms including macroalgae are generally only able to do so in single species or in closely related taxa with similar or identical defensive metabolites because the biochemical pathways they use to detoxify and/or excrete the defensive metabolites are relatively compound specific (Sotka and Whalen 2008; Sotka et al. 2009). *Plocamium* "*cartilagineum*" and *P. plumosa* are in separate taxonomic orders (Plocamiales and Ceramiales, respectively; Hommersand et al. 2009) which are not closely related and which were recently estimated to have been divergent for close to 400 million years (Yang et al. 2016). Although separate or convergent evolution of secondary metabolites is not common, it does occur (e.g., Daly 2004; Cutignano et al. 2012),

[1] This entity is genetically distinct from *P. cartilagineum* found elsewhere and certainly represents a separate species, but to date, no formal alternative name has been proposed (Hommersand et al. 2009).

and it is possible that defensive metabolites of *P.* "*cartilagineum*" and *P. plumosa* are similar enough to allow both to be eaten by *P. fissicauda. Paradexamine fissicauda* does not simply detoxify the halogenated secondary metabolite defenses of *P.* "*cartilagineum*" but also incorporates at least some of them into its tissues where they appear to deter predation on the amphipods by fish (Amsler et al. 2013). Although such sequestration of chemical defenses from the diet for use by the consumer as a defense against its own predators is known in terrestrial arthropods (Eisner and Meinwald 1995) and in marine systems from some opisthobranch molluscs (reviewed by Jormalainen and Honkanen 2008; Cimino and Ghiselin 2009; Hay 2009), we are not aware of other examples of this phenomenon in marine organisms.

While chemical defenses against amphipod herbivory are clearly very common in WAP macroalgae, we know specific compounds involved in that unpalatability in only a few cases. Purified phlorotannins from both *Desmarestia anceps* and *Cystosphaera jacquinotii* (but not several other brown algae) are deterrent in feeding bioassays with *Gondogeneia antarctica* (Iken et al. 2009) which is consistent with the deterrence observed in the hydrophilic crude extracts of these species (Table 17.1). Two halogenated monoterpenes, anverene and *epi*-plocamene D, from *Plocamium* "*cartilagineum*" have been shown to deter feeding by *G. antarctica* (Ankisetty et al. 2004). Antarctic *P.* "*cartilagineum*" produces a large number of halogenated secondary metabolites (Young et al. 2013), and it is likely that additional compounds contribute to the strong unpalatability of this species in most feeding bioassays. Similarly, it is possible that more than phlorotannins are responsible for feeding deterrence in *D. anceps* or *C. jacquinotii*, particularly if differential extraction is responsible for the differences discussed above between responses of *Cheirimedon femoratus* and *G. antarctica* to crude extracts of *D. anceps*.

Secondary metabolites are not the only molecules that can be involved in chemical defense. For example, reactive oxygen species (ROS) are known to be involved in defense against pathogens in a variety of macroalgae (Potin 2008; Thomas et al. 2014). Recent evidence suggests that this is probably true in Antarctic macroalgae and also that ROS has the potential to be involved in deterring predation by small grazers such as amphipods. Nine of 13 WAP macroalgae accumulated ROS within their tissues within 70 min of wounding, and four of five species tested released strong oxidants into the surrounding seawater within 1 min of wounding (McDowell et al. 2014a). *Palmaria decipiens*, the only macroalgal species tested that is eaten by *Gondogeneia antarctica*, increased ROS production at the site of grazing by *G. antarctica* (McDowell et al. 2014a). Although *P. decipiens* is consumed as fresh thallus by *G. antarctica*, as already noted, *Ascoseira mirabilis* is not (Table 17.1). *Ascoseira mirabilis* produces ROS upon wounding in the light but not in the dark (McDowell et al. 2016), and wounded *A. mirabilis* inhibits feeding of *G. antarctica* on *P. decipiens* disks within the same experimental container (McDowell et al. 2014b). Although hydrogen peroxide does not appear to be part of the ROS response in *A. mirabilis*, it is in other Antarctic macroalgae (McDowell et al. 2014a), and addition of hydrogen peroxide at concentrations comparable to those released by macroalgae also inhibits feeding by *G. antarctica* (McDowell et al. 2014b). ROS production in the light has been shown to be dependent on photosynthetic electron transfer in a temperate kelp species, and this is probably true in *P. decipiens* and *A. mirabilis* as well (McDowell et al. 2015). The light dependence of this potential defense against herbivores likely has ecological consequences as amphipods are active and probably feeding at night (authors' personal observations). This includes species such as *G. antarctica* moving off of chemically defended hosts to the palatable species *P. decipiens* at night, presumably to feed (Aumack et al. 2011a).

17.3.2 Macroalgal Palatability and Resistance to Fish

The species of fish most commonly observed in WAP macroalgal forests, *Notothenia coriiceps* (Fig. 17.1d), commonly has macroalgae in its guts when collected from macroalgal communities (e.g., Iken et al. 1997; Zamzow et al. 2011) and apparently selectively feeds on some macroalgal species including *Palmaria decipiens* (Iken et al. 1997). In feeding bioassays with fresh thallus material, *P. decipiens* is one of only four of 28 species tested that was consumed by *N. coriiceps* (Table 17.1). We have observed *N. coriiceps* taking bites out of *P. decipiens* in laboratory aquaria and observed *P. decipiens* individuals in nature with similar, apparent bites missing from the thallus (authors, unpublished). Of the other three subtidal WAP macroalgal consumed by the fish as fresh thallus, *Porphyra plocamiestris* is in a genus that is commonly palatable to animals, including humans, and is very thin and easily torn. It is not surprising that it is palatable to the fish. *Lambia antarctica* is palatable as extract to both amphipods and sea stars (Table 17.1) indicating no apparent chemical defenses although it was rejected as thallus by sea stars (Table 17.1). *Lambia antarctica* is not as physically tough as *P. decipiens* (Amsler et al. 2005) suggesting that *N. coriiceps* should be able to bite off pieces of its thallus as it does with *P. decipiens*. The fourth species the fish consumed as thallus, *Gigartina skottsbergii* (Table 17.1), is different. It was the physically toughest of 30 species tested for puncture resistance by Amsler et al. (2005). Although the fish swallowed the small pieces used in the feeding bioassays, we believe that it is very unlikely that they could remove pieces of thallus from an intact *G. skottsbergii* blade in nature.

Fewer extract feeding bioassays have been performed with *Notothenia coriiceps* than with the amphipod *Gondogeneia antarctica* or sea star *Odontaster validus* (Table 17.1). Of the 24 macroalgal species known to be unpalatable to *N. coriiceps*, extract bioassays have not been performed with seven. Of the other 17 species, extract bioassays indicate that chemical defenses can explain the unpalatability in 10. In four others, only one crude extract type has been bioassayed (Table 17.1), so it is possible that defensive compounds could have been present in the other extract. Only three species that were rejected as thallus have had both crude extract types tested and shown to be palatable to the fish including *Trematocarpus antarcticus* which similarly is unpalatable as thallus but not as either crude extract to amphipods and sea stars (Table 17.1).

The only purified compounds that have been shown to deter feeding of *Notothenia coriiceps* to date are phlorotannins. The strongest deterrent response was seen with phlorotannins from *Desmarestia menziesii*, but pure phlorotannins from *Cystosphaera jacquinotii* were also significantly rejected by the fish (Iken et al. 2009). It is important to note that these assays were done with isolated phlorotannins, not simply methanolic extracts that would be enriched in phlorotannins and are commonly what is used in "phlorotannin" feeding bioassays even though they also contain other compounds (Amsler and Fairhead 2006). Such methanolic extracts of *Desmarestia anceps* and *Ascoseira mirabilis* deterred *N. coriiceps*, but the deterrence was lost when the extracts were purified further with other solvents and a microcellulose column (Iken et al. 2009).

17.3.3 Macroalgal Palatability and Resistance to Sea Stars

The sea star *Odontaster validus* (Fig. 17.1c) is one of the most common benthic invertebrates throughout much of coastal Antarctica (Janosik and Halanych 2010) and is typically present in WAP macroalgal communities (Dearborn and Fell 1974). Although sea stars are usually not considered as macroalgal predators, *O. validus* was one of two species out of 20 WAP sea stars surveyed that had macroalgae in its guts (McClintock 1994). Although rare, we have personally observed *O. validus* individuals with their cardiac stomachs extended over macroalgal thalli with epiphytes, but it seems most probable that macroalgae reported within their guts came from detrital fragments. So although it appears unlikely that sea star predation would have been important in selecting for the production of chemical defenses, the fact that they do not bite into their prey as amphipods and fish do could mean that thallus toughness is not as important a factor in their food choices. Fresh thallus bioassays with them, therefore, could provide useful biophysical information relative to feeding deterrence observed in other predators. The feeding bioassay system our group uses with *O. validus* (Amsler et al. 2005) is relatively quick and, for crude extract bioassays, can use artificial foods leftover from preparation of foods for amphipod or fish bioassays, providing comparative information on chemical deterrence in a separate predator.

To date, more fresh thallus bioassays with different macroalgal species (35) have been done with *Odontaster validus* than with either amphipods or fish (Table 17.1). Of those 35 species, fresh thallus was consumed by *O. validus* in 13 of the species, a much higher acceptance rate than in fresh thallus assays with amphipods or fish (Table 17.1). Five of the 13 species accepted by the sea star were rejected as thallus by one or more of the amphipod species or fish and in several of these instances, crude extract bioassays with one or more of the other species did not explain the unpalatability (Table 17.1). The *O. validus* assay results in these instances would be consistent with the thallus unpalatability to amphipods or fish being biophysically based.

The two studies that have examined the palatability of lipophilic macroalgal extracts with *Odontaster validus* (Amsler et al. 2005; Núñez-Pons and Avila 2014) used not only different extraction solvent systems, as discussed previously (Sect. 17.3.1) for studies on amphipods by the two different research groups, but also markedly different bioassay procedures. It is probably not surprising that results differed in three of the four macroalgal species assayed in both studies (Table 17.1). One other difference between the two studies was in the composition and extract loading of the artificial foods. Amsler et al. (2005) coated extracts onto freeze-dried powder of a palatable, intertidal, filamentous Antarctic green alga (*Cladophora repens*) that was then incorporated into alginate-based foods. The "natural concentration" of the extract was calculated on a volumetric basis relative to the volume of the final artificial food pellet. The pellets were 5% dry *C. repens* powder in 2% alginate which based on Peters et al. (2005) should have contained approximately 2% protein. That protein content is within but at the lower end of the range observed for 36 species of WAP macroalgae (Peters et al. 2005). Núñez-Pons and Avila (2014) used artificial foods composed of cubes of non-Antarctic shrimp which were 12.4% protein, a level at the very top of the range of protein contents in WAP macroalgae (Peters et al. 2005). The lipophilic extracts were added on a dry weight basis to the outside of the shrimp cubes, and while some would have soaked in, most would have been retained near the surface which is the only place "tasted" by a nonbiting sea star predator. So while the foods themselves were higher in nutritional value – and, therefore, potentially more attractive foods – compared to those used by Amsler et al. (2005), the potentially unpalatable extracts added to them were probably presented to the sea stars at higher concentrations. Another difference between the studies is that the Amsler et al. (2005) assay placed food items onto the sea stars' chemosensory tube feet and recorded whether or not the food that had been "tasted" was moved to and retained at the mouth. Núñez-Pons and Avila (2014) placed the sea stars on top of the food cubes and recorded whether they were consumed in 24 hours, which captures the full complement of feeding behaviors ("tasting" followed by consumption) and is probably a more ecologically robust bioassay approach.

Purified phlorotannins from *Desmarestia menziesii* and *Himantothallus grandifolius* but not from five other brown macroalgal species significantly deter feeding by *Odontaster validus* (Iken et al. 2009). *Desmarestia menziesii* also produces a quinone derivative, menzoquinone, which deters *O. validus* feeding in bioassays (Ankisetty et al. 2004).

17.3.4 Macroalgal Palatability and Resistance to Sea Urchins in McMurdo Sound

As discussed in Sect. 17.1, macroalgae occur throughout coastal Antarctica although their biomass, percent cover of the bottom, and species richness all become lower south of the northern WAP (Wiencke and Amsler 2012; Wiencke et al. 2014). Only two species of fleshy macroalgae, *Phyllophora antarctica* and *Iridaea cordata*, occur throughout most of the year at the southernmost extent of macroalgal distribution (and seasonally open water) in McMurdo Sound (Ross Sea; Miller and Pearse 1991). The sea urchin *Sterechinus neumayeri* is a very common and important consumer in McMurdo Sound as well as throughout much of the rest of Antarctica (Pearse and Giese 1966; Pawson 1969; Saucède et al. 2014) and an obvious potential predator on *P. antarctica* and *I. cordata* in McMurdo Sound. Unfortunately, *S. neumayeri* does not reliably or reproducibly consume anything we have tested it with in laboratory aquaria, and so typical feeding bioassays are not practical with it because of the high variability caused by many animals not eating at all. To get around this limitation, Amsler et al. (1998) developed a "phagostimulation" bioassay which measured how long a small disk of algal thallus or crude extracts on feeding-stimulant-containing filter paper disks were retained at the urchins' mouths when forcibly placed there. While far from an ideal assay, the responses were quite strong which allowed for statistically robust outcomes. Thallus disks from both species were retained at the mouth less than half as long as a paper disk without feeding stimulants. Both lipophilic and hydrophilic extracts added to filter paper disks with a feeding stimulant were rapidly rejected, while solvent controls were retained at the mouth for several-fold longer times, indicating that the unpalatability of fresh thalli is chemically based (Amsler et al. 1998).

Even though *Sterechinus neumayeri* does not consume *Phyllophora antarctica* or *Iridaea cordata*, it prefers to cover itself with them over other potential cover objects if the algae are present (Amsler et al. 1999). Either the macroalgae or other objects used as cover help protect *S. neumayeri* from predatory sea anemones by acting as a detachable "shield" that the anemones' tentacles stick to, allowing the urchins to release the cover material and escape (Dayton et al. 1970; Amsler et al. 1999). The algae also benefit from this relationship because they are the preferred cover items for the urchins. Anchor ice tears macroalgal thalli from the rock substratum, and most of the algal biomass in the communities are held by urchins rather than being attached to the rock substratum (Dayton et al. 1969; Miller and Pearse 1991; Amsler et al. 1999). The algae so retained within the photic zone remain physiologically active (Schwarz et al. 2003) as well as reproductive (Amsler et al. 1999) and probably are responsible for most of the spore and gamete production which maintains the populations.

17.4 Macroalga-Invertebrate Interactions on the Western Antarctic Peninsula

As discussed in Sect 17.3.1, amphipods are the most abundant invertebrates associated with WAP macroalgae. As also discussed in Sect. 17.3.1, the vast majority of macroalgal species deter predation by amphipods using chemical defenses. This includes all of the large, ecologically dominant brown macroalgae as well as most of the common red macroalgae. Consequently, the vast majority of the WAP macroalgal biomass is not available or, at least, not at all preferable, as food to the dense amphipod assemblage. Although the amphipods are not consuming most of the large macroalgae, they do consume epiphytic microalgae, filamentous algal epiphytes, and emergent filaments from endophytic algae (Amsler et al. 2009b, 2012b; Aumack et al. 2011b, 2017) and thereby clearly benefit their macroalgal hosts by sustaining photosynthesis, gas exchange, and nutrient uptake capacities. In addition to consuming the smaller algae, the amphipods also benefit from associating with the chemically defended macroalgae by gaining a refuge from an important predator, the common benthic fish *Notothenia coriiceps* (Zamzow et al. 2010, 2011). Because macroalgae are the major structural component and primary producers in these communities and because amphipods are the most abundant invertebrates, we have previously described this association as a community-wide mutualism (Amsler et al. 2014). This relationship is presented diagrammatically in the upper half of Fig. 17.2. None of the individual components of this mutualism are unique to the WAP, and comparisons and contrasts of the WAP communities to specific lower-latitude communities are discussed at length by Amsler et al. (2014). However, the overall importance of these two groups of organisms – the macroalgae which dominate in terms of structure and primary productivity and the amphipods which are by far the most abundant animals – to these nearshore communities makes the interaction uniquely important compared to such interactions in lower-latitude communities (Amsler et al. 2014).

In addition to the "bottom-up" benefits to amphipods from gaining an associational refuge from fish predation, it is possible that the very high densities of amphipods associated with macroalgae result in part from "top-down" influences of predators on fish that prey on amphipods. *Notothenia coriiceps* is a benthic ambush predator with relatively small home ranges (e.g., Daniels 1982; North 1996; Campbell et al. 2008). Although it is common along the bottom beneath the macroalgal canopy, it is rarely observed by divers up off the bottom within the canopy (Casaux et al. 1990; authors' personal observations). The smaller *Harpagifer antarcticus* is found associated with small rocks in Antarctic macroalgal forests and is also an amphipod predator with a cryptic lifestyle (Daniels 1983; Casaux 1998). Several seal species including leopard seals have been observed with fish in their gut contents or scat (Hall-Aspland and Rogers 2004; Casaux et al. 2009), and members of our research group have observed and photographed leopard seals eating *N. coriiceps* in shallow water (A. Shilling, personal communication). It is reasonable to hypothesize that the cryptic behavior of the fish represents a nonconsumptive effect of seal predators and thereby reduces predation on amphipods and other macroalgal-associated invertebrates.

Another feature of these communities that is at least somewhat unique is that free-living filamentous algae are very uncommon throughout most of the year in the subtidal zone but filamentous algae are very common growing as endophytes within subtidal macroalgae (Peters 2003; Amsler et al. 2009b). Filamentous macroalgae are frequently observed in the WAP intertidal (e.g., Lamb and Zimmerman 1977; Marcías et al. 2017; Valdivia et al. 2019), indicating that there is nothing about

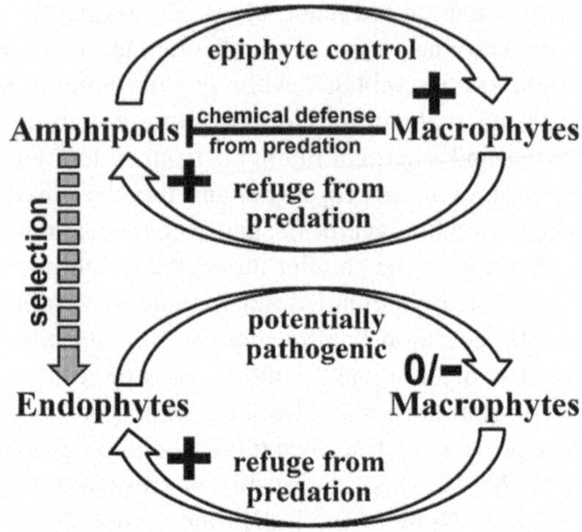

Fig. 17.2 Schematic representation of species interactions between algae and amphipods on the western Antarctic Peninsula. See text for details. + indicates a positive effect, 0 indicates no effect, and – indicates a negative effect. (From Amsler et al. (2014). Used with permission)

Antarctica that excludes free-living filamentous algae. Peters (2003) noted that escape from herbivory has long been hypothesized as a selective factor favoring endophytism (e.g., Kylin 1937) and postulated that the dense assemblages of amphipods on subtidal WAP macroalgae probably selected for this pattern of filamentous algae rarely being free-living but commonly being endophytic in the WAP subtidal. This is potentially problematic with respect to the idea of WAP macroalgae and amphipods being mutualists. While not always true (e.g., Gauna et al. 2009), filamentous algal endophytes are commonly pathogenic to their macroalgal hosts (e.g., Apt 1988; Correa and Sánchez 1996; Faugeron et al. 2000). If such pathogenicity is common in WAP endophytes and endophytism is driven by amphipod herbivory, could amphipods truly be viewed as beneficial to the larger macroalgae? Schoenrock et al. (2013) performed field experiments which showed that while endophytes could reduce growth and/or survival of some species of macrophytes when present in very high densities within the macroalgal thalli, this was not always true. In the most heavily impacted species in terms of growth, *Iridaea cordata*, there was no corresponding correlation between endophyte load and the density of either carposporangia or tetrasporangia, indicating no pathogenic impact on fecundity (Schoenrock et al. 2015b). In laboratory experiments on nine macroalgal species, almost no pathogenic impacts were seen on photosynthetic physiology, thallus toughness, or susceptibility to grazers, and the few significant effects observed were very mild (Schoenrock et al. 2015a). Consequently, although endophytes can sometimes be detrimental to their larger macroalgal hosts, overall, this almost certainly does not negate the benefit macroalgae gain from having the dense assemblage of associated amphipods (Fig. 17.2).

Although often not as abundant as amphipods, small gastropods are also commonly observed associated with WAP macroalgae (Richardson 1977; Picken 1979, 1980; Iken 1999; Amsler et al. 2015). Gastropods commonly consume microalgae and filamentous macroalgae (Purchon 1977; Santhanam 2018) and consequently could be analogous to amphipods in providing benefits to their macroalgal hosts by controlling epiphytic algae (see also Chaps. 12 and 13). They may be particularly important on the upper surfaces of large, blade-forming macroalgae such as *Himantothallus grandifolius* and *Gigartina skottsbergii* where amphipods are not as common as gastropods, at least during daylight hours (authors' personal observations). Amsler et al. (2019) performed a mesocosm experiment which maintained *H. grandifolius* blade sections either with no gastropods or with densities of gastropods corresponding to natural gastropod densities on *H. grandifolius*. A very similar mesocosm experiment had previously been done with or without natural densities of amphipods by Aumack et al. (2011b). The gastropods controlled epiphytic diatom densities on *H. grandifolius* blades (Amsler et al. 2019) but not to the extent that amphipods were observed to do so for *H. grandifolius* (Aumack et al. 2011b). No grazing marks from the gastropods were ever observed on the *H. grandifolius* thalli either in the mesocosms or in an accompanying laboratory experiment, but almost all of the gastropod species have radulae that are probably not able to rasp the tough thalli of *H. grandifolius* (Amsler et al. 2019). That is unlikely to be the case for the limpet *Nacella concinna* which was present in the experiment and has a very tough, minerally hardened, chitinized radula. This limpet is presumably deterred from consuming its *H. grandifolius* hosts by the same chemical defenses observed to deter feeding in other sympatric predators (Table 17.1).

The fish *Notothenia coriiceps* often consumes gastropods in addition to amphipods in macroalgal communities (Zamzow et al. 2011). Although small gastropods on chemically defended, branched macroalgae probably gain refuge from fish predators just as amphipods do, subjectively it is likely that a gastropod on a large bladed macroalga such as *Himantothallus grandifolius* is just as available to a predatory fish as it would be on bare rock. However, sea stars are also important predators of Antarctic gastropods (McClintock 1994), and *Odontaster validus* is chemically deterred from consuming *H. grandifolius* (Table 17.1). Because sea stars use their tube feet both for locomotion and for chemically sensing their prey (Hennebert et al. 2013), Amsler et al. (2019) hypothesized that *O. validus* might be less likely to be on *H. grandifolius* blades than on other possible substrata, giving gastropods on the macroalga somewhat of an associational refuge from predation. This hypothesis might be true, but was not supported in the one simple experiment performed (Amsler et al. 2019). Overall, compared to the macroalga-amphipod mutualism illustrated in Fig. 17.2, there are similarities in macroalga-gastropod interactions but they are not identical.

17.5 Overview

A very high percentage of Antarctic macroalgal species deter potential herbivores by elaborating chemical defenses. We are not aware of examples elsewhere in the world where such a high percentage of the macroalgal flora is chemically defended. In North Carolina (USA), few palatable macroalgal species are apparent in the summer, but this is because grazing fish remove palatable species that are apparent in the winter and spring (Hay 1986; Hay and Sutherland 1988; Duffy and Hay 1994).

Most of the Antarctic macroalgal flora are composed of perennial species that are present throughout all seasons (Wiencke and Clayton 2002; Wiencke et al. 2014) (see Chaps. 1, 11, 12). This high prevalence of chemical defenses against herbivory appears to be a fairly unique feature of Antarctic macroalgal forests. Chemical defenses against carnivory are also very common in Antarctic invertebrates (e.g., Amsler et al. 2001; Avila et al. 2008; McClintock et al. 2010; Moles et al. 2015), making Antarctica a powerful natural laboratory for the study of chemically mediated predator-prey relationships.

Macroalgal forests along the WAP are similar to temperate kelp forests in being dominated by large, perennial brown macroalgae (Wiencke and Amsler 2012). All of the overstory brown algal species that dominate these WAP forests are chemically defended from herbivory. Consequently, while being similar to kelp forests in some ways, their trophic dynamics differ greatly. Temperate kelps are usually palatable to abundant kelp forest consumers such as sea urchins, and the macroalgal biomass and numerical dominance are usually maintained by "top-down" factors controlling the herbivore populations (Steneck et al. 2002). Although top-down factors may have a role in allowing amphipods to be so abundant in WAP forests, as discussed above in the Sect. 17.4, the general unpalatability to herbivores of the Antarctic macroalgae suggests that their persistence and community dominance would exist even if only through this "bottom-up" factor.

As also discussed in Sect. 17.5, while the hypothesis of a community-wide mutualism between amphipods and macroalgae is well supported, there are similarities but also important differences in the interactions of macroalgal-associated gastropods and their hosts compared to the macroalga-amphipod relationship. While perhaps not as abundant, at least in terms of biomass, other crustaceans such as copepods, isopods, and ostracods also associate with canopy-forming macroalgae (Schram et al. 2016), and polychaetes commonly associate with holdfasts of the large macroalgae (Pabis and Sicinski 2010). The extent to which the relationships of these other macroalgal-associated invertebrates and their hosts are similar to or different from the macroalga-amphipod mutualism illustrated in Fig. 17.2 has yet to be examined.

Acknowledgments We are grateful to all our Antarctic field team members over the years, in particular to M. Amsler who also provided constructive comments on an earlier version of this chapter. The original version of this chapter also benefited from the constructive comments of Martin Thiel and an anonymous reviewer. Our Antarctic chemical ecology research has been generously supported by numerous awards from the NSF Antarctic Organisms and Ecosystem program, most recently by PLR-1341333 (CDA, JBM) and PLR-1341339 (BJB).

References

Amsler CD (2008) Algal chemical ecology. Springer-Verlag, Berlin, p xviii. 313

Amsler CD (2012) Chemical ecology of seaweeds. In: Wiencke C, Bischof K (eds) Seaweed biology: novel insights into ecophysiology, ecology and utilization. Springer-Verlag, Berlin, pp 177–188

Amsler CD, Fairhead VA (2006) Defensive and sensory chemical ecology of brown algae. Adv Bot Res 43:1–91

Amsler CD, McClintock JB, Baker BJ (1998) Chemical defense against herbivory in the Antarctic marine macroalgae *Iridaea cordata* and *Phyllophora antarctica* (Rhodophyceae). J Phycol 34:53–59

Amsler CD, McClintock JB, Baker BJ (1999) An antarctic feeding triangle: defensive interactions between macroalgae, sea urchins, and sea anemones. Mar Ecol Prog Ser 183:105–114

Amsler CD, Iken KB, McClintock JB, Baker BJ (2001) Secondary metabolites from Antarctic marine organisms and their ecological implications. In: McClintock JB, Baker BJ (eds) Marine chemical ecology. CRC, Boca Raton, pp 267–300

Amsler CD, Iken K, McClintock JB, Amsler MO, Peters KJ, Hubbard JM, Furrow FB, Baker BJ (2005) Comprehensive evaluation of the palatability and chemical defenses of subtidal macroalgae from the Antarctic Peninsula. Mar Ecol Prog Ser 294:141–159

Amsler CD, McClintock JB, Baker BJ (2008) Macroalgal chemical defenses in polar marine communities. In: Amsler CD (ed) Algal chemical ecology. Springer-Verlag, Berlin, pp 91–103

Amsler CD, Iken K, McClintock JB, Baker BJ (2009a) Defenses of polar macroalgae against herbivores and biofoulers. Bot Mar 52:535–545

Amsler CD, Amsler MO, McClintock JB, Baker BJ (2009b) Filamentous algal endophytes in macrophytic Antarctic algae: prevalence in hosts and palatability to mesoherbivores. Phycologia 48:324–334

Amsler MO, McClintock JB, Amsler CD, Angus RA, Baker BJ (2009c) An evaluation of sponge-associated amphipods from the Antarctic Peninsula. Antarct Sci 21:579–589

Amsler CD, McClintock JB, Baker BJ (2012a) Palatability of living and dead detached Antarctic macroalgae to consumers. Antarct Sci 24:589–590

Amsler CD, McClintock JB, Baker BJ (2012b) Amphipods exclude filamentous algae from the Western Antarctic Peninsula benthos: experimental evidence. Polar Biol 35:171–177

Amsler MO, Amsler CD, von Salm JL, Aumack CF, McClintock JB, Young RM, Baker BJ (2013) Tolerance and sequestration of macroalgal chemical defenses by an Antarctic amphipod: a 'cheater' among mutualists. Mar Ecol Prog Ser 490:79–90

Amsler CD, McClintock JB, Baker BJ (2014) Chemical mediation of mutualistic interactions between macroalgae and mesograzers structure unique coastal communities along the western Antarctic Peninsula. J Phycol 50:1–10

Amsler MO, Huang YM, Engl W, McClintock JB, Amsler CD (2015) Abundance and diversity of gastropods associated with dominant subtidal macroalgae from the western Antarctic Peninsula. Polar Biol 38:1171–1181

Amsler CD, Amsler MO, Curtis MD, McClintock JB, Baker BJ (2019) Impacts of gastropods on epiphytic microalgae on the brown macroalga *Himantothallus grandifolius*. Antarct Sci 31:89–97

Ankisetty S, Nandiraju S, Win H, Park YC, Amsler CD, McClintock JB, Baker JA, Diyabalanage TK, Pasaribu A, Singh MP, Maiese WM, Walsh RD, Zaworotko MJ, Baker BJ (2004) Chemical investigation of predator-deterred macroalgae from the Antarctic Peninsula. J Nat Prod 67:1295–1302

Apt KE (1988) Etiology and development of hyperplasia induced by *Streblonema* sp. (Phaeophyta) on members of the Laminariales (Phaeophyta). J Phycol 24:28–34

Arnold TM, Targett NM (2003) To grow and defend: lack of tradeoffs for brown algal phlorotannins. Oikos 100:406–408

Aumack CF, Amsler CD, McClintock JB, Baker BJ (2010) Chemically mediated resistance to mesoherbivory in finely branched macroalgae along the western Antarctic Peninsula. Eur J Phycol 45:19–26

Aumack CF, Amsler CD, McClintock JB, Baker BJ (2011a) Changes in amphipod densities among macroalgal habitats in day versus night collections along the Western Antarctic Peninsula. Mar Biol 158:1879–1885

Aumack CF, Amsler CD, McClintock JB, Baker BJ (2011b) Impacts of mesograzers on epiphyte and endophyte growth associated with chemically defended macroalgae from the western Antarctic Peninsula: a mesocosm experiment. J Phycol 47:36–41

Aumack CF, Lowe AT, Amsler CD, Amsler MO, McClintock JB, Baker BJ (2017) Gut content, fatty acid, and stable isotope analyses reveal dietary sources of macroalgal-associated amphipods along the western Antarctic Peninsula. Polar Biol 40:1371–1384

Avila C, Taboada S, Núñez-Pons L (2008) Antarctic marine chemical ecology: what is next? Mar Ecol 29:1–71

Baldwin IT, Halitschke R, Paschold A, von Dahl CC, Preston CA (2006) Volatile signaling in plant-plant interactions: "Talking trees" in the genomics era. Science 311:812–815

Barrera-Oro E, Moreira E, Seefeldt MA, Valli Francione M, Quartino ML (2019) The importance of macroalgae and associated amphipods in the selective benthic feeding of sister rockcod species *Notothenia rossii* and *N. coriiceps* (Nototheniidae) in West Antarctica. Polar Biol 42:317–334

Bousfield EL (ed) (1973) Shallow-water gammaridean amphipoda of New England. Comstock, Ithaca

Braeckman U, Pasotti F, Vázquez S, Zacher K, Hoffmann R, Elvert M, Marchant H, Buckner C, Quartino ML, Mác Cormack W, Soetaert K, Wenzhöfer F, Vanreusel A (2019) Degradation of macroalgal detritus in shallow coastal Antarctic sediments. Limnol Oceanogr 64:1423–1441

Bregazzi PK (1972) Life cycle and seasonal movements of *Cheirimedon femoratus* (Pfeffer) and *Tryphosella kergueleni* (Miers) (Crustacea: Amphipoda). Br Antarct Surv Bull 30:1–34

Breithaupt T, Thiel M (eds) (2011) Chemical communication in crustaceans. Springer-Verlag, New York, p 565

Brönmark C, Hansson L-A (eds) (2012) Chemical ecology in aquatic systems. Oxford University Press, Oxford

Brouwer PEM (1996) Decomposition in situ of the sublittoral Antarctic macroalga *Desmarestia anceps* Montagne. Polar Biol 16:129–137

Bucolo P, Amsler CD, McClintock JB, Baker BJ (2011) Palatability of the Antarctic rhodophyte *Palmaria decipiens* (Reinsch) RW Ricker and its endo/epiphyte *Elachista antarctica* Skottsberg to sympatric amphipods. J Exp Mar Biol Ecol 396:202–206

Bucolo P, Amsler CD, McClintock JB, Baker BJ (2012) Effects of macroalgal chemical extracts on spore behavior of the Antarctic epiphyte *Elachista antarctica* Phaeophyceae. J Phycol 48:1403–1410

Bullard SB, Hay ME (2002) Palatability of marine macro-holoplankton: nematocysts, nutritional quality, and chemistry as defenses against consumers. Limnol Oceanogr 47:1456–1467

Campbell HA, Fraser KPP, Bishop CM, Peck LS, Egginton S (2008) Hibernation in an Antarctic fish: on ice for winter. PLoS One 3:e1743

Casaux R (1998) The contrasting diet of *Harpagifer antarcticus* (Notothenioidei, Harpagiferidae) at two localities of the South Shetland Islands, Antarctica. Polar Biol 19:283–285

Casaux R, Mazzotta A, Barrera-Oro E (1990) Seasonal aspects of the biology and diet of nearshore nototheniid fish at Potter Cove, South Shetland Islands, Antarctica. Polar Biol 11:63–72

Casaux R, Baroni A, Ramon A, Carlini A, Bertolin M, DiPrinzio CY (2009) Diet of the leopard seal *Hydrurga leptonyx* at the Danco Coast, Antarctic Peninsula. Polar Biol 32:307–310

Cimino G, Ghiselin MT (2009) Chemical defense and the evolution of opisthobranch gastropods. Proc Calif Acad Sci 60:175–422

Clark GF, Stark JS, Palmer AS, Riddle MJ, Johnston EL (2017) The roles of sea-ice, light and sedimentation in structuring shallow Antarctic benthic communities. PLoS One 12:e0168391

Corbisier TN, Petti MAV, Skowronski RSP, Brito TAS (2004) Trophic relationships in the nearshore zone of Martel Inlet (King George Island, Antarctica): $\delta^{13}C$ stable-isotope analysis. Polar Biol 27:75–82

Correa JA, Sánchez PA (1996) Ecological aspects of algal infectious diseases. Hydrobiologia 326–327:89–95

Cutignano A, Villani G, Fontana A (2012) One metabolite, two pathways: convergence of polypropionate biosynthesis in fungi and marine molluscs. Org Lett 14:992–995

Daly JW (2004) Marine toxins and nonmarine toxins: Convergence or symbiotic organisms? J Nat Prod 67:1211–1215

Daniels RA (1982) Feeding ecology of some fishes of the Antarctic Peninsula. Fish Bull 80:575–585
Daniels RA (1983) Demographic characteristics of an Antarctic plunderfish, *Harpagifer bispinis antarcticus*. Mar Ecol Prog Ser 13:181–183
Dayton PK, Robilliard GA, DeVries AL (1969) Anchor ice formation in McMurdo Sound, Antarctica, and its biological effects. Science 163:273–275
Dayton PK, Robillard GA, Paine RT (1970) Benthic faunal zonation as a result of anchor ice at McMurdo Sound, Antarctica. In: Holgate MW (ed) Antarctic ecology, vol 1. Academic, New York, pp 244–258
De Broyer C, Jażdżewska A (2014) Chapter 5.17. Biogeographic patterns of Southern Ocean benthic Amphipods. In: De Broyer C, Koubbi P, Griffiths HJ, Raymond B, d'Udekem d'Acoz C, Van de Putte AP, Danis B, David B, Grant S, Gutt J, Held C, Hosie G, Huettmann F, Post A, Ropert-Coudert Y (eds) Biogeographic atlas of the Southern Ocean. Scientific Committee on Antarctic Research, Cambridge, 155–165
Dearborn JH, Fell FJ (1974) Ecology of echinoderms from the Antarctic Peninsula. Antarct J US 9:304–306
DeLaca TE, Lipps JH (1976) Shallow water marine associations, Antarctic Peninsula. Antarct J US 11:12–20
Dicke M, Takken W (eds) (2006) Chemical ecology: from gene to ecosystem. Springer, New York
Doyle SR, Momo FR, Brêthes J-C, Ferreyra GA (2012) Metabolic rate and food availability of the Antarctic amphipod *Gondogeneia antarctica* (Chevreux 1906): seasonal variation in allometric scaling and temperature dependence. Polar Biol 35:413–424
Duffy JE, Hay ME (1994) Herbivore resistance to seaweed chemical defense: the roles of mobility and predation risk. Ecology 75:1304–1319
Dunton K (2001) $\delta^{15}N$ and $\delta^{13}C$ measurements of Antarctic Peninsula fauna: trophic relationships and assimilation of benthic seaweeds. Am Zool 41:99–112
Eisner T, Meinwald J (eds) (1995) Chemical ecology: the chemistry of biotic interaction. National Academy Press, Washington, DC
Faugeron S, Martínez EA, Sánchez PA, Correa JA (2000) Infectious diseases in *Mazzaella laminarioides* (Rhodophyta): estimating the effect of infections on host reproductive potential. Dis Aquat Org 42:143–148
Gauna M, Parodi E, Cáceres E (2009) Epi-endophytic symbiosis between *Laminariocolax aecidioides* (Ectocarpales, Phaeophyceae) and *Undaria pinnatifida* (Laminariales, Phaeophyceae) growing on Argentinian coasts. J Appl Phycol 21:11–18
Hall-Aspland SA, Rogers TL (2004) Summer diet of leopard seals (*Hydrurga leptonyx*) in Prydz Bay, Eastern Antarctica. Polar Biol 27:729–734
Hay ME (1986) Associational plant defenses and the maintenance of species diversity: turning competitors into accomplices. Am Nat 128:617–641
Hay ME (1997) The ecology and evolution of seaweed-herbivore interactions on coral reefs. Coral Reefs 16:S67–S76
Hay ME (2009) Marine chemical ecology: chemical signals and cues structure marine populations, communities, and ecosystems. Annu Rev Mar Sci 1:193–212
Hay ME, Sutherland JP (1988) The ecology of rubble structures in the South Atlantic Bight: a community profile. US Fish Wildl Serv Biol Rep 85(7.20):1–67
Hay ME, Stachowicz JJ, Cruz-Rivera E, Bullard SB, Deal MS, Lindquist N (1998) Bioassays with marine and freshwater macroorganisms. In: Haynes KF, Millar JG (eds) Methods in chemical ecology. Volume 2: Bioassay methods. Kluwer Academic Publishers, Norwell, pp 39–141
Hennebert E, Jangoux M, Flamming P (2013) Functional biology of asteroid tube feet. In: Lawrence JM (ed) Starfish biology and ecology of the Asteroidea. John Hopkins Press, Baltimore, pp 24–36
Herms DA, Mattson WJ (1992) The dilemma of plants: to grow or defend. Q Rev Biol 67:285–335
Hommersand MH, Moe RL, Amsler CD, Fredericq S (2009) Notes on the systematics and biogeographical relationships of Antarctic and Sub-Antarctic Rhodophyta with descriptions of four new genera and five new species. Bot Mar 52:509–534

Huang YM, Amsler MO, McClintock JB, Amsler CD, Baker BJ (2007) Patterns of gammarid amphipod abundance and species composition associated with dominant subtidal macroalgae along the western Antarctic Peninsula. Polar Biol 30:1417–1430

Iken K (1999) Feeding ecology of the Antarctic herbivorous gastropod *Laevilacunaria antarctica* Martens. J Exp Mar Biol Ecol 236:133–148

Iken K, Barrera-Oro ER, Quartino ML, Casaux RJ, Brey T (1997) Grazing in the Antarctic fish *Notothenia coriiceps*: evidence for selective feeding on macroalgae. Antarct Sci 9:386–391

Iken K, Amsler CD, Amsler MO, McClintock JB, Baker BJ (2009) Ecological roles of phlorotannins in Antarctic brown algae. Bot Mar 52:547–557

Inderjit, Mallik AU (2002) Chemical ecology of plants: allelopathy in aquatic and terrestrial ecosystems. Birkhäuser, Basel, p 272

Janosik AM, Halanych KM (2010) Unrecognized Antarctic biodiversity: a case study of the genus *Odontaster* (Odontasteridae; Asteroidea). Integr Comp Biol 50:981–992

Jormalainen V, Honkanen T (2008) Macroalgal chemical defenses and their roles in structuring temperate marine communities. In: Amsler CD (ed) Algal chemical ecology. Springer-Verlag, Berlin, pp 57–89

Kang S, Kim S, Park H (2015) Transcriptome of the Antarctic amphipod *Gondogeneia antarctica* and its response to pollutant exposure. Mar Genomics 24:253–254

Kylin H (1937) Bemerkungen über die Entwicklungsgeschichte eniger Phaeophyceen. Acta Univ Lund 33:1–34

Lamb IM, Zimmerman MH (1977) Benthic marine algae of the Antarctic Peninsula. Ant Res Ser 5:130–229

Lane AL, Kubanek J (2008) Secondary metabolite defenses against pathogens and biofoulers. In: Amsler CD (ed) Algal chemical ecology. Springer-Verlag, Berlin, pp 229–243

Littler M, Littler D (1980) The evolution of thallus form and survival strategies in benthic marine macroalgae: field and laboratory tests of a functional form model. Am Nat 116:25–44

Ma WS, Mutka T, Vesley B, Amsler MO, McClintock JB, Amsler CD, Perman JA, Singh MP, Maiese WM, Zaworotko MJ, Kyle DE, Baker BJ (2009) Norselic Acids A-E, highly oxidized anti-infective steroids that deter mesograzer predation, from the Antarctic sponge *Crella* sp. J Nat Prod 72:1842–1846

Marcías ML, Deregibus D, Saravia LA, Campana GL, Quartino ML (2017) Life between tides: spatial and temporal variations of an intertidal macroalgal community at Potter Peninsula, South Shetland Islands, Antarctica. Estuar Coast Shelf Sci 187:193–203

Maschek JA, Baker BJ (2008) The chemistry of algal secondary metabolism. In: Amsler CD (ed) Algal chemical ecology. Springer-Verlag, Berlin, pp 1–24

McClintock JB (1994) Trophic biology of Antarctic echinoderms. Mar Ecol Prog Ser 111:191–202

McClintock JB, Baker BJ (eds) (2001) Marine chemical ecology. CRC, Boca Raton

McClintock JB, Amsler CD, Baker BJ (2010) Overview of the chemical ecology of benthic marine invertebrates along the western Antarctic Peninsula. Integr Comp Biol 50:967–980

McDowell RE, Amsler CD, Dickinson DA, McClintock JB, Baker BJ (2014a) Reactive oxygen species and the Antarctic macroalgal wound response. J Phycol 50:71–80

McDowell RE, Amsler CD, McClintock JB, Baker BJ (2014b) Reactive oxygen species as a marine grazing defense: H_2O_2 and wounded *Ascoseira mirabilis* both inhibit feeding by an amphipod grazer. J Exp Mar Biol Ecol 458:34–38

McDowell RE, Amsler MO, Li Q, Lancaster JR, Amsler CD (2015) The immediate wound-induced oxidative burst of *Saccharina latissima* depends on light via photosynthetic electron transport. J Phycol 51:431–441

McDowell RE, Amsler CD, Amsler MO, Li Q, Lancaster JR Jr (2016) Control of grazing by light availability via light-dependent, wound-induced metabolites: the role of reactive oxygen species. J Exp Mar Biol Ecol 477:86–91

Miller KA, Pearse JS (1991) Ecological studies of seaweeds in McMurdo Sound, Antarctica. Am Zool 31:35–48

Moles J, Núñez-Pons L, Taboada S, Figuerola B, Cristobo J, Avila C (2015) Anti-predatory chemical defences in Antarctic benthic fauna. Mar Biol 162:1813–1821

Müller-Schwarze D (ed) (2006) Chemical ecology of vertebrates. Cambridge University Press, Cambridge

North AW (1996) Locomotory activity and behaviour of the Antarctic teleost *Notothenia coriiceps*. Mar Biol 126:125–132

Núñez-Farfán J, Fornoni J, Valverde PL (2007) The evolution of resistance and tolerance to herbivores. Annu Rev Ecol Evol Syst 38:541–566

Núñez-Pons L, Avila C (2014) Deterrent activities in the crude lipophilic fractions of Antarctic benthic organisms: chemical defences against keystone predators. Polar Res 33:1

Núñez-Pons L, Rodríguez-Arias M, Gómez-Garreta A, Ribera-Siguán A, Avila C (2012) Feeding deterrency in Antarctic marine organisms: bioassays with the omnivore amphipod *Cheirimedon femoratus*. Mar Ecol Prog Ser 462:163–174

Obermüller B, Puntarulo S, Abele D (2007) UV-tolerance and instantaneous physiological stress responses of two Antarctic amphipod species *Gondogeneia antarctica* and *Djerboa furcipes* during exposure to UV radiation. Mar Environ Res 64:267–285

Pabis K, Sicinski J (2010) Polychaete fauna associated with holdfasts of the large brown alga *Himantothallus grandifolius* in Admiralty Bay, King George Island, Antarctic. Polar Biol 33:1277–1288

Pawson DL (1969) Echinoidea. Antarct Map Folio Ser 11:38–41

Pearse JS, Giese AC (1966) Food, reproduction and organic constitution of the common antarctic echinoid *Sterechinus neumayeri* (Meissner). Biol Bull 130:387–401

Peters AF (2003) Molecular identification, taxonomy and distribution of brown algal endophytes, with emphasis on species from Antarctica. Proc Int Seaweed Symp 17:293–302

Peters KJ, Amsler CD, Amsler MO, McClintock JB, Dunbar RB, Baker BJ (2005) A comparative analysis of the nutritional and elemental composition of macroalgae from the western Antarctic Peninsula. Phycologia 44:453–463

Picken GB (1979) Growth, production and biomass of the antarctic gastropod *Laevilacunaria antarctica* Martens 1885. J Exp Mar Biol Ecol 40:71–79

Picken GB (1980) The distribution, growth, and reproduction of the Antarctic limpet *Nacella (Patinigera) concinna* (Strebel) 1908. J Exp Mar Biol Ecol 42:71–85

Potin P (2008) Oxidative burst and related responses in biotic interactions of algae. In: Amsler CD (ed) Algal chemical ecology. Springer-Verlag, Berlin, pp 245–271

Puglisi MP, Becerro MA (eds) (2019) Chemical ecology: the ecological impacts of marine natural products. CRC, Boca Raton, p 400

Purchon RD (1977) Feeding methods and adaptive radiation in the Gastropoda. In: Purchon RD (ed) The biology of the mollusca, 2nd edn. Pergamon, Amsterdam, pp 41–99

Reichardt W, Dieckmann G (1985) Kinetics and trophic role of bacterial degradation of macroalgae in Antarctic coastal waters. In: Siegfried WR, Condy P, Laws RM (eds) Antarctic nutrient cycles and food webs. Springer-Verlag, Berlin, pp 115–122

Richardson MG (1971) The ecology and physiological aspects of Antarctic weed dwelling amphipods (Preliminary report, II). Brit Antarct Surv Rep N9/1971(-72)/H:1–16

Richardson MG (1977) The ecology including physiological aspects of selected Antarctic marine invertebrates associated with inshore macrophytes. PhD Dissertation. Department of Zoology, University of Durham

Rosenthal GA, Berenbaum MR (2012) Herbivores: their interactions with secondary plant metabolites. Ecological and evolutionary processes, 2nd edn. Academic Press, New York, p 493

Santhanam R (2018) Biology and ecology of edible marine gastropod molluscs. Apple Academic Press, Oakville

Saucède T, Pierrat B, David B (2014) Chapter 5.26. Echinoids. In: De Broyer C, Koubbi P, Griffiths HJ, Raymond B, d'Udekem d'Acoz C, Van de Putte AP, Danis B, David B, Grant S, Gutt J, Held C, Hosie G, Huettmann F, Post A, Ropert-Coudert Y (eds) Biogeographic atlas of the Southern Ocean. Scientific Committee on Antarctic Research, Cambridge, pp 213–220

Schoenrock KM, Amsler CD, McClintock JB, Baker BJ (2013) Endophyte presence as a potential stressor on growth and survival in Antarctic macroalgal hosts. Phycologia 52:595–599

Schoenrock KM, Amsler CD, McClintock JB, Baker BJ (2015a) A comprehensive study of Antarctic algal symbioses: minimal impacts of endophyte presence in most species of macroalgal hosts. Eur J Phycol 50:271–278

Schoenrock KM, Amsler CD, McClintock JB, Baker BJ (2015b) Life history bias in endophyte infection of the Antarctic rhodophyte, *Iridaea cordata*. Bot Mar 58:1–8

Schram JB, McClintock JB, Amsler CD, Baker BJ (2015) Impacts of acute elevated seawater temperature on the feeding preferences of an Antarctic amphipod toward chemically deterrent macroalgae. Mar Biol 162:425–433

Schram JB, Amsler MO, Amsler CD, Schoenrock KM, McClintock JB, Angus RA (2016) Antarctic crustacean grazer assemblages exhibit resistance following exposure to decreased pH. Mar Biol 163:106

Schwarz AM, Hawes I, Andrew N, Norkko A, Cummings V, Thrush S (2003) Macroalgal photosynthesis near the southern global limit for growth; Cape Evans, Ross Sea, Antarctica. Polar Biol 26:789–799

Sotka EE, Whalen KE (2008) Herbivore offense in the sea: the detoxification and transport of secondary metabolites. In: Amsler CD (ed) Algal chemical ecology. Springer-Verlag, Berlin, pp 203–228

Sotka EE, Forbey J, Horn M, Poore AGB, Raubenheimer D, Whalen KE (2009) The emerging role of pharmacology in understanding consumer-prey interactions in marine and freshwater systems. Integr Comp Biol 49:291–313

Steneck RS, Graham MH, Bourque BJ, Corbett D, Erlandson JM, Estes JA, Tegner MJ (2002) Kelp forest ecosystems: biodiversity, stability, resilience and future. Environ Conserv 29:436–459

Thomas F, Cosse A, Le Panse S, Kloareg B, Potin P, Leblanc C (2014) Kelps feature systemic defense responses: insights into the evolution of innate immunity in multicellular eukaryotes. New Phytol 204:567–576

Valdivia N, Pardo LM, Macaya EC, Huovinen P, Gómez I (2019) Different ecological mechanisms lead to similar grazer controls on the functioning of periphyton Antarctic and sub-Antarctic communities. Prog Oceanogr 174:7–16

von Salm JL, Schoenrock KM, McClintock JB, Amsler CD, Baker BJ (2019) The status of marine chemical ecology in Antarctica: form and function of unique high-latitude chemistry. In: Puglisi-Weening MP, Becerro MA (eds) Chemical ecology: the ecological impacts of marine natural products. CRC, Boca Raton, pp 27–69

Waters CM, Bassler BL (2005) Quorum sensing: cell-to-cell communication in bacteria. Annu Rev Cell Dev Biol 21:319–346

Wiencke C, Amsler CD (2012) Seaweeds and their communities in polar regions. In: Wiencke C, Bischof K (eds) Seaweed biology: novel insights into ecophysiology, ecology and utilization. Springer-Verlag, Berlin, pp 265–294

Wiencke C, Clayton MN (2002) Antarctic seaweeds. In: Synopsis of the Antarctic benthos, vol 9. ARG Gantner Verlag KG, Ruggell

Wiencke C, Amsler CD, Clayton MN (2014) Chapter 5.1. Macroalgae. In: De Broyer C, Koubbi P, Griffiths HJ, Raymond B, d'Udekem d'Acoz C, Van de Putte AP, Danis B et al (eds) Biogeographic atlas of the Southern Ocean. Scientific Committee on Antarctic Research, Cambridge, 66–73

Wyatt TD (ed) (2014) Pheromones and animal behavior: chemical signals and signatures. Cambridge Press, Cambridge

Yang EC, Boo SM, Bhattacharya D, Saunders GW, Knoll AH, Fredericq S, Graf L, Yoon HS (2016) Divergence time estimates and the evolution of major lineages in the florideophyte red algae. Sci Rep 6:21361

Young RM, von Salm JL, Amsler MO, Lopez-Bautista J, Amsler CD, McClintock JB, Baker BJ (2013) Site-specific variability in the chemical diversity of the Antarctic red alga *Plocamium cartilagineum*. Mar Drugs 11:2126–2139

Zamzow JP, Amsler CD, McClintock JB, Baker BJ (2010) Habitat choice and predator avoidance by Antarctic amphipods: the roles of algal chemistry and morphology. Mar Ecol Prog Ser 400:155–163

Zamzow JP, Aumack CF, Amsler CD, McClintock JB, Amsler MO, Baker BJ (2011) Gut contents and stable isotope analyses of the Antarctic fish, *Notothenia coriiceps* Richardson, from two macroalgal communities. Antarct Sci 23:107–116

Zenteno L, Cárdenas L, Valdivia N, Gómez I, Höfer J, Garrido I, Pardo LM (2019) Unraveling the multiple bottom-up supplies of an Antarctic nearshore benthic community. Prog Oceanogr 174:55–63

Chapter 18
Brown Algal Phlorotannins: An Overview of Their Functional Roles

Iván Gómez and Pirjo Huovinen

Abstract Phlorotannins are polyphenolic compounds, relatively hydrophilic, formed by polymers of phloroglucinol and found exclusively in brown algae. These molecules are located in vesicles denominated physodes (the soluble fraction) and also complexed to polysaccharides in the cell wall (the insoluble fraction). Well known as potential grazer deterrents, one of the most striking characteristics of these compounds, due to a number of hydroxyl groups, is their antioxidant potential, which opens promising perspectives for pharmaceutical and biotechnological uses. In Antarctic brown algae, especially endemic species of Desmarestiales, constitutively high levels of phlorotannins (up to 12% of dry weight) have been measured. Although translocation has not been conclusively confirmed, the differential allocation of phlorotannins in meristematic and reproductive tissues in some species suggests their involvement in chemical defenses protecting essential metabolic functions. Due to their UV-absorbing properties and peripheral localization in cells and tissues, phlorotannins have been related with the increased tolerance to UV radiation in various Antarctic brown algae. However, no induction of phlorotannins by UV has been demonstrated, which strongly supports the idea that these molecules are constitutive biochemical components of a suite of mechanisms against multiple stressors. Due to their structural role as primary compounds, phlorotannins are essential for various morpho-functional processes that in the case of Antarctic algae allow them to thrive under extreme conditions. Overall, the significance of phlorotannins in this group of algae has largely been recognized; however, fundamental aspects of their molecular expression, synthesis, and regulation still need to be addressed, especially considering the climate change-driven environmental scenarios.

Keywords Antioxidant activity · Brown algae · Phenolic compounds · Physodes · Secondary metabolites · UV-absorbing compounds

---

I. Gómez (✉) · P. Huovinen
Instituto de Ciencias Marinas y Limnológicas, Facultad de Ciencias, Universidad Austral de Chile, Valdivia, Chile

Research Center Dynamics of High Latitude Marine Ecosystems (IDEAL), Valdivia, Chile
e-mail: igomezo@uach.cl; pirjo.huovinen@uach.cl

I. Gómez, P. Huovinen (eds.), *Antarctic Seaweeds*,
https://doi.org/10.1007/978-3-030-39448-6_18

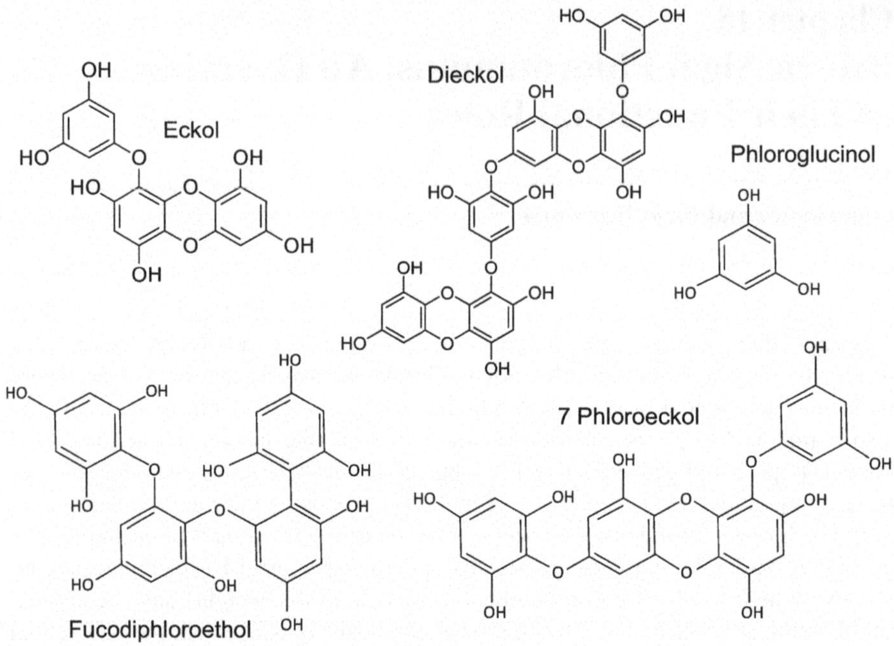

Fig. 18.1 Chemical structure of some common phlorotannins extracted from brown algae

18.1 Introduction

Brown algal phenolics belong to a single structural class, the phlorotannins, which are dehydropolymers of phloroglucinol (1,3,5-trihydroxybenzen). Depending on the degree of polymerization, phlorotannins present a wide range of molecular weights (between 140 and 40.000 Da; Ragan and Glombitza 1986). Thus, based on the number of phloroglucinol units, the profiling of phlorotannins can vary considerably among species (Steevensz et al. 2012). Within the phlorotannins found in brown algae, the most common are fucols, phlorethols, fucophlorethols, and eckols (Fig. 18.1), which vary depending on the type of chemical linkage, e.g., ether Aryl-O-Aryl linkages in phlorethols or dibenzodioxin linkages in eckols (Ragan and Glombitza 1986; La Barre et al. 2010). As secondary metabolites, these substances play a series of putative roles in the cell, mainly as anti-herbivory defense, antifouling activity, UV protectants, and antioxidants. Phlorotannins can be present as soluble substances sequestered in vesicle-denominated physodes and as insoluble fraction bound to polysaccharides in the cell wall (denominated the insoluble fraction) (Fig. 18.2). Due to this, these compounds are regarded also as primary compounds, essential during cell formation (Schoenwaelder 2002). This has been corroborated by histological studies indicating that phlorotannins

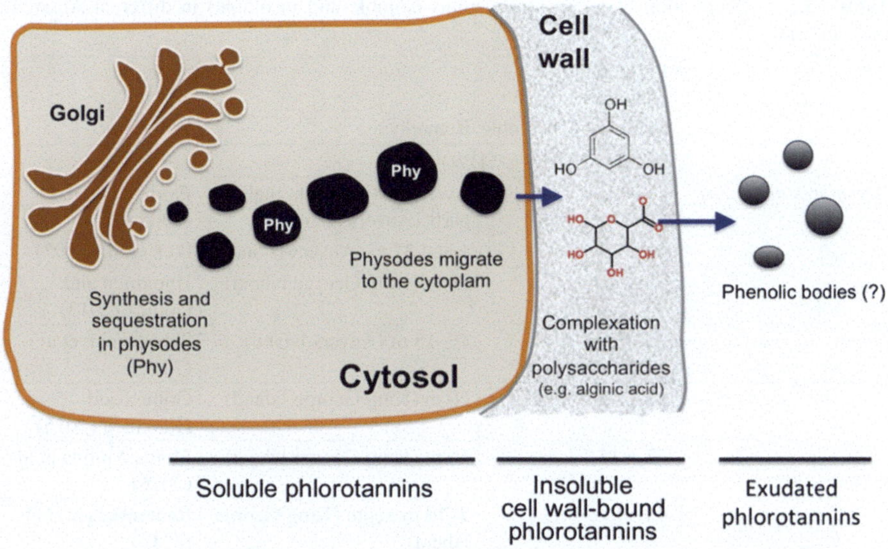

Fig. 18.2 Localization of physodes and the different functional forms of phlorotannins in brown algal cells. The relationship between exuded phlorotannins and extracellular phenolic bodies has not been conclusively established

participate actively in wall configuration of zygotes (Kevekordes and Clayton 1999), are concentrated at the periphery of multicellular embryos of *Fucus* (Schoenwaelder and Clayton 1998), and are actively synthesized during wound healing and sealing after amputation of thallus regions simulating herbivory (Lüder and Clayton 2004).

Because of their ubiquity, phlorotannins can be present in high concentrations in some species. For example, in Antarctic brown algae, particularly endemic species of Desmarestiales, constitutively high levels of phlorotannins have been measured, forming up to 12% of dry weight (Table 18.1). However, it has been well demonstrated that total phlorotannin content can vary significantly depending on a number of biotic and abiotic factors (Van Alstyne and Pelletreau 2000; Pavia and Toth 2000; Jormalainen et al. 2003; Gómez and Huovinen 2010).

Apart from their well-known role as herbivore deterrents (reviewed in Chap. 17 in this volume and by Iken et al. 2009), phlorotannins and various of their chemical derivatives apparently have other important functions, especially in oxidative metabolism and metal chelation (Stauber and Florence 1987), with far-reaching implications for ecophysiology of brown algae (La Barre et al. 2010).

Table 18.1 Concentration of total phlorotannins (soluble and insoluble) in different Antarctic brown algae

Species	Phlorotannin contents		Remarks	Reference
	Soluble	Insoluble		
	Percentage (mg g^{-1} DW)			
Desmarestia anceps	117 ± 3		Average of different thallus parts (Anvers Island)	Fairhead et al. (2005a)
	60 ± 19		1 and 37 m (Anvers Island)	Iken et al. (2007)
	60 ± 5.5		28 m (King George Island)	Huovinen and Gómez (2013)
	30		10–15 m (Anvers Island)	Schoenrock et al. (2015)
	80 ± 18	28 ± 4.9	20 m (King George Island)	Gómez and Huovinen (2015)
	52 ± 15		15 m (King George Island)	Flores-Molina et al. (2016)
	43–48	33	7–10 m depth (King George Island)	Rautenberger et al. (2015)
Desmarestia menziesii	52 ± 2		Average of different thallus parts (Anvers Island)	Fairhead et al. (2005a)
	30 ± 12		1 and 37 m (Anvers Island)	Iken et al. (2007)
	85 ± 5.6		Apical parts; 17 m (King George Island)	Huovinen and Gómez (2013)
	10		10–15 m (Anvers Island)	Schoenrock et al. (2015)
	55 ± 15		20 m (King George Island)	Gómez and Huovinen (2015)
	55–58	18	7–10 m depth (King George Island)	Rautenberger et al. (2015)
Ascoseira mirabilis	35 ± 25		1 and 37 m (Anvers Island)	Iken et al. (2007)
	15 ± 12		8 m (King George Island)	Huovinen and Gómez (2013)
	12 ± 3	50 ± 10	1 m (King George Island)	Huovinen and Gómez (2015)
	13 ± 3	15 ± 2.4	20 m (King George Island)	Gómez and Huovinen (2015)
	10–12	20–23	7–10 m depth (King George Island)	Rautenberger et al. (2015)
Himantothallus grandifolius	63 ± 8		1 and 37 m (Anvers Island)	Iken et al. (2007)
	120 ± 8.6		30 m (King George Island)	Huovinen and Gómez (2013)
	97 ± 20	49 ± 11	20 m (King George Island)	Gómez and Huovinen (2015)
	90–99	40	7–10 m depth (King George Island)	Rautenberger et al. (2015)

(continued)

Table 18.1 (continued)

Species	Phlorotannin contents		Remarks	Reference
	Soluble	Insoluble		
Cystosphaera jacquinotii	26 ± 16		1 and 37 m (Anvers Island)	Iken et al. (2007)
	29 ± 5.2		25 m (King George Island)	Huovinen and Gómez (2013)
	75 ± 19	10 ± 3	15–20 m (King George Island)	Huovinen and Gómez (2015)
Desmarestia antarctica	42 ± 5.7		20 m (King George Island)	Huovinen and Gómez (2013)
Adenocystis utricularis	52 ± 4.4		Intertidal zone (King George Island)	Huovinen and Gómez (2013)

18.2 Synthesis and Cellular Localization of Phlorotannins: Dual Functions as Secondary Metabolites and Structural Compounds

Phlorotannins are synthesized via the acetate-malonate pathway through a type III polyketide synthase (Herbert 1989; Meslet-Cladiere et al. 2013), and in terms of their chemical properties, they differ from the condensed tannins of vascular plants (Arnold and Targett 2002). It has been suggested that some fatty acids associated with the Acetyl-CoA, a key intermediate in the polyketide pathway, could be regarded as precursors of phlorotannins (Steinhoff et al. 2011). However, metabolism of phenolics in algae has been much less studied, and key aspects of biosynthesis and regulation are unknown.

Due to their reactivity, phlorotannins may easily form complexes with macromolecules, and they are sequestered in physodes through the formation of covalent, hydrogen, or ionic bonds (the soluble fraction) (Fig. 18.2). Traditionally physodes or phenolic precursors are thought to be synthesized in the chloroplast or at the chloroplast membrane, and various authors have described an osmiophilic material being released from chloroplasts (Evans and Holligan 1972; Feldmann and Guglielmi 1972; Pellegrini 1980). An alternative explanation suggests that phenolic material is produced in the chloroplast endoplasmic reticulum (CER), which may play a role in the transport of phenolic precursors to cell vacuoles and physodes (Pellegrini 1980; Clayton and Beakes 1983; Kaur and Vijayaraghavan 1992) or may directly give rise to physodes (Feldmann and Guglielmi 1972; Oliveira and Bisalputra 1973). When the physodes make contact with the plasmalemma, phlorotannins are released and polymerized in the apoplast (cell wall) forming complexes with polysaccharides, e.g., alginic-acid-bound phlorotannins (Schoenwaelder and Clayton 1999) (Fig. 18.2). Some studies have identified peroxidases in the cell wall of *Ascophyllum nodosum* suggesting that phlorotannins excreted from the cells may be modified through the activity of these enzymes (Vilter 1995). In fact, the process of oxidative condensation and the linkage to alginic acids in the apoplast is apparently driven by vanadium-dependent haloperoxidases (Potin and Leblanc 2006; Salgado et al. 2009). Vreeland and Laetsch (1988) proposed that phenolic cross-linking of alginate may occur in early wall formation in *Fucus* and that peroxidases may be involved in the catalysis of phenolic condensation into alginate. During early phases of growth, physode movements to regions of active wall formation from the cell periphery to the rhizoid tip and to the impending plane of cytokinesis are dependent on interactions with the cytoskeleton (Schoenwaelder and Clayton 1999). Hence, actin microfilaments may be acting as a general circulatory system moving physodes around the cell, with microtubules directing physodes (and probably other wall components) to cell-wall deposition sites, both in the primary wall and at cross-walls (Kevekordes and Clayton 1999; Schoenwaelder and Clayton 1999). Although the exudation of phlorotannins has been reported (Ragan and Glombitza 1986; Toth and Pavia 2002; Koivikko et al. 2005), no clear evidence exists that they are related with phenolic bodies described in embryos of *Durvillaea antarctica* (Kevekordes and Clayton 1999). For example, it was demonstrated that extracellular excretion of phenolic compounds in *Eisenia bicyclis* and *Ecklonia kurome* corresponded to monomeric bromophenols, while phloroglucinol or polymeric phlorotannins were not detected (Shibata et al. 2006). On the other hand, phlorotannins are strong chelators of metals, and thus, they are thought to participate in exudation-based detoxification mechanisms: they may sequester metal ions in physodes (Smith and Harwood 1986), and through exudation processes, these metal-complexing compounds may decrease the metal concentration or alter its speciation in the surrounding water (Gledhill et al. 1999). As physodes are more abundant in peripheral cell layers, their role as a filter stopping metals from entering the inner cells has been proposed (reviewed by Schoenwaelder 2002).

18.3 Phlorotannins as UV-Screening Substances

In contrast to terrestrial plants, where natural levels of UV radiation do not necessarily result in damage (Paul and Gwynn-Jones 2003; Hideg et al. 2013), aquatic organisms, especially subtidal seaweeds, can be impaired when they are exposed to high solar radiation (Bischof et al. 2006a). Thus, synthesis and accumulation of UV-screening substances is a common photoprotective strategy observed in several groups of living organisms (Karentz et al. 1991; García-Pichel and Castenholz 1993; Cockell and Knowland 1999). In fact, various of these compounds have been isolated and tested as bioactive substances for use in skin care, cosmetics, and pharmaceutical products (reviewed in Pangestuti et al. 2018). Due to their chemical properties, diverse phenolics are regarded as general anti-stress agents, including UV protection and antioxidant activity, and apart from brown algae, they have been reported in green (Menzel et al. 1983; Pérez-Rodriguez et al. 1998, 2001; Gómez et al. 1998; Ross et al. 2005) and red algae (Athukorala et al. 2003; Yildiz et al. 2011; Heffernan et al. 2014; Cruces et al. 2018).

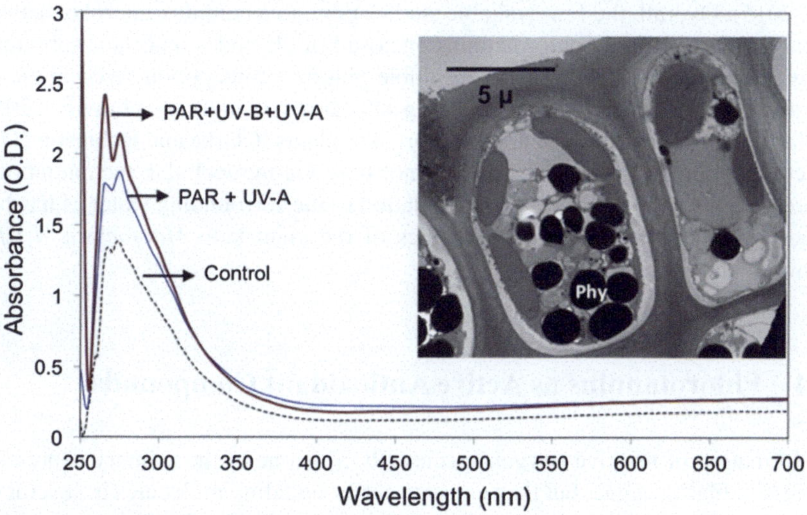

Fig. 18.3 Absorption of ethanol-water extract of apical fronds of *Lessonia spicata* exposed to UV radiation for 24 h

Phlorotannins have been correlated with an increased tolerance to UV radiation (Pavia et al. 1997; Swanson and Fox 2007; Gómez and Huovinen 2010; Steinhoff 2010). In general, phlorotannins absorb at wavelengths between 200 and 300 nm, i.e., well in the UV-C range and at shorter UV-B wavelengths. When phlorotannins are removed using polyvinylpolypyrrolidone (PVPP), absorption of algal extract decreases by 70% in the range between 280 and 300 nm (Pavia et al. 1997). Although the effectiveness of phlorotannins as UV-screening compounds is higher at the UV-C range, absorption spectra of the different fractions of phlorotannins (physodes, cell-wall-bound and excreted phlorotannins) may be different. In fact, the absorption peak of the trihydroxy-coumarin from *Dasycladus vermicularis* suffers a shift when excreted to the seawater (Pérez-Rodriguez et al. 2001). Protection against UV radiation by phlorotannins in early developmental phases has been demonstrated in *Fucus* embryos and spores of *Laminaria* (Schoenwaelder et al. 2003; Henry and Van Alstyne 2004; Roleda et al. 2010). In the kelp *Lessonia spicata*, 24-h exposure to UV radiation increases the synthesis of phlorotannins, compared to control without UV (Fig. 18.3). In this species, the UV-mediated increase in phlorotannins can minimize photodamage of key physiological processes and cellular components, such as photosynthesis and DNA (Gómez and Huovinen 2010). Interestingly, differences between species, season, and morpho-functional processes have raised the question whether the photoprotective role of phlorotannins in brown algae can be regarded as a constitutive or inducible mechanism. For example, in *Lessonia*, the induction of phlorotannins by UV radiation has been shown to occur only during the period when sporophytes actively grow (Gómez and Huovinen 2010). However, in *Fucus vesiculosus*, no UV induction of soluble phlorotannins was found, which was related with a lack of upregulation of *pksIII* genes (Creis et al. 2015). Overall, the few available studies point to a complex interplay between the induction of soluble phlorotannins enclosed in physodes and their subsequent deposition in the cell-wall matrix. Insoluble phlorotannins polymerized in the cell wall are regarded as primary UV-shielding substances (Gómez and Huovinen 2010), similar to cell-wall-bound phenolics reported in plants (Clarke and Robinson 2008). Moreover, photoprotection is conferred not only via intracellular accumulation of phlorotannins but also as a result of exudation to the surrounding water as has been suggested for adult thalli and propagules of the giant kelp *Macrocystis pyrifera* (Swanson and Druehl 2002).

18.4 Phlorotannins as Active Antioxidant Compounds

The formation of reactive oxygen species (ROS) is one of the primary expressions of stress in marine algae, but they can also act as signaling molecules in several cellular reactions (revised in Bischof and Rautenberger 2012). Phlorotannins are known to act, not only as photoprotective substances but also as highly efficient ROS scavengers (Nakai et al. 2006; Wang et al. 2009; Heffernan et al. 2014). Due to the presence of various interconnected rings (up to eight) in their chemical structure, phlorotannins are regarded as potent antioxidants scavenging different types of ROS, e.g., superoxide anions (O_2^-), peroxides, singlet oxygen ($^1O_2$), and hydroxyl radicals ($^{\bullet}OH$) (Ahn et al. 2007; Koivikko et al. 2007). Thus, the hydroxyl groups present in phlorotannins act as reducing agents, hydrogen donors, and singlet oxygen (reviewed in Michalak 2006). It has been postulated that the relationship between increased levels of ROS and phlorotannin induction in brown algae can follow the methyl jasmonate signal transduction pathway, a plant defense-related pathway reported commonly in plants during high ROS production (Arnold et al. 2001; Küpper et al. 2009). The evidence gained during the last decades appears to indicate that operation of efficient and rapid ROS scavenging mechanisms based on phenols can be regarded as an important physiological adaptation in seaweeds when they are exposed to different environmental stressors, e.g., high solar irradiation, metal pollution, or high temperature (Aguilera et al. 2002; Contreras et al. 2009; Cruces et al. 2017).

18.4.1 Phlorotannins and UV-Induced Oxidative Stress

UV radiation is a primary factor inducing ROS in seaweeds, mostly by increasing the activity of peroxidases and NADPH oxidase (Mackerness et al. 2001). During exposure to high levels of UV radiation, the xanthophyll cycle is inhibited, which increases ROS production and impedes an effective dissipation of excess absorbed excitation energy of photosynthetically active radiation (PAR) (Dring 2005; Bischof et al. 2006b; Lesser 2012). Thus, the increased electron fluxes result in a direct photoreduction of oxygen in the Mehler reaction in PSI and lower activity of Rubisco oxygenase, which exacerbate formation of ROS (e.g., superoxides) (Badger et al. 2000). In cold temperate/Arctic kelp gametophytes, formation of ROS could be demonstrated to occur predominantly in the peripheral cytoplasm or in plasmatic vesicles (Müller et al. 2012). Under UV stress, a relationship between the content of phlorotannins and antioxidant activity in various species of brown algae has been reported.

Temperature can modify the phlorotannin response to UV radiation. In the sub-Antarctic *Lessonia spicata*, *Durvillaea antarctica*, and *Macrocystis pyrifera* from the coast of Chile, a rapid induction of soluble phlorotannins triggered by UV radiation ameliorated the effects of oxidative stress on photochemical processes after short-term thermal stress (Cruces et al. 2012). Interestingly, under elevated temperatures >20 °C, the UV induction of phlorotannins ceases, and lipid peroxidation increases in *D. antarctica* and *L. spicata*, suggesting that the ROS scavenging potential of these sub-Antarctic species has a geographic component associated with prevalent UV levels and temperature (Cruces et al. 2013). This can have important consequences for stress tolerance in species living at the limits of their geographic distribution or those exposed to changing conditions of temperature and solar radiation (e.g., polar, intertidal species). In fact, at a molecular level, oxidative stress caused by UV radiation in the Arctic kelp *Saccharina latissima* is higher at 2 °C, which is reflected in upregulation of genes encoding for different ROS defense mechanisms, especially antioxidant enzymes, but not phlorotannins (Heinrich et al. 2015). On the other hand, UV radiation does not directly regulate the expression of genes involved in phlorotannin metabolism in the intertidal *Fucus vesiculosus*, suggesting that UV induction of these substances relies on other processes or their accumulation represents a constitutive metabolic strategy (Creis et al. 2015). Because phlorotannins act also as primary, structural cell components, their accumulation depends on cellular cycles and biomass formation. Hence, the constitutive nature of phlorotannins confers side-by-side advantages to brown algae under multiples stress factors, including UV radiation (Arnold and Targett 2003; Gómez and Huovinen 2010).

18.4.2 Phlorotannins and Their Interaction with Metals

Although the relationship between phlorotannins of brown algae and metals is not fully understood, increasing evidence indicates that metal tolerance of seaweeds can be associated with both internal and external metal-complexing ligands (Andrade et al. 2010; Connan and Stengel 2011). Decreased levels of soluble phenolic compounds in seaweeds (e.g., *Scytosiphon lomentaria* and *Ulva compressa*) have been reported in copper-impacted sites (Ratkevicius et al. 2003; Contreras et al. 2005). Metals are redox active and also participate in many reactions generating ROS. The importance of phenolic compounds as key antioxidant agents during metal exposure has been recognized in plants (reviewed by Sakihama et al. 2002). UV radiation is known to induce or enhance the toxicity of certain organic contaminants (phototoxicity) (Huovinen et al. 2001), and the presence of various metals has been reported to have antagonistic effects on seaweeds (Andrade et al. 2006). In copper-impacted areas, seaweeds (e.g. *Ulva compressa*) have been shown to develop oxidative stress, and decreased levels of soluble phenolic compounds have been reported (Ratkevicius et al. 2003; Contreras et al. 2005). Adverse effects of copper in the brown alga *Laminaria digitata* were buffered by protective mechanisms regulated by lipid peroxide derivatives (Ritter et al. 2008). Proteins potentially involved in the control of copper-mediated oxidative stress in the brown alga *Scytosiphon gracilis* were identified recently (Contreras et al. 2010). Species-specific antioxidant activity of the soluble phlorotannins and its response to environmental stress (UV radiation, metals) has been shown in three Pacific kelps (Huovinen et al. 2010). Here, inorganic nitrogen was shown to mitigate the adverse effects of copper: the impact of the interaction of copper, nitrate, and UV radiation was species-specific, *Lessonia spicata* showing the strongest responses in photosynthetic activity and *Durvillaea antarctica* the strongest response in phlorotannins and their antioxidant activity. *Macrocystis pyrifera* accumulated threefold more copper in its tissues than the other kelps, but its photosynthetic activity was twofold less inhibited by copper than in *D. antarctica*, suggesting higher metal tolerance of *M. pyrifera*, which was partly explained by the decreased accumulation of copper in the algal tissues in the presence of nitrate (Huovinen et al. 2010).

Whether phlorotannins react increasing their ROS scavenging activity after exposure to metals is not well known. When algae are exposed to metal stress, increased exudation of organic compounds, including probably phlorotannins, may retain free metals in form of extracellular complexing ligands. On the other hand, detoxification of intracellular metals via algal exudates may also increase (Andrade et al. 2010). *Lessonia spicata* from uncontaminated sites has been shown to have capacity to rapidly respond to copper exposure by producing organic ligands that, due to their complexing capacity in the water, can rapidly attenuate the level of labile copper (Andrade et al. 2010), thus affecting its bioavailability.

18.5 Phlorotannins in Antarctic Seaweeds

An important feature of various endemic Antarctic brown algae is their high content of phlorotannins. Different studies have reported total phlorotannin contents in Antarctic brown algae ranging between 1% and 12% DW. In the case of insoluble phlorotannins, values are between 1% and 5% DW (Table 18.1). Although some cold-temperate genera (e.g., *Fucus, Ascophyllum*) can contain high concentrations of phlorotannins (>10% DW), normally the maximal values detected in Antarctic brown algae are higher than most of reported values from temperate, cold-temperate, and Arctic species (Connan et al. 2004; Dubois and Iken 2012; Cruces et al. 2013; Generalíc-Mekiníc et al. 2019). The high concentrations of phlorotannins in some endemic Antarctic brown algae can be also evidenced by the abundance of physodes

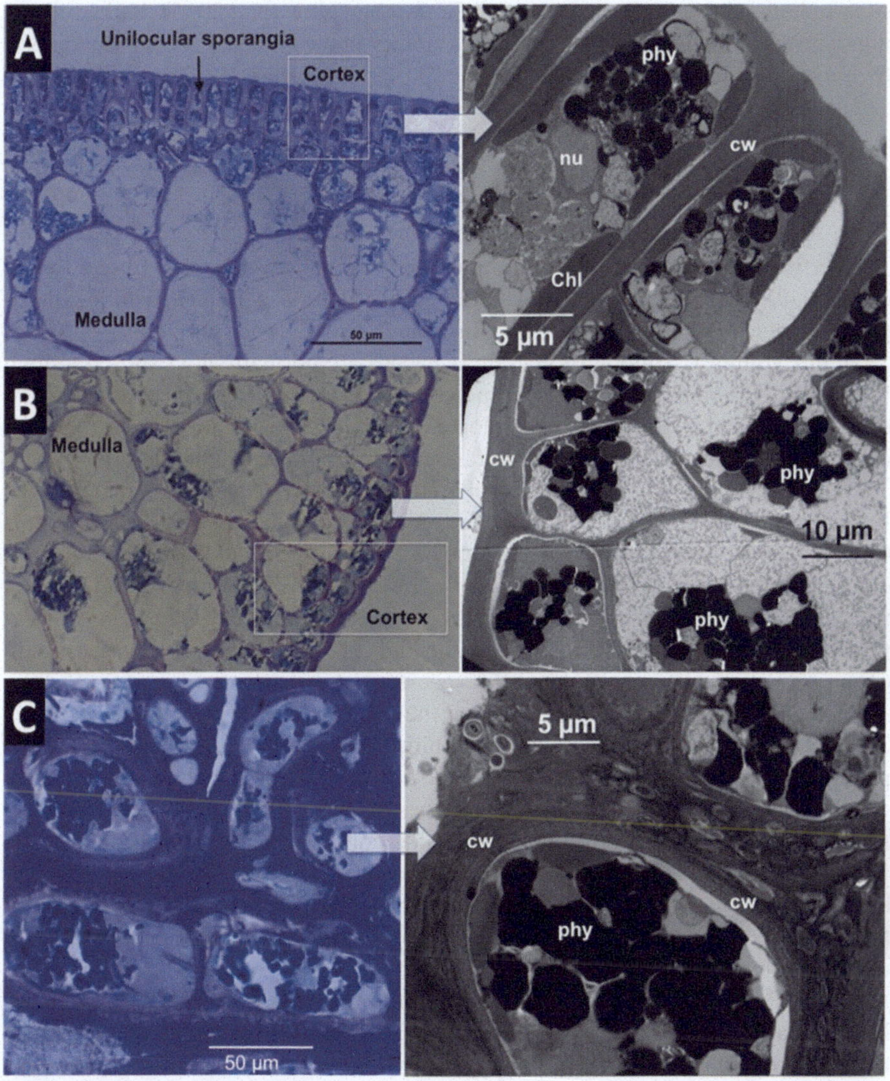

Fig. 18.4 Ultrastructure and localization of phlorotannin-containing physodes in Antarctic brown algae. (**a**) *Desmarestia anceps*; (**b**) *Phaeurus antarcticus*; (**c**) *Halopteris obovata*. Left: cross sections stained with toluidine blue; right: transmission electron microscopy of cortical cells indicating the presence of physodes (*phy*). Cell wall (*cw*) and chloroplasts (*chl*) are indicated

in the outer cell layers (Fig. 18.4). Herbivory is among the factors that may contribute to these high levels of phlorotannins (see Chap. 17 in this volume); however, other intrinsic factors related to biomass formation can also play an important role. Whether these high phlorotannin levels found in endemic brown algae living between 5 and 30 m depth can also be associated with photoprotection against high solar radiation has also been evaluated (Gómez and Huovinen 2015). In the following sections of the present chapter, the variability in the contents of phlorotannins is revised in the context of abiotic stress.

18.5.1 Depth Patterns in Phlorotannin Contents

The vertical distribution of endemic Antarctic brown algae can range from 1 to 2 m down to 40 m or greater depths. This broad distribution is related with a suite of photobiological adaptations operating in a range of different light fields (see Chap. 11 in this volume). Although Antarctic seaweeds are normally not exposed to high UV levels, seasonal and oceanographic conditions can increase the eventual incidence of harmful irradiances (the biological impact of this factor on different processes related with algal distribution is revised in Chap. 11 by Gómez and Huovinen). The hypothesis that phlorotannins of Antarctic seaweeds can also be related with the light acclimation strategies has been tested in algae collected at different depths (Fairhead et al. 2005a; Huovinen and Gómez 2013; Gómez and Huovinen 2015). In eight brown algae collected along a depth gradient in King George Island, species such as *Desmarestia anceps*, *Cystosphaera jacquinotii*, and *Himantothallus grandifolius* collected at depths >20 m showed the highest phlorotannin concentrations, in contrast to shallow water or intertidal species such as *Adenocystis utricularis* or *Ascoseira mirabilis*, which in general had the lowest values (Huovinen and Gómez 2013). However, when intraspecific variability of phlorotannins is examined, this pattern can be different. In fact, Gómez and Huovinen (2015) analyzed the contents of phlorotannins in conspecifics of *Ascoseira mirabilis*, *Desmarestia anceps*, *D. menziesii* and *Himantothallus grandifolius* collected from 5/10, 20 to 30 m at King George Island, and found that variation with depth was species-specific. For example, in *A. mirabilis*, no changes with depth were detected, while in *D. anceps* and *D. menziesii*, values increased in algae collected at 10 m depth compared to 20 or 30 m. Similar results have been reported in *D. anceps* from Anvers Island, West Antarctic Peninsula, where higher phlorotannin contents were measured in shallower locations (3–12 m) compared to samples collected between 18 and 30 m depth (Fairhead et al. 2005a). Although many factors can preclude a conclusive comparison between different studies (e.g., time of collection, study site, and the characteristics of the depth gradient or differences in depths between samples), the results appear to indicate that (a) endemic Antarctic brown algae from depth >20 m in general show constitutively high levels of phlorotannins, (b) phlorotannin contents and their vertical variability mirror differences in life adaptations developed to cope with multiple abiotic or biotic variables, and (c) phlorotannins form part of a trade-off between shade adaptation marked by high photosynthetic efficiencies at low light and tolerance to high solar stress. Thus, phlorotannins act as multifunctional substances that can be "mobilized" in any situation that poses a threat to the algae (Gómez and Huovinen 2015) (see Sect. 18.5.3).

18.5.2 Phlorotannin Allocation in Antarctic Seaweeds

Although the Antarctic is devoid of kelps, which resemble plants in being structurally complex, many Antarctic Desmarestiales and *Ascoseira* and *Cystosphaera* show a complex thallus anatomy with morpho-functional processes analogous to those described in Laminariales and Fucales (see Chap. 11 in this volume). In this context, it has been commonly observed that concentrations of phlorotannins vary strongly among different thallus parts (Van Alstyne et al. 1999; Connan et al. 2004; Iken et al. 2007), stimulating researchers to propose diverse hypothesis explaining whether this observed variability is related with functional processes at an organismal level. Whether the unequal allocation of phlorotannins to different thallus regions is related with putative benefits for the alga, e.g., protection of metabolic performance, reproductive output, or, in general, to guarantee the algal fitness during environmental stress, is a relevant question. The optimal defense theory (ODM; Rhoades 1979) is one of the ecological models used to explain the differential distribution of phlorotannins in the brown algal thalli, suggesting that chemical defenses are produced in direct proportion to the risk, i.e., the phenolic compounds would be produced at a direct expense of other functions (Pavia et al. 2002). High concentrations of phlorotannins may be expected when an environmental pressure, e.g., grazing, is high (inducible response) on thallus parts that make an important contribution to the whole fitness (e.g., meristematic or reproductive regions) or during seasonal periods when algae are especially vulnerable. Results in *Fucus* and *Ecklonia* (Steinberg 1985; Yates and Peckol 1993) and the sub-Antarctic kelps *Lessonia spicata* and *Macrocystis pyrifera* (Pansch et al. 2008) indicate that phlorotannins could vary as predicted by the ODT. As has been reported for *Ascophyllum nodosum*, production of phlorotannins can be highly costly at the expense of growth (Pavia et al. 1999). Thereby, it has been proposed that due to these costs, synthesis and accumulation of phenolics could indicate inducible rather than constitutive defenses (Rhoades 1979), which has been confirmed in some studies of simulated herbivory (Lüder and Clayton 2004). However, studies carried out in some Antarctic species indicate that phlorotannin allocation not necessarily confers chemical defense consistent with the ODM assumptions. For example, regarding the "value" of different thallus parts in relation with perennial and annual growth strategies, *D. anceps* did not show differences in phlorotannin contents between thallus parts (Fairhead et al. 2005a; Iken et al. 2007); however, there were marked differences in the toughness, and the chemical defenses in primary stems/stipes were much higher than the laterals supporting the ODT model (Fairhead et al. 2005b). In contrast, *D. menziesii* and *Ascoseira mirabilis* had higher phlorotannin concentrations in the holdfasts compared to the branch or lamina regions. Due to that holdfasts were regarded here as the most valuable thallus part conferring attachment, the patterns in these species appear to meet well the ODT (Iken et al. 2007). It must be emphasized that the deterrent role of Antarctic phlorotannins against grazers and microbia or as antifouling agents is species specific and probably depends on the type of predominant phlorotannins and other not well-known qualitative properties of this compounds (reviewed in Iken et al. 2009). The results agree with longitudinal profiles determined in the cold-temperate kelps *Laminaria hyperborea* and *Laminaria digitata* (Connan et al. 2006) and *Lessonia spicata* (Gómez et al. 2016), where valuable regions and basal parts such as haptera or holdfasts and meristematic tissues allocated the highest phlorotannins compared to the fronds, which can be regarded as transient structures. These patterns can be associated with various longitudinal profiles of physiological performance normally described for various kelps (Van Alstyne et al. 1999; Gómez et al. 2005; Gruber et al. 2011).

In the case of photoprotective responses, it could be reasonable to argue that valuable thallus regions, e.g., reproductive tissues, should be protected when they are exposed to UV radiation (Holzinger et al. 2011). This hypothesis has also been tested in two Antarctic brown algae by exposing reproductive and vegetative thallus pieces to UV radiation during a short-term period (Huovinen and Gómez 2015). In the brown alga *Cystosphaera jacquinotii*, the reproductive structures (receptacles containing conceptacles) showed higher UV tolerance than its vegetative blades, whereas in *Ascoseira mirabilis*, high UV tolerance was demonstrated in both vegetative and reproductive tissues. Interestingly, the reproductive structures of both species of brown algae had higher levels of soluble phlorotannins than the vegetative tissues, and thus, allocation and proportions of soluble and insoluble, cell-wall-bound phlorotannins could be related with the observed patterns of UV tolerance of the different tissues. Observations of tissue cross sections under violet-blue light excitation using epifluorescence microscopy confirmed a high allocation of phenolic compounds (as blue autofluorescence) in *C. jacquinotii*, especially in its reproductive structures (Fig. 18.5a). The study is among the first approaches to address the defense strategies that Antarctic macroalgae exploit to protect their reproductive structures. It is likely that the allocation of chemical defenses and UV-absorbing compounds in reproductive tissues is a widespread strategy to ensure the viability of spores and gametes during their maturation. For example, blue autofluorescence, indicating the presence of phenolics compounds in reproductive tissues (carporangia) of the Antarctic red algae *Trematocarpus antarcticus* (Fig. 18.5b), suggests that not only phlorotannins but also other phenolics can be allocated providing protection to reproductive tissues.

It has been suggested that phlorotannins can be remobilized between tissues with different metabolic demand and age (Arnold and Targett 2000). In Laminariales, compounds as mannitol, amino acids, and other low-molecular-weight compounds are transported through specialized cells to power meristematic growth through translocation processes (Küppers and Kremer 1978; Gómez and Huovinen 2012). Since phlorotannins are structural components in cells, it may be intuitively suggested that these compounds or some key precursors may be mobilized along the thallus. The abundance of low-molecular-weight phlorotannins (<1200 Da) in various species of brown algae (Steevensz et al. 2012) supports also the idea that these compounds might be rapidly remobilized. For example, accumulation of phlorotannins in response to artificial wounding in the kelp *Ecklonia radiata*, including the presence of physodes in medullary sieve elements (Lüder and Clayton 2004), suggests that these compounds can be "transported" along the thallus. In fact, some

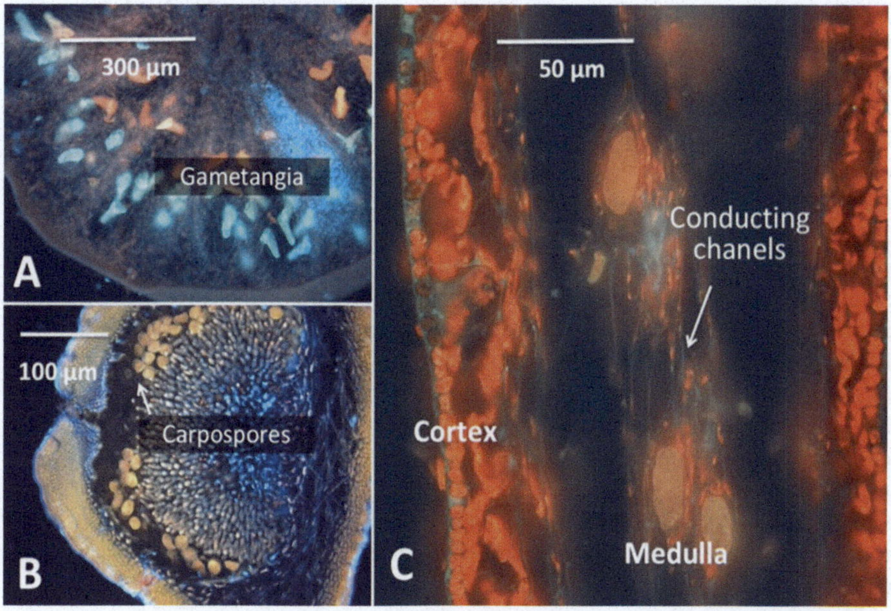

Fig. 18.5 Blue autofluorescence of phenolic compounds under violet-blue light excitation in (**a**) reproductive receptacles of the brown alga *Cystosphaera jacquinotii* and (**b**) carposporangium in the Antarctic red alga *Trematocarpus antarcticus*; (**c**) cross section of non-reproductive lamina of brown alga *Ascoseira mirabilis* indicating presence of phenolic compounds in medullar "conducting channels"

Antarctic species such as *Ascoseira mirabilis* appear to have the anatomical prerequisites as this alga shows "conducting channels" in its medulla (Fig. 18.5c), which are suggested to have putative functions in remobilization of substances (Clayton and Ashburner 1990). Overall, phlorotannins are important structural elements in the algal thallus, and since growth is localized in specific regions, a trade-off between phlorotannin synthesis, mobilization, and growth might be defined in these species in a similar way as in terrestrial vascular plants.

18.5.3 Phlorotannins in Response to UV Radiation

Experimental evidence has pointed to a relatively high tolerance of Antarctic seaweeds to UV radiation in the short-term, which at least partly can be related to their chemical defense mechanism based on phenolic substances. Indeed, Antarctic algae exposed for 2 h to UV radiation at 2 °C showed very low inhibition of photosynthesis measured as maximal quantum yield of fluorescence (F_v/F_m), which can reach up to 35% in algae collected from depth >20 m. Even in algae growing at 30 m, inhibition of chlorophyll fluorescence did not exceed 10–15% (Huovinen and Gómez

2013). In the case of the brown algae, almost all have high levels of phlorotannins, which due to their UV-absorbing characteristics are the main candidates conferring photoprotection in these species (Huovinen and Gómez 2013; Gómez and Huovinen 2015; Núñez-Pons et al. 2018). However, testing these properties experimentally is not an easy task. In fact, manipulative studies conducted in algae attaining high concentrations of phenols, e.g., *Desmarestia anceps*, have not demonstrated induction in phlorotannins in response to UV (Fairhead et al. 2006; Gómez and Huovinen 2015; Flores-Molina et al. 2016). In contrast, some species with relatively low concentrations, such as *Ascoseira mirabilis*, show a slightly UV-mediated induction of soluble phlorotannins (Rautenberger et al. 2015). As this species is normally found at shallower depths (1–10 m), the results suggest that it can become exposed to harmful solar radiation in summer, thus activating the synthesis of phlorotannins. In all, photoprotection against excess solar radiation, e.g., via UV shielding, is a collateral function in Antarctic seaweeds as these molecules form part of an integral defense machinery operating in response to multiple stressors in the polar environment such as herbivores, antifouling, and changes in temperature or simply they are synthesized to supply of structural elements during cell growth (as insoluble phlorotannins). These multiple functional roles are explained by their high antioxidant capacity, the most important chemical property of phlorotannins. In fact, soluble phlorotannins have been positively correlated with the high antioxidant potential determined in extracts of various species of Antarctic brown algae. This positive correlation is observed in algae exposed to different conditions of UV radiation and temperatures (Gómez and Huovinen 2015; Flores-Molina et al. 2016). Interestingly, it has been demonstrated that UV effects on photosynthesis in Antarctic macroalgae are modified by temperature: when algae are incubated at 7 °C, i.e., 5 °C above the field temperature, inhibition of photosynthesis decreases, and recovery increases, suggesting that, e.g., the PSII repair cycle is more effective at elevated temperature resulting in a higher UV tolerance, at least in the short-term (Rautenberger et al. 2015). However, the relationship between the phlorotannin contents and the antioxidant potential of extracts does not change with temperature (Fig. 18.6), reinforcing the idea that in these species, phlorotannins are not UV-inducible compounds. This raises questions related to the role of these compounds and their physiological consequences under changing environmental conditions. For example, it is known that during oxidative stress, Antarctic brown algae can active their enzymatic machinery (e.g., superoxide dismutase, SOD), whose efficiency varies in response to environmental gradients (Bischof and Rautenberger 2012). Thus, the operation of complementary mechanisms of ROS detoxifying less affected by, e.g., temperature or UV radiation could be favored. In this context, it has been reported recently that under high solar stress conditions, algae display a suite of complementary and consecutive protective mechanisms based on energy dissipative downregulation of photosynthesis, rapid pigment acclimation and PSII repair mechanisms, synthesis of phenolics with specific UV absorption characteristics, and complementary ROS scavenging mediated by antioxidant enzymes and phenols (Cruces et al. 2017).

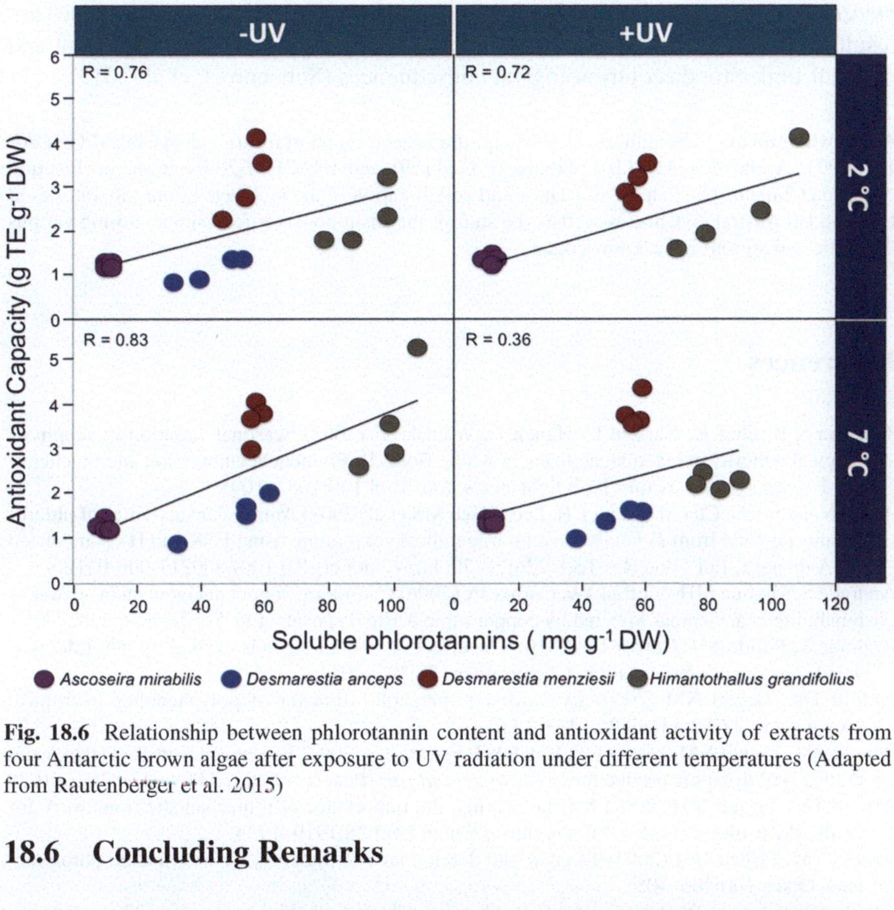

Fig. 18.6 Relationship between phlorotannin content and antioxidant activity of extracts from four Antarctic brown algae after exposure to UV radiation under different temperatures (Adapted from Rautenberger et al. 2015)

18.6 Concluding Remarks

Antarctic macroalgae are equipped with a suite of anti-stress mechanisms to cope with the harsh polar environment, probably much more sophisticated and efficient than previously thought. However, in the case of the putative roles of phlorotannins in stress tolerance, many gaps still persist, especially those related with their synthesis, the action of different forms of phlorotannins, and their turnover and regulation. In the case of Antarctic algae, no data on gene expression exist, which precludes an understanding of crucial aspects related with their multifunctional roles.

In Antarctic brown algae, processes related with synthesis and functional dynamics of phlorotannins take place under constantly low temperatures of 0–2 °C, and phlorotannin induction can be sensitive to changes in temperate and cold-temperate species. This raises questions related with the effects of shifts in temperature on the anti-stress properties of these substances in endemic Antarctic algae, where induction has not been reported.

Finally, the role of phlorotannins in response to new and emergent stressors, e.g., pollutants and acidification (OA), has been very little studied in Antarctic algae. Although recent experimental essays carried out in *Desmarestia anceps* and *D.*

menziesii indicated that a combination of low pH and elevated temperature does not result in marked effects in phlorotannin content, the shifts can have important and not well-understood ecophysiological consequences (Schoenrock et al. 2015).

Acknowledgments The authors acknowledge the financial support from Conicyt Chile (FONDAP 15150003, Anillo-PIA ART1101, Fondecyt 1161129) and INACH T-20-09 from the Instituto Antártico Chileno. The helpful assistance and collaboration of the members of our laboratories at Universidad Austral de Chile as well as the staff of the Instituto Antártico Chileno during various Antarctic expeditions are acknowledged.

References

Aguilera J, Bischof K, Karsten U, Hanelt D, Wiencke C (2002) Seasonal variation in ecophysiological patterns in macroalgae from an Arctic fjord. II. Pigment accumulation and biochemical defence systems against high light stress. Mar Biol 140:1087–1095

Ahn GN, Kim KN, Cha SH, Song CB, Lee J, Heo MS et al (2007) Antioxidant activities of phlorotannins purified from *Ecklonia cava* on free radical scavenging using ESR and H_2O_2-mediated DNA damage. Eur Food Res Tech 226:71–79. https://doi.org/10.1007/s00217-006-0510-y

Andrade S, Medina MH, Moffett JW, Correa JA (2006) Cadmium–copper antagonism in seaweeds inhabiting coastal areas affected by copper mine waste disposals. Env Sci Tech 40:4382–4387

Andrade S, Pulido MJ, Correa JA (2010) The effect of organic ligands exuded by intertidal seaweeds on copper complexation. Chemosphere 78:397–401

Arnold TM, Targett NM (2000) Evidence for metabolic turnover of polyphenolics in tropical brown algae. J Chem Ecol 26:1393–1410

Arnold TM, Targett NM, Tanner CE, Hatch WI, Ferrari KE (2001) Evidence for methyl jasmonate-induced phlorotannin production in *Fucus vesiculosus* (Phaeophyceae). J Phycol 37:1026–1029

Arnold TM, Targett NM (2002) Marine tannins: the importance of a mechanistic framework for predicting ecological roles. Mini review. J Chem Ecol 28:1919–1934

Arnold TM, Targett NM (2003) To grow and defend: lack of tradeoffs for brown algal phlorotannins. Oikos 100:406–408

Athukorala Y, Lee K-W, Song C, Ahn C-B, Shin T-S, Cha Y-J, Shahidi F, Jeon Y-J (2003) Potential antioxidant activity of marine red alga *Grateloupia filicina* extracts. J Food Lipids 10:251–265

Badger MR, von Caemmerer S, Ruuska S, Nakano H (2000) Electron flow to oxygen in higher plants and algae: rates and control of direct photoreduction (Mehler reaction) and rubisco oxygenase. Philos Trans R Soc Lond Ser B Biol Sci 355(1402):1433–1446. https://doi.org/10.1098/rstb.2000.0704

Bischof K, Gómez I, Molis M, Hanelt D, Karsten U, Lüder U, Roleda MY, Zacher K, Wiencke C (2006a) Ultraviolet radiation shapes seaweed communities. Rev Environ Sci Biotechnol 5:141–166

Bischof K, Rautenberger R, Brey L, Perez-Llorens JL (2006b) Physiological acclimation to gradients of solar irradiance within mats of the filamentous green macroalga *Chaetomorpha linum* from southern Spain. Mar Ecol Prog Ser 306:165–175

Bischof K, Rautenberger R (2012) Seaweed responses to environmental stress: reactive oxygen and antioxidative strategies. In: Wiencke C, Bischof K (eds) Seaweed biology: novel insights into ecophysiology, ecology and utilization, ecological studies, vol 219. Springer-Verlag, Berlin, pp 109–132

Clarke LJ, Robinson SA (2008) Cell wall-bound ultraviolet-screening compounds explain the high ultraviolet tolerance of the Antarctic moss, *Ceratodon purpureus*. New Phytol 179:776–783

Clayton MN, Beakes GW (1983) Effects of fixatives on the ultrastructure of physodes in vegetative cell of *Scytosiphon lomentaria* (Scytosiphonaceae, Phaeophyta). J Phycol 19:416

Clayton MN, Ashburner CM (1990) The anatomy and ultrastructure of "conducting channels" in *Ascoseira mirabilis* (Ascoseirales, Phaeophyceae). Bot Mar 33:63–70

Cockell CS, Knowland J (1999) Ultraviolet radiation screening compounds. Biol Rev 74:311–345

Connan S, Goulard F, Stiger V, Deslandes E, Ar Gall E (2004) Interspecific and temporal variation in phlorotannin levels in an assemblage of brown algae. Bot Mar 47:410–416

Connan S, Delisle F, Deslandes E, Ar Gall E (2006) Intra-thallus phlorotannin content and antioxidant activity in Phaeophyceae of temperate waters. Bot Mar 49:39–46

Connan S, Stengel DB (2011) Impacts of ambient salinity and copper on brown algae: 2. Interactive effects on phenolic pool and assessment of metal binding capacity of phlorotannin. Aquat Toxicol 104:1–13

Contreras L, Moenne A, Correa JA (2005) Antioxidant responses in *Scytosiphon lomentaria* (Phaeophyta) inhabiting copper-enriched coastal environments. J Phycol 41:1184–1195

Contreras L, Mella D, Moenne A, Correa JA (2009) Differential responses to copper induced oxidative stress in the marine macroalgae *Lessonia nigrescens* and *Scytosiphon lomentaria* (Phaeophyceae). Aquat Toxicol 94:94–102

Contreras L, Moenne A, Gaillard F, Potin P, Correa JA (2010) Proteomic analysis and identification of copper stress-regulated proteins in the marine alga *Scytosiphon gracilis* (Phaeophyceae). Aquat Toxicol 96:85–89

Creis E, Delage L, Charton S, Goulitquer S, Leblanc C, Potin P, Ar Gall E (2015) Constitutive or inducible protective mechanisms against UV-B radiation in the brown alga *Fucus vesiculosus*? A study of gene expression and phlorotannin content responses. PLoS One 10(6):e0128003. https://doi.org/10.1371/journal.pone.0128003

Cruces E, Huovinen P, Gómez I (2012) Phlorotannin and antioxidant responses upon short term exposure to UV radiation and elevated temperature in three South Pacific kelps. Photochem Photobiol 88:58–66

Cruces E, Huovinen P, Gómez I (2013) Interactive effects of UV radiation and enhanced temperature on photosynthesis, phlorotannin induction and antioxidant activities of two sub-Antarctic brown algae. Mar Biol 160:1–13

Cruces E, Rautenberger R, Rojas-Lillo Y, Cubillos VM, Ramirez-Kushel E, Gómez I (2017) Physiological acclimation of *Lessonia spicata* to diurnal changing PAR and UV radiation: differential regulation among down-regulation of photochemistry, ROS scavenging activity and phlorotannins as major photoprotective mechanisms. Photosynth Res 131:145–157. https://doi.org/10.1007/s11120-016-0304-4

Cruces E, Flores MR, Díaz MJ, Huovinen P, Gómez I (2018) Phenolics as photoprotective mechanism against combined action of UV radiation and temperature in the red alga *Gracilaria chilensis*? J Appl Phycol 30(2):1247–1257. https://doi.org/10.1007/s10811-017-1304-2

Dring M (2005) Stress resistance and disease resistance in seaweeds: the role of reactive oxygen metabolism. Adv Bot Res 43:175–207

Dubois A, Iken K (2012) Seasonal variation in kelp phlorotannins in relation to grazer abundance and environmental variables in the Alaskan sublittoral zone. Algae 27(1):9–19. https://doi.org/10.4490/algae.2012.27.1.009

Evans LV, Holligan MS (1972) Correlated light and electron microscope studies on brown algae. II. Physode production in *Dictyota*. New Phytol 71:1173–1180

Fairhead VA, Amsler CD, McClintock JB, Baker BJ (2005a) Variation in phlorotannin content within two species of brown macroalgae (*Desmarestia anceps* and *D. menziesii*) from the Western Antarctic Peninsula. Polar Biol 28:680–686. https://doi.org/10.1007/s00300-005-0735-4

Fairhead VA, Amsler CD, McClintock JB, Baker BJ (2005b) Within-thallus variation in chemical and physical defenses in two species of ecologically dominant brown macroalgae from the Antarctic Peninsula. J Exp Mar Biol Ecol 322:1–12

Fairhead VA, Amsler CD, McClintock JB, Baker BJ (2006) Lack of defense or phlorotannins induction by UV radiation or mesograzers in *Desmarestia anceps* and *D. menziesii* (Phaeophyceae). J Phycol 42:1174–1183

Feldmann G, Guglielmi MG (1972) Les physodes et les corps irisants du *Dictyota dichotoma* (Hudson) Lamouroux. Comptes Rendus de l'Academie des. Sciences 275:751–754

Flores-Molina MR, Muñoz P, Rautenberger R, Huovinen P, Gómez I (2016) Stress tolerance to UV radiation and temperature of the endemic Antarctic brown alga *Desmarestia anceps* is mediated by high concentrations of phlorotannins. Photochem Photobiol 92:455–466. https://doi.org/10.1111/php.12580

García-Pichel F, Castenholz RW (1993) Occurrence of UV-absorbing, mycosporine-like compounds among cyanobacterial isolates and an estimate of their screening capacity. Appl Environ Microbiol 59:170–176

Generalíc-Mekiníc I, Skroza D, Šimat V, Hamed I, Câgalj M, Popovíc-Perkovíc Z (2019) Phenolic content of brown algae (Phaeophyceae) species: extraction, identification, and quantification. Biomol Ther 9:244. https://doi.org/10.3390/biom9060244

Gledhill M, Nimmo M, Hill SJ, Brown MT (1999) The release of copper-complexing ligands by the brown alga *Fucus vesiculosus* (Phaeophyceae) in response to increasing total copper levels. J Phycol 35:501–509

Gómez I, Huovinen P (2010) Induction of phlorotannins during UV exposure mitigates inhibition of photosynthesis and DNA damage in the kelp *Lessonia nigrescens*. Photochem Photobiol 86:1056–1063

Gómez I, Huovinen P (2012) Morpho-functionality of carbon metabolism in seaweeds. In: Wiencke C, Bischof K (eds) Seaweed biology: novel insights into ecophysiology, ecology and utilization, ecological studies, vol 219. Springer-Verlag, Berlin, pp 25–46

Gómez I, Huovinen P (2015) Lack of physiological depth patterns in conspecifics of endemic Antarctic brown algae: a trade-off between UV stress tolerance and shade adaptation? PLoS One 10(8):e0134440

Gómez I, Pérez-Rodríguez E, Viñegla B, Figueroa FL, Karsten U (1998) Effects of solar radiation on photosynthesis, UV-absorbing compounds and enzyme activities of the green alga *Dasycladus vermicularis* from southern Spain. J Photochem Photobiol B Biol 47:46–57

Gómez I, Ulloa N, Orostegui M (2005) Morpho-functional patterns of photosynthesis ad UV sensitivity in the kelp *Lessonia nigrescens* (Laminariales, Phaeophyta). Mar Biol 148:231–240

Gómez I, Véliz K, Español S, Huovinen P (2016) Spatial distribution of phlorotannins and its relationship with photosynthetic UV tolerance and allocation of storage carbohydrates in blades of the kelp *Lessonia spicata*. Mar Biol 163:110. https://doi.org/10.1007/s00227-016-2891-1

Gruber A, Roleda MY, Bartsch I, Hanelt D, Wiencke C (2011) Sporogenesis under ultraviolet radiation in *Laminaria digitata* (Phaeophyceae) reveals protection of photosensitive meiospores within soral tissue: physiological and anatomical evidence. J Phycol 47:603–614

Heffernan N, Smyth TJ, Soler-Villa A, Fitzgerald RJ, Brunton NP (2014) Phenolic content and antioxidant activity of fractions obtained from selected Irish macroalgae species (*Laminaria digitata*, *Fucus serratus*, *Gracilaria gracilis* and *Codium fragile*). J Appl Phycol 27:519. https://doi.org/10.1007/s10811-014-0291-9

Heinrich S, Valentin K, Frickenhaus K, Wiencke C (2015) Temperature and light interactively modulate gene expression in *Saccharina latissima* (Phaeophyceae). J Phycol 51:93–108

Henry BE, Van Alstyne KL (2004) Effects of UV radiation on growth and phlorotannins in *Fucus gardneri* (Phaeophyceae) juveniles and embryos. J Phycol 40:527–533

Herbert RB (ed) (1989) The biosynthesis of secondary metabolites, 2nd edn. Chapman and Hall, New Yorkk

Hideg É, Jansen MAK, Strid A (2013) UV-B exposure, ROS, and stress: inseparable companions or loosely linked associates? Trends Plant Sci 18:107–115. https://doi.org/10.1016/j.tplants.2012.09.003

Holzinger A, Di Piazza L, Lütz C, Roleda MY (2011) Sporogenic and vegetative tissues of *Saccharina latissima* (Laminariales, Phaeophyceae) exhibit distinctive sensitivity to experimentally enhanced ultraviolet radiation: photosynthetically active radiation ratio. Phycol Res 59:221–235

Huovinen P, Gómez I (2013) Photosynthetic characteristics and UV stress tolerance of Antarctic seaweeds along the depth gradient. Polar Biol 36:1319–1332

Huovinen P, Gómez I (2015) UV Sensitivity of vegetative and reproductive tissues of three Antarctic macroalgae is related to differential allocation of phenolic substances. Photochem Photobiol 91:1382–1388. https://doi.org/10.1111/php.12500

Huovinen PS, Soimasuo MR, Oikari AOJ (2001) Photoinduced toxicity of retene to *Daphnia magna* under enhanced UV-B radiation. Chemosphere 45:683–691

Huovinen P, Leal P, Gómez I (2010) Impact of interaction of copper, nitrogen and UV radiation on the physiology of three south Pacific kelps. Mar Freshw Res 61:330–341

Iken K, Amsler CD, Hubbard JM, McClintock JB, Baker BJ (2007) Allocation patterns of phlorotannins in Antarctic brown algae. Phycologia 46(4):386–395. https://doi.org/10.2216/06-67.1

Iken K, Amsler CD, Amsler MO, McClintock JB, Baker BJ (2009) Field studies on deterrent roles of phlorotannins in Antarctic brown algae. Bot Mar 52:547–557

Jormalainen V, Honkanen T, Koivikko R, Eränen J (2003) Induction of phlorotannin production in a brown alga: defense or resource dynamics? Oikos 103:640–650

Karentz D, McEuen FS, Land MC, Dunlap WC (1991) Survey of mycosporine-like amino acid compounds in Antarctic marine organisms: potential protection from ultraviolet exposure. Mar Biol 108:157–166

Kaur I, Vijayaraghavan MR (1992) Physode distribution and genesis in *Sargassum vulgare* C. Agardh and *Sargassum johnstonii* Setchell & Gardner. Aquat Bot 42:375–384

Kevekordes K, Clayton MN (1999) Shedding of the zygote wall by *Durvillaea potatorum* (Durvillaeales, Phaeophyta) embryos. Eur J Phycol 34:65–70

Koivikko R, Loponen J, Honkanen T, Jormalainen V (2005) Contents of soluble, cell-wall-bound and exuded phlorotannins in the brown alga *Fucus vesiculosus*, with implications on their ecological function. J Chem Ecol 31:195–212

Koivikko R, Loponen J, Pihlaja K, Jormalainen V (2007) High-performance liquid chromatographic analysis of phlorotannins from the brown alga *Fucus vesiculosus*. Phytochem Anal 18(4):326–332

Küpper FC, Gaquerel E, Cosse A, Adas F, Peters AF et al (2009) Free fatty acids and methyl jasmonate trigger defense reactions in *Laminaria digitata*. Plant Cell Physiol 50:789–800

Küppers U, Kremer BP (1978) Longitudinal profiles of carbon dioxide fixation capacities in marine macroalgae. Plant Physiol 62:49–53

La Barre S, Potin P, Leblanc C, Delage L (2010) The halogenated metabolism of brown algae (Phaeophyta), its biological importance and its environmental significance. Mar Drugs 8:988–1010. https://doi.org/10.3390/md8040988

Lesser M (2012) Oxidative stress in tropical marine ecosystems. In: Abele D, Vazquez-Medina JP, Zenteno-Savin T (eds) Oxidative stress in aquatic ecosystems. Blackwell Publishing Ltd, Oxford, pp 9–19

Lüder UH, Clayton MN (2004) Induction of phlorotannins in the brown macroalga *Ecklonia radiata* (Laminariales, Phaeophyta) in response to simulated herbivory – the first microscopic study. Planta 218:928–937

Mackerness AHS, John CF, Jordan B, Thomas B (2001) Early signaling components in ultraviolet-B responses: distinct roles for different reactive oxygen species and nitric oxide. FEBS Lett 489(2–3):237–242

Menzel D, Kazlauskas R, Reichelt J (1983) Coumarins in the siphonalean green algal family Dasycladaceae Kützing (Chlorophyceae). Bot Mar 29:23–29

Meslet-Cladiere L, Delage L, Leroux CJ, Goulitquer S, Leblanc C, Creis E, Gall EA et al (2013) Structure/function analysis of a type III polyketide synthase in the brown alga *Ectocarpus siliculosus* reveals a biochemical pathway in phlorotannins monomer biosynthesis. Plant Cell 25:3089–3103. https://doi.org/10.1105/tpc.113.111336

Michalak A (2006) Phenolic compounds and their antioxidant activity in plants growing under heavy metal stress. Polish J Environ Stud 15(4):523–530

Müller R, Desel C, Steinhoff FS, Wiencke C, Bischof K (2012) UV-radiation and elevated temperatures induce formation of reactive oxygen species in gametophytes of cold temperate/Arctic kelps (Laminariales, Phaeophyceae). Phycol Res 60:27–36

Nakai M, Kageyama N, Nakahara K, Miki W (2006) Phlorotannins as radical scavengers from the extract of *Sargassum ringgoldianum*. Mar Biotechnol 8:409–414

Núñez-Pons L, Avila C, Romano G, Verde C, Giordano D (2018) UV-protective compounds in marine organisms from the Southern Ocean. Mar Drugs 16(9):336. https://doi.org/10.3390/md16090336

Oliveira L, Bisalputra T (1973) Studies in the brown alga *Ectocarpus* in culture. I. General ultrastructure of the sporophytic vegetative cells. J Submicrosc Cytol 5:107–120

Pangestuti R, Siahaan EA, Kim SK (2018) Photoprotective substances derived from marine algae. Mar Drugs 16(11):399. https://doi.org/10.3390/md16110399

Pansch C, Gómez I, Rothäusler E, Véliz K, Thiel M (2008) Species-specific defense strategies of vegetative versus reproductive blades of the Pacific kelps *Lessonia nigrescens* and *Macrocystis integrifolia*. Mar Biol 155:51–62. https://doi.org/10.1007/s00227-008-1006-z

Paul N, Gwynn-Jones D (2003) Ecological roles of solar UV radiation: towards an integrated approach. Trends Ecol Evol 18:48–55

Pavia H, Toth G (2000) Inducible chemical resistance to herbivory in the brown seaweed *Ascophyllum nodosum*. Ecology 81:3212–3225

Pavia H, Cervin G, Lindaren A, Åberg P (1997) Effects of UV-B radiation and simulated herbivory on phlorotannins in the brown alga *Ascophyllum nodosum*. Mar Ecol Progr Ser 157:139–146

Pavia H, Toth G, Åberg P (1999) Trade-offs between phlorotannin production and annual growth in natural populations of the brown seaweed *Ascophyllum nodosum*. J Ecol 87:761–771

Pavia H, Toth GB, Åberg P (2002) Optimal defense theory: elasticity analysis as a tool to predict intraplant variation in defenses. Ecology 83:891–897

Pellegrini L (1980) Cytological studies on physodes in the vegetative cells of *Cystoseira stricta* Sauvageau (Phaeophyta, Fucales). J Cell Sci 41:209–231

Pérez-Rodríguez E, Gómez I, Karsten U, Figueroa FL (1998) Effects of UV radiation on photosynthesis and excretion of UV-absorbing compounds of *Dasycladus vermicularis* (Dasycladales, Chlorophyta) from southern Spain. Phycologia 37:379–387

Pérez-Rodríguez E, Aguilera J, Gómez I, Figueroa FL (2001) Excretion of coumarins by the Mediterranean green alga *Dasycladus vermicularis* in response to environmental stress. Mar Biol 139:633–639

Potin P, Leblanc CL (2006) Phenolic-based adhesives of marine brown algae. In: Smith AM, Callow JA (eds) Biological adhesive. Springer, Berlin, pp 105–124. https://doi.org/10.1007/978-3-540-31049-5_6

Ragan MA, Glombitza K-W (1986) Phlorotannins, brown algal polyphenols. Prog Phycol Res 4:130–241

Ratkevicius N, Correa JA, Moenne A (2003) Copper accumulation, synthesis of ascorbate and activation of ascorbate peroxidase in *Enteromorpha compressa* (L.) Grev. (Chlorophyta) from heavy metal-enriched environments in northern Chile. Plant Cell Environ 26:1599–1608

Rautenberger R, Huovinen P, Gómez I (2015) Effects of increased seawater temperature on UV-tolerance of Antarctic marine macroalgae. Mar Biol 162:1087–1097

Rhoades DF (1979) Evolution of plant chemical defense against herbivores. In: Rosenthal GA, Janzen DH (eds) Herbivores: their interaction with secondary plant metabolites. Academic Press, New York, pp 1–55

Ritter A, Goulitquer S, Salaün JP, Tonon T, Correa JA, Potin P (2008) Copper stress induces biosynthesis of octadecanoid and eicosanoid oxygenated derivatives in the brown algal kelp *Laminaria digitata*. New Phytol 180(4):809–821. https://doi.org/10.1111/j.1469-8137.2008.02626.x

Roleda MY, Lüder UH, Wiencke C (2010) UV-susceptibility of zoospores of the brown macroalga *Laminaria digitata* from Spitsbergen. Polar Biol 33:577–588

Ross C, Küpper FC, Vreeland V, Waite JH, Jacobs RS (2005) Evidence of a latent oxidative burst in relation to wound repair in the giant unicellular chlorophyte *Dasycladus vermicularis*. J Phycol 41:531–541

Sakihama Y, Cohen MF, Grace SC, Yamasaki H (2002) Plant phenolic antioxidant and prooxidant activities: phenolics-induced oxidative damage mediated by metals in plants. Toxicology 177:67–80

Salgado LT, Cinelli LP, Viana NB, de Carvalho RT, de Souza Mourão PA et al (2009) A vanadium bromoperoxidase catalyzes the formation of high-molecular-weight complexes between brown algal phenolic substances and alginates. J Phycol 45:193–202. https://doi.org/10.1111/j.1529-8817.2008.00642.x

Shibata T, Hama Y, Miyasaki T, Ito M, Nakamura T (2006) Extracellular secretion of phenolic substances from living brown algae. J Appl Phycol 18:787–794. https://doi.org/10.1007/s10811-006-9094-y

Schoenrock KM, Schram JB, Amsler CD, McClintock JB, Angus RA (2015) Climate change impacts on overstory *Desmarestia* spp. from the western Antarctic Peninsula. Mar Biol 162:377–389. https://doi.org/10.1007/s00227-014-2582-8

Schoenwaelder MEA (2002) The occurrence and cellular significance of physodes in brown algae. Phycologia 41:125–139

Schoenwaelder MEA, Clayton MN (1998) Secretion of phenolic substances into zygote wall and cell plate in embryos of *Hormosira* and *Acrocarpia* (Fucales, Phaeophyta). J Phycol 34:969–980

Schoenwaelder MEA, Clayton MN (1999) The presence of phenolic compounds in isolated cell walls of brown algae. Phycologia 38:161–166

Schoenwaelder MEA, Wiencke C, Clayton MN, Glombitza KW (2003) The effect of elevated UV radiation on *Fucus* spp. (Fucales, Phaeophyta) zygote and embryo development. Plant Biol 5:366–377

Smith KL, Harwood JL (1986) The subcellular localization of absorbed copper in *Fucus*. Physiol Plant 66:692–698

Stauber JL, Florence TM (1987) Mechanisms of toxicity of ionic copper and copper complexes in algae. Mar Biol 94:511–519

Steevensz AJ, MacKinnon SL, Hankinson R, Craft C, Connan S, Stengel DB, Melanson JE (2012) Profiling phlorotannins in brown macroalgae by liquid chromatography–high resolution mass spectrometry. Phytochem Anal 23(5):547–553

Steinberg PD (1985) Feeding preferences of *Tegula funebralis* and chemical defences of marine brown algae. Ecol Monogr 5:333–349

Steinhoff F (2010) Phlorotannins as UV-protective substances in early developmental stages of brown algae, PhD thesis, University of Bremen

Steinhoff FS, Graeve M, Wiencke C, Wulff A, Bischof K (2011) Lipid content and fatty acid consumption in zoospores/ developing gametophytes of *Saccharina latissima* (Laminariales, Phaeophyceae) as potential precursors for secondary metabolites as phlorotannins. Polar Biol 34:1011. https://doi.org/10.1007/s00300-011-0960-y

Swanson AK, Druehl LD (2002) Induction, exudation and the UV protective role of kelp phlorotannins. Aquat Bot 73:241–253

Swanson AK, Fox C (2007) Altered kelp (Laminariales) phlorotannins and growth under elevated carbon dioxide and ultraviolet-B treatments can influence associated intertidal food webs. Glob Chang Biol 13:1696–1709

Toth GB, Pavia H (2002) Lack of phlorotannin induction in the kelp *Laminaria hyperborea* in response to grazing by two gastropod herbivores. Mar Biol 140:403–409

Van Alstyne KL, Pelletreau KN (2000) Effects of nutrient enrichment on growth and phlorotannin production in *Fucus gardneri* embryos. Mar Ecol Prog Ser 206:33–43

Van Alstyne KL, McCarthy JJ, Hustead CL, Kearns LJ (1999) Phlorotannin allocation among tissues of northeastern Pacific kelps and rockweeds. J Phycol 35:483–492. https://doi.org/10.1046/j.1529-8817.1999.3530483.x

Vilter H (1995) Vanadium-dependent haloperoxidases. In: Sigel H, Sigel AM (eds) Vanadium and its role in life. Metal ions in biological systems, vol 31. Dekker, New York, pp 325–352

Vreeland V, Laetsch WM (1988) Role of alginate self-associating subunits in the assembly of *Fucus* embryo cell walls. In: Varner JE (ed) Self assembling architecture. Alan R. Liss, New York, pp 77–96

Wang T, Jónsdóttir R, Ólafsdóttir G (2009) Total phenolic compounds, radical scavenging and metal chelation of extracts from Icelandic seaweeds. Food Chem 116:240–248

Yates JL, Peckol P (1993) Effects of nutrient availability and herbivory on polyphenolics in the seaweed *Fucus vesiculosus*. Ecology 74:1757–1766

Yildiz G, Vatan Ö, Çelikler S, Dere Ş (2011) Determination of the phenolic compounds and antioxidative capacity in red algae *Gracilaria bursa-pastoris*. Int J Food Prop 14:496

Index

I. Gómez, P. Huovinen (eds.), *Antarctic Seaweeds*,
https://doi.org/10.1007/978-3-030-39448-6

W

The manufacturer's authorised representative in the EU is Springer Nature Customer Service Centre GmbH, Europaplatz 3, 69115 Heidelberg, Germany. If you have any concerns regarding our products, please contact ProductSafety@springernature.com

Printed and bound by CPI Group (UK) Ltd, Croydon, CR0 4YY
30/09/2026
02225556-0011